Humid Tropical Environments

Humid Tropical Environments

Alison J. Reading
Russell D. Thompson
Andrew C. Millington

First published 1995

Blackwell Publishers Ltd
108 Cowley Road
Oxford OX4 1JF
UK

Blackwell Publishers Inc.
238 Main Street
Cambridge, Massachusetts 02142
USA

British Library Cataloguing in Publication Data
A CIP catalogue record for this book is available from the British Library.

Library of Congress Cataloging-in-Publication Data
Reading, Alison J.
Humid tropical environments/Alison J. Reading, Russell D. Thompson, Andrew C. Millington.
p. cm.–(The natural environment)
Includes bibliographical references (p.) and index.
ISBN 0–631–17287–4 (acid-free paper). – ISBN 0–631–19174–7 (pbk. acid-free paper)
1. Tropics–Environmental conditions. 2. Tropics–Climate.
3. Tropics–Geography. I. Thompson, Russell D. II. Millington, A. C. III. Title. IV. Series.
GE160.T73R43 1994 94–27803
363.7'00913–dc20 CIP

Typeset in Sabon and Helvetica
by Keyword Typesetting Services Ltd, Wallington, Surrey
Printed in Great Britain by T.J. Press Ltd, Padstow, Cornwall

This book is printed on acid-free paper

Contents

Preface

Interest and scientific investigation in the tropics, by both expatriate and local scientists, has increased dramatically during the second half of this century. Our improved understanding of the physical landscape has been accompanied by a realization that the humid tropics are distinctive in many ways, yet are at the same time inextricably linked with more poleward regions of the globe. Furthermore, areas currently considered tropical may have developed under very different conditions and, conversely, landscapes outside the tropics may have developed under tropical regimes. Studies are now revealing details of the climatic and denudational history of the tropics and are providing indications of future conditions, taking into account the possibility and extent of human-induced global climate change. Scientific as well as humanitarian attention and concern about development issues within the nations of the low latitudes have grown enormously during the past two decades. The inhabitants of tropical regions have an intimate (and often precarious) relationship with their physical environment and any attempt to secure a sustainable relationship hinges upon an understanding of the physical environment as well as the needs and resources of the human environment.

As a result of our improved awareness of the importance of the tropics many students of geography and allied disciplines now study the physical characteristics and environmental issues of the low latitude regions as part of their degree course. At present, however, there is a noticeable imbalance in the amount of information available about the dry tropics and savannas compared with the amount dealing with the humid tropics.

Many of the books and other publications available which examine humid tropical environments are rather specialized. A few, however, deserve special mention for the contribution they have made to their particular specialism. These include books by Tricart (1972), Thomas (1974) and Faniran and Jeje (1983), widely used by students of geomorphology; Nieuwolt (1977) and Jackson (1990), which contain large amounts of valuable information for students of the atmosphere and hydrosphere; and Richards (1964) and Greenland and Lal (1977), which provide a foundation of information for students of biogeography and pedology.

Increasingly, geographers are realizing the merit of a holistic approach to pure and applied research. It is impossible, for example, to understand patterns of erosion without an appreciation of climatic factors and unwise to make assumptions about vegetation without also considering soils. This book is an attempt to provide a holistic approach to the study of the physical environment of the humid tropics. It does so by systematically introducing some of the major physical characteristics of the environment and by stressing their interrelationships. This book also acknowledges the importance of the human dimension within the humid tropics and addresses some of the pressing environmental concerns of the region.

In a single volume of limited length it is impossible to cover all aspects of the physical landscape. Subjects have therefore been carefully selected according to their importance, as measured by their spatial extent or widespread effect. These criteria have meant that, for example, high altitude parts of the humid tropics have received limited attention, despite their uniqueness. However, those parts of the humid tropics within which the majority of people live and which contain the majority of natural resources have been discussed more fully. Undoubtedly other authors would have chosen to include, expand and exclude the available material differently.

The book begins by establishing a context for the information which is to follow. First, various definitions of the humid tropics are examined and this is followed by an evaluation of our understanding of the humid tropics in a historical context. Chapters 2–7 systematically discuss various aspects of the humid tropical physical environments and chapter 8 explores the utilization of the environment as a resource. Chapter 9 is essentially an overview, within which physical–human interactions are examined more fully and prospects for the future are evaluated.

This book would not have been produced without the help of a number of friends, colleagues and family members and the authors would like to thank them all for their assistance, patience and encouragement. George Dardis at Anglia Polytechnic University, Martin Thorpe at University College Dublin and Ross Reynolds and Ian Fenwick at the University of Reading deserve a special thankyou for the loan of articles and their helpful and generous comments on the text. Judith Fox, who produced almost all the artwork, deserves special mention for her patience and understanding during the protracted preparation of the book. Finally the authors would like to thank the numerous individuals who have helped develop our interest and shape our ideas on the humid tropics: these include some of the authors of publications listed in the bibliography and also the people we have met with and talked to whilst undertaking fieldwork in the humid tropics.

Acknowledgements

We are grateful to the following for permission to reproduce figures, commonly in modified form:

For figure 1.1, adapted with permission from *Ecological Aspects of Development in the Humid Tropics*, copyright 1982 by the National Academy of Sciences, Courtesy of the National Academy Press, Washington, DC; for figure 1.2, W. Köppen, *Das Geographische System der Klimate*, p. 44, Gebrüder Borntraeger, Stuttgart; figure 1.4, I. J. Jackson, *Climate, Water and Agriculture*, copyright 1989 Longman Group; for figure 1.5, *Tropical Geomorphology Newsletter*, Singapore; for figure 1.6, UNEP; for figure 2.2, D. F. Hayward and J. S. Oguntoyinbo, *Climatology of West Africa*, copyright 1987 Hutchinson; for figures 2.3(a) and 2.5(c), H. J. Critchfield, *General Climatology*, 4th edn, 1983, p. 99, reprinted by permission of Prentice Hall, Englewood Cliffs, NJ; for figures 2.3(b), 2.4 and 2.5(a), (b), R. G. Barry and R. J. Chorley, *Atmosphere, Weather and Climate*, copyright Routledge; for figures 2.7 and 2.12, R. D. Thompson et al., *Processes in Physical Geography*, copyright 1992 Longman Group; for figures 2.10, 2.11 and 2.13, R. G. Barry and R. J. Chorley, *Atmosphere, Weather and Climate*, copyright Routledge; for figures 2.14(c), (d), Royal Meteorological Society; for figures 3.1, 3.2, 3.3, 3.4 and 3.8, D. F. Hayward and J. S. Oguntoyinbo, *Climatology of West Africa*, copyright 1987 Hutchinson; for figures 3.6, 3.7 and 3.9, I. J. Jackson, *Climate, Water and Agriculture*, copyright 1989 Longman Group; for figure 3.10(a), J. Lockwood, *The Physical Geography of the Tropics*, copyright 1974 Oxford University Press, Kuala Lumpur; for figure 3.10(b), S. Nieuwolt, *Journal of Tropical Geography*, 27 (1968); for figure 4.2, H. Eswaran, H. Ikwara and J. Kimble, *Proceedings, International Symposium on Red Soils*, copyright 1986 Elsevier; for figure 4.7, A. C. Millington; for figure 4.9, A. Young, *Tropical Soils and Soil Survey*, 1976, Cambridge University Press; for figure 4.10, *The Philippine Agriculturist*, Laguna; for figure 4.14, Soil Science Society of America, Madison; for figure 4.15, A. C. Millington et al., *Zeitschrift für Geomorphologie*; 52 (1985), Gebrüder Borntraeger; for figure 5.1, H. Walter, *Vegetation of the Earth and Ecological Systems of the*

Geobiosphere, copyright 1979 Springer-Verlag; for figures 5.4 and 5.13, K. A. Longman and J. Jenik, *Tropical Forest and its Environment*, copyright 1987 Longman Group; for figures 5.12 and 5.14, J. P. Schulz, *Ecological Studies on Rain Forest in Northern Suriname*, copyright 1960 Elsevier; for figure 5.17(a), M. Dantas and J. Phillipson, *Journal of Tropical Ecology*, 5 (1989), Cambridge University Press; for figure 5.17(b), A. Martínez–Yrízar and J. Sarukhán, *Journal of Tropical Ecology*, 6 (1990), Cambridge University Press; for figure 5.19, P. B. Tomlinson and M. N. Zimmermann, *Tropical Trees as Living Systems*, Cambridge University Press; for figure 5.25(b), E. Mayr and R. J. O'Hara, *Evolution*, 40 (1986); for figure 6.5, J. Budel, *Zeitschrift für Geomorphologie*, 1 (1957), Gebrüder Borntraeger; for figure 6.6, A. Faniran and L. E. Jeje, *Humid Tropical Geomorphology*, copyright 1983, Longman Group; for figures 6.13 and 6.14, D. Ford and P. W. Williams, *Karst Geomorphology and Hydrology*, copyright 1989 Chapman and Hall; for figures 6.15 and 6.16, P. W. Williams, *Zeitschrift für Geomorphologie*, 29 (1985), Gebrüder Borntraeger; for figure 7.6, J. Tricart, *Landforms of the Humid Tropics, Forests and Savannas*, copyright 1972 Longman Group; for figure 7.7, reprinted from 'A global survey of sediment yield', by M. B. Jansson, from *Geografiska Annaler*, *Series A, 70* (1988), 81–98, by permission of Scandinavian University Press; for figure 7.11, reprinted from 'The Rufiji River, Tanzania. Hydrology and sediment transport' by P. H. Temple and A. Sundborg, from *Geografiska Annaler*, *Series A*, 54 (1972), 345–68, by permission of Scandinavian University Press; for figure 7.12, R. P. D. Walsh; for figure 8.2, C. D. Ollier, *Weathering*, copyright 1984 Longman Group; for figure 8.3, M. J. McFarlane, *Journal of African Earth Sciences*, 12, 267–82, copyright 1991 Elsevier; for figure 8.4, O. Malm et al., *Ambio*, 19 (1990), 11–15; for figure 8.5, I. Burton, R. W. Kates and G. F. White, *The Environment as a Hazard*, 1978, p. 20, Guildford Publications, New York; for figure 8.6, *World Map of Natural Hazards* with permission from Munich Reinsurance Company, 1988; for figure 8.8, A. Goudie, *The Nature of the Environment*, p. 11, copyright 1989 Blackwell Publishers; for figure 8.10, A. Rapp and R. Nyberg, *Ambio*, 20 (1991), 210–18; for figure 8.13, C. F. Jordan, *Amazonian Rain Forest, Ecosystem Disturbance and Recovery*, copyright 1987 Springer-Verlag; for figure 8.14, A. T. Grove and F. M. G. Klein, *Rural Africa*, 1979, Cambridge University Press; for figure 8.16, from 'Slash and burn impacts on a Costa Rican wet forest' by J. Ewel et al., *Ecology*, 62 (1981), 827, copyright 1981 by the Ecological Society of America, reprinted by permission; for figure 8.17, *Southeast Asian Studies*; for figure 8.20, *Association for Tropical Biology*; for figures 8.21 and 8.23, C. F. Jordan, *Amazonian Rain Forest. Ecosystem Disturbance and Recovery*, copyright 1987 Springer-Verlag; for figure 9.2, A. C. Millington, *Conservation in Africa: People, Policies and Practice*, copyright 1988 Cambridge University Press; for figures 9.6, 9.7 and 9.8, H. C. Brookfield and Y. Byron, first published in *Global Environmental Change*, 1 (December 1990) 45, 49, 50 and 51, and reproduced here with the permission of Butterworth-Heinemann, Oxford; for figure 9.11, J. Gribbin, *New Scientist*, 110 (1508) (1986); for figure 9.12, J. Gribbin, *New Scientist*, 116 (1588) (1987); for figure 9.13, UNEP; for figure 9.14, Commonwealth Secretariat.

We are grateful to the following for permission to reproduce tables, commonly in a modified form:

For table 5.8, I. Deshmukh, *Ecology and Tropical Biology*, copyright 1986 Blackwell Scientific Publications; for table 6.8, H. A. Viles, *Biogeomorphology*, p. 332, copyright 1988 Blackwell Publishers; for table 7.4, reprinted from 'Comparison of surface runoff and soil loss from runoff plots in forest and small scale agriculture in the Usumbara Mountains, Tanzania', by L. Lundgren from *Geografiska Annaler, Series A*, 61 (1980), 113–48, by permission of Scandinavian University Press; for tables 7.5 and 7.6, *Zeitschrift für Geomorphologie*, 36 (1980), 176–202, Gebrüder Borntraeger; for table 8.1, from R. Edwards and K. Atkinson, *Ore Deposit Geology*, copyright 1986 Chapman and Hall.

We are also grateful to the following for the loan of photographs and permission to include them in this book:

Dr E. Cater (6.8), Mr J. B. Reading (8.5), Dr A. Brass (8.6), Mr Jack Tyndale-Biscoe (8.7), Dr M. Thorp (8.3(b)), Professor I. Douglas (8.9).

Every effort has been made to trace all the copyright holders of illustrations used in this book. However, if any have been overlooked the authors and publishers will be pleased to make the necessary arrangements at the first opportunity.

ONE

Introduction

1.1 Defining the Humid Tropics

Sensu stricto, the 'tropics' refer to parts of the world which lie between the Tropic of Cancer (23.5 °N) and the Tropic of Capricorn (23.5 °S) (figure 1.1). These latitudinal boundaries correspond to the outer limits of the areas where the sun can lie at zenith. The tropics therefore receive large amounts of solar radiation throughout the year and as a result seasonal fluctuations in temperature are minimal and there is no distinct winter season.

For most geographical and ecological purposes, the above astronomical definition is too rigid and does not adequately delimit a region with distinctive physical or biological characteristics. Numerous other schemes for defining and subdividing the tropics have therefore been proposed by climatologists, geomorphologists and biologists. These, however, are also problematic. Difficulties of delimiting the outer boundary of the tropics and defining different types of tropics are particularly pronounced, especially across continental areas. Each scientific discipline also has different requirements of a classification system. For example, botanists will require the tropics to be distinguished by particular vegetation assemblages, climatologists will perceive the tropics as areas where specific atmospheric conditions prevail and geomorphologists will prefer that boundaries describe areas where physical processes take place with a certain intensity or magnitude. It is unusual for the boundaries of the various classification systems to correspond exactly.

Through its control on the solar radiation receipt, latitude is a major factor affecting climatic conditions and climate. Directly or indirectly, it is a major factor affecting the nature of landform processes, vegetation, soils, agriculture and even economic development. Boundaries based on climatic parameters or indices therefore make up the major group of classification schemes. Any boundary, however, is inevitably arbitrary since the atmosphere is a global entity with complex latitudinal interconnections. Furthermore, actual climatic conditions often change very gradually over

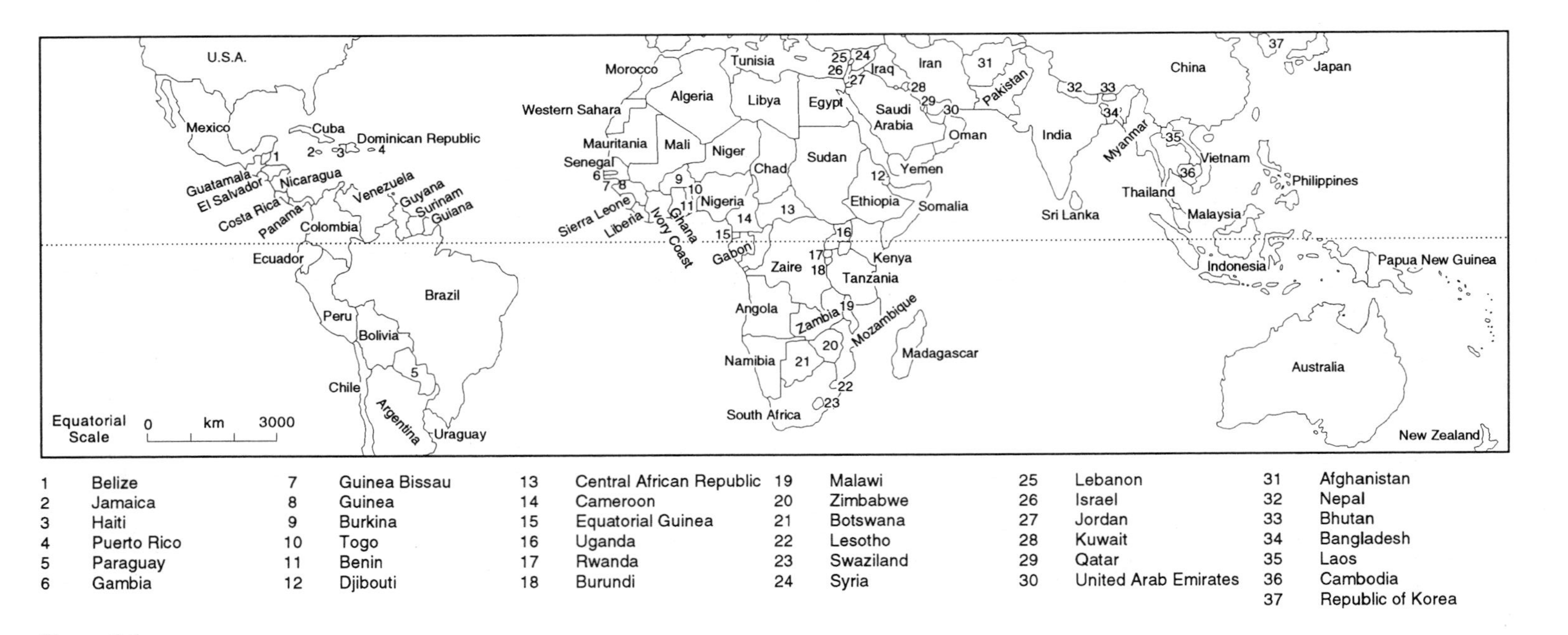

Figure 1.1
Countries which may be regarded as tropical (at least in part) (after Savage et al., 1982).

long distances and are likely to vary from year to year. Boundaries should therefore be regarded as representing transition zones (Nieuwolt, 1977).

Some classifications are based directly upon climatic parameters such as temperature and rainfall (e.g. Köppen, 1936; Garnier, 1958; Gourou, 1966; Tricart, 1972). Köppen's classification (figure 1.2) is the most widely recognized and forms the basis from which most other classifications have developed (Mink, 1983). Köppen's thermal requirement for a tropical climate is for an average mean temperature for the coldest month of 18 °C. This criterion excludes the cooler highlands (areas above 900 m), which make up around 25 per cent of the land surface within the tropics. However, in common with the tropical lowlands, these highlands receive high amounts of solar radiation and do not have a thermally depressed (winter) season. Nieuwolt (1977) includes the high altitude areas by transposing temperatures to sea-level equivalents, based upon a 5–6 °C per thousand metre elevation relationship. With this modification, Quito, Ecuador (2818 m above sea level (a.s.l.)), becomes as tropical as, for example, Madras, India (7 m a.s.l.). However, as Barrow (1987) points out, the highlands temperatures may be sufficiently depressed to affect biotic activity. High and low altitude tropical areas are therefore not directly comparable in botanical terms.

Since insolation and temperature are relatively uniform at any tropical site, differences in the amount and temporal distribution of available water, principally in the form of precipitation, largely account for regional and seasonal differences in tropical regions (Savage et al., 1982). It is therefore generally accepted that moisture is a valid criterion for differentiating between different types of tropics. However, the units used and the limits imposed vary enormously from scheme to scheme. Several schemes use mean annual rainfall totals to distinguish different types of tropics but there is no consensus on the required amount or its seasonal distribution. For example, Köppen (1936) and Nieuwolt (1977) use figures of 450–600 mm per year to divide the tropics into humid and dry parts, Gourou (1966) places the dry–wet boundary as low as 400 mm per year, while Tricart (1972) raises it to 750–800 mm per year.

Some classification systems conceptualize tropicality in hydrometeorological terms and differentiate the tropics based upon the relationship between water inputs (i.e. precipitation) and water outputs (i.e. evaporation). Garnier (1958), for example, defines humid areas as ones in which actual evapotranspiration equals potential evapotranspiration such that soil moisture is not depleted below a critical level. For areas which fulfil the tropical requirements of a minimum mean monthly temperature of 18 °C and a mean vapour pressure of at least 20 mb Garnier (1958) uses the number of months in which this condition is satisfied to differentiate tropical types. Holdridge (1967) notes that at higher elevations, because atmospheric temperatures are lower, the

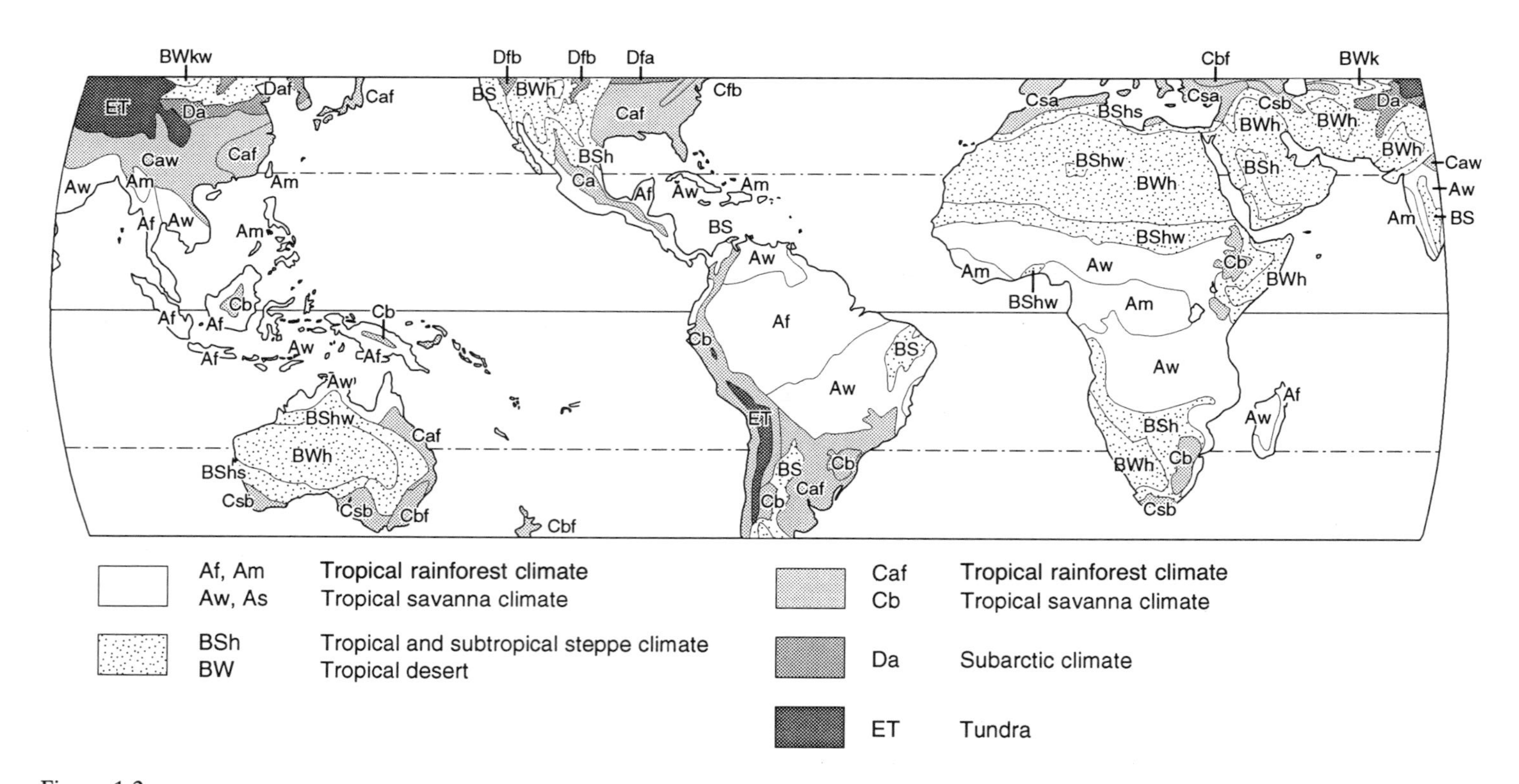

Figure 1.2
Köppen's (1936) classification of climates between 40°N and 40°S: humid tropical area includes places under Af, Am, Aw, As and the southern parts of Caf in the northern hemisphere and the northern parts of Caf in the southern hemisphere.

amount of evapotranspiration is also lower per unit time. Thus, less rainfall is required at lower temperatures to maintain humidity. Holdridge's classification incorporates a temperature criterion, biotemperature (equation (1.1)), to divide the tropics into thirty-eight bioclimates or 'life zones' (see section 5.2).

$$\text{Biotemperature } (^\circ\text{C}) = \frac{\text{sum of unit period temperatures } (^\circ\text{C})}{\text{number of unit periods (e.g. days, weeks etc.)}} \tag{1.1}$$

Holdridge differentiates seventeen humid tropical bioclimates in seven temperature-limited altitudinal zones (see table 5.3). Each zone takes its name from the mature plant formations found in the area (figure 1.3).

It should be noted that not all classification systems are quantitatively specific. Gourou, for example, describes the humid tropics simply as receiving 'rain enough for agriculture to be possible without irrigation' (Gourou, 1966, p. 1). However, he does then go on to quantify the amount of rainfall needed to satisfy this requirement on an annual basis. Barrow (1987) is concerned largely with moisture in terms of water resources and agricultural production. He divides the tropics into (a) humid (wet or equatorial), (b) subhumid (semi-humid or wet–dry) and (c) dry (semi-arid and arid), according to whether water is usually abundant, sometimes problematical or usually difficult respectively.

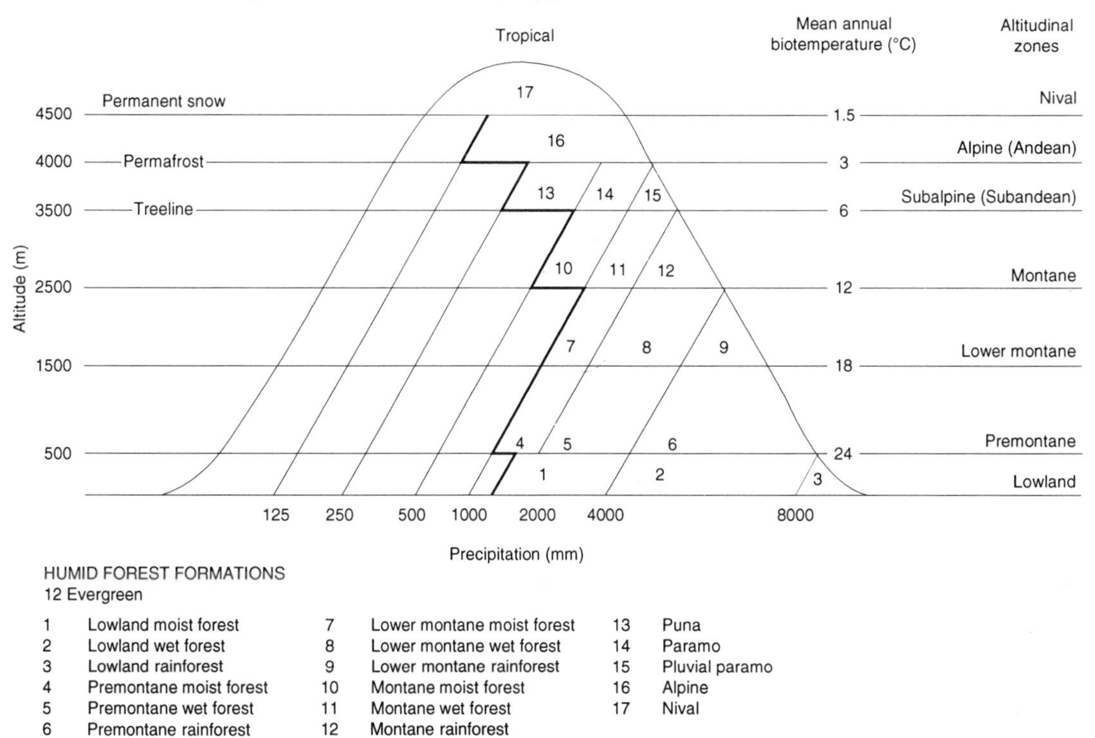

Figure 1.3
Humid tropical life zones (after Holdridge, 1967).

Most classification schemes distinguish different types of humid tropical areas but, again, there is little consistency in the number of subdivisions defined. Köppen (1936) identifies three climatic types: tropical rainforest climate (Af); tropical monsoon rainforest climate (Am); and tropical savanna climate (Aw) (figure 1.2). In most places these types correspond to Tricart's (1972) zoning categories, namely constantly humid; constantly humid with a short dry season; and seasonally humid with a long dry season. Tricart (1972) is more specifically concerned with areas which exhibit humid tropical weathering patterns and profiles. He adds a fourth subhumid category to encompass those areas (e.g. southern China up to 30 °N, eastern Brazil down to 30 °S) within which geomorphological processes, especially weathering patterns, are very similar to those of the humid tropics proper.

Other classifications have been directed towards agriculture. For example, Fosberg et al. (1961) recognize two types of humid tropics: constantly humid and seasonally humid. Jackson (1989) further divides these according to their seasonal rainfall regime (figure 1.4), since this is particularly critical for vegetation development and agriculture (section 3.6.1).

Chang and Lau (1982) have attempted a synthesis of classification schemes for the humid tropics based on the following criteria: (a) a minimum mean monthly temperature of 18 °C; (b) a wet month defined as having average rainfall of 100 mm or more; and (c) a half wet month defined as having average rainfall of 60–100 mm. Using these criteria they divide the humid tropics into three classes: (a) wet tropics (more than 9.5 equivalent wet months); (b) moist tropics (7–9.5 equivalent wet months); and (c) wet and dry tropics (4.5–7 equivalent wet months). This classification is similar to Köppen's scheme but incorporates almost all monsoonal climates. It also offers a convenient framework for discussing agriculture (Mink, 1983).

The number, variation and degree of sophistication of classification systems can appear bewildering. For single subject studies, it is usually possible to find or adapt a scheme which adequately describes the areas under investigation. This text, however, is concerned with many aspects of the humid tropical environment and a single classification scheme which accurately defined all the areas to be discussed could not be found. Therefore, no single system was imposed. The working definition used throughout the text was that the regions dealt with were regarded, by consensus, as tropical owing to their consistently high receipt of solar radiation, heat and moisture (the latter two reduced to sea-level equivalents). In combination, these climatological parameters produce regions which support vegetation, soil, patterns of landform development, animal life, agriculture and economic development distinctive from other parts of the earth surface. Linear boundaries were considered relatively unimportant since they rarely exist as physical features in the landscape.

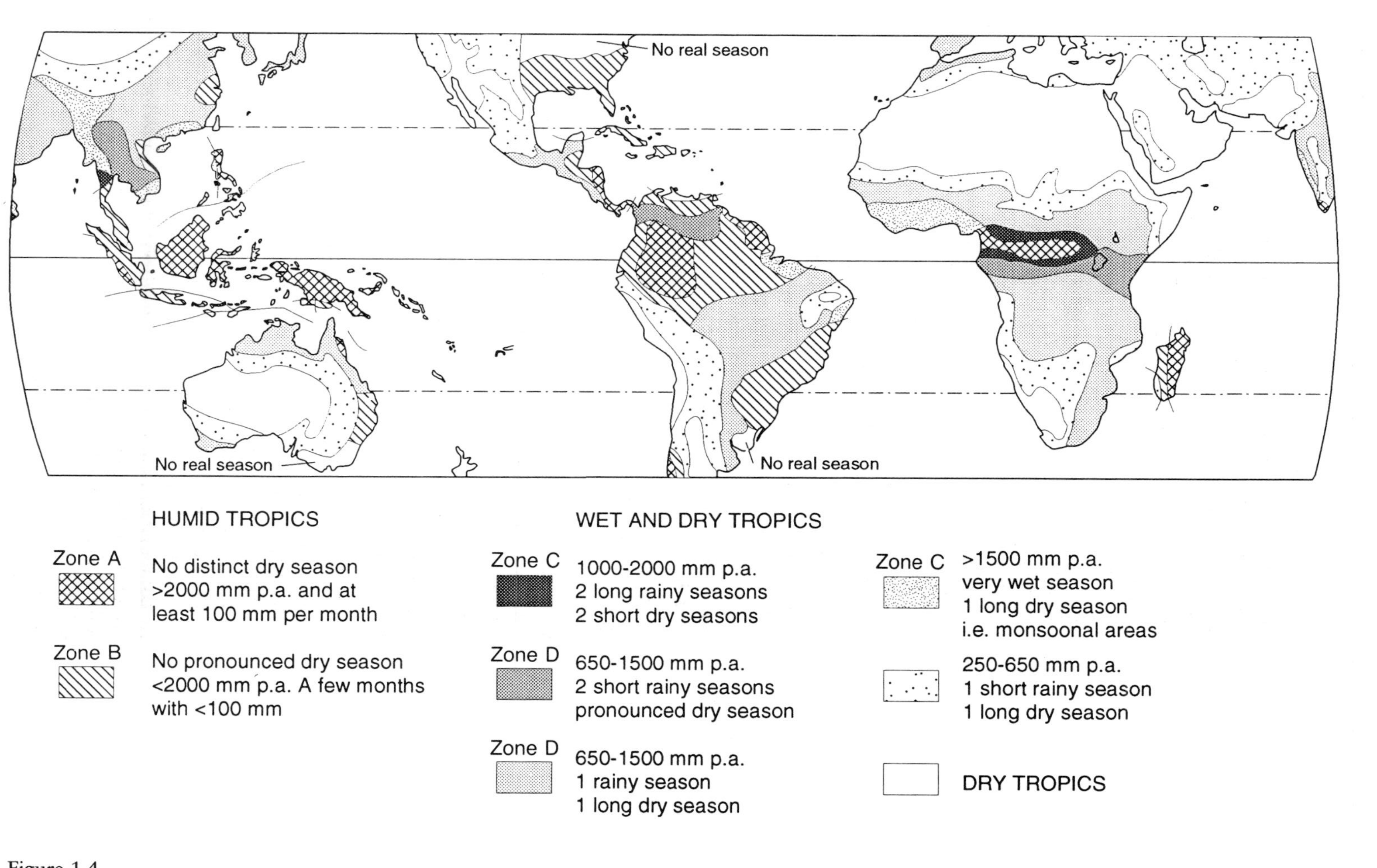

Figure 1.4
Classification of the tropics based on the seasonal distribution of rainfall (after Jackson, 1989, p. 60, figure 3.10).

The subject matter of this text also requires a temporal context. On a geological time scale tectonic activity may have significantly altered climatic and physiographic conditions (Thomas, 1994). Some of the features discussed may be relic features, representing artifacts of tropical conditions which have now disappeared. Although they lie outside the humid tropics at present, they have been included since they owe their origin or characteristic form to humid tropical conditions. Consideration has also been given to the fact that some features present in areas currently regarded as humid tropics are likely to have at least partly been formed under cooler and/or drier climatic regimes.

It is also important to recognize the importance of shorter term changes in climate. The boundary of the humid tropics will fluctuate regularly as climatic conditions vary from year to year. Furthermore, along the margins of humid tropical lands, the frequency and intensity with which humid tropical conditions have been interrupted due to climatic variability may be significant in determining physical forms.

1.2 The Importance of the Humid Tropics

Depending upon the classification used, the tropics extend to cover approximately 35–40 per cent of the earth's land surface. Following Köppen's scheme, only around 4 per cent is sufficiently wet to support rainforest (i.e. Af and Am categories) but up to 20 per cent can be regarded as humid tropical if more seasonal areas (i.e. Aw and As) are included. The largest continental expanses of humid tropical land extend between around 10 °N and 10 °S of the equator; they include the Amazon Basin in South America; southeast Mexico and Central America; the Congo Basin, Mozambique, Madagascar and the Guinea coast of Africa; the subcontinent and large islands of southeast Asia; northeast Australia and New Guinea. Over oceanic areas, the humid tropics extend poleward to over 20 ° and include the islands of the Caribbean and Indian Ocean (e.g. Antilles and Mascarene Islands). In the Pacific the region extends to islands as far north as Hawaii and as far south as New Caledonia (Mink, 1983).

Over such an extensive area it is not surprising to find that differences in land-surface age, tectonics and structure have produced striking variations in ecosystems and landforms and in the dominance of physical and biotic processes. There are also marked seasonal and longitudinal variations in climatic conditions within the humid tropical region, the result of variations in the global atmosphere and enhanced by the distribution of land and ocean. Relief is another important control on climatic conditions, since it moderates temperatures and modifies rainfall (section 2.6.2).

A detailed understanding of these diverse and complex environments is essential to both pure and applied aspects of science. First, investigations during the second half of this century have confirmed that the tropics provide the key to the understanding of much of biological and earth science (Douglas and Spencer, 1985). For example, low latitudes do not suffer great seasonal contrasts in the type and intensity of physical processes and, although the legacy of past climatic change in low latitudes is now recognized as having been much more significant than once believed, most areas escaped the dramatic changes brought about by periglacial and glacial conditions during the Pleistocene. This has helped geomorphologists to unravel the nature of geomorphic processes and ecologists to better understand the complexities of ecosystem functioning.

The tropics are also becoming increasingly important in their own right, especially in economic and political terms. There are at least sixty-three countries and major islands with territories lying within the humid tropics (table 1.1). These countries contain around 40 per cent of the world's population, i.e. over 14 billion people, and enormous reserves of biotic and mineral reserves (chapter 8). However, almost all are currently categorized as underdeveloped by the United Nations and rely on agriculture to sustain both their local and national economies. Agriculture is, of course, particularly sensitive to environmental conditions. A detailed understanding of the physical environment is therefore an essential prerequisite to the sustainable development of humid tropical regions. The key to success here lies in the effective application of the available knowledge in the context of technological, social and political constraints.

1.3 Perceptions of the Humid Tropics

Humid tropical environments have been expertly managed by their indigenous populations for thousands of years. Some of the earliest civilizations were located in the tropics (Ooi Jin Bee, 1983). The Mayan civilization, for example, which made its appearance in central America before the Christian era and survived until the Spanish conquest, is known to have operated sophisticated irrigation systems for its crops and extensive sewage systems for its cities. The peoples' adaptation to and manipulation of their physical environments could only occur through an appreciation of the nature and dynamics of physical systems.

Modern Western science and explanation, however, originated in more northerly latitudes. The humid tropics were 'discovered' by European explorers in the late fifteenth century, and by the nineteenth century many explorers and naturalists were recording their impressions of the humid tropics for the benefit of their temperate-bound kinsmen. The thoughts and ideas of the early scientists and agriculturalists were invariably influenced by preconceptions based upon their knowledge of temperate landscapes. Also, since relatively few humid tropical areas were accessible to Europeans, they were unlikely to become acquainted

Table 1.1 Major countries within the humid tropics

Country	*Land area (million ha)*	*Humid tropical (per cent)* [a]
Africa		
Burundi	2.8	32
Cameroon	47.5	35
Central African Republic	62.3	25
Congo	34.2	100
Equatorial Guinea	2.8	100
Gabon	26.8	85
Ghana	23.9	30
Guinea	24.9	95
Ivory Coast	32.2	40
Kenya	58.3	18
Liberia	11.1	100
Malagasy Republic	58.7	
Mauritius	0.2	
Mozambique	78.3	
Nigeria	92.4	10
Rwanda	2.6	35
Sao Tome	0.1	100
Sierra Leone	7.2	100
Tanzania	94.0	5
Togo	5.6	33
Uganda	23.6	10
Zaire	234.5	85
Asia		
Brunei	0.6	100
Burma	67.8	
India	215.3	
Indonesia	190.4	95
Kampuchea	18.1	35
Laos	23.7	
Peninsular Malaysia	143.1	
Sabah	7.6	100
Sarawak	12.5	
Papua New Guinea	46.2	100
Philippines	30.0	36
Sri Lanka	6.6	25
Thailand	51.4	20
Vietnam	33.3	25
Central America		
Belize	2.3	15
Costa Rica	5.1	93
Dominican Republic	4.9	
El Salvador	2.1	
Guatamala	10.9	
Haiti	2.8	
Honduras	11.2	
Mexico	98.6	
Nicaragua	13.0	30
Panama	7.6	100
Puerto Rico	0.9	

Table 1.1 cont.

Country	*Land area (million ha)*	*Humid tropical (per cent)* [a]
South America		
Bolivia	109.9	8[b]
Brazil	851.2	45
Colombia	114.0	72
Ecuador	28.4	55[b]
French Guiana	9.1	100
Guyana	21.5	70
Peru	128.5	56[b]
Suriname	16.3	85
Trinidad	0.5	90
Venezuela	91.2	57[b]
Oceania		
Australia	253.9	
Fiji	1.8	
New Caledonia	1.9	
New Hebrides	1.5	
Samoa	0.3	
Solomon Islands	2.8	

[a]Where known
[b]Tropical tundra accounts for an additional 5 per cent.

Source: Adapted from Savage et al. 1982, based on information from Persson, 1974, and Brown et al., 1980

with a comprehensive range of humid tropical environments. The fact that the humid tropics are extremely diverse as well as being very different from areas at higher latitudes has not been widely appreciated until relatively recently. The pattern and nationality of early scientific investigations closely follow the spread of colonialization and, to the present day, the majority of physical scientists interested in the humid tropics live, or at least were educated in, temperate regions (e.g. figure 1.5).

One of the earliest perceptions of the humid tropics was that it was an area of lush, fertile forest. This view was initiated by accounts of the richness and diversity of the flora and fauna which were interpreted as great wealth and fertility. The apparent ubiquity of the rainforest was also reinforced by the large amounts of attention it received by botanists and biologists. The forests were often regarded as invincible obstacles to development which would grow up again as fast as they were cut down. Nourishing the forests were soils, commonly referred to in terms of their inexhaustible fertility. During the late eighteenth century, colonialists looking to exploit this apparent richness found to their cost that these observations were misfounded.

Physiographically, the humid tropics are extremely varied and it is impossible to visualize a mode of tropical landscape applicable throughout the area (Gupta, 1984). However, the majority of early geomorphology was carried out in parts of Africa and South America which lie within the ancient cratons. As a result, the humid tropics were frequently described as flat plainlands whose

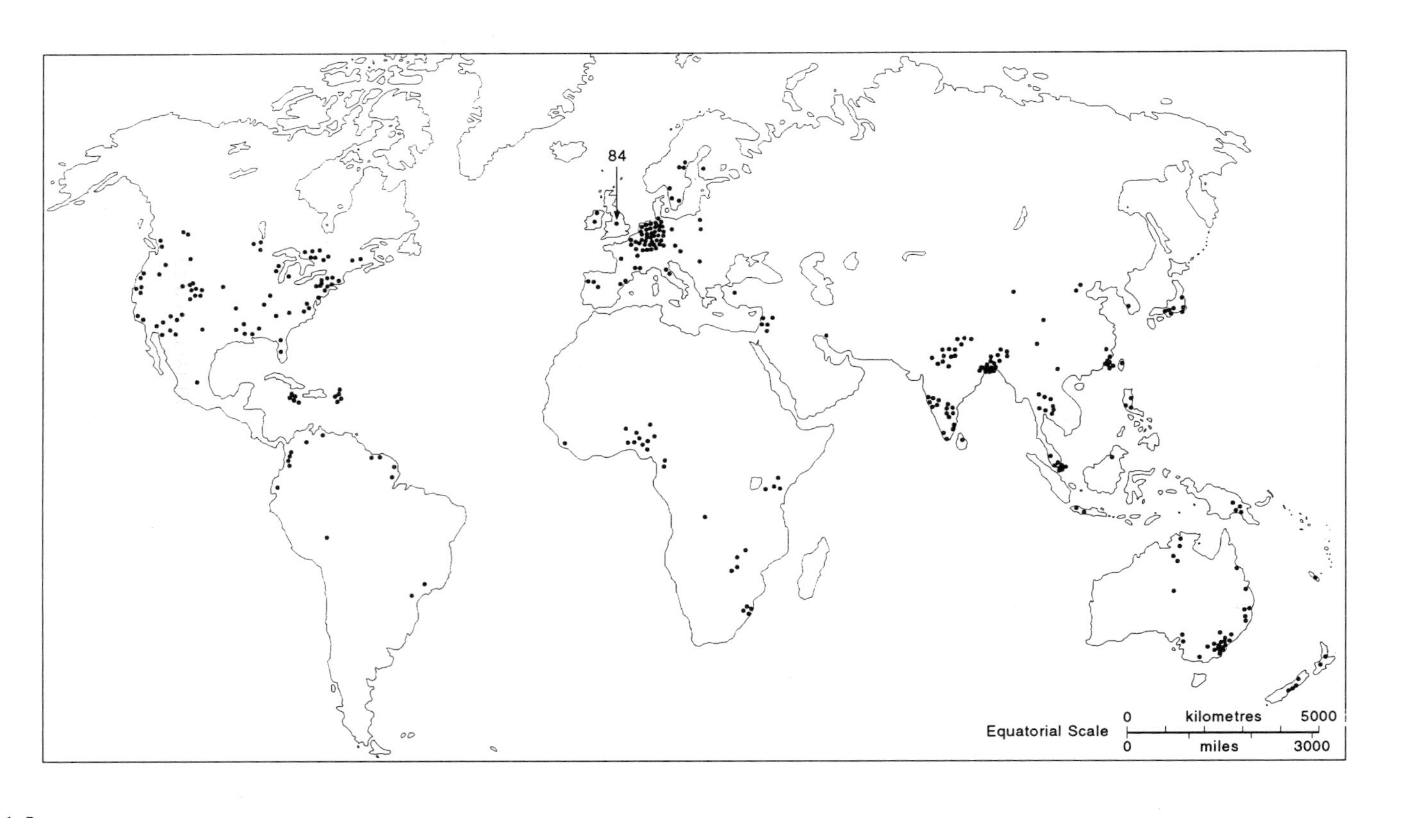

Figure 1.5
Distribution of 'tropical scientists' (after Tropical Geomorphology Newsletter*).*

monotony is only sporadically interrupted by 'extraordinary isolated hills' (Thomson, 1882, p. 69). The great depth to which rock had 'rotted' and the high percentage of clay present in the soils were also widely discussed.

The classic perception of humid tropical weather is that days and nights are similarly hot and humid and that rainfall occurs in heavy localized downpours. Unlike temperate precipitation, tropical rainfall does often occur as a result of localized convective disturbances (Riehl, 1979). Rainfall is therefore generally more intense and localized but there is considerable variation in both time and place in the distribution of these tropical rainstorms. A major exception to this pattern is the longer duration but less intense rainfall associated with the monsoon. The apparent uniformity and simplicity of weather in the tropics was reinforced by meteorologists from higher latitudes who noted the absence of active fronts and steep pressure gradients which are responsible for the majority of weather in temperate regions (section 2.3). Seasonal variations in rainfall were explained simply by the existence of an equatorial trough within which air ascends and gives rise to a belt of precipitation which migrates following the movement of the sun. Thus, poleward margins of the tropics received a single wet season, corresponding to the summer solstice, and at the equator the double wet season was explained by the presence of the sun overhead at each of the equinoxes.

Another perception of the humid tropics is of an environment where disease and unhealthiness is commonplace. Gourou (1966), for example, states: 'We who live in temperate lands find it difficult to realize that the water in streams, and even the soil, may swarm with dangerous germs, and that myriads of blood-sucking insects may inject deadly microbes into the skin' (Gourou, 1966, p. 7). This unhealthiness was related to climate since the constant high temperatures, high humidity and abundance of water surfaces fed by the rains were seen as favourable for the persistence of pathogenic complexes. In the early days of European contact with the tropics there were certainly high death rates associated with tropical diseases and general unsanitary conditions. Lewis and Berry (1988) argue that these factors were often compounded by Europeans overdressing, which resulted in attacks of heat exhaustion and pneumonia. Environmental disease, affecting indigenous and introduced human populations and animals, is undoubtedly an important problem in the tropics (figure 1.6), but its prevalence and distribution relate to a wide range of social and cultural factors and especially poverty.

Huntington (1915, 1945) and his followers used the association between tropicality and unhealthiness to explain the fact that no tropical country has achieved a high state of economic development in modern times. Tropical diseases were seen as problems of the 'uncivilized' countries (Gourou, 1966, p. 13). Climatic determinism also popularized the belief that tropical peoples were in some way inferior and lazy. Modern day reactions against such associations

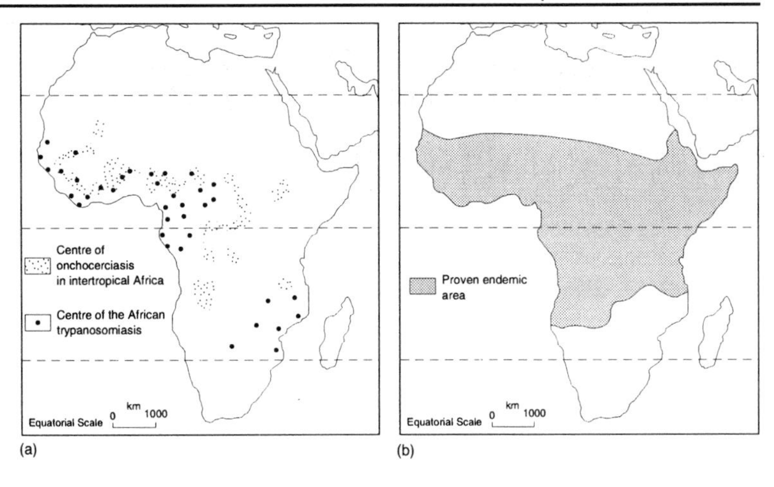

Figure 1.6
Distribution, within Africa, of selected environmental diseases (after Environmental Development Action, 1981): (a) onchocerciasis and trypanosomiasis; (b) endemic yellow fever; (c) leprosy; (d) cerebrospinal meningitis.

have tended to cause the pendulum to swing the other way. Climate has frequently been ignored or scantly regarded in examinations of tropical underdevelopment.

The incomplete understanding of the diversity and complexity of tropical environmental systems and crude generalizations, the result of a skewed information base, have resulted in the formulation and persistence of serious misconceptions. By providing an overview of some of the most important aspects of the humid tropical physical environment, it is hoped that this text will help dispel some of the misconceptions and qualify some of the preconceptions which undergraduates of geography and allied disciplines frequently hold.

TWO

The Humid Tropical Atmosphere

2.1 Introduction

In this chapter we analyse atmospheric circulation systems operating within the humid tropics, especially the role of the Hadley cell in general and intertropical convergence zone (ITCZ) changes in particular. We examine atmospheric disturbances at every scale, from localized convective cells to more extensive hurricanes, in order to emphasize the factors which control rapidly changing weather systems and the pronounced climatic variability which characterize the region. The mechanisms of large-scale monsoon circulations are discussed in detail, particularly in terms of the seasonal characteristics of the distinctive monsoonal climatic regimes. The chapter ends with a discussion of small-scale weather patterns associated with terrain factors and we examine the role of the southern oscillation in extreme tropical weather events.

2.2 Atmospheric Circulation Systems

Circulation systems in the atmosphere over the humid tropics are primarily controlled by the well-known Hadley cell regime (figure 2.1(a)). This thermally directed cell has two major low level components, namely equatorial low pressure convergence (the ITCZ) and subtropical high pressure divergence at about 30° north and south, with associated subsidence or trade wind inversions. These convergence and divergence zones in both hemispheres are associated with a poleward flow aloft and equatorward flow at the surface, which is represented by the trade winds.

The weather and climate of the humid tropics are dominated by the equatorial limb of the Hadley cell which, lying between the trade wind regions of the two hemispheres, is a distinctive surface trough which is nearly continuous around the world when plotted

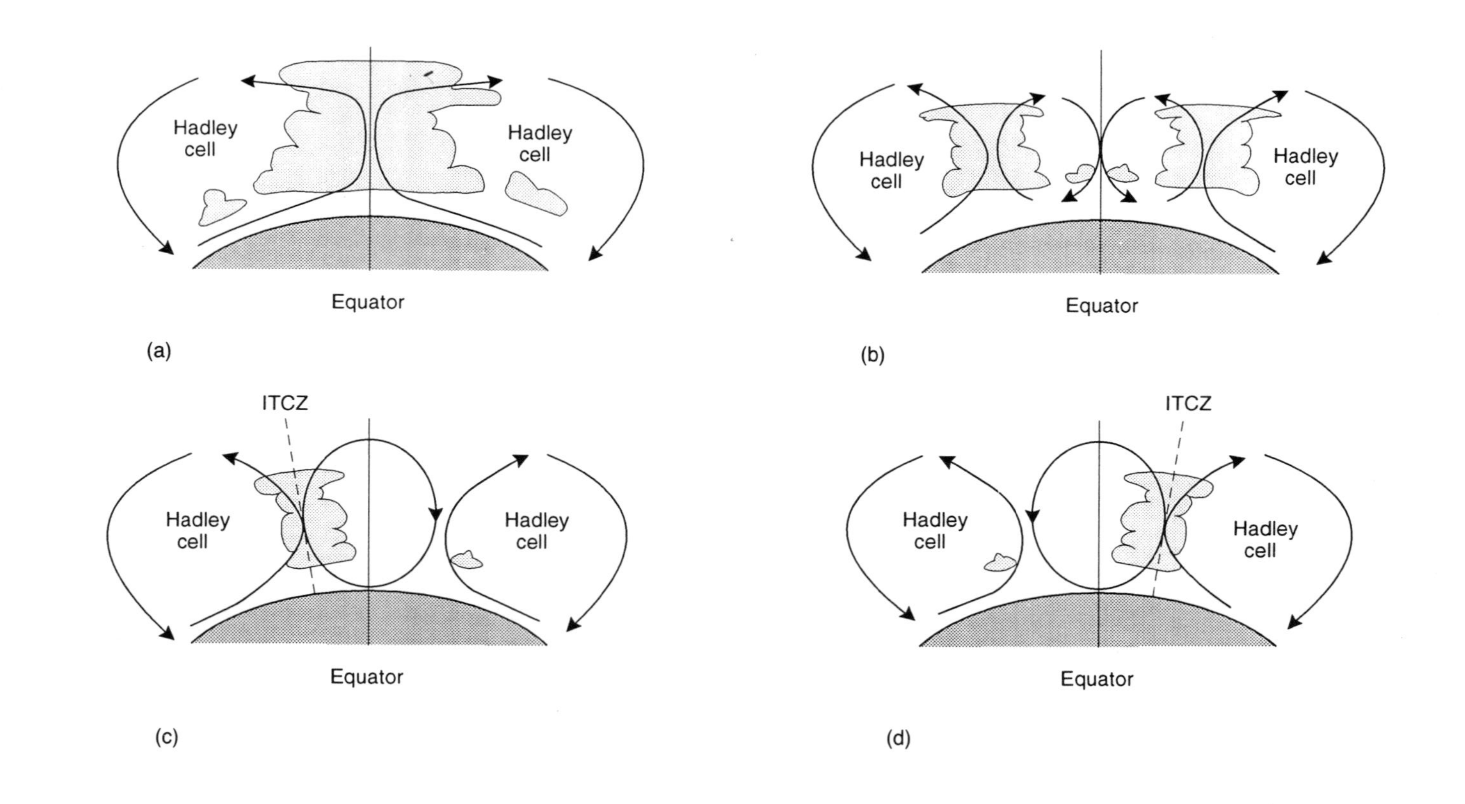

Figure 2.1
Models of the equatorial meridional circulation (after Chang, 1972): (a) a Hadley cell on each side of the equator; (b) two equatorial cells separating two Hadley cells; (c) one equatorial cell separating two Hadley cells, with an ITCZ in the northern hemisphere; (d) one equatorial cell separating two Hadley cells, with an ITCZ in the southern hemisphere.

on mean annual pressure charts. Usually, it can be traced upwards to the mid-troposphere as a quasi-vertical axis of symmetry. In the 1930s, the streamline convergence of the trades was recognized as the intertropical front (ITF). Initially it was assumed that the front was thermodynamically similar in structure to the polar front and synoptically was amenable to the same methods of analysis (Chang, 1972).

The ITF can be recognized over continental areas such as West Africa and southern Asia where, in summer, hot dry continental tropical air converges with cooler, humid equatorial air. Under these conditions, sharp temperature and moisture gradients occur but the resultant discontinuity or front (i.e. the West African intertropical discontinuity (ITD) in figure 2.2) rarely produces weather disturbances of the mid-latitude type (Barry and Chorley, 1987). Elsewhere, pronounced temperature and density changes are rare along the ITF and its behaviour is fundamentally different from that of the polar front.

In 1945, the terms 'intertropical trough' (ITT) and 'intertropical convergence zone' (ITCZ) were preferred owing to the recognition of wind field convergence in tropical weather systems. The former term was defined as a low pressure equatorial trough between the northeast and southeast trades where they meet at a small angle without pronounced convergence.

The ITT is oriented approximately west to east and is usually located within 350 km of the equator. When the trades meet at a pronounced angle, the ITT becomes the ITCZ, a zone of convergence about 80–300 km wide and usually more than 5° of latitude from the equator. However, in practice it is difficult to differentiate between the ITT and the ITCZ so that the latter term is now commonly used to denote both features (Chang, 1972). The ITCZ is characterized by great variability in both its structure and location. Structural contrasts are associated with the varying strength of the opposing trade winds since weak convergence makes it difficult to locate the ITCZ, where cloud development is limited to isolated, shallow cumulus clouds and scattered altocumulus clouds, with little precipitation. As the convergence intensifies, cumulonimbus clouds become more common with violent turbulence and heavy rainfall. At times, two or more vigorous convergence zones can be identified with subsidence and fair-weather clouds located between vigorous cumulus convection.

The classical model of ITCZ structure is represented by the basic circulation of the Hadley cell (figure 2.1(a)), where the equatorial convergence zone is the common limb of upward vertical movement between the Hadley cells in the two hemispheres. Fletcher (1945) suggested that radiational cooling at the cloud tops is responsible for subsidence–divergence along the equator, forming two new circulation cells and two new convergence zones in the hemispheres (figure 2.1(b)). Asnani (1968) proposed two further refinements of the classical model with a single equatorial cell separating the Hadley circulations (figures 2.1(c) and 2.1(d)). The ITCZ is now

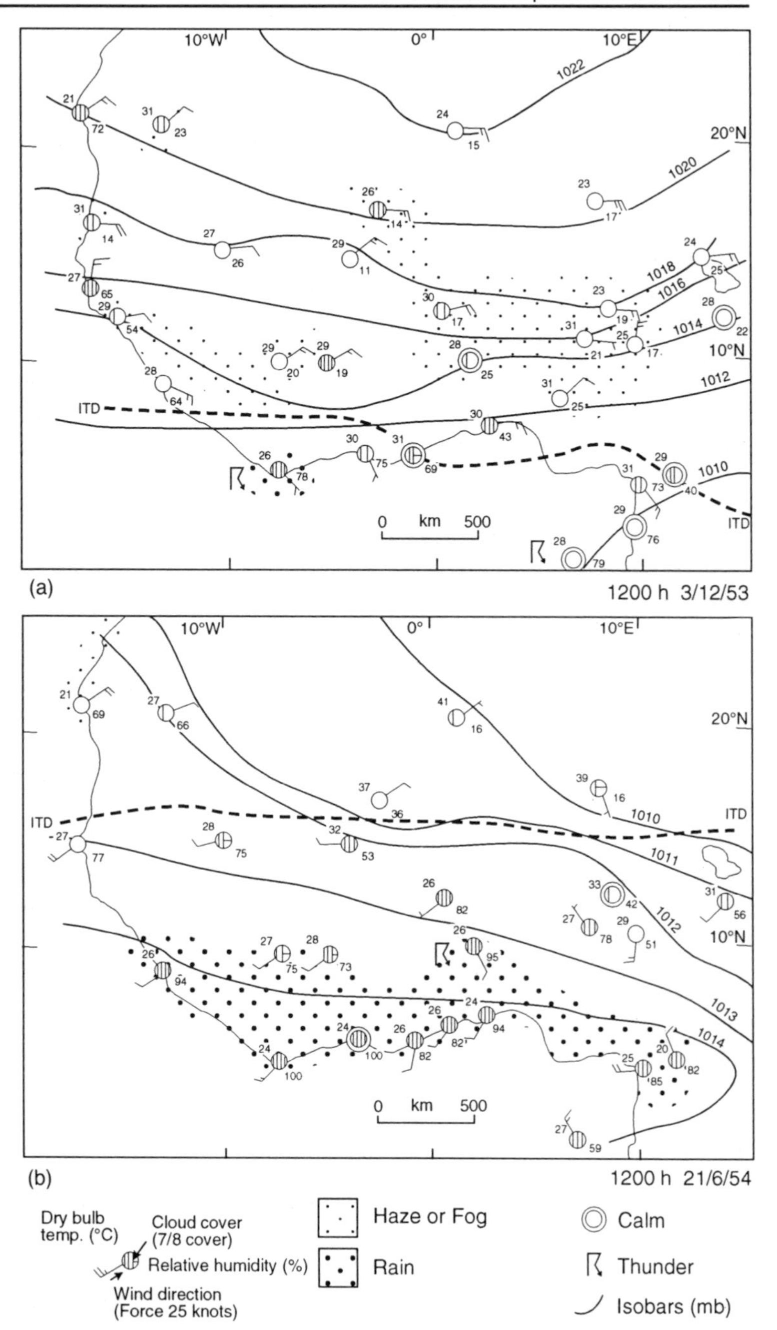

Figure 2.2
Circulation systems and prevailing weather over West Africa (a) in December and (b) in June (after Hayward and Oguntoyinbo, 1987).

located to the north and south of the equator, associated with heavy cloud conditions and precipitation; conversely, on the descending limb of the equatorial cell, pronounced subsidence accounts for the absence of clouds.

It is now apparent that the equatorial trough is not a region of continuously rising air with associated cloud and rain. Convergence is intermittent in time and space and maximum convergence and ascent is often found several degrees equatorward of the trough (Jackson, 1989). Satellite observations have confirmed that a complete cloud cover does not occur in the region and that a series of fairly well-developed cloud clusters (convergence) and cloudless areas (subsidence) is more common. Also, the convergent areas of rain and cloud have very varied structures and range from linear to oval and circular formations. Furthermore, it is evident that these cloud clusters are commonly associated with westward-moving disturbances, where areas of convergence can grow or decay over periods of a few days (as discussed in later sections).

The position of the ITCZ is generally assumed to vary seasonally in direct response to changes in location of maximum solar heating and the zone of seasonal maximum temperature (i.e. the so-called thermal equator). Consequently, in July, the ITCZ is likely to be located in the northern hemisphere, at around 25 °N over hot continental southeast Asia and about 5–10 °N over the comparatively cooler Pacific and Atlantic Oceans (figure 2.3(a)). In January, when the thermal equator moves towards the Tropic of Capricorn, the ITCZ is located mainly in the southern hemisphere where (because of the smaller land masses and reduced continental heating) it is likely to be around 15 °S over land and close to the equator over the oceans. However, figure 2.3(a) reveals that over the eastern Pacific Ocean (off Peru) and equatorial Atlantic Ocean (Gulf of Guinea) the ITCZ remains in the northern hemisphere, well to the north of the sun's zenithal latitude.

It is apparent from figure 2.3(b) that the location of the ITCZ is more complex than the seasonal distributions discussed above. In the southwestern parts of the Pacific and Atlantic Oceans, satellite observations reveal the presence of two semi-permanent convergence zones in February, both north and south of the equator, related to long waves in the middle and upper troposphere. These zones are not found at this time in the eastern South Atlantic and South Pacific owing to the occurrence of stabilizing cold ocean currents. Also, the convergence observed in the western South Pacific in February is now recognized as an important discontinuity. It is a zone of maximum cloudiness termed the South Pacific convergence zone (SPCZ), which extends from the eastern tip of Papua New Guinea to about 30 °S, 120 °W (Barry and Chorley, 1987).

The significance of the latitudinal migrations of the ITCZ in the prevailing weather conditions in the tropics is clearly demonstrated by the situation in West Africa (figure 2.2). Across West Africa the ITCZ (sometimes referred to as the intertropical discontinuity (ITD)) separates dry, tropical continental (TC) air in the north from humid tropical maritime (TM) air in the south. The northern TC airmass, centred over the Sahara, has marked seasonal thermodynamic changes whereas the southerly (TM) air mass possesses more uniform thermodynamic properties. In December, the ITD is located close to the Gulf of Guinea coast and the greater part of the

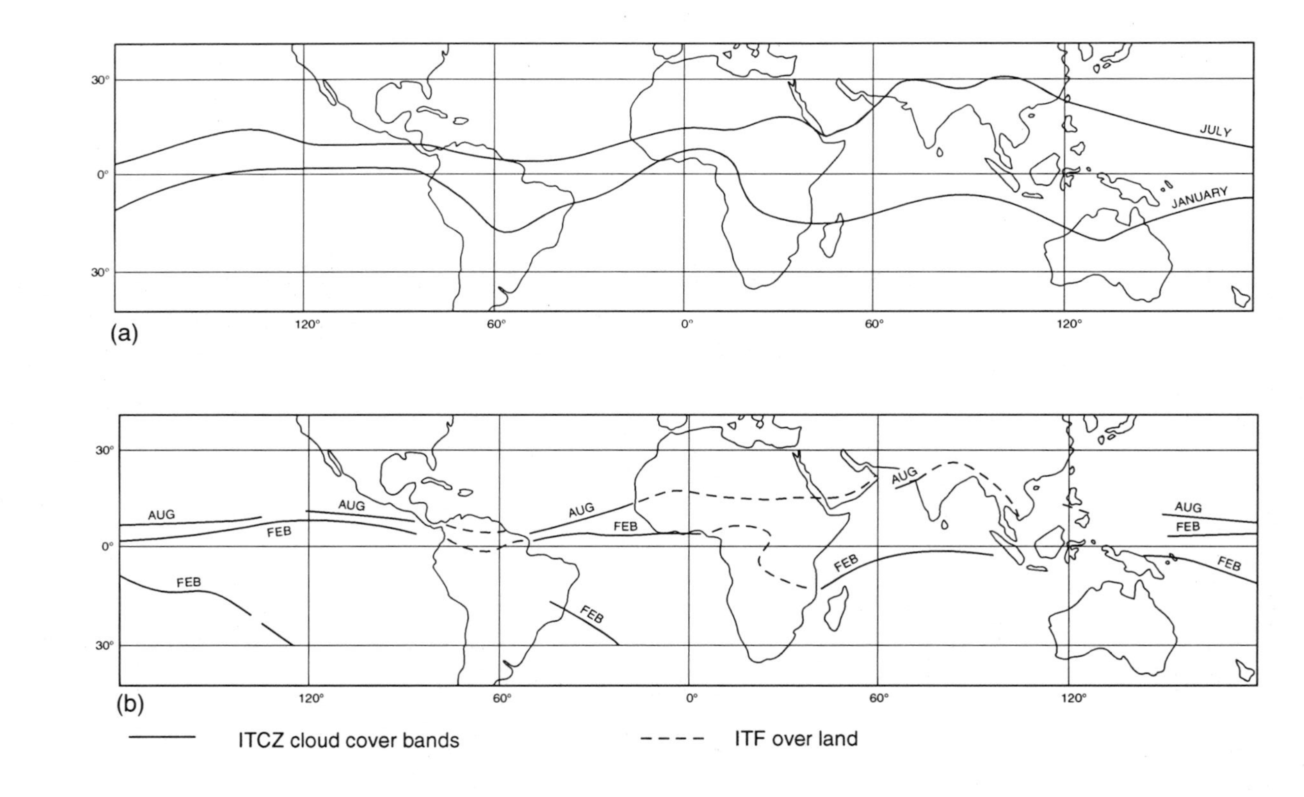

Figure 2.3
Mean positions of the ITCZ (a) in January and July (after Henderson-Sellers and Robinson, 1987) and (b) in February and August (after Barry and Chorley, 1987).

region is dominated by the hot (>26 °C), dry (<26 per cent relative humidity) and dusty Saharan air mass. Cloud and rain (between 30 and 50 mm) are confined to the more humid coastal 'strip' associated with the humid airflow from the Gulf. By June, the ITD has migrated to about 15 °N (it can reach 25 °N) and the TM air mass now extends over the southern part of the region. Cloud, high humidities and heavy rainfall occur as far north as 10 °, with monthly rainfall exceeding 1000 mm in the extreme southwest corner. To the north of the ITD, dry clear conditions continue, with temperatures now in excess of 37 °C.

2.3 Weather Disturbances

Our knowledge of humid tropical weather systems has been greatly hindered by the dearth of synoptic stations, particularly over the long term. For example, in Nigeria, which has a relatively good rainfall record (records commenced as early as 1891 at Lagos racecourse), synoptic stations were not established until 1941. By 1981 the number of stations had increased from fifteen to thirty, but these represented some 923,773 km^2, an area three times as large as Britain. For West Africa as a whole, there were 192 fully operational synoptic stations in 1981, of which thirty-nine measured solar radiation and fifty-six recorded upper air data, mostly wind speed and direction from balloon ascents. The distribution of these stations, however, was extremely patchy. For example across Liberia's 111,400 km^2 only two stations operated and both these were located near the capital Moravia (Hayward and Oguntoyinbo, 1987).

Until the 1940s, the controls of humid tropical climates were considered to be simple and obvious, especially compared with the more complex higher latitudes. A basic model of 'trade wind weather' was established with simple weather processes operating in response to persistent insolation and rainfall regimes. The long-recognized tropical cyclone was assumed to be the only disturbance of note in the region. Furthermore, its significance was underplayed because of its infrequent nature and the fact that many of these systems were 'missed' since they were remote from land-based observing stations (Thompson, 1981; Reading, 1990). Weather observations during the Second World War indicated the inadequacy of the trade wind model and revealed the complexity of humid tropical climates with distinctive, rapidly evolving weather systems and considerable climatic variations, even over extensive oceans. The use of operational weather satellites since the 1960s has greatly improved the understanding of tropical weather disturbances. At the same time, there has been a significant expansion of the regular observational network in the tropics and the initiation of concerted large-scale field experiments. For example, two specific experiments of the World Meteorological Organization's Global Atmosphere Research Program (GARP) in 1974 and 1979 carried out some 5000 upper air soundings over the equatorial Atlantic and Indian Oceans (Reynolds, 1985).

It is now generally accepted that within the humid tropics there are five categories of weather disturbances, which can be classified according to their space and time scales (Barry and Chorley, 1992). The smallest disturbances, with a life time of only a few hours, are represented by individual cumulus clouds (figure 2.4) which are deep convective systems some 1–10 km in diameter which develop from dynamically induced convergence in the trade wind boundary layer. Weak convective cells produce so-called 'street' alignments of cumulus clouds which occur roughly parallel to the air flow. With more pronounced convective mixing, the clouds can be organized into polygonal or honeycomb-like structures, particularly when cold air moves over a warmer sea surface (Barry and Chorley, 1992). The most vigorous convective towers are confined to regions where sea surface temperatures exceed 26 °C over tropical oceans (Zangvil, 1975). They can extend over 20,000 m with updraughts reaching 14 m s^{-1} and are accompanied by violent thunderstorms.

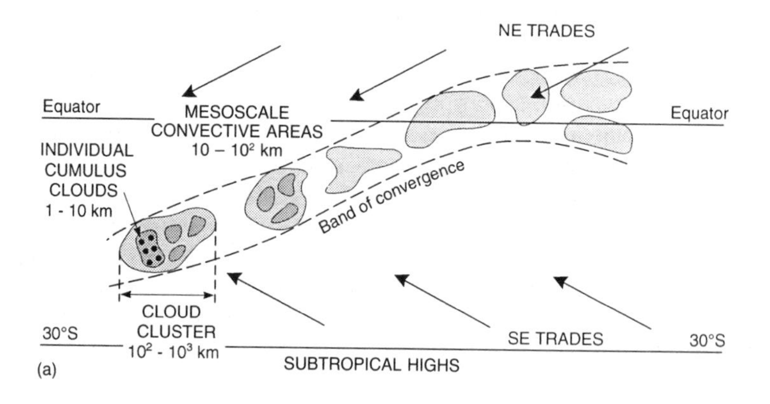

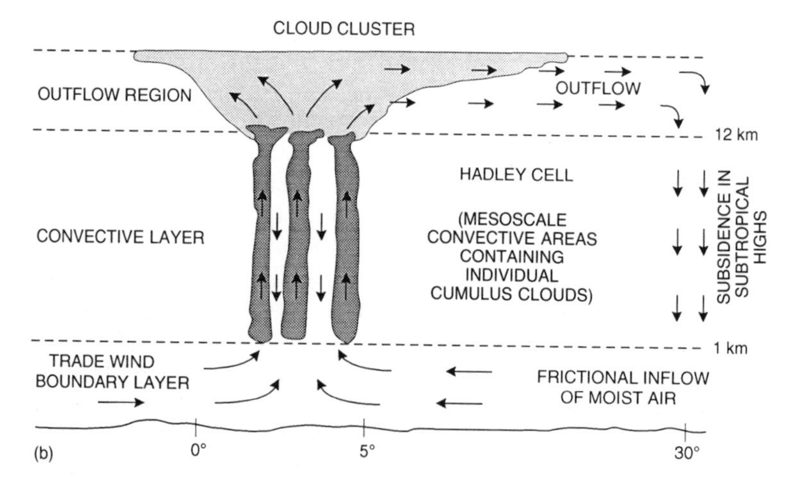

Figure 2.4
Mesoscale/subsynoptic weather disturbances showing (a) the spatial distribution and (b) the vertical structure of the convective elements which form cloud clusters (after Barry and Chorley, 1987).

The grouping of individual cumulus cells forms the second disturbance category which Barry and Chorley (1987) term meso-scale convective areas (MCAs) (figure 2.4), up to 100 km across where vigorous convergence and deep cumulonimbus clouds alternate with strong subsidence–divergence and cloud-free areas. The grouping of several MCAs represents the third category of a cloud cluster (figure 2.4), some 100–1000 km in diameter. They represent concentrations of cyclonic vorticity at low levels and anticyclonic vorticity at high levels, with a reversed flow of strong subsidence between the convective areas. Cloud clusters are commonplace features in the humid tropics, with ten to fifteen such groupings recorded per month, and can persist over several days. They are essentially subsynoptic-scale features with a somewhat arbitrary definition since they appear as amorphous cloud areas on satellite images, extending over areas ranging from 2° to 12° squares (Barry and Chorley, 1987).

Despite these subsynoptic-scale characteristics, the generation and maintenance of clusters have been recognized as functions of the synoptic-scale environment, including exceptionally strong meridional wind shear, strong low level positive relative vorticity, substantial latent heat flux 'fuelling' and small amounts of vertical wind shear in the troposphere. Most clusters display a similar thermodynamic evolution by warming the upper troposphere and cooling the lower troposphere during the growth of convection, along either highly organized north–south aligned disturbance lines or more loosely arranged clusters (Reynolds, 1985).

The fourth category of weather disturbance in the region is the synoptic-scale development of low pressure troughs/waves and well-organized cyclonic systems. These major disturbances are discussed in the next section. The final category is represented by the planetary waves, with wavelengths extending up to 40,000 km, which are observed in the upper troposphere. They do not appear to control weather mechanisms in the tropics directly but may interact with lower tropospheric systems through alternating high level cyclonic or anticyclonic vorticity, which is conducive to low level divergence and convergence respectively.

2.4 Synoptic-scale Weather Disturbances

2.4.1 Easterly waves

Within the ITCZ, the Coriolis force is too weak to generate circular atmospheric motions and to organize major synoptic-scale disturbances comparable to mid-latitude depressions. Instead, tropical convection becomes associated with marked wave-like undulations in the form of distinctive low pressure troughs and high pressure ridges, with wavelengths between 2000 and 4000 km long. These waves have a 'life span' of one to two weeks and can travel some

6°–7° longitude per day (Barry and Chorley, 1987). The first wave-type disturbance identified in the humid tropics was the easterly wave, which appeared to be a regular feature in the Caribbean and central Pacific Ocean. Even though recent satellite imagery reveals that these simple wave disturbances are less common than originally suggested, they are conspicuous, if shallow and weak, troughs of low pressure athwart the trade winds. These waves move westwards at speeds of 5–7 m s^{-1} in the deep trade wind flow of the southern limb of the Azores and north Pacific subtropical high pressure systems. Furthermore, they are clearly recognized on the weather map by the poleward tongueing/troughing of the isobars or streamlines (figure 2.5).

The origin of easterly waves is not clear but they appear to develop over tropical oceans, like the Caribbean, where the trade wind inversion is weak or absent during summer and autumn (Barry and Chorley, 1987). This inversion results from adiabatic warming in the subsiding air of the subtropical anticyclone and is best developed and closest to the surface (i.e. below 1500 m) at the outer, poleward limits of the Hadley cells. However, it persists throughout most of the trade wind belt, although it rises to above 2000 m at the equator, particularly over the Atlantic Ocean. Also, it is lowest (*c.* 500 m) over the cold waters of the Benguela and Canary currents around 15 °S and 15 °N respectively. The absence of this inversion within the ITCZ allows surface heating and lapse rate steepening to develop into unstable, freely convected systems, with considerable amounts of latent heat release and low pressure troughs forming in the lower troposphere. Another factor associated with the formation of these waves is the penetration of cold fronts into low latitudes, especially in the area between two subtropical anticyclones where the equatorward part of the front tends to fracture, forming a westward-travelling wave (Barry and Chorley, 1987).

A distinctive feature of easterly waves is the weather sequence observed before, during and after the passage of the trough axis (figure 2.5). Ahead of the axis, the trade wind inversion is particularly low which inhibits free convection and produces stable, fine and hazy weather. Figure 2.5 also shows that this area is a zone of strong divergence as the air moves equatorward and curves anticyclonically, with vertical contraction of the air column and descending, drying air. Behind the wave axis, the inversion is much higher and free convection can take place with deep cumulus congestus and cumulonimbus clouds developing in the unstable, moist air with moderate or heavy thundery showers. Figure 2.5 also reveals that the rear of the trough is a zone of strong convergence as the air moves polewards and curves cyclonically, with vertical expansion of the air column, and ascending air. Apart from this characteristic weather sequence, easterly waves appear to act as parent vortices for the generation of tropical cyclone seedlings (see section 2.4.2) although only about 10 per cent of these seedlings survive to become mature hurricanes.

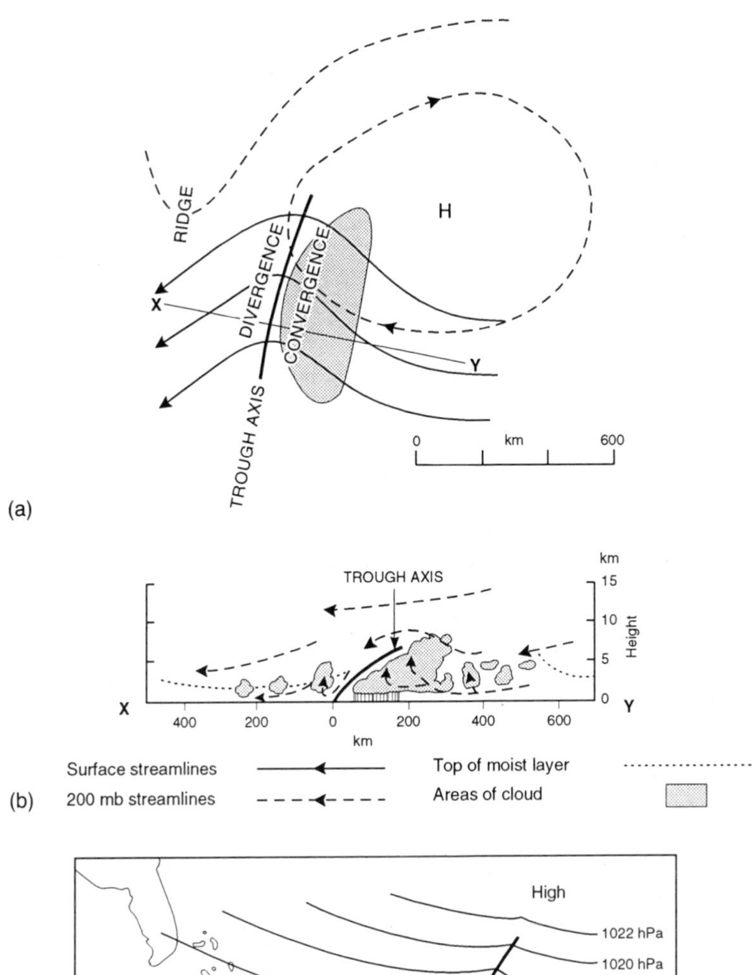

Figure 2.5
A model of (a) the areal structure and (b) the vertical structure of an easterly wave (after Barry and Chorley, 1987); (c) the pressure distribution and trough axis in an easterly wave in the Caribbean with winds converging at B and diverging at A (after Henderson-Sellers and Robinson, 1987).

2.4.2 Tropical cyclones

These weather systems are intense, circular low pressure vortices which are given different regional names: hurricanes in the western North Atlantic, cyclones in the Indian Ocean, typhoons in the west Pacific and even willy-willies in northern Australia. The study of tropical cyclones has been confused by terminological discrepancies, which have limited the strict comparison of data on cyclone frequency and movement. Consequently, Kerr (1976) urged that the name tropical cyclone should be used as a generic term which should include tropical disturbances of all intensities, based on the World Meteorological Organization classification in table 2.1.

Table 2.1 The World Meteorological Organization classification of tropical cyclones

Term	*Wind speed*
Tropical depression	Winds up to 33 knots (17.1 $m\,s^{-1}$) or Beaufort force 7
Tropical storm, moderate	Winds between 34 and 47 knots (17.2–24.4 $m\,s^{-1}$) or Beaufort force 8 or 9
Tropical storm, severe	Winds between 48 and 63 knots (24.5–32.6 $m\,s^{-1}$) or Beaufort force 10 or 11
Hurricane (or local synonym)	Winds in excess of 64 knots (32.7 $m\,s^{-1}$) or Beaufort force 12

Knowledge and fear of tropical cyclones have existed for centuries. In the southwest Pacific for example, Polynesians often referred to these storms in their myths and legends. The Cook Islanders called the backing northeast wind the terrible 'Maoake', which we now realize is associated with the violent circulation of a cyclone approaching from the west.

Some of the earliest written accounts of tropical cyclones come from the western North Atlantic region. Records extend back to the discovery of the West Indies by Christopher Columbus in the 1590s and, compared with other cyclone-prone regions, are particularly comprehensive in the pre-charted years. The main reasons for this are first the more or less continual occupation, by literate European settlers, of the numerous islands and islets which make up the West Indies. Second, the distribution of the numerous islands and islets which make up the West Indies archipelago mean that it is unlikely that a cyclone will pass through the Caribbean without affecting land or encountering a ship. Third, a great deal of time and effort has been spent 'discovering' references to cyclones in historical documents. By the nineteenth century European settlers, traders and missionaries throughout cyclone-prone regions were meticulously recording violent storms (along with other climatological phenomena) in diaries, estate records, ships' logs and religious reports. Since the Second World War, improvements in the observational network and satellite monitoring have guaranteed that few, if any, cyclones anywhere in the world are undetected. Analysis of historical data has revealed significant variations in both the tracks and frequency of storms (figure 2.6). Improvements in data quality and an increased understanding of the tropical atmosphere has aided more quantitative studies of cyclone behaviour (e.g. Thompson, 1986; Reading, 1990). However, our knowledge of the causal factors involved in determining global and intra-regional frequency distributions is severely lacking.

Tropical cyclones have distinctive characteristics. They are generally small sized, about one-third the size of middle latitude

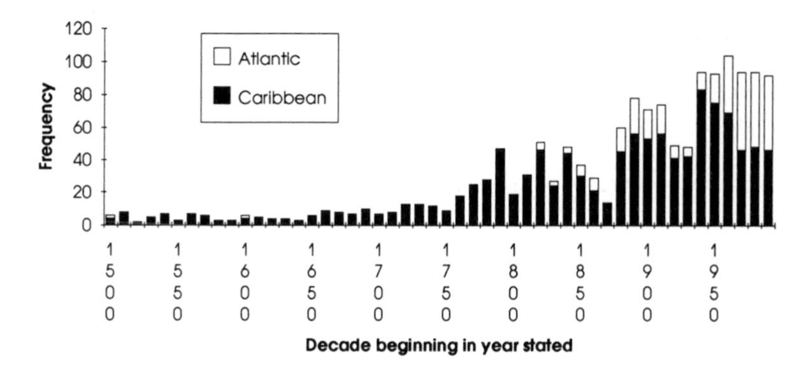

Figure 2.6
Tropical cyclone frequency in the western North Atlantic. Improvements in surveillance methods are responsible for the apparent increase in cyclone frequency up to around 1900. However, the relative peaks and troughs are thought to represent real changes in levels of cyclone activity.

depressions, although the actual storm diameter normally varies from 500 to 800 km taking the entire cloud envelope into account. The central pressure averages about 950 mb, no lower than some extreme temperate lows, but it can be in the range 860–900 mb. Hurricane Gilbert which devastated the Caribbean in September 1988, for example, had a record low pressure of 882 mb. An extreme horizontal pressure gradient is associated with this deep vortex; e.g. Hurricane Bebe in Fiji during October 1972 recorded 'eye' wall pressure gradients close to 1 mb km^{-1}. This resulted in wind speeds approaching 100 m s^{-1}; Hurricane Gilbert's wind gusted to 97 m s^{-1}.

The cloud structure in a tropical cyclone is cylindrical in form, extending from a low base almost to the Tropopause. This cloud 'cylinder' widens out in its upper reaches and consists mainly of cumulonimbus towers massed together. Also, it is characterized by spiralling bands of stratocumulus and heavy cumulus, which enter at low levels, and cirrus/cirrostratus which spread out at high levels (figure 2.7). The 'eye' of the cylinder is a central well or funnel of subsidence and adiabatic warming of high level air drawn down into the heart of the vortex. It is mainly cloud-free air (apart from some broken, thin cloud below 1500 m) which extends from the base to the top of the system, with a diameter of some tens of kilometres. Typically the 'eye' of hurricanes averages between 10 and 25 km across.

Most of the energy of a large tropical cyclone is concentrated in a ring within 100 km radius of the centre and, in this zone, the winds attain maximum force. The strongest winds are found in a ring encircling the storm centre at a distance of some 24 km from the 'eye'. Inside the ring, wind speeds decrease rapidly to the relatively calm conditions of the 'eye'. Outside the high velocity ring, wind speeds also decrease progressively towards the periphery of the storm (Thompson, 1986). The central ring is characterized by the thickest cloud and heaviest precipitation, which can exceed 500 mm day^{-1}. For example, Hurricane Bebe in Fiji in 1972 deposited 755 mm of rain between 23 and 24 October as it passed over the mountains some 27 km northwest of Suva.

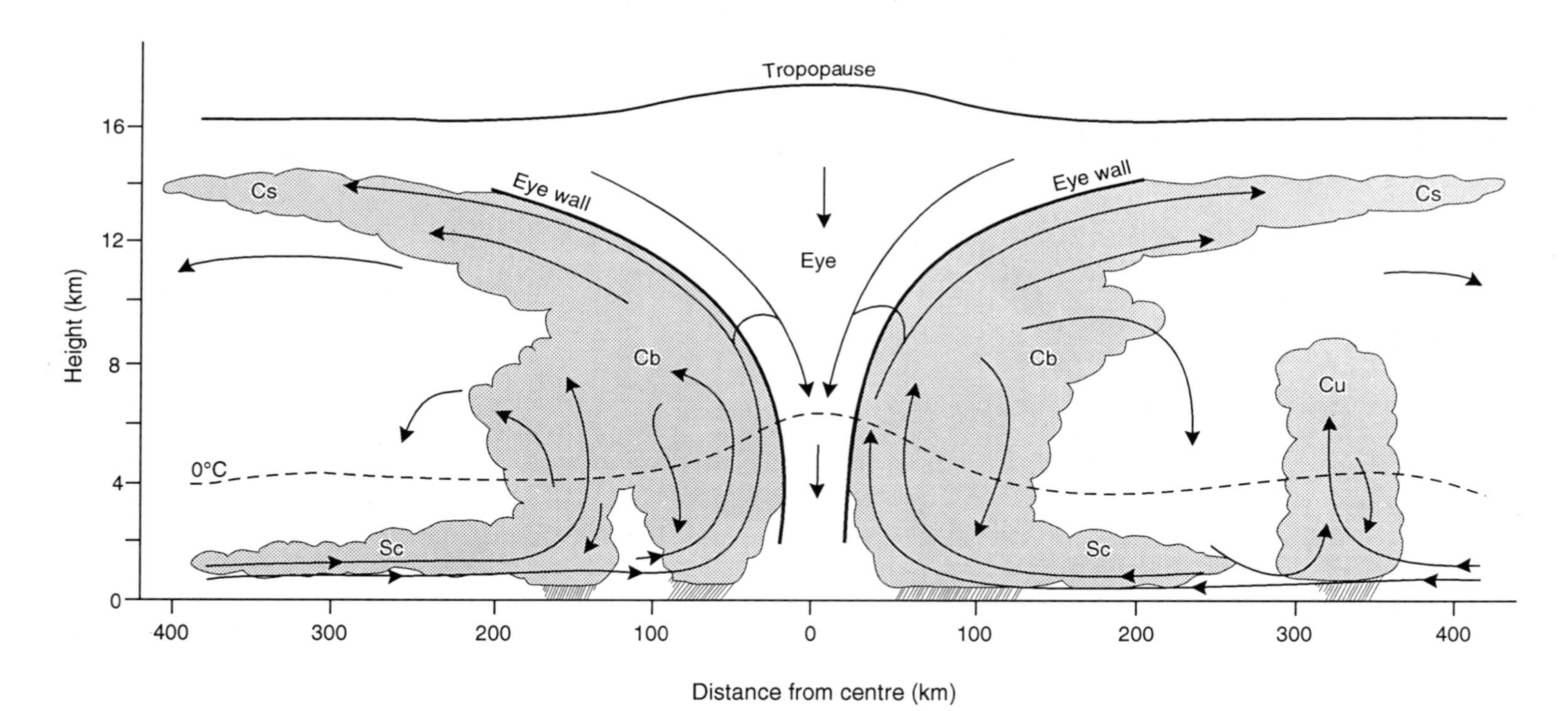

Figure 2.7
Suggested vertical structure of a mature hurricane (after Thompson et al., 1986). Air flow patterns, freezing level, precipitation and common cloud types are indicated: Cs, cirrostratus; Cu, cumulus; Cb, cumulonimbus; Sc, stratocumulus.

The hurricane-force winds and excessive rainfall are coupled with storm surges (called Loka in Fiji), which can reach heights of 15 m. Loka form over low-lying coasts when the abnormally low atmospheric pressure initiates a strong, positive sea surge (e.g. the sea level rises by approximately 1 cm for every 1 mb fall in pressure), drawing water up to some 30 cm and, with wind assistance, up to 4 m above normal sea level. Surges are particularly destructive along low-lying coasts and their destructive potential is increased in estuaries when river floodwater meets and checks this sea water surge or when the wind impels the water against the coastline. Then, the water piles up to scour low-lying areas as a formidable wall.

Hurricanes usually move slowly at some 5–7 $m s^{-1}$ and initially the systems are steered from east to west in the trade wind flow (figures 2.8 and 2.9). Indeed, the tracks of most hurricanes are largely dictated by the deep circulation of the subtropical anticyclones in which they are embedded. Hurricanes tend to recurve poleward around the western margins of these highs and eventually enter the circulation of the westerlies. They tend to decay as they move across the colder waters of high latitudes or over land, where the supply of latent heat is reduced considerably and the systems degenerate into weaker extra-tropical depressions. Alternatively, they can become regenerated in middle latitudes when they become incorporated into frontal depressions.

Figure 2.8
Typical tracks of West Indian tropical cyclones: the 1989 season in the western North Atlantic (after Case and Mayfield, 1990). Western North Atlantic cyclones originate off West Africa, around the Cape Verde Islands or in the western Caribbean Sea. They have irregular tracks but almost all move around the periphery of subtropical high pressure cells, i.e. they move roughly westwards in the northern hemisphere. The frequency and spatial distribution of cyclones varies considerably from decade to decade. In the Caribbean, cyclone frequencies were relatively high during the 1890s and 1900s and between the 1930s and 1950s. Frequencies were low during the 1910s and since around 1960 (especially in the eastern Caribbean). There is some evidence of increased cyclone activity in some parts of the Caribbean during the late 1980s.

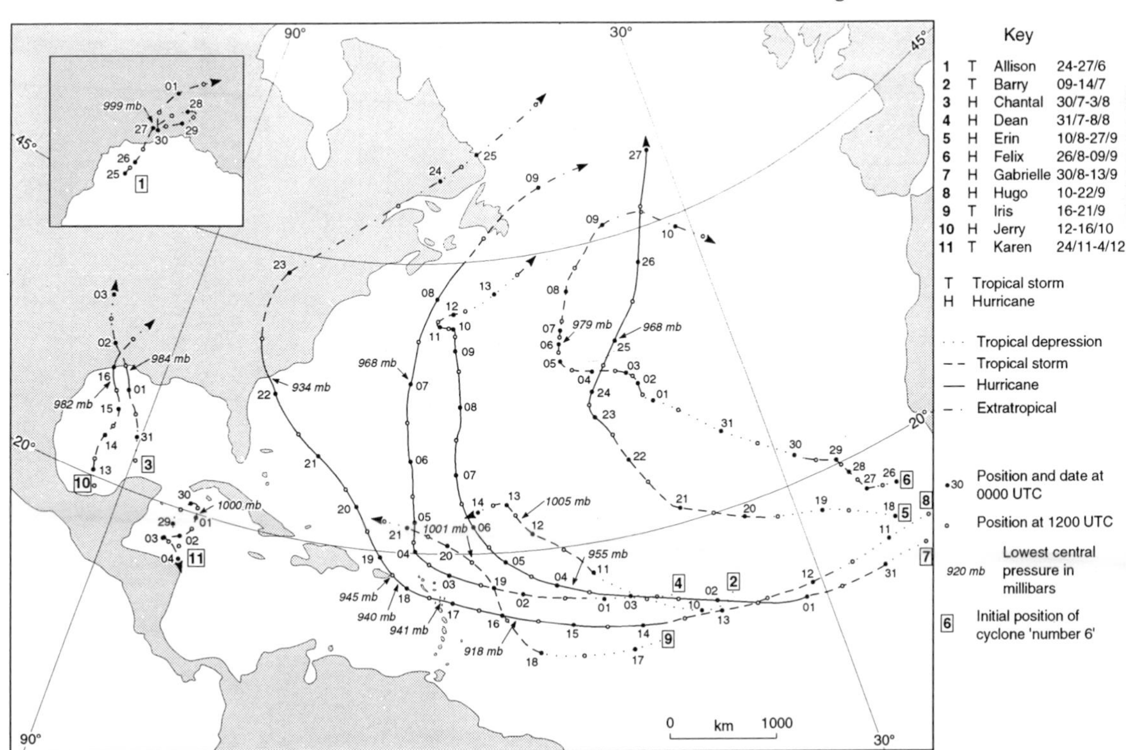

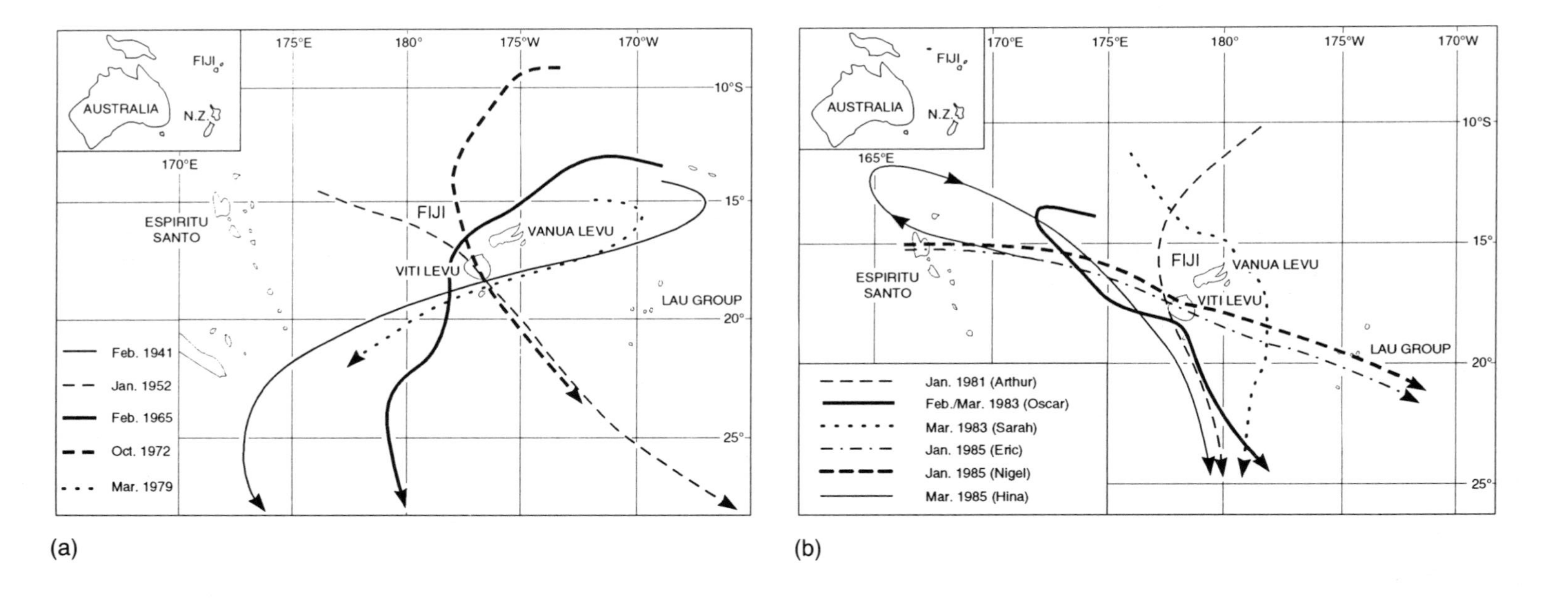

Figure 2.9
Severe hurricanes in Fiji: (a) 1940–79; (b) 1980–5 (after Thompson, 1986). Cyclones originate outside the Fijian region, generally between 10 °S and 15 °S. Fijian cyclones have an irregular track but most move eastwards or southeastwards during at least part of their life. Cyclone frequencies over Fiji are highly variable. For example, the average frequency interval is four years. However, three occurred in 1875 and in 1985. Between 1941 and 1979 only five severe cyclones were recorded. Between 1981 and 1985 there were six.

The development of hurricanes is not fully understood but there are basic requirements for cyclogenesis in the humid tropics which Gray (1979) refers to as 'primary climatological genesis parameters'. Furthermore, hurricanes have a composite origin and require the complete synchronization of the various parameters involved. When this coupling is missing, then tropical cyclone seedlings generated by parent vortices (such as the easterly waves discussed earlier) will not survive to become more mature and violent hurricanes. Tropical cyclones are in part thermal systems since nearly all form over extensive ocean areas (with low frictional retardation) when surface temperatures exceed 26 °C, integrated to a maximum depth of 60 m (Reynolds, 1985), especially in late summer and early autumn when ocean waters are at their warmest. Tropical cyclones therefore occur in distinctive 'seasons'. In the western North Atlantic this is between July and October. In the South Pacific the hurricane 'season' is between December and April. Very few cyclones occur outside the season. One notable exception, however, was the very intense Hurricane Bebe which tracked across Fiji in October 1972 (figure 2.9(a)).

Oceanic thermal energy represents the major energy source of a tropical cyclone which develops a warm core in association with excessive rates of evaporation and subsequent condensation/latent heat release. High surface sea temperatures (figure 2.10) are necessary for the lapse rate steepening and deep free convection inherent in any hurricane, which are initiated in close clusters of large cumulonimbus towers (discussed earlier). There appears to be a strong correlation between the seasonal location of the ITCZ thermal energy concentration and cyclones, as evidenced by the fact that no tropical cyclones occur in the South Atlantic (where the ITCZ never lies south of 5 °S) or in the southeast Pacific (where the ITCZ remains north of the equator) (Barry and Chorley, 1987). In these areas, sea surface temperatures are always too low and the vertical wind shear is very often unfavourable.

The latitudinal control of cyclogenesis is associated with the fact that cyclones rarely form near the equator, where the Coriolis parameter is zero and the vertical component of the planetary vorticity is absent. This means that any semblance of organized balanced rotational motion (so vital to sustain a circulation) does not occur normally within 5 ° of the equator (Reynolds, 1985). It has long been recognized that the formation of deep convection cells generates a sudden and massive release of latent heat of condensation which represents the basic energy source for cyclone maturity. However, the scale of this localized convection was a problem in determining the growth of a system hundreds of kilometres in diameter. It now appears that the transfer of energy from the individual cumulus-scale to the synoptic-scale circulation of a mature cyclone occurs through the organization of some 100–200 cumulonimbus towers into interconnected spiral bands, although the process is not fully understood (Barry and Chorley, 1987).

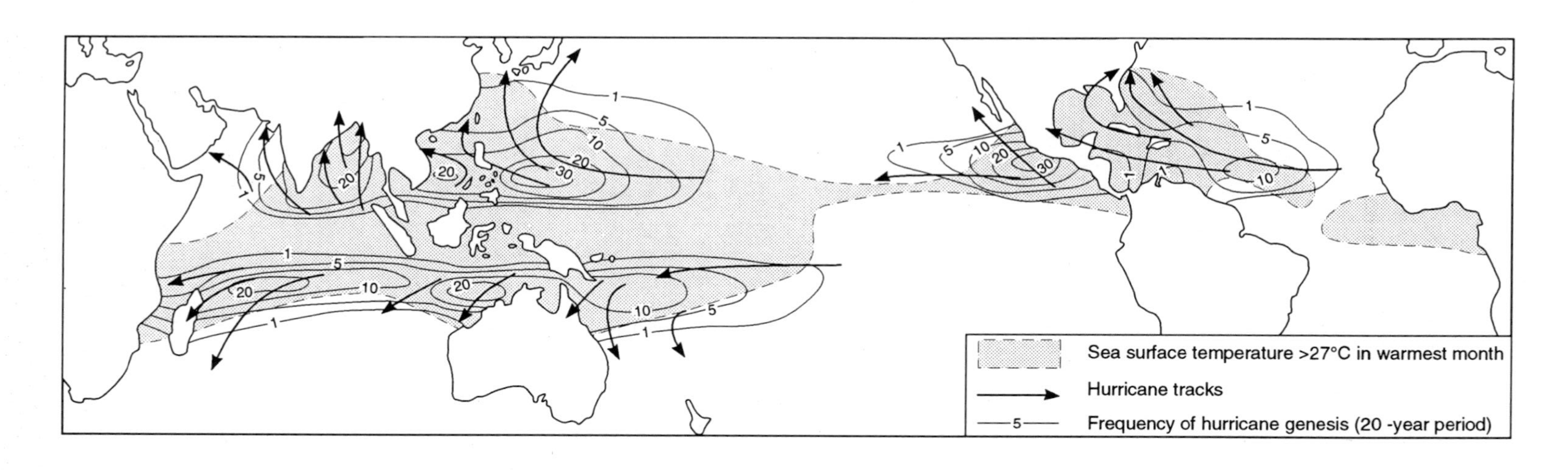

Figure 2.10
Frequency of hurricane genesis, areas with sea surface temperatures above 27°C in the warmest month and principal hurricane tracks (after Barry and Chorley, 1987).

Tropical cyclogenesis is also associated with weak vertical shear of the horizontal wind in the basic air circulation. With strong wind shear, the latent heat of condensation released by the developing convection will be dissipated in the upper atmosphere, since the heat energy in these layers is now advected in a direction different from that in the lower layers of heat release. Under these conditions, there could be no vertical concentration of heat energy and a surface vortex fails to develop in any area beneath strong jet stream wind shear. Therefore hurricane formation will only occur in latitudes equatorward of the subtropical jet stream. Furthermore, with the exception of the western North Pacific and western North Atlantic Oceans, strong wind shear is found in all regions poleward of 20 ° latitude. In the north Indian Ocean, weak vertical wind shear only occurs between 20 ° and 23 ° north in August and 10 ° to 15 ° north in late spring and autumn. These restrictions limit cyclogenesis in the oceans concerned, and in areas where strong wind shear is present throughout the year (e.g. the western South Atlantic and Central Pacific Oceans), hurricanes will not develop.

The final requirement for tropical cyclone genesis is associated with a mechanism necessary to couple the low level convergence with a divergent flow above 12 km or the 200 mb pressure level in order to sustain the intense free convection and storm generation. Such high level outflow (figure 2.7) is normally provided by an upper tropospheric anticyclone cell but can also occur on the eastern limb of an upper trough in the westerlies. The divergence maintains the ascent and low level inflow which is necessary to generate energy continually from latent heat release, which is transformed into kinetic energy. These requirements are essential for the maintenance of hurricane intensity and, as soon as one factor declines, the storm begins to decay. As was noted earlier, degeneration occurs quite quickly when the tracking takes the vortex over a cool sea surface and especially over land, where friction increases and the supply of water vapour (the major energy source) is reduced. Rapid decay can also occur when cold, polar air enters the system or when the upper level outflow becomes detached from the surface vortex, which now fills since the inflow at lower levels exceeds the outflow aloft.

2.5 The Monsoon Circulation

The best known large-scale circulation in the humid tropics is associated with the seasonal reversal of airflow (of at least 120 ° change in wind direction) across southern Asia. This wind regime reversal (and associated changes in precipitation patterns) is known as the monsoon and indeed the change influences an immense area which extends well outside tropical latitudes (figure 2.11). The mechanism of airflow/precipitation change has been traditionally explained in terms of the seasonal changes in land-mass heating/cooling in response to the seasonal migration of solar heating and

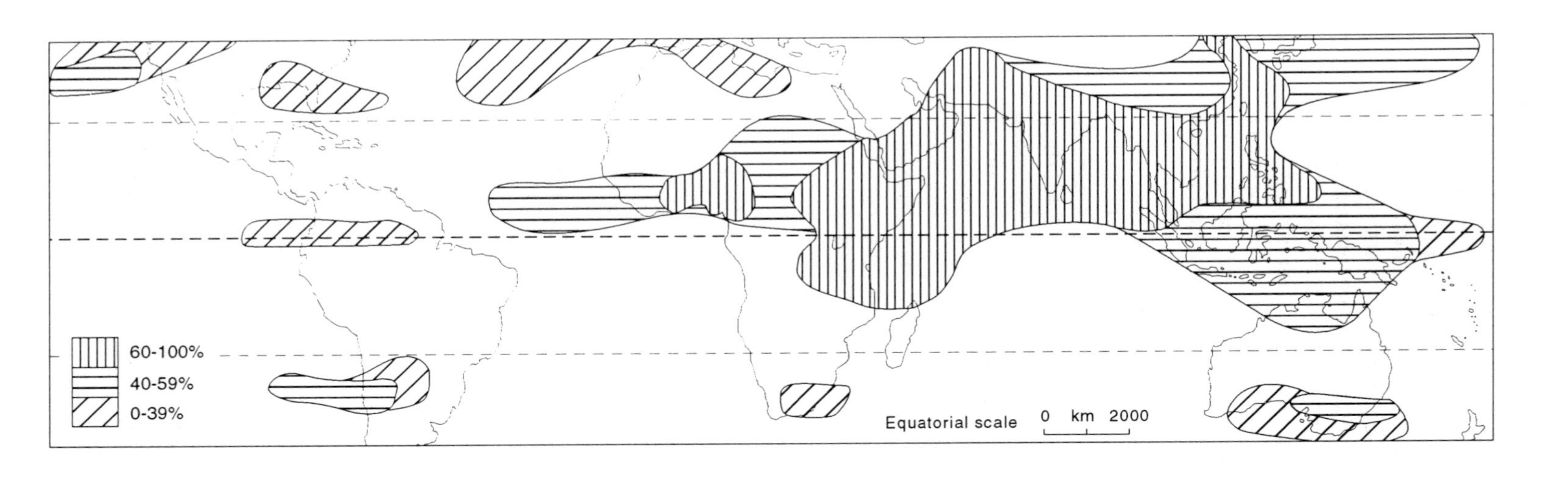

Figure 2.11
Monsoon regions experiencing a seasonal surface wind change of at least 120°, showing the frequency of the prevailing octant (Barry and Chorley, 1987).

the so-called thermal equator or ITCZ discussed earlier. A simple circulation change was envisaged in the form of a gigantic sea-breeze mechanism (figure 2.12), associated with summer thermal low pressure over the continent (with converging southerly onshore winds) and a winter thermal high pressure, with diverging northerly offshore winds. The resultant climatic regime of hot wet summers and cool dry winters remains a conspicuous feature of monsoon circulations. However, the traditional simplicity of these circulation changes has now been replaced by a more complex mechanism involving the interaction of the upper and lower troposphere and the role of the Tibetan (Qinghai-Xizang) plateau on jet stream flow.

Since the significance of these complex associations varies seasonally, which dictates the prevailing climatic regime, it is pertinent to examine the circulations of the winter and summer seasons individually. The winter monsoon (figure 2.13(a)) is characterized by outflowing northeasterly surface winds and a vigorous westerly airflow in the upper troposphere, with a

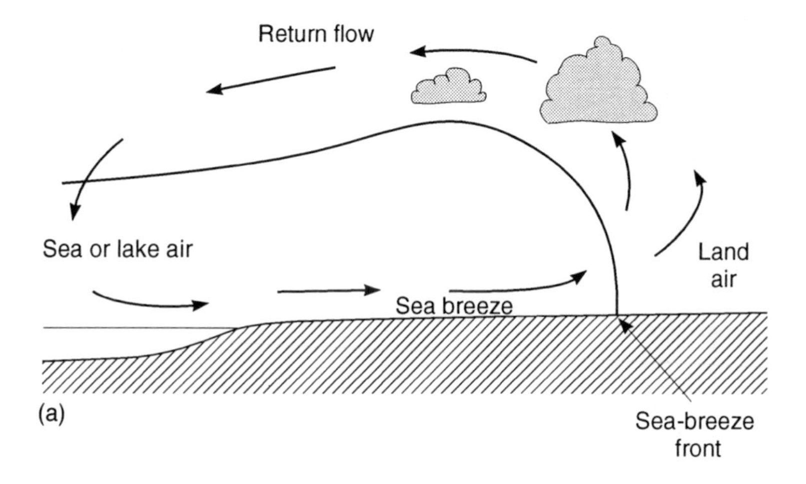

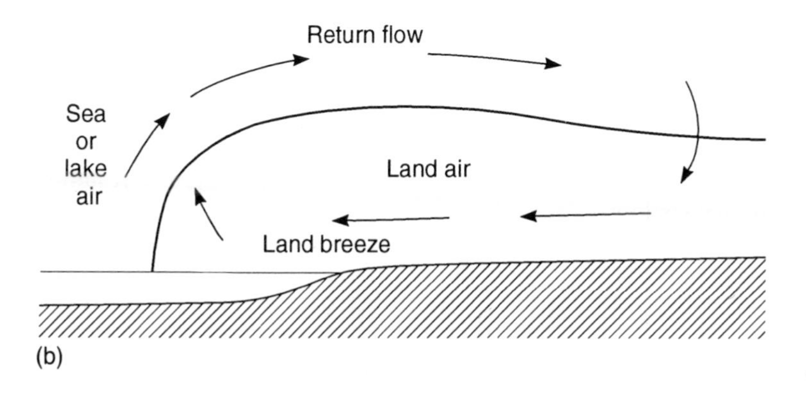

Figure 2.12
Land and sea-breeze circulation: (a) daytime; (b) night-time (after Thompson et al., 1986).

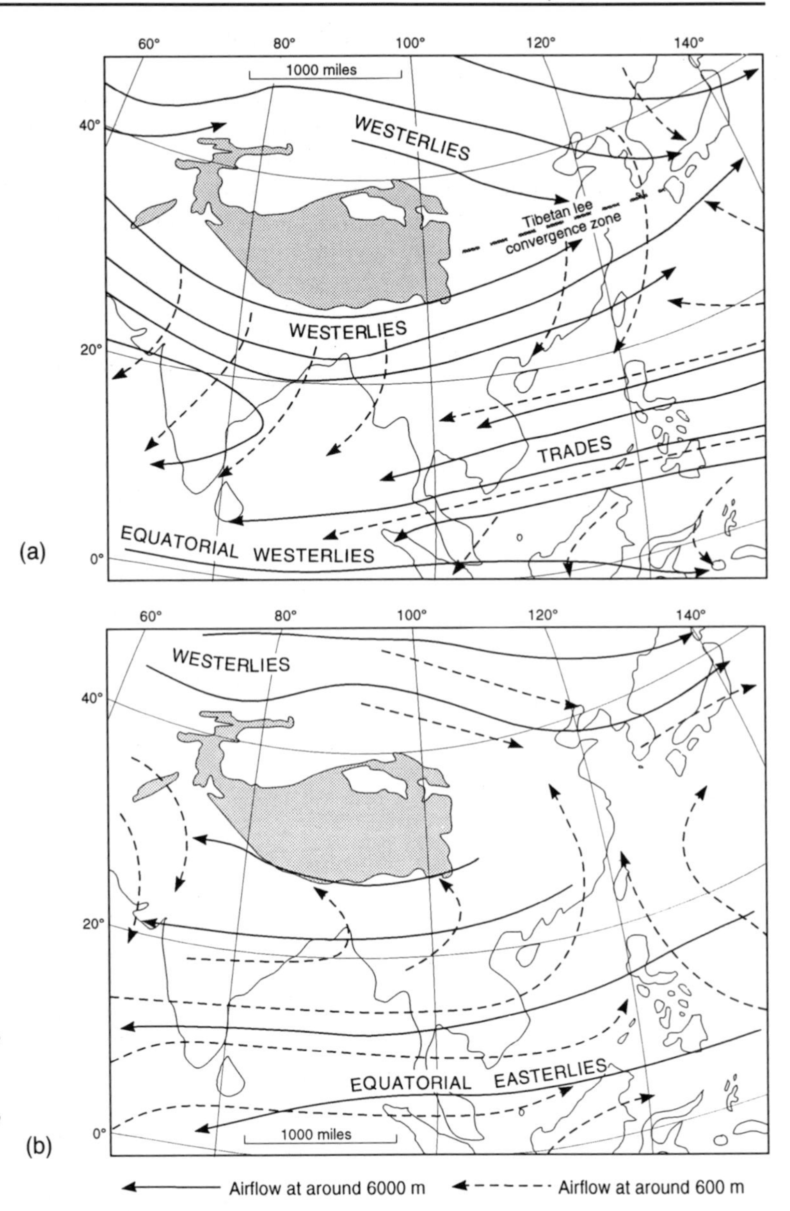

Figure 2.13
Characteristic airflow of (a) the winter monsoon and (b) the summer monsoon over India and southeast Asia (after Barry and Chorley, 1987). Names refer to the winds aloft.

conspicuous subtropical jet stream embedded at 200 mb over northern India. At the surface, a north to south pressure gradient extends almost to the equator, from a shallow layer of cold, anticyclonic air situated over Siberia. Consequently, the resulting surface airflow is from the north or northeast, and over northern India this airflow is reinforced by the subsiding and diverging air from the subtropical anticyclone. This deep subsidence is associated with convergence to the south of the subtropical jet in the upper westerlies. The resultant surface airflow is commonly known as the 'land trades' and is northerly over most of northern India. It becomes northeasterly over Burma and Bangladesh and easterly over peninsular India (Barry and Chorley, 1987).

The dominance of the offshore winter monsoon is responsible for mainly dry weather over the central part of the Indian subcontinent. The only exceptions are the southern tip of the peninsula, which lies close to the ITCZ and is frequently affected by the onshore equatorial westerlies, and in the north of India and Pakistan. Winter precipitation in these northern regions is associated with the passage of so-called 'western disturbances', which represent the occluded stages of wave depressions which originate along the Mediterranean front. These depressions are steered over the region by the subtropical jet stream and, between December and April on average, northern India is frequented by nineteen such systems. They typically appear as troughs in the upper westerlies which occasionally extend to the surface as closed cyclonic circulations which deepen over the subcontinent. This deepening was originally explained in terms of the sudden initiation of convergence provided by the inflow of warm, moist air from the Bay of Bengal. However, more recent explanations note that it is caused primarily by the marked diffluence at the 300 mb level (Chang, 1972).

Some of these depressions continue eastwards and become regenerated over China (at about 30 °N, 105 °E) in the zone of jet stream confluence, in the so-called Tibetan lee convergence zone (figure 2.13(a)). Barry and Chorley (1987) note that the mean axis of the subtropical jet stream over China (at 12 km) correlates closely with the distribution of winter precipitation in excess of 600 mm. Furthermore, in the rear of these depressions, there are invasions of very cold polar continental airstreams (e.g. the buran blizzards of Mongolia and Manchuria). These cold waves are comparable with the northers in the southern USA and can extend over the coast as far as Hong Kong (22.5 °N).

At the end of the winter (normally in March), the increasing solar radiation and rising air temperatures initiate a thermal surface trough which reaches its maximum intensity in May. Despite the development of onshore coastal winds at this time and hot, unstable weather conditions, rainfall is limited and is restricted to squall lines of localized thunderstorms in the Ganges delta of India, known as nor'westers. In the northwest, lapse rate steepening and associated free convection generate violent squalls and dust storms called 'andhis'. It appears that the generation of these storms depends on upper-air divergence in the waves of the subtropical jet stream, especially when they are coupled with low level convergence of maritime air from the Bay of Bengal. China does not experience India's hot, squally pre-summer monsoon weather, since the subsiding air from the upper westerlies maintains the offshore northeasterly airflow at the surface. The weather remains cool even in the south, although spring is wetter then winter due to persistent cyclogenesis in the Tibetan lee convergence zone (figure 2.13(a)) discussed earlier. For example, the three months from March to

May contribute up to a third of China's annual precipitation (Barry and Chorley, 1987).

The onset or 'burst' of the summer (or so-called southwest) monsoon depends on the shift northwards (in May and June) of the upper westerlies (figure 2.13(b)) and the subtropical high pressure cells. This anticyclone now becomes centred some 2000 m above the Tibetan plateau (4000 m elevation), which acts as a significant high level heat source. An upper tropospheric easterly jet stream now forms on the southern limb of the high and dominates southern Asia, at an elevation between 13 and 16 km, equatorwards of 15 °N (figure 2.13(b)). This jet extends from the South China Sea across India into the Arabian Sea and the southeastern African Sahara. The equatorial or monsoon trough, associated with the ITCZ, now intensifies at about 25 °N over India. It establishes a strong north–south pressure gradient which is reinforced by a similar steep lateral temperature gradient and, in turn, accentuates the development of the easterly jet. The jet stream acts as an important air scavenger since the flow of air away from southern Asia at high levels is necessary to maintain the flow toward the region at low levels.

This upper–lower tropospheric coupling is essential for the formation of deep cumulus clouds (with excessive precipitation) and the transfers of sensible and latent heat to increase the store of potential energy within the upper troposphere. Some of this energy is converted into kinetic energy which rapidly strengthens the upper airflow and intensifies the scavenging role of the easterly jet stream. The conversion often accelerates spectacularly within five to ten days during late May which is the time when the onset of the monsoon takes place over the Indian Ocean at about 5 °N. On average, the monsoon first appears over Sri Lanka and the southeast Bay of Bengal during late May and reaches southern India in early June. It advances northwards in a series of surges, characterized by vigorous rain-bearing low pressure vortices on their leading edges. The advance occurs along two main branches across the Arabian Sea and Bay of Bengal which meet over central India in late June and enter Kashmir in mid July (Hamilton, 1987).

In India, the southwest monsoon arrives in late May in the extreme south and progressively advances northwards to reach the Thar desert by early July (figure 2.14(a)). However, the advance is not regular since the monsoon can retreat or surge forward with renewed vigour. At Delhi, the average onset date is 2 July but between 1901 and 1950 the actual date ranged from 17 June to 20 July, with a 7.8 day standard deviation (Chang, 1972). The monsoon retreats from northwestern India in late September (figure 2.14(b)). The withdrawal is much slower in the south where stagnation of the monsoon trough gives the east coast (south of 15 °N) maximum rainfall amounts in November. On average, it retreats from southernmost parts by early December. Mean rainfall at Bombay (figure 2.14(c)) shows the dramatic onset of the monsoon, which deposits over 90 per cent of the annual total between June and September. In late May, daily rainfall averages

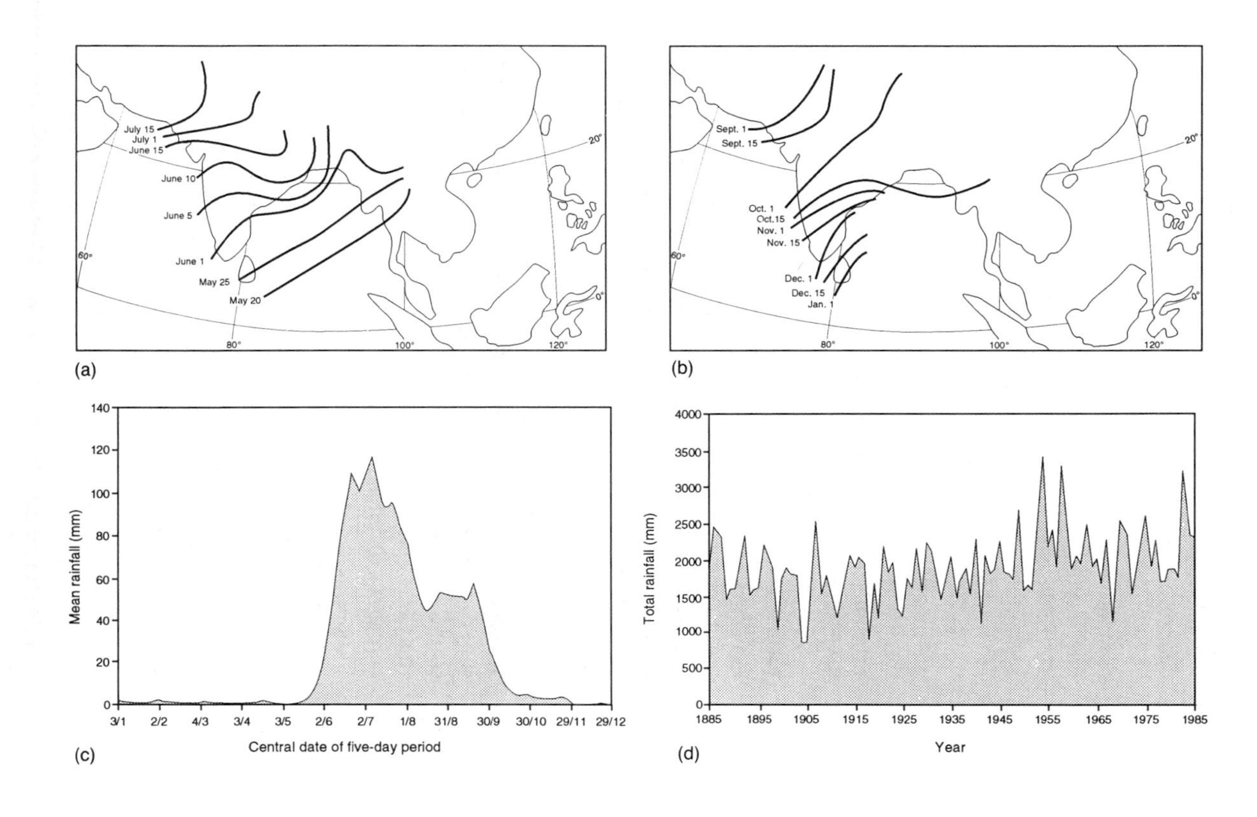

Figure 2.14
Characteristics of the Indian monsoon: (a) onset dates; (b) withdrawal dates; (c) five-day mean rainfall at Bombay; (d) variability of southwest monsoon rain at Bombay, 1885–1985 ((a), (b) after Chang, 1972; (c), (d) after Hamilton, 1987).

2 mm per day which increases to 22 mm by early July. Rainfall declines rapidly during August but trough stagnation causes a secondary peak in September. Monsoon rainfall is characterized by pronounced inter-annual fluctuations. It is apparent from figure 2.14(d) that since 1885 seasonal totals at Bombay have ranged from under 600 mm (1918) to over 3400 mm (1954). The current trend is towards wetter conditions since eighteen of the twenty driest years occurred before 1950 (Hamilton, 1987).

Throughout southeast Asia the summer monsoon weather is complex and variable, even though it is characterized overall by hot, humid and wet conditions. For example, Rangoon (16 °N, 96 °E) averages over 350 mm rainfall per month between June and September, which represents about 75 per cent of the annual total. However, during a normal wet monsoon season, conspicuous 'breaks' occur in the rainfall deposition when the trough moves north over the sub-Himalayan region, leaving central India and Burma mainly dry. One or two such breaks occur in an average rainfall season, most commonly in August, and the typical duration is between three and five days (Reynolds, 1985). During the break period, heavy rainfall is concentrated along and near the foot of the Himalayas from the Punjab to Assam, whereas central India experiences drought conditions when breaks last more than a week. The break is initiated by a change in mid-latitude circulation from a high zonal index to a low index/meridional flow. This introduces an extension of a large-amplitude upper air trough and westerlies of jet stream strength from mid-latitudes into Indo-Pakistan. This weakens the Tibetan anticyclone and facilitates a northward shift of the monsoon trough to the southern fringe of the Himalayas. At the same time, the easterly jet is strengthened and moves northwards along with a shift in the rainfall belt. The westerly and easterly jets are now within 12 ° latitude compared with a 32 ° distance during an active monsoon period (Chang, 1972).

Monsoon depressions provide further variability in the summer monsoon weather, especially at the head of the Bay of Bengal and along the Ganges valley. These depressions are distinctive, if small, low pressure vortices, roughly circular in shape with diameters of 250–500 km; a cyclonic circulation extends up to 4–8 km, with its axis tilting southeast where temperatures are lowest (Hamilton, 1987). Monsoon depressions appear on synoptic charts as weak tropical cyclones (table 2.1), with wind speeds below 17 m s^{-1}. They are prevented from developing into hurricanes during the monsoon season by the prevailing, strong vertical wind shear between the low level westerlies and the tropical easterly jet. They are fairly common disturbances, occurring about twice a month between June and September, and have a 'life time' of between two and five days. During this time they are responsible for daily precipitation totals averaging 100–200 mm, mainly as convective rains in the southwest quadrant of the depression (Barry and Chorley, 1987). The depressions usually track west-northwest across India, towards the region of maximum surface convergence, steered by the upper easterly airflow. They normally fill over the Rajasthan desert or recurve into the Punjab and break up in the Himalayas. The mechanisms of development for monsoon depressions remain unclear but it appears that they form in an active monsoon trough over the Bay of Bengal, associated with lapse rate steepening and a potent Coriolis force. Also, the latent heat release associated with condensation in this organized free convection is necessary for the generation and maintenance of the disturbances, although most are not simple, warm-cored systems. A further requirement is the coupling of divergence in an upper tropospheric ridge in the easterlies with the convergence in a low level trough, which intensifies the inflowing airflow and strengthens the cyclonic vorticity.

The complexity and variability of the circulation changes and associated weather patterns of the summer monsoon are evident from the discussion on monsoon breaks and depressions. With the onset of autumn, the weather becomes more settled over northern India, with the gradual transition to winter monsoon conditions. However, the Bay of Bengal now experiences a peak development of tropical cyclones in October–November, which is associated with a rainfall maximum in southeast India. Cyclogenesis in the Bay is aided by the confluence at the 500 mb level of an easterly airflow from the Pacific with the equatorial westerlies (Barry and Chorley, 1987). During October over the rest of India, and indeed most of southern Asia, the thermal contrasts between land and sea are reduced with the movement south of the thermal equator/ITCZ. The summer circulation breaks down and the westerlies of mid-latitudes begin to advance southwards. By late October, the westerly jet stream has returned south of the Tibetan Plateau, replacing the easterly jet, and reinforces the winter monsoon circulation (figure 2.13(a)) associated with anticyclonic subsidence over northern India. The so-called surface land trades now return often within a week or so, giving cool dry weather over most of southern and eastern Asia. The winter monsoon has now set in and the atmospheric circulation/climatic regimes discussed earlier are repeated. However, it is not until the end of December that the summer monsoon circulation finally withdraws from the southern tip of the Indian subcontinent and Sri Lanka (figure 2.14(b)).

2.6 Other Weather and Climate Controls

The weather and climatic elements in the humid tropics are mainly controlled by the circulation systems and atmospheric disturbances discussed in the preceding four sections. However, small-scale weather patterns are also important and are responsible for a variety of localized and contrasting climatic regimes from time to time. These elements are conspicuous in the humid tropics, both in spatial and temporal terms. They are initiated primarily by terrain factors associated with land–sea thermal differences in coastal zones and orographic displacement over mountainous regions. In both cases, they represent tertiary-scale systems which are superimposed on the major, larger-scale synoptic and monsoonal circulations.

2.6.1 Land and sea breezes

These circulations represent miniature monsoonal regimes (figure 2.12) due to the fact that, in coastal environments, the day-time heating of tropical air over land can be up to five times that over adjacent water surfaces. Consequently, the vertical expansion of the heated air column over the land is associated with low level convergence. This is linked to onshore winds and sea breezes at the surface and a compensatory counterflow of offshore winds at about 1200 m elevation (figure 2.12(a)). At night, the more rapid cooling of the land surface reverses the pressure distribution, with vertical contraction over the land and air divergence. This now causes offshore winds or land breezes at the surface and a compensatory onshore airflow aloft (figure 2.12(b)). The sea-breeze component has a conspicuous timing in that it is initiated in the early morning (*c.* 10:00 hours), reaches a maximum velocity of 6–16 m s^{-1} during the early afternoon (*c.* 15:00 hours) and terminates around 20:00 hours. The counterflow aloft in the tropics usually occurs around 1200 m with maximum sea-breeze wind speeds recorded at an elevation of 200–400 m (Barry and Chorley, 1987).

The arrival of the sea breeze inland is associated with a significant fall in temperature, an increase in relative humidity, an increase in wind speed and a reversal of wind direction. The inland penetration (normally between 20 and 60 km) of maritime air continues behind a conspicuous migrating sea-breeze front (figure 2.12(a)) which encourages the uplift of unstable air. This leads to the development of cumulus clouds and short, sharp showers. There is an early afternoon rainfall intensity over large tropical islands in the western Pacific compared with nocturnal maxima over small islands, a time when sea–air temperature differences are greatest (Barry and Chorley, 1987). The clouds move seawards on the return counterflow aloft but do not survive in the subsiding air over the cooler sea. Thus, a line of cumulus clouds during the day often indicates the presence of the coastline beneath it. The oppositely directed land breeze at night is associated with stable, cloudless conditions over the land and a flow of colder 'continental' air seawards. The resultant airflow is represented by a shallow circulation with only about half the speed of the day-time sea breeze. However, the land-breeze velocity may often be reinforced by localized topographic effects, such as the downslope katabatic airflow along transverse coastal valleys. This leads to cool, clear, gusty weather conditions along the coastline.

2.6.2 The orographic factor

Pronounced surface relief features have a significant role to play in the humid tropics where the air masses are particularly unstable and buoyant. This so-called orographic factor is responsible for enhanced precipitation amounts over exposed windward slopes and reduced amounts in the rain-shadow area over the leeward slopes. For example, in the Hawaiian Islands, mean annual precipitation exceeds 1000 mm on mountain summits around 1500 m elevation (i.e. Mt Waialeale on Kauai) whereas the leeward slopes record less than 500 mm over wide areas. Interestingly on Hawaii (and other tropical islands affected by the trade wind inversion, discussed in section 2.4), maximum rainfall totals are recorded at 900 m elevation whereas the 4200 m summits of Mauna Loa and Mauna Kea (rising above this inversion) receive between 250 and 500 mm (Barry and Chorley, 1987).

It should be noted that the orographic enhancement of recorded precipitation is most conspicuous over mountainous islands in the humid tropics during the low-sun season. At this time, the ITCZ is relatively inactive and weather disturbances are generally absent. Then, the prevailing trade winds provide a regular 'flush' of moist airstreams over windward slopes which accentuates the rain-shadow effect over the leeward slopes. Fijian rainfall patterns reveal this enhancement since, in the low-sun season from May to October, the southwest to northeast trending mountains (exceeding 600 m on the main islands of Viti Levu and Vanua Levu) lie athwart the persistent southeast trade winds. At this time, a distinctive 'wet' zone characterizes the south and east slopes of the mountains whereas a pronounced 'dry' zone dominates the north and west slopes. Consequently, Lautoka (on the northwest coast of Viti Levu) now experiences a three to five month dry season, with rainfall averaging about 60 mm per month at this time (figure 2.15). Conversely, Suva (on the southeast coast of the same island) does

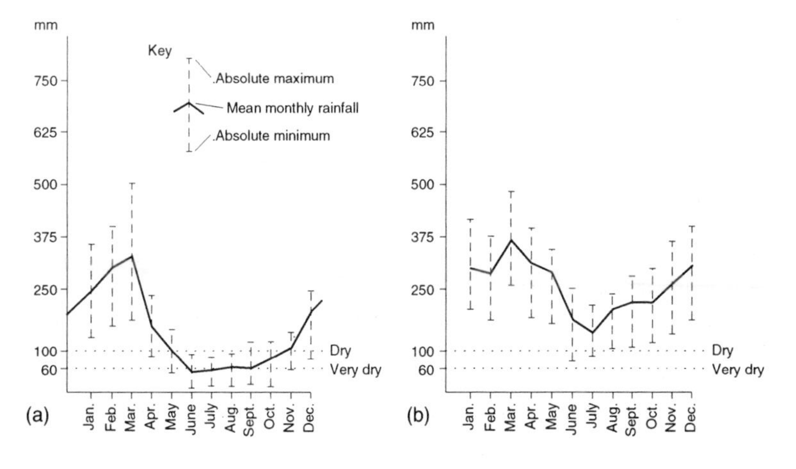

Figure 2.15
Mean monthly rainfall over Fiji: (a) at Lautoka, average number of rainy days 99; (b) at Suva, average number of rainy days 240 (after Thompson, 1986).

not normally experience a dry season (figure 2.15) since the onshore trades are responsible for monthly rainfall totals well in excess of 100 mm (which represents the main criterion for a dry season in Fiji). Also the number of raindays at Suva is roughly 2.5 times that over Lautoka to emphasize the orographic effect in the region.

Figure 2.15 also reveals that during the high-sun season between December and March, monthly mean rainfall at both Lautoka and Suva averages more than 250 mm and can exceed 400 mm. At this time, there is no distinctive wet–dry zonation since the SPCZ is very intense and is located just to the north of Fiji at about 13 °S (Thompson, 1986). Consequently, the proximity of this very active convective zone plus regular cyclogenesis causes converging, humid onshore winds to dominate the entire coastline of the main islands, with pronounced orographic enhancement over the mountainous interior. Over the year, this forced convection dominates the rainfall distribution over the islands. For example, figure 2.16 illustrates the mean annual rainfall over Viti Levu which exceeds 5000 mm over the mountains, approximates 3000 mm in the wet zone and is less than 2000 mm over the drier north and west coastlines.

The monsoon circulation is also influenced by the orographic effect where the moist onshore winds are channelled by topographic 'funnels' (like the head of the Bay of Bengal) towards the high ground. Pronounced orographic uplift follows this airstream convergence which results in some of the heaviest annual rainfall totals ever recorded. For example, over the Khasi Hills in Assam, mean annual totals at 1400 m elevation exceed 11,000 mm with a

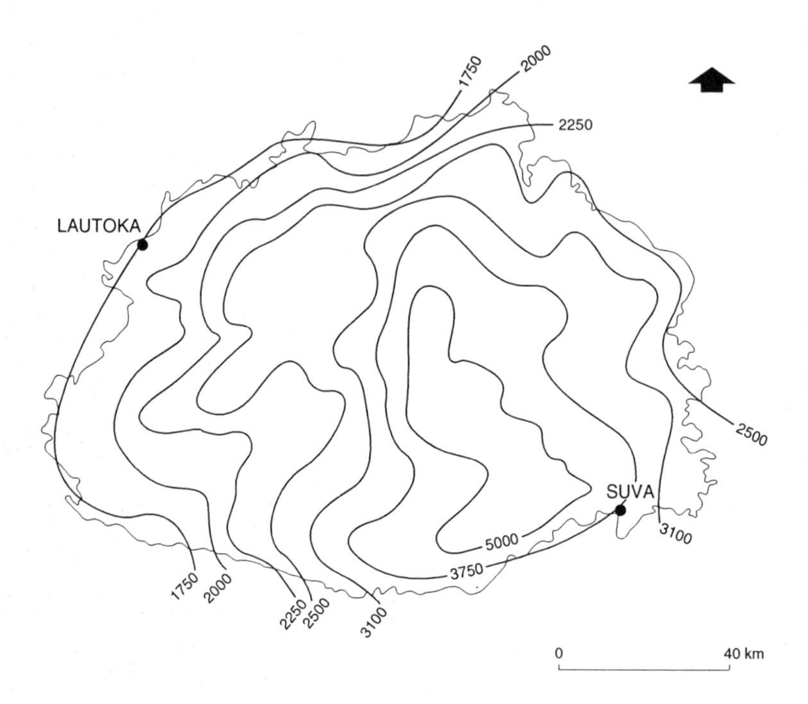

Figure 2.16
Viti Levu, Fiji: mean annual rainfall (mm) (after Thompson, 1986).

record 24,400 mm measured at Cherrapunji in 1974. However, despite these dramatic examples, it is apparent that, throughout the monsoon region, topography plays a secondary role to synoptic activity and large-scale dynamics in determining rainfall distribution in the area (Barry and Chorley, 1987).

2.7 The Role of El Niño and the Southern Oscillation in Extreme Weather Events

When strong trade winds blow, frictional drag on the ocean transports warm surface water away from the source environment and this divergence is driven by the Coriolis force. It then permits an upwelling of cold benthos water to the surface like, for example, the Peruvian current off South America (figure 2.17(a)). Furthermore, the replacement of warm water by a cold supply has important atmospheric consequences. Lapse-rate steepening, instability and free convection over warm water are replaced by conductive chilling, a strong surface temperature inversion and stability.

In the tropical Pacific basin, the cold Peruvian current is replaced at periodic intervals by the incursion of a weak, warm ocean current that flows south along the coasts of Ecuador and Peru. This disruption is traditionally referred to as El Niño which is named after the Christ boy/child since it commonly occurs during the Christmas season. However, the El Niño event is now associated with the Pacific-wide climatic changes that are coincident with the more irregular occurrences of an exceptionally strong and warm current. These events are responsible for a more large-scale readjustment of the tropical atmospheric and oceanic regimes (figure 2.17), which are now regarded as part of a long-term climatic variation in the Pacific basin known as the southern oscillation (SO).

The climatic readjustment between these alternating warm and cold occurrences was first described by Sir Gilbert Walker in 1928. Indeed, the so-called 'Walker circulation' represents the normal state of the SO when upwelling, cold deep water is conspicuous along the coasts of Ecuador and Peru (figure 2.17(a)). This results in a 'normal' atmospheric–oceanic circulation mode, with a distinctive west to east longitudinal cell across the Pacific. There now exists a typical trans-ocean pressure gradient at sea level of 5–10 mb (Bigg, 1990), between low pressure in the west and high pressure in the east (figure 2.17(a)). Strong, surface easterly trade winds feed the low level cyclonic convergence over eastern Australia, Fiji and Indonesia. This leads to the formation of large convectional systems, heavy rainfall and flooding in these areas. At about the 200 mb level in the troposphere (figure 2.17(a)), there is a distinctive westerly counterflow feeding upper-air convergence over northwestern South America. The associated subsidence and low

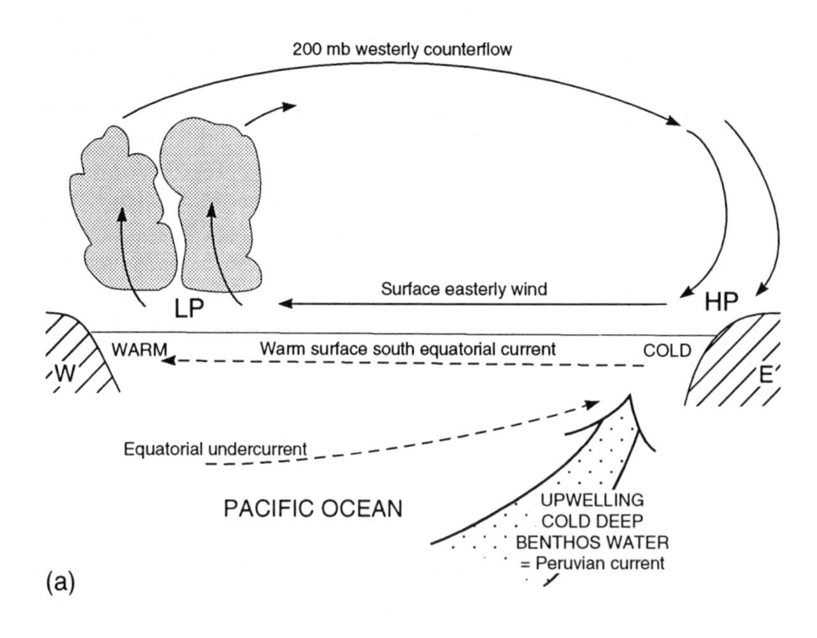

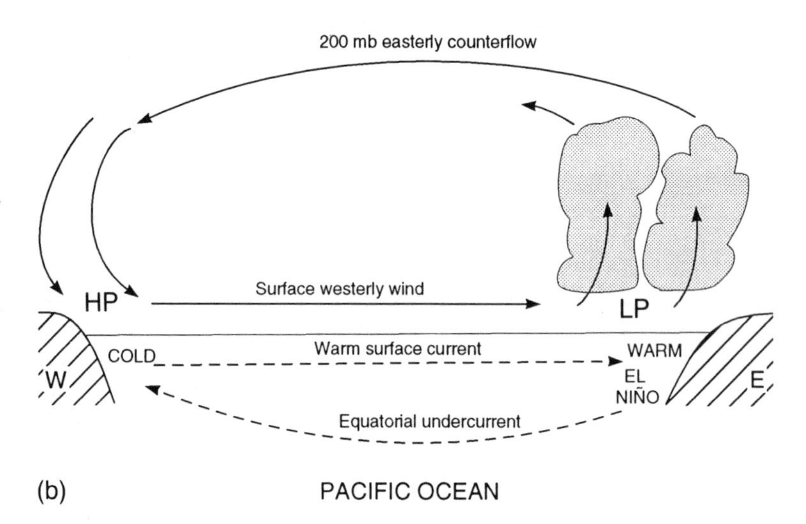

Figure 2.17
Schematic representation of the southern oscillation, with (a) the normal or Walker circulation and (b) El Nino/ENSO: HP, high pressure/low-level divergence (arid with drought); LP, low pressure/low-level convergence (heavy rain with flooding).

level anticyclonic divergence over Peru and Ecuador accentuates the aridity of these desert coasts.

It is apparent that this normal cell and circulation mode is disturbed periodically during weak Hadley cell activity. Then, there is a replacement of cold water off South America by a southward flow of warm water, as equatorial currents, to about 6 °S along the Peruvian coast. This replacement occurs at irregular intervals every two to seven years as the El Niño. Then, coastal surface water temperatures suddenly rise by about 4 °C although El Niño normally takes from three to six months to reach local peak intensity after onset (Reynolds, 1985). Furthermore, the cessation of upward-flowing cold deep water and its nutrients causes economic chaos with a massive fish-kill and the loss of dependent sea birds (which is catastrophic for local guano industries).

El Niño is a very well known phenomenon and, apart from its economic consequences, it is responsible for torrential rainfall and massive flooding over the normally arid coastal deserts of northern Peru and Ecuador. This anomalous weather results from the reversal and weakening of the normal Walker circulation. Thermal low pressure convergence develops over the warm El Niño waters, off northwestern South America (figure 2.17(b)), which accentuates lapse-rate steepening, free convection and precipitation totals. The reversed Walker circulation is linked to stronger Hadley cell activity, which increases the intensity of the surface trade winds and prevents warm water from flowing across the Pacific. This eventually terminates El Niño and, when the colder water is re-established in the eastern Pacific, the Hadley cell now weakens and conditions are restored for the eventual, periodic return of the warm equatorial currents as El Niño. The development and decay of El Niño characteristically take place over a period of twelve to eighteen months. It occurs roughly every three to four years, but, as Bigg (1990) has shown, the interval can vary between two years and a decade.

The reversed Walker circulation or ENSO episode (figure 2.17(b)) is also responsible for pronounced anticyclogenesis over the southwest Pacific, including the eastern periphery of Australia. This development, and resultant westerly surface flow, is necessary to maintain cyclogenesis in the eastern Pacific which in turn leads to an easterly counterflow at high altitudes which is a product of the SO (figure 2.17(b)). The persistent high pressure subsidence in the southwest Pacific appears to introduce severe drought conditions into a normally humid tropical environment including an impact on the Indian summer monsoon through a weakening of land–sea circulation systems. For example, the 1983 El Niño was a very intense episode and triggered massive drought conditions and forest fires in Australia, which were the worst recorded this century (Pearce, 1988). Similarly, the 1987 drought in Fiji proved to be the worst one in at least the last century and has been attributed to the strong ENSO episode of that year (Reddy, 1989). During this episode a broad area of anomalously high pressure persisted over the islands continuously from April to October 1987. The resultant subsidence counteracted the normal orographic effect (discussed in section 2.6.2) and weakened the occasional frontal bands. Consequently, monthly rainfall totals recorded at Suva over the period were between 8 and 48 per cent of normal.

Conversely, 1988 experienced the disappearance of El Niño and the return of a strengthened, normal Walker circulation (termed La Niña, or girl child), which was responsible for unusually heavy rainfall in eastern Australia. At the same time, catastrophic flooding occurred in the Sudan, Bangladesh and southern China, with a spectacular millet harvest in the African Sahel for the first time for twenty years and above average rainfall in India (Thompson, 1989). It is apparent that strong teleconnections appear between these south Pacific ENSO events and extreme weather regimes around the globe, well outside the humid tropics (see Bigg, 1990). However, at the present time, we can only suggest that the occurrence of sea surface temperature anomalies seems to coincide with disrupted atmospheric patterns worldwide and that Rossby-wave positions (and the generation of surface lows and highs) appear to be very sensitive to such anomalies (Henderson-Sellers and Robinson, 1987).

THREE

Humid Tropical Climates

3.1 Introduction

The climate of the humid tropics represents the long-term aggregate response to the complexities of the circulation systems and weather disturbances discussed in chapter 2. This particularly applies to temperature and precipitation where studies over a long period of time reveal important temporal and spatial variations. Rainfall seasonality and inter-annual variability are distinctive climatic features in the region. Furthermore, the majority of people living in the humid tropics depend on agriculture for their livelihood and are thus controlled by the vagaries of weather systems and short-term climatic extremes, especially drought and famine when the monsoon rains fail (section 2.5). Most countries of the humid tropics fall into the category of 'underdeveloped'. They have poor and fragile economies which operate in complex, variable climatic regimes with the emphasis on traditional rather than technological responses to the adversity of the climate.

The analysis of humid tropical climates suffers from the same dearth and short-term nature of observations as was noted in the last chapter for synoptic stations. However, the basic elements of climate (notably temperature and rainfall) have been recorded over a much longer time period (compared with synoptic data). The first rainfall station in West Africa, for example, was established in Dakar (the colonial Gold Coast) in 1855. This station functioned permanently after 1862, apart from gaps between 1863–8 and 1883–92 (Hayward and Oguntoyinbo, 1987), and must represent the longest 'run' of rainfall observations in the humid tropics. Consequently, the variability of West African rainfall regimes is the best understood in the region (along with many other climatic elements) and this explains the bias in the literature (and this chapter). However, many of the few hundred rainfall stations established in the region by the late 1930s were located in schools and institutions which failed to maintain continuous and accurate readings. During the 1940s, official government synoptic and rainfall stations were set up, with more standardized siting, instrumental and observational controls and increased data accuracy and reliability. This was associated with the greatly increased understanding of the complexity of humid tropical weather disturbances discussed in section 2.3. By 1981, in West Africa, 192 fully-operational synoptic stations were operating together with 204 additional rainfall stations, although their distribution was very patchy, as revealed by the Liberia example in section 2.3.

A meaningful and accurate classification of the humid tropical climates has proved difficult to achieve. Some of the numerous attempts to classify the humid tropics, most using climatic characteristics as a major defining feature, have already been discussed in chapter 1. In recent decades, the desire to classify the humid tropics in broad average climatic terms has largely disappeared. It has been replaced with a need to understand fully the dynamics and temporal/spatial variability of the major elements concerned. It is these subjects which will be discussed in the following sections.

This chapter examines the major climatic elements of the humid tropics. It begins with a discussion of radiation balance and energy fluxes, temperature, humidity/cloud cover and evaporation/evapotranspiration. Considerable emphasis is placed on spatial and temporal patterns of rainfall since, in the absence of pronounced annual temperature changes, rainfall becomes the determinant of season and a major control of physical processes and agricultural activities.

3.2 Radiation and Energy Balances

The collection of radiation and energy flux data is limited in the humid tropics since the capital investment required for sensors and integrators/printers or data loggers is not available in most Third World economies. Indeed, the majority of these stations rely on easily and cheaply measured basic elements like temperature and especially precipitation. However, it should be noted that solarimeters are being increasingly introduced into the region, to expand the coverage of solar radiation monitoring at least. The global radiation data for the tropics are mainly in the form of generalized, average annual totals of direct plus diffuse surface-received solar radiation (SW ↓), which are readily available in the literature (e.g. Sellers, 1965; Trewartha, 1968). These data confirm that, because of the greater obliquity of the solar beam with increasing latitude, the average annual solar radiation value in the tropics (7400 mJ m^{-2} $year^{-1}$) is about two and a half times that in polar regions. However, there is considerable variation within the region associated in particular with the degree of cloudiness and the related reflection of solar radiation back to space. The cloudiest parts of the tropics are found in the equatorial belt where the ITCZ is more active (section 2.2). Here, the annual total solar radiation is relatively low, ranging from 5000 mJ m^{-2} $year^{-1}$ in the cloudiest Zaire basin to 6000 mJ in Amazonia and in less cloudy Indonesia. This increased insolation with progressively reduced cloud cover is also evident in the tropical marine regions of the Caribbean (7500 mJ m^{-2} $year^{-1}$) and the monsoon lands of India (7000 mJ), where cloud-free low-sun seasons are conspicuous. The highest values worldwide (up to 9200 mJ m^{-2} $year^{-1}$) are observed in the cloud-free arid tropics of the Sahara-Arabian desert.

Regional global radiation data are rare in the humid tropics, although the increased distribution of solarimeters in West Africa since the 1940s (from three in 1952 to thirty-nine in 1981) has facilitated more accurate representations in this area. For example, figure 3.1 illustrates West Africa's global radiation in January (low-sun season) and July (high-sun season). The former data range between 12 and 22 mJ m^{-2} day^{-1} for the humid tropical stations, with the lowest values recorded in the cloudy coastal belt. A particularly steep gradient of change is evident at 8 °–9 °N with increasing cloudiness, although anomalous high values do occur, associated possibly with the less cloudy uplands like the Jos Plateau of Nigeria (Hayward and Oguntoyinbo, 1987). It is interesting to note from figure 3.1(b) that the solar radiation recorded in the July

Figure 3.1
West Africa's global radiation in January (a) and July (b) and net radiation in January (c) and July (d) (after Hayward and Oguntoyinbo, 1987). All values are in millijoules per square metre per day.

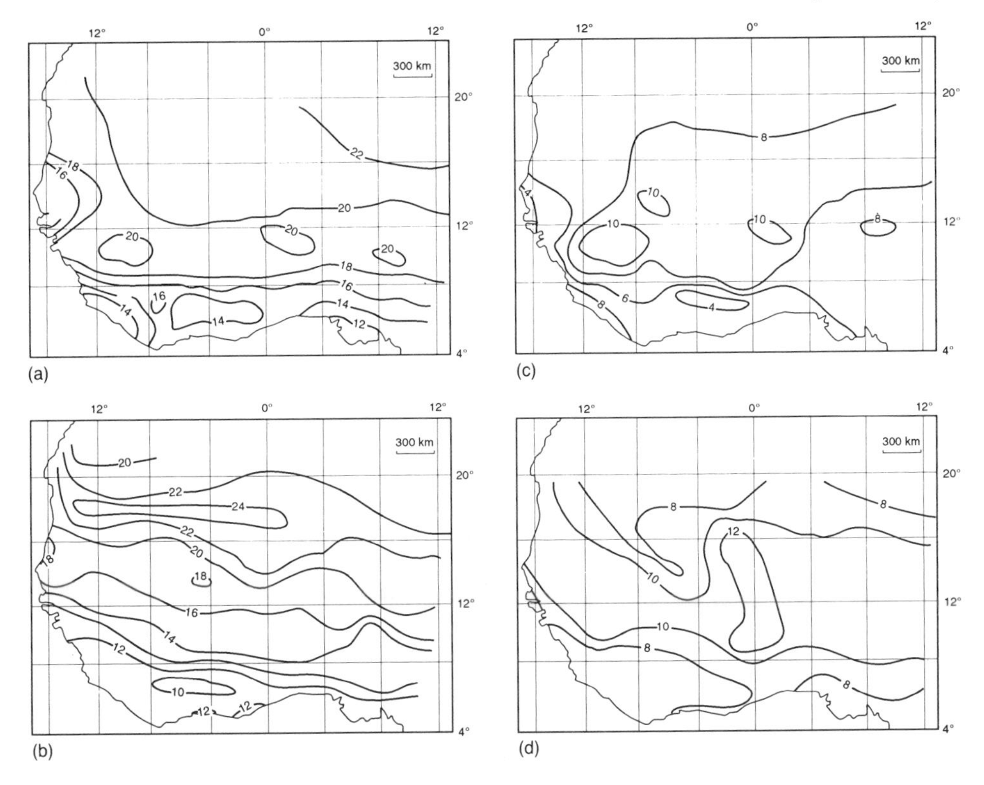

high-sun season at the coastal stations is considerably less (between 2 and 3 $mJ\,m^{-2}\,day^{-1}$) than that observed in January. Despite the increased solar elevation, there is more cloud cover in the high-sun season associated with increased ITCZ activity (section 2.2). For example, over Liberia, cloud cover averaged 4 octas in January compared with 7 octas in August. The higher albedo (70–90 per cent) and increased cumulus cloud-top reflectivity is responsible for the decrease in solar radiation at the surface, which in July ranges between 10 and 12 $mJ\,m^{-2}\,day^{-1}$ (figure 3.1(b)). Inland, the high-sun solar radiation has increased slightly at this time, compared with the low-sun value.

The outgoing short-wave radiation component (SW ↑) of the radiation balance is controlled by the surface albedo, which varies according to vegetation and soil types. For example, the wet, dark tropical soils have albedos around 10 per cent, rain-soaked forests and bushland average 12 per cent and tropical crops (like rice and sugar cane) average 13 per cent. Consequently, the surface of the tropics is a conspicuous absorber of solar radiation, compared with the more reflective, arid tropics where albedos range between 25 and 45 per cent. Specific data for the long-wave infrared components, both outgoing (LW ↑) and incoming (LW ↓), are rarely measured in the developed mid-latitudes let alone the developing humid tropics. However, LW ↓ monitoring will be essential worldwide in the future to determine the actual magnitude of the enhanced greenhouse effect and the proposed global warming (chapter 9).

Global annual calculations of LW ↑ (Sellers, 1965; Trewartha, 1968) are all within 11 per cent across the earth's surface with the greatest losses recorded in the hot subtropical deserts, where atmospheric absorption is weakest. Equatorial values average 5900 $mJ\,m^{-2}\,year^{-1}$, associated with a high emissivity from warm surfaces which is tempered by the persistent cloud cover. Regional LW ↑ data are not generally available but important variations will be associated with the different vegetation and soil types and their variable emissivities. Interestingly, LW ↓ data are available for West Africa (Ojo, 1970) where, in the cloudy humid coastal zones, LW ↓ values averaged 7.75 $mJ\,m^{-2}\,day^{-1}$ over the year. Indeed, the annual range here was less than 2 per cent which emphasizes the persistent role of water vapour and cloud droplets, and the associated counter radiation.

The net exchange, between the four incoming and outgoing components of the radiation balance discussed so far, is termed net radiation (Rn):

$$Rn = SW\downarrow - SW\uparrow - LW\uparrow + LW\downarrow \qquad (3.1)$$

For the humid tropics, insolation exceeds the long-wave radiation loss to represent a radiative energy surplus. This is an important heat source for the global energy balance which initiates the ITCZ and the atmospheric global circulation. The best available regional data again come from West Africa, where figures 3.1(c) and 3.1(d) illustrate net radiation values for January and July. The maps indicate a modest if expected Rn maximum during the high-sun season (figure 3.1(d)) both along the coast and at inland stations, with respective values of 7–8 and 14 $mJ\,m^{-2}\,day^{-1}$. Furthermore, this radiative energy concentration correlates with the ITCZ season and hurricane formation over the oceans (section 2.4.2). Relatively low Rn values along the cloudy, humid coasts (compared with the more cloud-free inland areas) are also evident in the low-sun season, where almost twice as much Rn is recorded inland. In both seasons, the gradients of change towards the coast are not as pronounced as those for global radiation (figures 3.1(a) and 3.1(b)), apart from the January 'bunching' along 8 °N between longitudes 0 ° and 6 °W in central parts of Ghana and the Ivory Coast (figure 3.1(c)). This is due to the important role of cloud cover in maximizing the LW ↓ component of Rn which partly compensates for the insolation (SW↓)) deficiency of these cloudy sites. Furthermore, the inland–coastal Rn differences in July are moderated by the increasing dust-loading of the inland atmosphere from wind deflation in the more arid tropics to the north and east. The resultant increased atmospheric turbidity from this dust 'veil' scatters an appreciable amount of solar radiation back to space with greatly increased SW ↑ (Hayward and Oguntoyinbo, 1987).

The net radiation received in an area is 'consumed' by a number of important non-radiative fluxes, particularly by the turbulent cascade of energy away from the earth's surface associated with conduction and convection. The resultant global energy balance is represented by the equation

$$\mathrm{Rn} = H + \mathrm{LE} = 0 \tag{3.2}$$

where H is the sensible heat flux (i.e. the direct transfer by conduction/convection) and LE is the latent heat flux associated with the energy stored/released in the evaporation–transpiration/condensation processes. Obviously the energy balance of an area will be controlled by the availability of surface water in every form. Consequently, since the humid tropics are pluvial with freely evaporating and transpiring surfaces, in general terms the energy balance will be dominated by LE.

Indeed, latitudinal values from Sellers (1965) reveal (table 3.1) that the area between the equator and 10 ° north and south has an energy balance where LE utilizes about 70 per cent of Rn over land and sea. Also, over the land the balance is consumed by H, whereas over the seas ocean current tansfers (Af) account for the majority of the balance. In the former area, the Bowen ratio ($\beta = H/\mathrm{LE}$) ranges between 0.4 and 0.5, to confirm the LE dominance. Between latitudes 10 ° and 20 ° north and south, the LE control of energy balance over the oceans continues, ranging from 83 to 92 per cent (table 3.1). However, over the land masses, LE/H relationships vary

Table 3.1 Energy balance of the humid tropics

	Oceans				Land			
Latitude zone	*Rn*	*LE*	*H*	*Δf*	*Rn*	*LE*	*H*	*β*
0–10 °N	100	70	3	27	100	67	33	0.5
0–10 °S	100	73	3	24	100	69	31	0.4
10–20 °N	100	83	5	12	100	41	59	1.5
10–20 °S	100	92	4	4	100	56	44	0.8

H, LE and *f* values are expressed as a percentage of Rn.

Source: Adapted from Sellers, 1965

in both hemispheres. The zone 10°–20°S extends the LE control into the subhumid part of the tropics, although it is much reduced ($\beta = 0.8$). Conversely, the zone 10°–20°N (being more continental) is increasingly resistant to vapour flux and H now dominates, with a β of 1.5 to confirm the resistance. This energy balance is typical of Köppens Am and Aw climate zones where dry seasons are conspicuous, to greatly suppress evapotranspiration rates.

At a regional scale, particularly a more local level, the energy balance is controlled by the nature of the earth's surface together with specific ground properties. The surface energy balance is now expressed as

$$\text{Rn} = H + \text{LE} + G = 0 \tag{3.3}$$

where Rn, LE and H are as before (equation (3.2)) and G is a measure of the conduction of heat to or from the underlying ground. The latter flux depends on a wide range of soil properties (moisture content, porosity, diffusivity etc.) which vary in any environment. In the context of the humid tropics, the main variables will be the water content of the soil and the amount of surface cover in the form of leaf litter. For example, the litter in grasslands and especially the rain forest will act as a thermal insulator, keeping G very small (of the order of 2 per cent of Rn). Furthermore, in the Aw/Am dry season, the desiccation of the litter cover will act as a vapour barrier to reduce LE considerably, forcing H into a larger mode of Rn dissipation. On the other hand, saturated soils, paddy fields and tropical swamps introduce continuous demands by evapotranspiration, so that LE normally exceeds the available Rn at all times. The necessary energy requirements are now supplemented by H, which has to be advected from surrounding drier areas (with negative β).

3.3 Temperature

The above radiation balance data emphasize that only a small proportion (30 per cent) of Rn is available as a source of air heating in the humid tropics. Most of the radiative energy is utilized in the

LE flux which does not influence surface air temperatures. Therefore, temperatures in the region are moderated and over well-watered surfaces it is rare for shade temperatures to exceed 34 °C, compared with 50 °C and above recorded in the arid tropics where *H* dominates. Indeed, the annual average temperatures (adjusted to sea level) in the humid tropics normally range between 24 and 30 °C at individual stations. Furthermore, these stations are characterized by small (less than 2 °C) seasonal and annual temperature ranges, due to the modest Rn differences discussed in the last section. For example, Belem (Brazil) has a mean annual temperature range of 1.6 °C and Jaluit (Marshall Islands) has a range of about 0.5 °C. As a contrast to these negligible annual ranges, the diurnal range appears quite considerable and can regularly equal 6 °C. However, this range is smaller in the cloudy, wet season with reduced SW ↓ by day and increased LW ↓ by night. Conversely, in the cloud-free dry season, the nights are particularly cool (following maximum LW ↑) and indeed the 12 °C diurnal range then (figure 3.2(a)) represents the 'winter' of the tropics in terms of sensitivity of the native people in traditional buildings.

The general uniformity of temperature in the tropics is interrupted by elevation, which can cause large temperature changes over short distances. The rate at which temperature falls with increasing altitude, the lapse rate, is also highly variable but local. The main factor controlling the lapse rate is cloudiness but

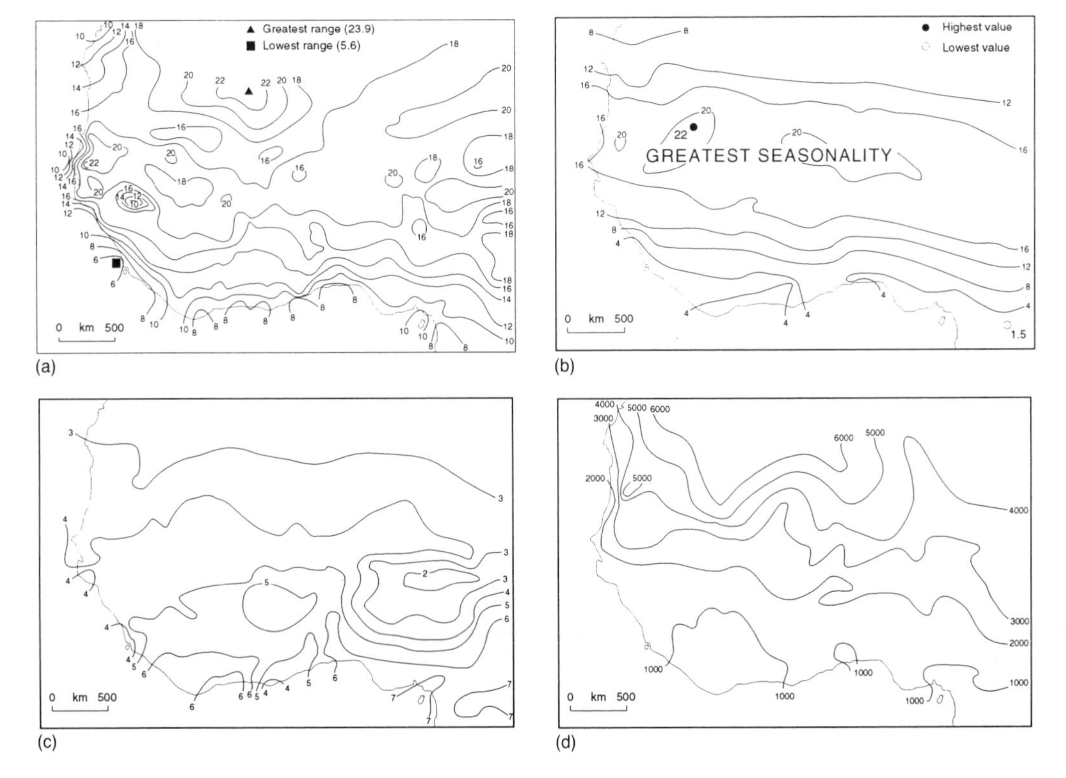

Figure 3.2
Elements of West African climates: (a) mean diurnal temperature range (°C); (b) mean annual range of vapour pressure (mb); (c) mean annual cloud cover (oktas); (d) mean annual potential evaporation (mm) (after Hayward and Oguntoyinbo, 1987).

local topographical conditions (particularly aspect) are also important. Excessive heat in the humid tropics is also moderated by offshore cold ocean currents which represent the upwelling of benthos water assumed (by Pople and Mensah, 1971) to be due to the increased evaporation, salinity and density of the subsiding surface water. This upwelling is seasonal in the humid tropics; for example, it occurs between July and October along the West African coast from western Nigeria to the Ivory Coast. At this time, mean sea temperatures average about 5 °C lower than those in March–April although extreme 10 °C differences have been recorded. Figure 3.3 clearly illustrates the relationship between air and sea surface temperatures in the Gulf off Guinea of Lomé and Abidjan and confirms the pronounced high-sun season cooling resulting from cold-water upwelling.

There are four distinctive temperature regimes in the humid tropics, as illustrated in figure 3.4. First, in the equatorial low-elevation type, which is represented by Belem (Brazil) (figure 3.4(a)), temperatures are uniformly high and extremely constant around 27 °C throughout the year, with a mean annual range of 1.6 °C. The extreme maxima do not correspond to the seasonal SW ↓ maxima but are controlled more by cloud cover and humidity concentrations. For example, the time of greatest heating correlates with relatively cloud-free conditions (enhanced SW ↓) and very high

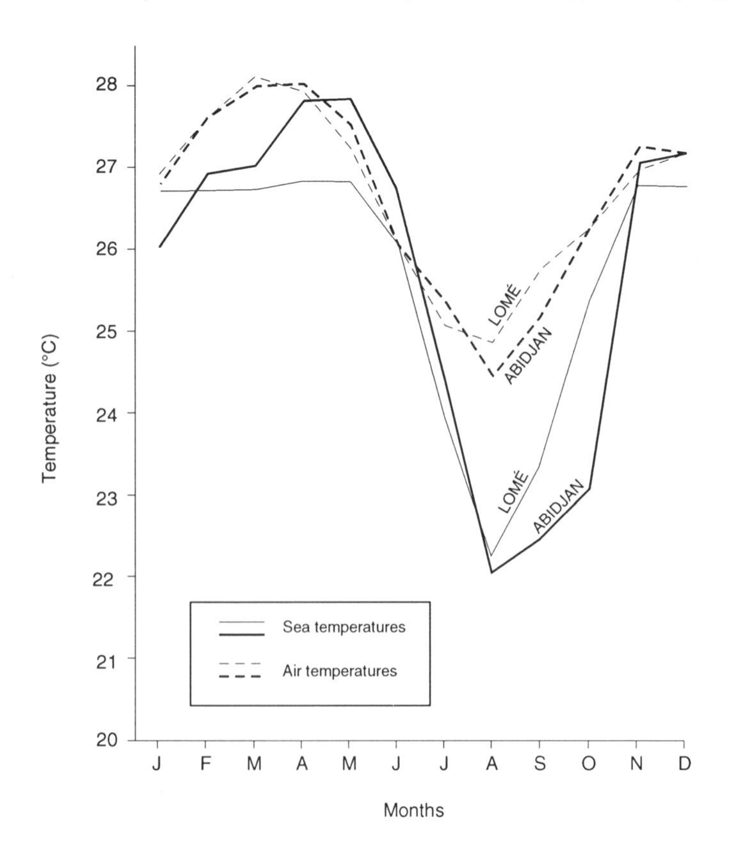

Figure 3.3
Air and sea temperatures off Lomé (6 °N, 1 °E) and Abidjan (5 °N, 4 °W), West Africa (after Hayward and Oguntoyinbo, 1987).

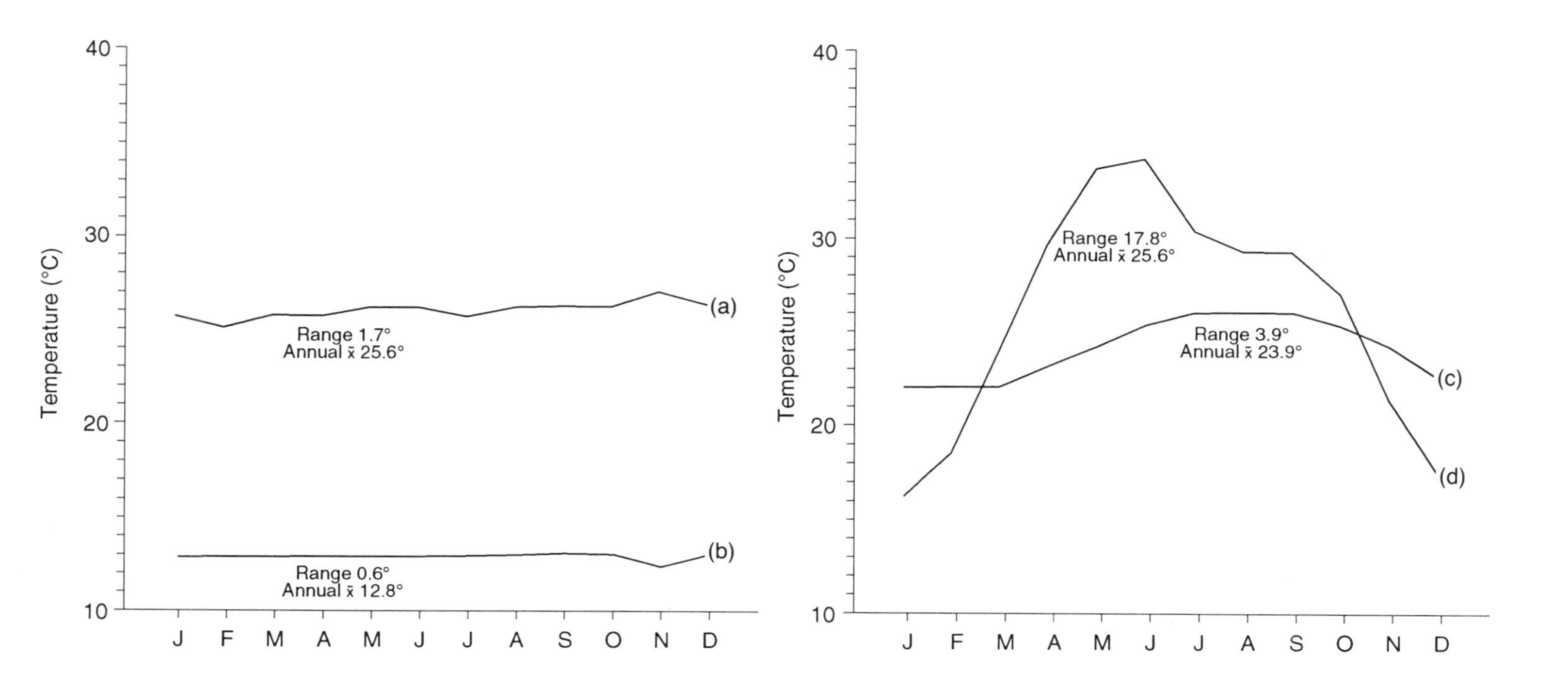

Figure 3.4
Monthly temperatures at four humid tropical stations: curve (a), Belem, 1 °S, 48 °W, altitude 13 m; curve (b), Quito, 0 °, 79 °W, altitude 2949 m; curve (c), Honolulu, 21 °N, 158 °W, altitude 12 m; curve (d), Jaipur, 27 °N, 76 °E, altitude 438 m (after Hayward and Oguntoyinbo, 1987).

humidity (increased LW ↓), which is prolonged when the cooling rains are delayed. At Lagos (Nigeria) the mean monthly temperatures peak at 27.8 °C in March (the end of the dry season) and decrease rapidly to 23.9 °C in July during the rainy season. The second temperature regime is the equatorial high elevation type, which is represented by Quito (Ecuador) (figure 3.4(b)). At an elevation of 2949 m in the Andes, the equatorial heating is moderated considerably by lapse-rate cooling (equal to 6 °C per 1000 m ascent). In fact, temperatures at elevations between 1800 and 3200 m resemble those of temperate regions, since the mean annual range here is between 12 and 18 °C. For example, the so-called 'tierra fria' of the high Mexican plateau is more desirable for human settlement and the population density here is four times that of the 'tierra caliente' on the hot, steamy coastal lowlands (Miller, 1971). Similar altitudinal zonations occur in Brazil, Peru, Bolivia, Ecuador and Colombia. Quito (figure 3.4(b)) has a mean annual temperature of 13 °C and the monthly variations are negligible (even less than those at low levels), well within 0.5 °C. Bogota (Colombia) at an elevation of 2754 m has similar thermal characteristics with a mean annual temperature of 14 °C and an mean annual range of less than 1.0 °C.

The third temperature regime is the tropical marine type (figure 3.4(c)) which lies in the trade wind belt around 20 ° north and south, outside the true equatorial zone. Hawaii and Fiji are good examples of this type where seasonal changes in ITCZ/SPCZ location and intensity (section 2.6.2) are responsible for conspicuous rainfall and temperature distributions. Consequently, mean temperatures are conspicuously higher in the ITCZ/SPCZ 'season'; for example, at Honolulu (figure 3.4(c)) the January monthly mean is 21.7 °C compared with 25.6 °C in July, the middle of this 'season'. At Suva (Fiji) the July mean temperature equals 23 °C compared with 27 °C in February. Thus the 4 °C annual range at both these stations is more than twice that recorded in the equatorial low elevation locations like Belem and Jaluit (discussed above).

The final temperature regime is the monsoon type (figure 3.4(d)) where very distinctive seasonal changes occur. Indeed this region only displays humid tropical characteristics during the high-sun season when temperatures exceed 35 °C, well above those in the equatorial lowlands. However, the low-sun values are considerably lower, for example 16.1 °C at Jaipur in India (figure 3.4(d)). Consequently, the mean annual temperature range is 17.8 °C at this station, with this peak summer heating of 33.9 °C recorded in June before the onset of the monsoon rains. This compares with mean temperatures some 4 °C lower during the rainy season a month later, when temperatures fall due to increased cloud cover, decreased Rn and increased evaporative cooling.

It should be noted that monsoon-type temperatures are controlled primarily by latitude, altitude and continentality. For example Madras (India) is 14 ° closer to the equator, some 430 m lower and is coastal, compared with Jaipur (figure 3.4(d)). This results in less extreme heating and cooling over the year and an annual temperature range between 32.2° and 24.4°C, which is 10°C less than that at Jaipur. In the dry, cool low-sun season, and especially at inland locations, the diurnal range is very large and can exceed 25°C. This is due to cloudless conditions and reduced humidity levels which are conducive to excessive rates of LW ↑. For example, in the Punjab at this time, day-time maxima reach 27°C while air frosts are common occurrences at night.

At the same time, the increased cloud cover, humidity and LW ↓ in the coastal regions are responsible for greatly reduced diurnal ranges, e.g. less than 5°C at Bombay, where air frosts are unknown.

3.4 Humidity and Cloud Cover

So far in this chapter, we have emphasized the roles played by humidity and cloud cover in the humid tropics, especially in radiation balance and temperature distributions. Humidity levels are always high in the area mainly because the LE flux dominates the energy balance (section 3.2) with persistent rates of *in situ* evapotranspiration. This is aggravated in the high-sun, wet season by the pronounced convergence of equatorial air. It should be reiterated that the high water vapour content increases LW ↓ and keeps the nocturnal temperatures high, accentuating the human discomfort. In the wet season, this nocturnal high humidity is a considerable problem since the very muggy saturated conditions cannot promote evaporation and the associated cooling of the perspiring body. Increasing altitude and cooler air with a reduced moisture capacity makes the humidity level more conducive to European settlement. The 'tierra fria' (discussed in the last section) represents such a favourable environment in many Andean–Latin American countries.

Vapour pressures normally exceed 25 mb in the humid tropics, although lower values characterize the reduced temperatures of the cloudy high-sun season. For example, along the Gulf of Guinea coast, April (pre-rainy season) is the time of highest vapour pressure (30 mb) compared with rainy August (26 mb), with a mean annual range of 4 mb (figure 3.2(b)). Away from the coast, the vapour pressure gradient is steep since, in January in West Africa, vapour pressure reduces from 30 mb at 6°N to less than 10 mb north of 10°N. However, at this time, the gradient is exaggerated by the desiccating harmattan wind off the Sahara (Hayward and Oguntoyinbo, 1987). In monsoon climates, seasonal ranges of vapour pressure are much more pronounced. For example, at Calcutta (India) dry-season January values average 15 mb compared with 34 mb in July, the wet season (Nieuwolt, 1977).

Relative humidity levels are always high in the humid tropics, generally exceeding 80 per cent since the maritime air is always close to saturation point. The mean annual range is small, less than

10 per cent along the Gulf of Guinea coast, with the highest levels (>90 per cent) recorded in February during the low-sun season when saturation point is close in the comparatively cooler air. It should be noted that relative humidity levels have a distinctive diurnal rhythm following the temperature distribution. This is particularly so in the high-sun season when afternoon relative humidities are some 8 per cent lower than those during the cooler morning. The 'tierra fria' effect on air temperatures and vapour pressure discussed above extends naturally to relative humidity values. At Mali in Guinea, for example (1464 m elevation), no month has a mean value above 70 per cent, which reinforces the human comfort factor.

The humid tropics are notoriously cloudy areas (figure 3.2(c)) but pronounced seasonal variations are distinctive, associated with the changing location and intensity of the ITCZ. The high-sun, wet season in the humid tropics is generally overcast (averaging 6 octas of cloud) compared with the reduced cloud cover (average 4 octas) in the low-sun, dry season when free convection is reduced. For example, along the Gulf of Guinea coast, the probability of an overcast sky is 40–50 per cent in the dry season compared with 70–80 per cent in the wet season (Hayward and Oguntoyinbo, 1987). The diurnal changes in cloud amount are also considerable and vary in terms of coastal versus inland locations. For example, the former stations experience maximum cloud cover during the early morning and clearer conditions in the early afternoon. In contrast, inland stations record maximum cloud cover in the late afternoon, associated with the progressive penetration inland of the sea-breeze front (figure 2.12) and its related forced convection (sections 2.6.1 and 3.6.4).

3.5 Evaporation and Evapotranspiration

Vapour flux data are limited, irregular and unreliable in the humid tropics and those available are mainly in a potential form acquired from simple, inexpensive Piché evaporimeters. However, these sensors are particularly sensitive to wind velocity and are known to under-record evaporation in humid areas.

Compared with the arid tropics (where evaporation can exceed 500 mm per month when water is available), evaporation in the humid tropics is not excessive and averages about 100 mm per month. This is due to the more overcast conditions and reduced Rn already discussed, although it has been noted that, in relative terms, LE is the dominant energy flux (section 3.2). The most complete, if questionable, set of evaporation data comes from West Africa where some eighty-seven stations (mostly using Piché evaporimeters) provide data of varying reliability and accuracy (Hayward and Oguntoyinbo, 1987). Figure 3.2(d) reveals that mean annual potential evaporation is about 1000 mm along the Gulf of Guinea coast. Furthermore, it is apparent that the atmospheric evaporative power increases rapidly northwards and especially inland. Mean potential rates in these cloud-free hotter areas are up to six times greater than those along the cloudy coast (when, of course, surface water is available).

The complexities of evaporative demand, especially in forested areas, and the associated problems of its measurement are frequently discussed at length in the literature (e.g. Geiger, 1965; De Bruin, 1983; Jackson, 1989). A number of approaches to the problems of measuring evaporation have been developed (table 3.2); all have advantages and disadvantages and some are inherently more suitable for tropical situations than others. For example, the actual volume of water loss measured using standard evaporation pans consistently produces values higher than those derived by other methods (Balek, 1983). Ayoade (1976) recorded average rates of 1500 mm over the year from three evaporation pans in coastal Nigeria. These were 50 per cent higher than those based on Piché measurements. The excessive evaporation from pans is largely attributed to their design and aluminium construction, which allows heating of the water during the day. Evaporation from pans constructed of thermally inert fibreglass may be up to 15 per cent lower.

Seasonal evaporation changes are most conspicuous in subhumid and semi-arid areas when the cloud-free low-sun season is the time of highest evaporation (with increased Rn). Conversely, the increased cloud cover and reduced Rn of the high-sun, wet season are associated with decreased evaporation. Spatial distributions, however, are complicated by the surface water characteristics, especially salinity and temperature. For example, cold ocean currents are responsible for the reduced evaporative power of the lower atmosphere. Evaporation rates in high altitude regions are also reduced.

Evapotranspiration is a much more pertinent element in the humid tropics since vegetation is extensive and freely transpiring, to represent a vital part of vapour flux. Actual rates are normally obtained from lysimeter measurements but very few of these exist in developing countries. Furthermore, the accuracy of those in use is somewhat suspect due to design weaknesses (especially with regard to aerodynamic roughness) and inadequate natural vegetation representations.

Table 3.2 Measuring evaporation and transpiration

Direct monitoring methods	*Empirical formulae*
Atmometers and evaporation pans	Calculation of water vapour flux, e.g. profile method
Lysimeters	Calculation of PET, e.g. Penman (1948), Monteith (1965), Van Bavel (1966), Priestly and Taylor (1972)
Tracers (for transpiration only)	Water balance approach

In most parts of the world, observed or actual rates of evapotranspiration generally fall below theoretical maximum or potential rates since water is rarely given up or taken into the atmosphere freely. However, in the humid tropics where there is a large latent heat flux and abundant moisture, the absolute 'demand' or rate of evapotranspiration is often high, at least during wet season months. Actual evapotranspiration may approach potential rates and in some circumstances even exceed it (e.g. owing to topographical irregularities). In wetter parts of the tropics annual water losses by evapotranspiration depend primarily upon mean sunshine duration, i.e. the available energy for the vapour flux to take place. De Bruin (1983) sites a maximum figure of 1700 mm per annum for a selection of humid tropical stations below 600 m, in general agreement with those obtained by Kayane (1971) for monsoonal Asia. More seasonal parts of the tropics are thought to lose between 700 and 1200 mm moisture per annum by evapotranspiration, the actual amount being largely determined by the length and intensity of the dry season when actual evapotranspiration rates fall well below potential values.

Most theoretical calculations of potential evapotranspiration (PE) are derived from Penman's (1948) formula (table 3.2), originally prescribed to model PE from a (hypothetical) extended surface of short green crop actively growing, completely shading the ground, of uniform height and well watered. The original Penman formula is unsuitable for situations of high wind speed, low radiation input, low rainfall intensities of long duration and a negative sensible heat flux due to meso-scale advection (Thom and Oliver, 1977; Monteith, 1981). Such conditions are rare in the tropics and several authors have demonstrated that the formula can give satisfactory results in the humid tropics (e.g. Brutsaert, 1965; Edwards et al., 1981; Bruijnzeel, 1983; Gunston and Batchelor, 1983). Doorenbos and Pruitt (1977) recommend a slightly modified version for use in irrigation estimates and De Bruin (1983) favours the simplified Priestly–Taylor modification (table 3.2) for practical purposes, since its parameters (principally sunshine duration or incoming solar radiation) can be easily measured directly in the field or calculated remotely from satellite data.

Most PE data available for the humid tropics are derived using Thornthwaite's (1948) approach, which depends on relatively easily obtained monthly mean data (temperature, precipitation and hours of daylight). Jackson (1989), however, argues that although the approach appears to work well in temperate continental latitudes like the USA, it may introduce serious errors under tropical conditions. Such PE estimates for southern Nigeria average about 1200 mm per year, based on Obasi (1972), Walker (1962) and Hayward and Oguntoyinbo (1987). Ojo (1969) revealed seasonal changes in PE in West Africa which correlated with seasonal ITD changes (section 2.2 and figure 2.2). For example, the Gulf of Guinea coast averaged 100 mm PE in the low-sun season, compared with a much reduced 50 mm PE in the more overcast high-sun season. Furthermore, Hayward and Oguntoyinbo (1987) suggest that the concentration of PE isopleths between coastal Senegal/ Mauritania and the interior at this time is clearly related to the ITD.

Figure 3.5 illustrates PE values for selected humid tropical stations, based on the Penman formula (Nieuwolt, 1977). Equatorial Kuala Lumpur (Malaysia) has consistent PE over the year, around 125 mm per month (figure 3.5), with a very small annual deviation. Conversely, monsoonal Bijapur (India) has distinctive seasonal variations with maximum PE (*c.* 180 mm) in May, prior to the onset of the southeast monsoon (figure 3.5). During this cloudy wet season PE rates decline significantly, but minimum rates are confined to the cool dry season, when they approach 65 mm per month. Kasama (Zambia) experiences the lowest PE rates of the three regions considered (figure 3.5), reflecting its subhumid continental location. However, lowest rates occur in the low-sun season (May to August) and the annual range is considerably greater than in the equatorial regions.

Nullet and Giambelluca (1990) demonstrate that, at least in island locations, none of the available theoretical methods of estimating PE adequately represent real conditions. They compare winter season PE rates calculated from a variety of established formulae with atmometer and adjusted open pan results for sites at various altitudes on Haleakala mountain, Maui, Hawaii. They reveal that the theoretical methods consistently underestimated PE (cf. the relationship usually assumed) since they fail to account for additional sources of energy provided by heat advection from oceanic and land sources at low elevations and from the entrainment of dry air at altitudes near the trade wind inversion (table 3.3). In summer, they suggest the situation would be reversed, with pan data overestimating true values.

The field hydrological balance method of estimating evapotranspiration is both time consuming and expensive but may provide one of the most accurate estimates. Sengele (1981) demonstrates its utility for the forested Loweo catchment in the Zaire Congo. Annual evapotranspiration losses were here calculated at 79 per

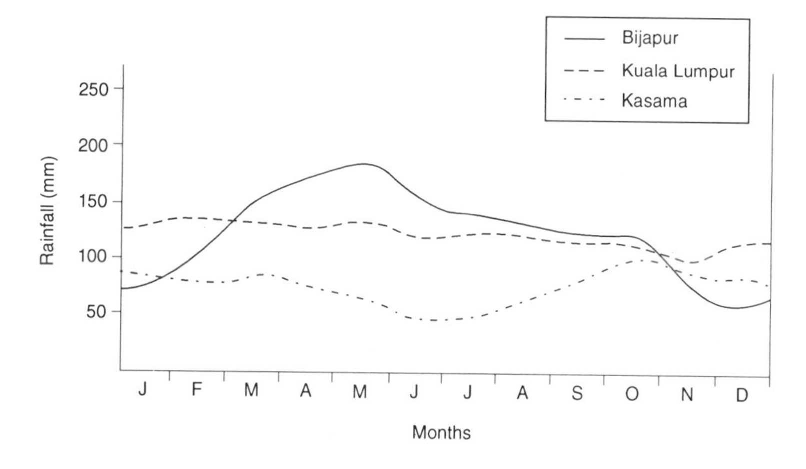

Figure 3.5
Mean potential evapotranspiration (according to the Penman formula) for Kuala Lumpur, 3 °N, 42 °E; Bijapur, 17 °N, 75 °E; Kasama, 10 °S, 31 °E (after Nieuwolt, 1977).

Table 3.3 Daily evaporation rates (mm day^{-1}) at six sites on Mt Haleakala, Maui, Hawaii, during winter 1987–8

	Station					
Method of calculation	*1*	*2*	*3*	*4*	*5*	*6*
Penman (1948)	2.4	2.8	2.8	2.1	2.9	2.5
Van Bavel (1966)	2.2	2.6	2.5	1.9	2.7	2.2
Monteith (1965)	2.2	2.6	2.6	2.0	2.8	2.3
Priestly and Taylor (1972)	2.1	2.9	2.8	2.1	3.1	2.6

Source: Nullet and Giambelluca, 1990

cent of annual rainfall (1280 mm), but important diurnal and seasonal variations were noted. Sinusoidal daily oscillations in water discharge with peak flow around midnight and minimum flow around noon were correlated with periods of least and minimum transpiration respectively.

3.6 Rainfall

Except in the highest areas (which comprise a very small part of the tropics), precipitation consists almost entirely of rainfall. It varies widely with respect to both time and place. Other climatic factors in the tropics are more uniform, and hence rainfall is the main factor used in the delimitation of tropical climates (Nieuwolt, 1977). The characteristics of the rainfall regime in the humid tropics influence all parts of the physical and human landscape. The availability of water in the landscape directly affects rates and types of rock weathering and soil formation and ultimately, therefore, landform development and vegetation cover. Rainfall is also particularly important in relation to agriculture.

3.6.1 Seasonality

In the absence of pronounced temperature changes over the year in the humid tropics, rainfall becomes the determinant of season. The characteristic convectional rains and wet seasons reach a maximum after the passage of the zenithal sun, about April and November on the equator. Furthermore, it is well known that one of these maxima is more pronounced than the other. For example, the 'greater rains' and 'smalls' of Ghana and the 'long rains' (Uua) and 'short rains' (Nthwa) of Kenya. In most cases, the so-called spring maximum rainfall (following the northward shift of the zenithal sun) is more substantial than the autumn maximum, following its southward return. For example, Lagos (Nigeria) records 990 mm in May to July compared with 406 mm in September to November. The equatorial rainfall regime does not extend far on either side of the equator; within a few degrees occur two dry seasons of equal length following the solstices which generally become more significant. Beyond 5° north and south, the dry seasons became less equal in duration and intensity.

It is now recognized that the traditional, seasonal north–south migrations of the zenithal sun/ITCZ (and associated convective rainfall belts) are gross oversimplifications. Instead, as section 2.2 confirms, the equatorial trough is much more complex in terms of structure and location which results in pronounced variability of the associated cloud cover and rainfall regimes. The trough is no longer regarded as a simple zone of continuously rising air and condensation. Furthermore, its position varies seasonally in direct response to the location of the so-called thermal equator, which is influenced by the degree of continental heating and cold, stabilizing ocean currents. For example, the ITCZ fails to penetrate very far south over northwest Brazil in the high-sun season because of the control of the Atlantic subtropical highs.

Consequently, Recife (8 °S) and indeed the entire Brazilian coastal zone from 5 °S to 13 °S receives little rainfall (i.e. between 50 and 150 mm per month) in the period January to March (figure 3.6(c)). This compares with Belem (1 °S) which records 190–450 mm of rainfall per month in the same period under the stagnating trough (figure 3.6(e)). Recife only receives rainfall of note between March and July (figure 3.6(c)) owing to the passage of weather disturbances moving equatorwards from higher latitudes (Jackson, 1989). From September to December, the rainfall at both Belem and

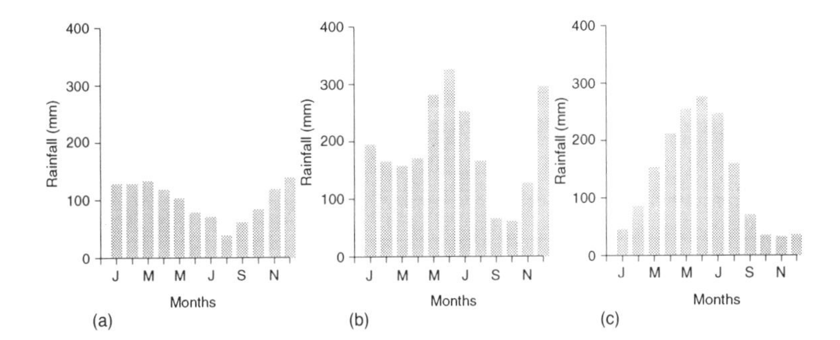

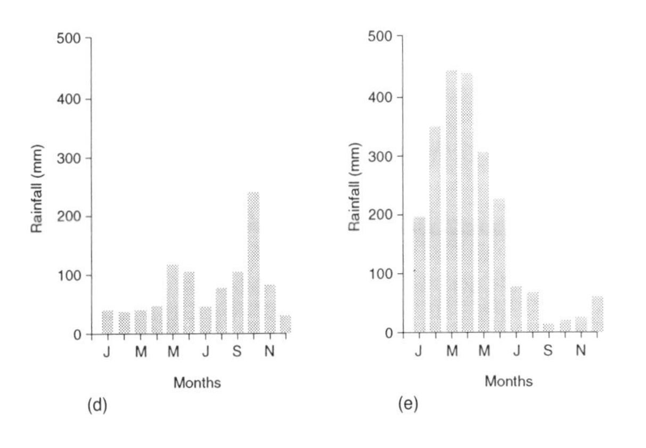

Figure 3.6
Rainfall seasonality at five South American/Caribbean stations: (a) Rio de Janeiro, 23 °S, 43 °W; (b) Georgetown, 7 °S, 58 °W; (c) Recife, 8 °S, 35 °W; (d) Kingston, 18 °N, 76 °W; (e) Belem, 1 °S, 48 °W (after Jackson, 1989).

Recife (figure 3.6(c)) is considerably less than 50 mm and the former station is particularly dry at this time. This indicates that the influence of anticyclonic subsidence now extends well into the Amazon basin on the equator in the low-sun season when the ITCZ has retreated into the North Atlantic (figure 2.3).

Section 2.3 also reveals that further complexities are introduced into the humid tropics as a result of the variable frequency of weather disturbances at every scale. Consequently, as in the Caribbean, it is often impossible to relate rainfall regimes to the movement of the ITCZ. At Georgetown (7 °S), for example (figure 3.6(b)), double rainfall peaks are evident around both summer solstices (with the convergence zone located in the vicinity). However, at Kingston (18 °N), Riehl (1954) suggested that the double peaks in May and October (figure 3.6(d)) are associated more with the occurrence of disturbances, especially tropical cyclones in the latter month. Further continental examples from Africa and southeast Asia (Jackson, 1989) confirm the role of asymmetrical ITCZ movement and associated rainfall anomalies throughout the humid tropics. In addition, additional complexities were introduced by tropical and extratropical disturbances, particularly at the synoptic scale (sections 2.4 and 2.5).

It is generally accepted that the 'classic' double rainfall maximum of the humid tropics, without a conspicuous dry season, is modified in a number of ways. The influence of the subtropical highs is paramount, especially since subsidence is less pronounced over the western parts of the tropical oceans and adjacent land masses, compared with eastern parts. This reduced subsidence is coupled with more disturbed trade winds as the stabilizing upper air inversion (section 2.4.1) is weakened. This promotes wetter conditions throughout the year and extends the tropical rains polewards. For example, the rainfall regime at Rio de Janeiro (23 °S) shows maxima around March and December (figure 3.6(a)). However, appreciable amounts are recorded in every month, despite the increasing remoteness of the convergence zone.

The conspicuous seasonal variations of rainfall in the monsoon climates are discussed in section 2.5, in particular with the Indian case study in figure 2.14. However, the rainfall regimes are not always dominated by a pronounced single high-sun peak as at Bombay (figures 2.14(c) and 3.7(a)) which is the classic, wet monsoon pattern. For example, Mandalay (figure 3.7(b)) is characterized by a double peak in May and September, with the latter maximum thought to be associated with jet stream fluctuations and the nature of the southeast Asian monsoon (Jackson, 1989). Minicoy (figure 3.7(c)) in the southwest of India experiences a double peak over the year with the main maximum in June following the dramatic onset of the southwest monsoon. A lesser but secondary peak occurs in October as the monsoon trough and related disturbances stagnate over southern India (figure 2.14(b)). Finally, Kalat (figure 3.7(d)) in the extreme northwest of the subcontinent reveals a rainfall regime on the extreme periphery of

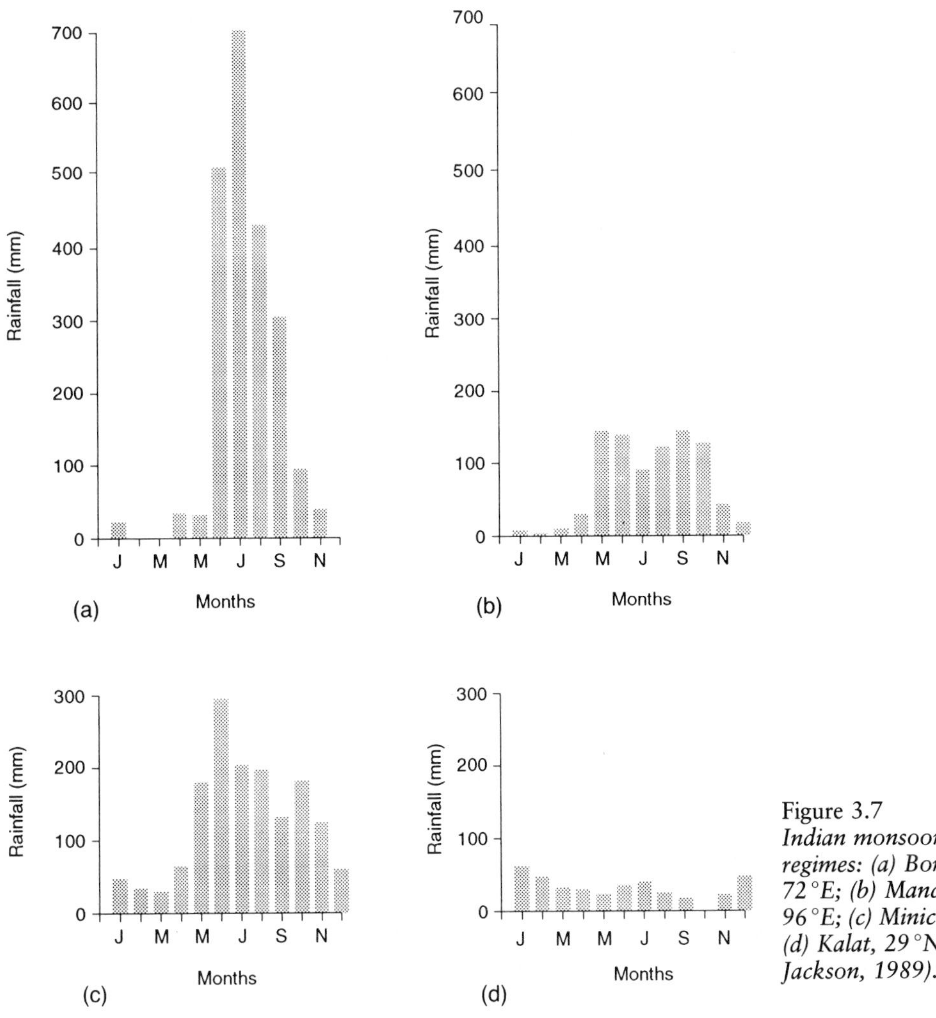

Figure 3.7
Indian monsoon rainfall regimes: (a) Bombay, 19 °N, 72 °E; (b) Mandalay, 22 °N, 96 °E; (c) Minicoy, 8 °N, 72 °E; (d) Kalat, 29 °N, 66 °E (after Jackson, 1989).

monsoon activity. In fact, the winter rainfall maximum here emphasizes the role of the so-called 'western disturbances' (section 2.5). These systems represent the occluded stages of wave depressions originating along the Mediterranean front which are steered over northern India by the subtropical jet stream.

A detailed classification of seasonal variations of rainfall regimes in the humid tropics has been attempted by Jackson (1989). It is condensed and illustrated in figure 1.4 (p. 7) as four distinctive rainfall zones which can be roughly correlated with the map of mean ITCZ positions in section 2.2 (figure 2.3). Zone A represents the 'textbook' humid tropics with annual rainfall totals in excess of 2000 mm. All months record at least 100 mm of rain and, while modest seasonal variations can occur, there is no true dry season. Likewise in zone B, there is no conspicuous dry season, but annual rainfall totals are now less than 2000 mm and a drier season does occur with a few months experiencing less than 100 mm. Jackson

(1989) suggests that it is a transitional zone between the humid and wet and dry tropics, which is common on the poleward margins of zone A. It also characterizes the windward (eastern) coasts of continents and islands where it may extend over a wide range of latitudes in the disturbed trade winds discussed above (figure 3.6(a)). Zone C represents the wet–dry tropics and is simply an arbitrary division which includes rainfall variability characterized by an increasing dominance of a dry season. This culminates in typical monsoon situations, where one season of exceptionally heavy rain is followed by a pronounced dry season (figure 3.7). However, the zone includes regimes with annual rainfall totals in excess of 1500 mm which are associated with one or two rainy seasons. Zone D represents a subhumid division with annual rainfall between 650 and 1500 mm, and only those areas with a mean annual rainfall approaching the latter figure can qualify as part of the humid tropics. Further complexity in this zone is introduced by the number and duration of the rainy and dry seasons (p. 7, figure 1.4).

3.6.2 Variability

The rainfall seasonality discussed in the last section exerts an overall control on water availability in the humid tropics. However, this distinctive, temporal variability (based on mean monthly rainfall data) does not reveal the certainty of the seasonal rainfall event. Furthermore, fluctuations and deviations from these meaningless average totals (characterizing a seasonal regime) are a conspicuous feature of the region, in spite of the general misconception concerning the permanently wet nature of the humid tropics. The degree of variability determines the reliability and probability of rainfall which, coupled with the seasonality variance, have a direct influence over agricultural systems in the humid tropics. However, it should be noted here that in the analysis of variability (as with annual means) the time period considered is important and even critical at certain stages of the growth season of a crop.

In the context of rainfall seasonality discussed above, variability of the start and finish of the rainy season is of paramount importance. Since the agricultural calendar is normally based on expected rainfall, these temporal variations may have serious effects on the cultivation of crops and feeding of livestock. For example, a late onset of rainfall will often result in a loss of seeds/plant failure while earlier than usual rainfall will encourage farmers to sow early and a following dry interval may destroy the crop. Livestock losses are also considerable since animals are moved in anticipation of the rains which are delayed. An indication of the variability of the length of alternating wet and dry seasons is evident in data from Kinshasa, Zaire. Between 1930 and 1959, the start of the wet season ranged from 26 August to 12 October, and in 80 per cent of years the length of the dry season was between 103 and 139 days (Jackson, 1989).

Assessments of rainfall variability require reasonably long periods of rainfall data which are not common in the humid tropics, even in colonial West Africa. Furthermore, most of the data available are in the form of average values and extreme maximum and minimum totals (table 3.4) which can only give a crude indication of variability since the occurrence of a few extremely wide-ranging values can be misleading. It is apparent that a measure is needed which considers all relevant values to obtain an estimate of variability together with reliability and probability assessment. A basic measure of relative variability was attempted for West Africa by Hayward and Oguntoyinbo (1987), although the necessary rainfall data for a sufficiently large number of stations (119) were only available for the period 1951–60. The coefficient of variability V was determined as

$$V = 100s/m \tag{3.4}$$

where s is the standard deviation and m is the mean rainfall. Figure 3.8 illustrates the rainfall variability calculated from this equation, with least variability in the coastal lands bordering the Gulf of Guinea (where the ITCZ has a more regular occurrence). Pronounced increases occur towards the equatorial trough periphery in the arid northeast and especially towards the northwest coast.

However, even though figure 3.8 provides a more realistic representation of rainfall variability (especially compared with table 3.4), it is still based on a measure of the standard deviation. This expression of rainfall variability, along with the arithmetic mean, suffers from the disadvantages of characteristically skewed

Table 3.4 Rainfall variability (mm) for two stations in Ghana in terms of (a) mean, (b) absolute maximum and (c) absolute minimum values

	Axim			*Kumasi*		
Month	*(a)*	*(b)*	*(c)*	*(a)*	*(b)*	*(c)*
January	49	230	2	17	124	0
February	73	281	0	59	150	1
March	112	311	30	149	268	55
April	122	401	20	143	259	26
May	336	892	145	180	296	68
June	510	1190	119	234	396	62
July	177	729	10	126	446	16
August	52	362	1	74	400	3
September	84	428	9	176	356	50
October	147	537	3	202	394	94
November	201	489	2	98	220	28
December	110	326	13	31	127	0
Year	1973	3312	1210	1491	2343	1086

Source: Hayward and Oguntoyinbo, 1987

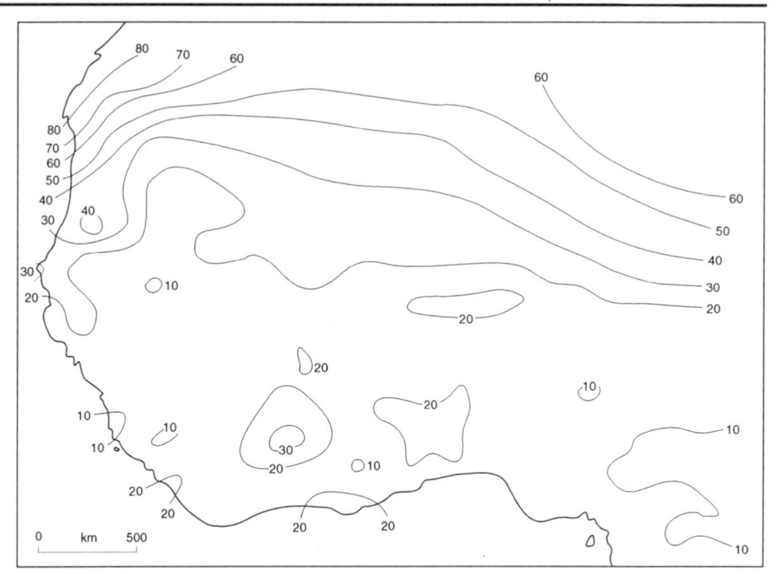

Figure 3.8
Rainfall variability in West Africa, 1951–60 (after Hayward and Oguntoyinbo, 1987).

distributions associated with the variable and unpredictable origins of tropical rainfall. For example, in parts of the humid tropics, much of the rainfall is derived from very infrequent, torrential downpours related to short-lived and unreliable weather disturbances, like hurricanes and monsoon depressions. Since the frequency of these disturbances varies over the years (section 2.3), this can lead to a much higher variability of rainfall and conspicuously skewed distributions. Conversely in an area where such irregular and extreme storms do not occur, the rainfall distribution is characterized by a more normal distribution (e.g. Singapore).

In the parts of the humid tropics experiencing periodic extreme rainfall events and skewed distributions, the calculation of percentiles is much more relevant than a measure of standard deviation. Furthermore, these values are particularly important in probability calculations (Jackson, 1989). Table 3.5 shows the tenth and ninetieth percentiles for Kinshasha, Zaire (P_{10} and P_{90} respectively). Since P_{10} represents 10 per cent of years less than this value, it is apparent that for November (as an example) 80 per cent of years will have between 135.0 and 334.8 mm rainfall. Rainfall probability data are particularly useful for agriculture and water resources in the tropics, although specific needs will vary according to individual crop type, evaporative demand and soil type. In particular, according to Jackson (1989), the period required for probability calculations will vary as will the critical rainfall values in these periods. Furthermore, he stressed that the problem of skewed distributions decreases with the increasing length of the time period concerned and with increased rainfall totals. Most probability studies have been carried out in the semi-arid regions of East Africa. Little attention has been paid to such analysis in the humid tropics where rainfall totals are well in

Table 3.5 Lower Zaire (Kinshasa) monthly and annual rainfall (1930–59): tenth and ninetieth percentiles and extremes (mm)

	Min	P_{10}	P_{90}	*Max*
January	1.7	32.9	227.1	320.6
February	48.6	67.0	234.2	329.8
March	58.0	79.6	317.3	428.9
April	58.9	112.4	326.9	378.6
May	22.2	42.7	217.8	280.0
June	0.0	0.0	22.7	37.6
July	0.0	0.0	1.1	34.0
August	0.0	0.0	17.8	24.4
September	1.6	5.0	76.3	100.4
October	19.5	52.6	206.8	281.6
November	84.3	135.0	334.8	347.7
December	47.7	01.3	275.8	326.6
Year	1824.0	1170.0	1655.0	1824.0

Source: Griffiths, 1972

excess of 1500 mm, where seasonality is a more vital issue and where probability and reliability studies have a much reduced practical application. In more marginal subhumid areas, these studies have a far more relevant role to play in agricultural production and water resource management. Table 3.6 reveals rainfall probability at eleven recording stations in Tanzania where the long-term probability of receiving 800 mm rainfall per annum varies between 32 and 96 per cent. These values are compared with the probability of receiving this amount in at least four out of five years and the probability of two successive years receiving less than this amount. The former comparison represents a direct association whereas the latter comparison reveals a conspicuous, if expected, inverse relationship. Here the percentage probability of experiencing two successive dry years increases with the smallest long-term probabilities.

Table 3.6 Annual rainfall probability – Tanzania

	Percentage probability of receiving:		
Station	*800 mm 4 out of 5 years*	*800 mm in two successive years*	*<800 mm*
Kingolwira SE	61	36	16
Lugoba Mission	83	79	3
Marios SE	52	23	23
Kikeo Mission	94	96	<1
Morogoro	68	48	10
Kilondeni	76	65	6
Tungi SE	49	18	26
Melela	32	3	46
Maneromango	65	44	12
Utondwe Salt Works	59	32	16
Kisiju	96	98	<1

Source: Jackson, 1989

3.6.3 Intensity, duration and frequency

It is generally assumed that precipitation in the humid tropics is characterized by torrential downpours of short duration which occur most afternoons (chapter 1). Cilaos, La Réunion, holds the record for the highest daily total of 1870 mm on 16 March 1952. According to Miller (1971), 'the daily incidence of rain is strikingly regular at any given place, often so regular that appointments are made for "after the rain" as one might make arrangements for"after tea". The time of this daily maximum ... on land is nearly always between midday and midnight and usually about 3 or 4 pm; i.e. following the greatest heat and convectional action'. More objective analysis of tropical rainfall regimes since the 1950s has confirmed that much of the rain is derived from storms or severe convectional showers lasting between 3 and 6 hours and producing intensities in excess of 25 mm h^{-1}, the critical value (according to Hudson, 1971) for runoff and soil erosion (chapter 4). For example, Swami (1970) calculated that throughout Nigeria 50–60 per cent of total rainfall was associated with showers in excess of 25 mm on 20–25 per cent of total raindays.

Rainfall intensity studies in New Guinea (Jackson, 1989) reveal (figure 3.9) that 34 per cent of the deposition occurs at intensities in excess of 25 mm h^{-1} and a further 12 per cent occurs at an intensity between 20 and 25 mm, i.e. approaching the erosive threshold. Hudson (1971) gave a general value of 40 per cent for tropical rainfall intensities exceeding 25 mm h^{-1}, compared with only 5 per cent for temperate areas. He also indicated that intensities of 150 mm h^{-1} can occur regularly and that a rate of 340 mm h^{-1} over a few minutes has been recorded. Furthermore, these high intensity rainfalls are concentrated into a relatively few but comparatively large storms. Figure 3.9(b) reveals that 47 per cent of the total rainfall occurs in single storms of more than 25 mm, including some 20 per cent concentrated in storms in excess of 50 mm. Figure 3.9(c) shows that the 50 per cent of total rainfall in storms exceeding 25 mm is produced by only 10 per cent of the storms and that 80 per cent of the total is attributed to 28 per cent of the storms involved.

The short duration and intense concentration of tropical rainfall is apparent from data for Bogor, Indonesia (Mohr and Van Baren, 1959). Out of an annual rainfall total of 4230 mm, 62 per cent was attributed to some sixty-five showers of 20–100 mm, 21 per cent to sixty showers of 10–20 mm and 17 per cent to showers of less than 10 mm. The length of the showers was not quantified by these authors although they did state that 22 per cent of Bogor's annual rainfall occurred in cloud bursts, which represented falls of at least 1 mm min^{-1} for not less than five minutes (i.e. forty times as much rain in this way compared with temperate regions). Also Henry (1974) stated that rainstorms (in excess of 10 mm) over southeast Asia and South/Central America had a life cycle of one to three hours and more than 60 per cent of the daily rain was deposited in a

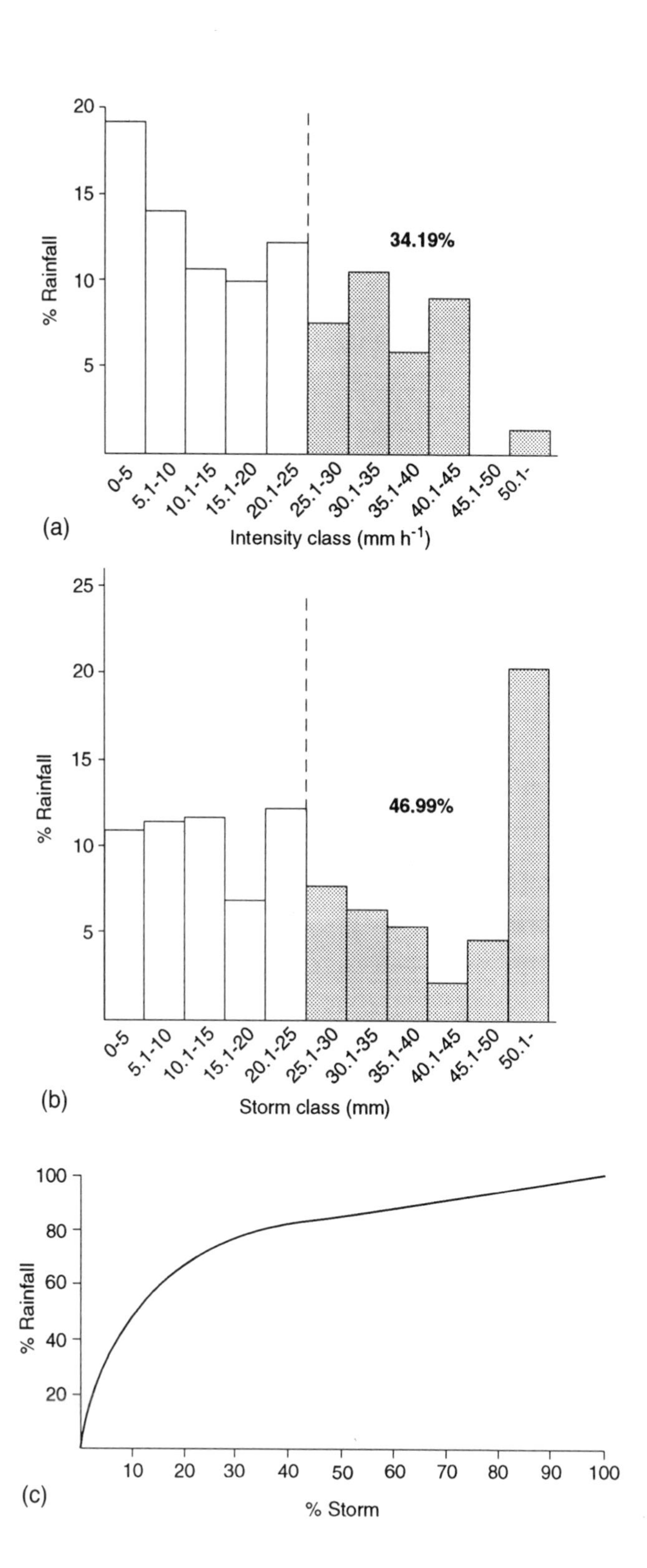

Figure 3.9
Rainfall intensities in New Guinea: (a) percentage in intensity classes; (b) percentage according to storm size; (c) cumulative percentage for total rainfall and number of storms (after Jackson, 1989).

1 hour period. Similarly, at Bidar, India (Rakhecha et al., 1985), up to 56 per cent of the daily rainfall could be experienced in one hour and 73 per cent in two consecutive hours.

The frequency of intensive, short-duration rainfall events is of paramount importance in water management schemes in the humid tropics, especially in the design of reservoirs and flood-control systems and the operation of river navigation. A major consideration is the return period of an event although the critical length of this period depends on a variety of physical and socioeconomic factors. For example, protection against flood damage may dictate that the control system involved should be able to withstand an event with a very long return period. Conversely, the cost of investment in protective measures must be balanced against the actual cost of flood damage which should demand a much lower return period. This cost–benefit ratio is a major factor in the developing regions of the humid tropics with limited financial resources.

A common assessment of the maximum rainfall expected over a period of time utilizes the Gumbal equation:

$$x_T = x(\text{mean}) + k(Tn)\text{Sx} \qquad (3.5)$$

where x_T is the extreme rainfall estimates for the return period T, x(mean) is the mean maximum rainfall, k is a frequency factor, Sx is the standard deviation of the extreme rainfall and n is the actual number of extreme values. Applied to Ghana rainfall data, Axim (mean annual rainfall 1973 mm) may expect to record up to 2568 mm in a ten year period and 3094 mm in fifty years (Hayward and Oguntoyinbo, 1987). A more detailed analysis of rainfall return periods for various durations is given in table 3.7. The expected rainfall obviously increases in total with increasing duration and longer return periods, from 23 mm over a ten minute duration every two years to 143 mm over twenty-four hours every fifty years.

Table 3.7 Lower Zaire (Kinshasa) rainfall for various durations and return periods (mm)

Return period (years)	*Duration (min)* 10	20	30	40	50	60	70	80	90	24 h
2	23.3	37.5	46.5	56.2	62.0	66.1	67.3	69.5	69.8	
10	30.6	49.3	61.2	74.4	82.4	87.9	89.5	92.5	92.7	117
25										132
50										143

Source: after Griffiths, 1972

3.6.4 Spatial and diurnal variations

From section 2.6 it is apparent that small-scale weather patterns are responsible for a variety of localized and contrasting rainfall regimes. They are conspicuous in the humid tropics, both in spatial and temporal/diurnal terms, and are initiated primarily by terrain factors associated with land–sea thermal contrasts and orographic displacement/forced convection. Section 2.6 reveals that during the day (and mostly in the afternoons) the sea-breeze front (figure 2.12(a)) penetrates inland up to 60 km as an advancing band of cumulus clouds and short, sharp showers. The afternoon rainfall intensity over large tropical islands in the Pacific approaches 175 mm h^{-1} (Barry and Chorley, 1992) which moves seawards on the return counterflow aloft (figure 2.12(a)).

Diurnal variability of rainfall is a well-documented occurrence in the humid tropics although the variations depend upon location, time of year and precipitation type. For example, away from the Gulf of Guinea in Ghana, maximum rainfall tends to occur in the late afternoon/early evening, especially in the high-sun season of strong free convection. However, during the main rains at Tamale, rainfall is concentrated at night and towards dawn, and at Navrongo, in mid-morning and early evening (Hayward and Oguntoyinbo, 1987). Such variability over short distances makes it difficult to generalize about diurnal rainfall regimes over large areas of the humid tropics. Furthermore, it is generally (if erroneously) assumed that two main diurnal regimes exist in the region. The first is a maritime type which is characterized by a rainfall maximum at night and early morning whereas the second is a continental regime with an afternoon maximum (Jackson, 1989).

There are many examples suggesting that these simple regimes do operate, i.e. the former type is evident at Abidjan, Ivory Coast, where 54 per cent of the rain falls at night and indeed 61 per cent of the daytime deposition is recorded before noon (Drochon, 1976). Conversely, inland Kuala Lumpur (Malaysia) has a more continental regime where the rainfall peaks in the late afternoon (figure 3.10(a)) during the four main rainfall seasons. However, it is apparent that there is considerable evidence to support the fallacy of the rigid acceptance of this coastal–inland division. For example, figure 3.10(b) reveals that Singapore records a maritime regime during the southwest monsoon between May and September and a continental one over the rest of the year. It is also important to consider the significance of weather disturbances (section 2.3) in rainfall deposition, which are generally immune from diurnal controls due to their varied nature and occurrence in time and space. However, it is apparent that the intensity or arrival of a disturbance may show a distinctive rhythm and hence might produce rainfall concentrations at certain times of day. For example, at Freetown, Sierra Leone, maximum rainfall deposition occurs at night/early morning, especially in the period preceding the ITCZ 'main' rains. At this time, the bulk of the rainfall is derived from line squalls of individual cumulus clouds (figure 2.4). These originate well to the east of Sierra Leone (even over Nigeria) and most frequently reach the Freetown area at night (Hayward and Oguntoyinbo, 1987).

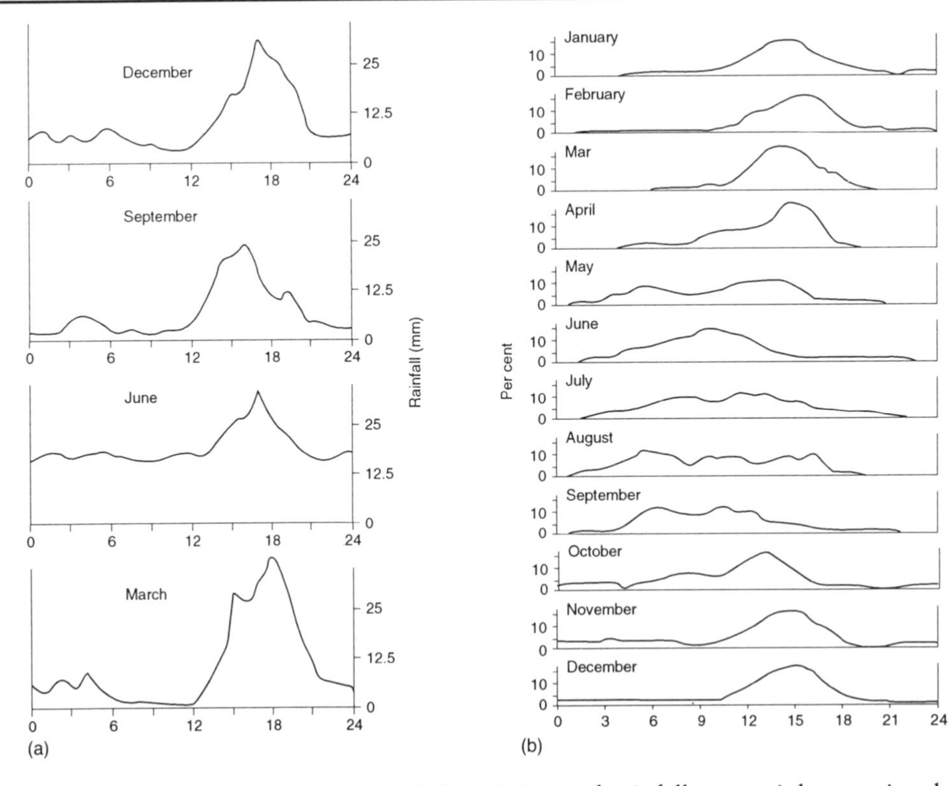

Figure 3.10
Diurnal variation of rainfall: (a) Kuala Lumpur, 3 °N, 101 °E (after Lockwood, 1974); (b) Singapore, 1 °N, 103 °E (after Jackson, 1989).

Conspicuous spatial variations of rainfall are mainly associated with the distribution of high ground in the humid tropics (and indeed all climatic zones) and localized forced convection. This so-called orographic factor is discussed in section 2.6.2 where the associated pronounced spatial variations over Hawaii, Papua New Guinea and particularly Fiji (figures 2.15 and 2.16) are most distinctive. However, in addition to the control of these fixed relief influences, spatial contrasts are also related to the localized nature of individual tropical rainstorms, with pronounced horizontal gradients in intensity and amount. Figure 2.4 illustrates the spatial distribution of the convective elements forming cloud clusters which represent the major meso-scale, subsynoptic weather disturbance in the humid tropics. The system is characterized by cells of vigorous convergence and torrential rainfall which alternate with areas of strong divergence–subsidence which are cloud/rain-free over a region up to 1000 km in diameter. It should be noted that rainfall varies spatially in all climatic regions owing to localized free/forced convection (altitude and urban 'oases') and perturbation contrasts within extra-tropical disturbances.

There is considerable evidence of the patchy, fickle nature of tropical rain events and the pronounced catch variations over a few kilometres. For example, in Dar es Salaam, Tanzania, on 11 March 1969 total rainfall for the day over the city varied between 2.5 and 15 mm (Jackson, 1989). Four separate storm cells were located, each depositing more than 10 mm independently of the others, which emphasizes the complexity of tropical disturbances. Further studies in Tanzania (tables 3.8 and 3.9) compared monthly and annual rainfall for two stations some 3.2 and 8 km apart. Table 3.8 shows that large differences exist for individual months, especially April and March, and that over the year the difference totalled 229 mm (some 14 per cent) greater at station B. Table 3.9 confirms that, even though the average rainfall is within 2 per cent, the catch is considerably different for individual years. Very large differences (exceeding 22 per cent), which are opposite in sign, are evident for 1948 and 1953.

It is apparent that for periods of months, seasons or years, significant local rainfall variations exist in the humid tropics which are quite independent of average totals. At any time, one locality may record rainfall well below average whereas over an area a few kilometres away the deposition could be well above average. These

Table 3.8 Monthly rainfall (mm) at two stations 3.2 km apart near Dar es Salaam, April 1967 to March 1968

	Station A	*Station B*
April	315.2	437.6
May	193.4	236.0
June	31.8	44.3
July	131.2	97.3
August	47.5	59.7
September	89.1	83.1
October	49.7	48.3
November	206.0	180.7
December	134.5	172.4
January	2.6	2.18
February	90.5	2.8
March	145.8	222.2

Total for April 1967 to March 1968
Station A 1437.3
Station B 1666.5

Total for April 1967 to March 1969
Station A 2593.3
Station B 2793.9

Source: after Jackson, 1989

Table 3.9 Annual rainfall (mm) 1946–54 for two Tanzanian stations 8 km apart (altitude difference 15 m)

	A	*B*	*A – B*
1946	754.6	859.8	–105.2
1947	973.1	1248.4	–275.3
1948	966.5	1420.9	–454.3
1949	790.1	619.8	+153.7
1950	1069.6	1223.3	153.7
1951	1188.7	1127.0	+61.7
1952	755.1	565.4	+189.7
1953	1108.7	859.8	+248.9
1954	930.7	869.8	+61.0
Average, 1942–66	943.9	924.8	

A, Fatemi Sisal Estate; B, Mgude Sisal Estate.
Source: Jackson, 1989

pronounced spatial differences have a profound effect on local agricultural practices (in terms of irrigation) and can be of sufficient magnitude to control crop yields over small distances.

3.7 Water Balance

The relationship between seasonal variations of rainfall and evapotranspiration can be assessed by a simple water balance technique introduced by Thornthwaite and Mather (1955). These assessments are useful in the humid tropics where the basic data used are readily available, especially for PE estimations (section 3.5). However, they are generalized oversimplifications, since soil moisture capacity is assumed at all times and places to be 102 mm (Thornthwaite, 1948). Also, the seasonal utilization and recharge of this water ignores soil type (particularly infiltration capacity) and plant-rooting characteristics. The final disadvantage (which is emphasized in section 3.5) is that PE estimates based on the Thornwaite approach have limited applications in humid tropical environments. In very basic terms, rainfall is classed as 'surplus' when it exceeds PE over a period of time, so that soil moisture reserves are rapidly and fully recharged and the so-called 'field capacity' is achieved. Any further rainfall is regarded as a surplus which mainly leaves the system through surface runoff or interflow. Conversely, a water deficit occurs when PE exceeds rainfall over a period of time, so that soil moisture reserves are now fully utilized and the so-called 'wilting point' is reached.

Despite the oversimplification of water surplus and deficit representations, together with the problems associated with the pronounced spatial and temporal variations in rainfall already discussed (section 3.6), water balance data clearly illustrate the average water regimes for a region. Figure 3.11 shows two distinctive patterns at (a) Vila, New Hebrides, and (b) Port Moresby, Papua New Guinea, which are associated primarily with contrasting climatic conditions within a general humid tropical framework. Vila's rainfall exceeds 2100 mm over the year, and although a maximum occurs in the high-sun/ITCZ season between December and April, the mean monthly rainfall always exceeds the PE. Consequently, Vila records a continuous water surplus which is particularly pronounced in the high-sun months between January and April. This influences agricultural practices since irrigation is not necessary and crops with a high water demand (like bananas) can be grown.

On the other hand, Port Moresby (figure 3.11(b)) only receives about half of Vila's rainfall (annual mean 1180 mm). Also, even though the former station is characterized by a similar concentration of rainfall in the December to April high-sun period, the seasonal contrasts are more pronounced than at Vila. For example, there is a much reduced low-sun total rainfall due to the orographic factor discussed in sections 2.6.2 and 3.6.4. Indeed, between the end

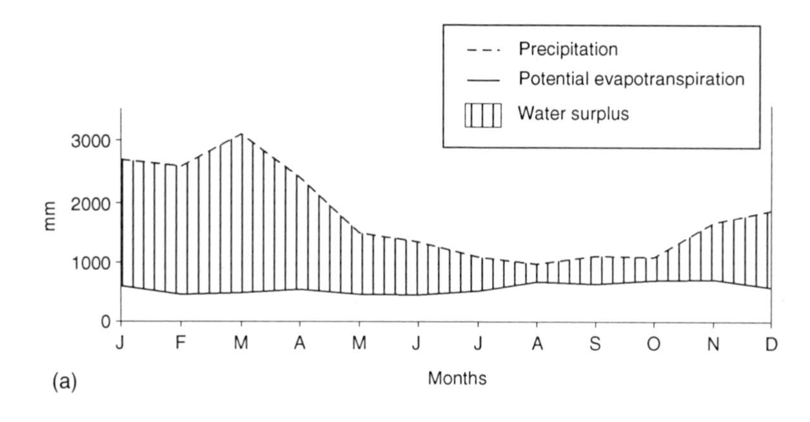

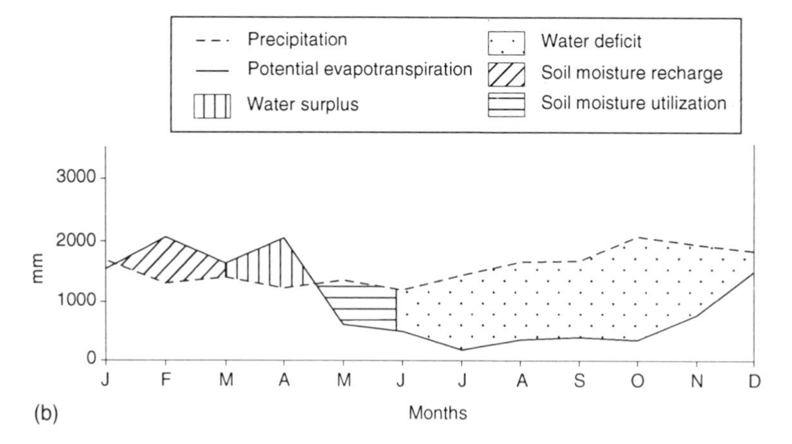

Figure 3.11
Water balance regimes: (a) Vila, New Hebrides, 15 °S, 168 °E; (b) Port Moresby, Papua New Guinea, 9 °S, 147 °E.

of June and January, Port Moresby records considerably less rainfall than PE and experiences a serious water deficiency. Now, crop desiccation is only avoided by extensive irrigation schemes or the use of drought-tolerant species. At Port Moresby, rainfall only exceeds PE between February and April but initially (i.e. in February and March), 102 mm of this excess is used to recharge the soil moisture. Consequently, a true water surplus only occurs in the month of April. However, although PE exceeds rainfall in May, the water deficit is delayed until early June by the utilization of soil moisture.

FOUR

Humid Tropical Soils

4.1 Introduction

Many of the most widely distributed soils in the humid tropics are quite distinct from those found in other parts of the world; they are characterized by red colours, the accumulation of iron and aluminium, kaolinite, and low cation exchange capacity and high rates of base leaching. Such soils dominate large tracts of all humid tropical regions except Australasia and in total account for almost 66 per cent of the humid tropics. Other soil types are related to specific parent materials. In this chapter we will focus on three aspects of these soils: (i) their distribution; (ii) contemporary soil-forming processes (pedogenesis) and their relationships to geomorphological processes; and (iii) how the different soils are managed by farmers.

It is important at the outset to adopt a soil classification scheme to structure this discussion. Soil classification anywhere, but perhaps more so in the humid tropics, is a tangled mess of national and international schemes in which the different classes only partially correspond. We will use the reports of the FAO/UNESCO Soils Map of the World (and its subsequent modifications) in this chapter (FAO/UNESCO, 1977a, b, c, d, e, f, 1988).

4.2 Soil-forming Processes in the Humid Tropics

The main controls on pedogenesis in the humid tropics are the amount of rainfall and its seasonal distribution, topographic position, altitude and lithology.

High annual rainfall, common throughout most of the humid tropics, means that processes such as base leaching and clay translocation are commonplace and most soils have been formed by these processes. However, this generalization is not true for all of the humid tropics and in some circumstances downward movement of material in the soil is rare. For instance, whilst there may be a water surplus throughout the year in the equatorial regions (figure

3.11(a)), the monsoonal humid tropics exhibit long seasonal droughts with little rainfall and high rates of evaporation (cf. chapter 3). Under such conditions downward movements of water, cations and clay in the profile are mainly restricted to the wet season.

Even in regions with a high water surplus free drainage may be impeded, especially in the low-lying undulating terrain that is common on continental shields (e.g. dambos and bolis: Boast, 1990; Millington et al., 1985) and in coastal regions (cf. chapter 6). Soils in these environments are either permanently or seasonally waterlogged and exhibit the typical properties of reduced, gleyed soils rather than freely draining pedalfers. Research into the changes in physical and chemical properties after submergence have been conducted in the context of swamp rice cultivation (e.g. Ponnamperuma, 1972; Sanchez, 1976; Young, 1976). Obstacles to free drainage are also provided by indurated ferricrete and alcrete horizons in ferralitic soils, and silica hard pans in andosols.

Recent work in East Africa and Papua New Guinea has shown that low temperatures act as a significant climatic control on pedogenesis at high altitudes (section 1.1). In these environments chemical weathering rates are reduced and organic matter accumulates. Consequently soils akin to those found in temperate and alpine regions have been recorded in such diverse places as the Andes, the Ruwenzori and Viphya Massifs in East Africa, and the Central Ranges of Papua New Guinea. In some circumstances organic matter accumulates at such a rate that blanket peat forms and at very high altitudes soils exhibit solifluction features which are indicative of periglacial conditions.

The nature of the parent material mainly influences soil development in geologically young environments, and is a much less obvious control on old continental shields. Four soils in particular are strongly influenced by parent materials: andosols are restricted to volcanic ashes, regosols and arenosols to sandy substrates, and dystric nitisols to basic rocks on continental shields.

4.3 Soil Classification

Three main groups of soils are commonly found in the humid tropics – freely draining pedalfers, soils with impeded drainage, and immature stony soils (leptosols). Pedalfers are characterized by free, or at the most slightly impeded, drainage. They generally contain few exchangeable bases, carbonates or soluble salts owing to intense leaching and consequently exhibit an acid reaction. Andosols, however, are an exception on many of these criteria although they are still classed as pedalfers. Most soils with impeded drainage are either hydromorphic (swamp) or alluvial soils occurring in poorly drained situations. They exhibit gleying, mottling and the slow breakdown and subsequent accumulation of organic material. Leptosols are the least common of the three main soil types and include both soils with very shallow profiles in mountain regions and shallow soils developed over duricrusts.

One of the greatest problems facing the student of tropical soils is the necessity to swop between soil classifications. The multiplicity of classification systems is partly a function of the differing colonial histories of humid tropical countries, partly due to the national systems that many countries have established, and partly due to the proliferation of 'global' classification systems. The FAO has exerted much effort towards regional and international soil correlation and this has resulted in the FAO/UNESCO Soils Map of the World. To simplify the discussion of soils in this chapter the FAO/UNESCO system will be used. A succinct account of the different systems applied to the tropics can be found in Sanchez (1976) and a table showing the correspondence between some of the main classifications is provided in table 4.1.

Table 4.1 Correspondence between the main soil classification schemes used in the humid tropics

USDA soil taxonomy	*Revised FAO Legend*	*French classification*	*Brazilian classification*
Oxisols	Ferralsols	Sols ferraltiques fortement désaturés typiques ou humifères	Latossolos, Terra Roxa legítima
Ultisols	Acrisols and dystric nitisols	Sols ferraltiques, lessivés	Podzólico Vermelho-Amarelo distrófico
Inceptisols	Various	Sols peu évolués	Solos com horizonteB
Aquepts	Gleysols	Sols hydromorphes	Solos hidromorficos
Andepts	Andosols	Andosols	Solos com horizonte B, incipiente
Tropepts	Cambisols	Sols brunifiés tropicaux	
Entisols	Various	Sols minéraux bruts	
Fluvents	Fluvisols	Regosols	
Psamments	Arenosols and regosols		Regossolos Areias e Quartzisosas
Lithic subgroups	Leptosols	Sols lithiques	Litossolos
Alfisols	Lixisols, Eutric nitisols and planosols	Sols ferringineaux tropicaux lessivés	Podzolico Vermelho-Amarelo equivalente eutrófico, Terra Roxa Estruturada, Planossolos
Histosols	Histosols	Sols organiques	Solos orgânicos
Spodosols	Podzols	Podsols	Podzols
Mollisols	Rendzinas, phaeozems	Podsols	Brunizens
Vertisols	Vertisols	Vertisols	Grumusols
Aridisols (saline only)	Solonchaks	Sols halomorphes	Solonchak

Source: adapted from Sanchez, (1976)

4.4 Soil Distribution

Information on the distribution of soils in the humid tropics has been taken from the FAO/UNESCO maps (figure 4.1) and accompanying tables. It can be seen from figure 4.1 and table 4.2 that the humid tropics are dominated by two, closely related, soil types – ferralsols (34.1 per cent of the area) and acrisols/alisols (17.3 per cent). Four other types of soil which are related to acrisols and ferralsols – cambisols, lixisols, nitisols and plinthisols – account for 5.2 per cent, 9.6 per cent, 5.5 per cent and 0.2 per cent of the humid tropics respectively. These six soils, commonly known as red soils, cover almost two-thirds of the humid tropics although their distribution is not consistent between regions. They account for over 70 per cent of the humid tropics in South America, Africa and southeast Asia, a little over half of Central America, the Caribbean and south Asia, but only 28.5 per cent of the Australasian humid tropics. The dominant types of soil with the 'red soil group' also differ. Ferralsols dominate in South America and Africa, although in the former acrisols/alisols and lixisols are also important, whilst in Africa lixisols and nitisols are the next most common soil types. Acrisols/alisols are the dominant red soils in southeast and south Asia. Australasia is dominated by cambisols, whilst in Central America and the Caribbean acrisols/alisols, cambisols and lixisols are almost equally important.

Fluvisols and gleysols, which are commonly found together, account for 6.2 per cent of the humid tropics. They are particularly important in southeast and south Asia where they account for 11.9 and 9.2 per cent of the land area respectively but are rare in Australasia. Arenosols account for 5.9 per cent of the humid tropics, a proportion which can be increased to 7.7 per cent by including regosols – a closely related soil type. They are particularly important in Africa and South America where they account for 14.7 and 6.6 per cent of the land area. Leptosols account for 5.6 of the humid tropics and are shallow, stony soils which are mainly found in mountainous terrain. They are important in South and Central America (figure 4.1). Of the remaining soils histosols are important in southeast Asia, and andosols account for 6.9 per cent of Central America and the Caribbean.

A number of soils have distributions which are marginal to the humid tropics and are not considered in this chapter.

1. Vertisols are found in wet savanna environments in all six regions (table 4.2) and account for 5.6 per cent of the humid tropics.
2. Planosols, which are found in the subtropical margins of the humid tropical zone in South and Central America, Africa and Australia, in total account for 1.3 per cent by area.
3. Podzols, with only minor occurrences in the tropics, are restricted to southeast Asia and Australasia and account for 0.3 per cent by area.

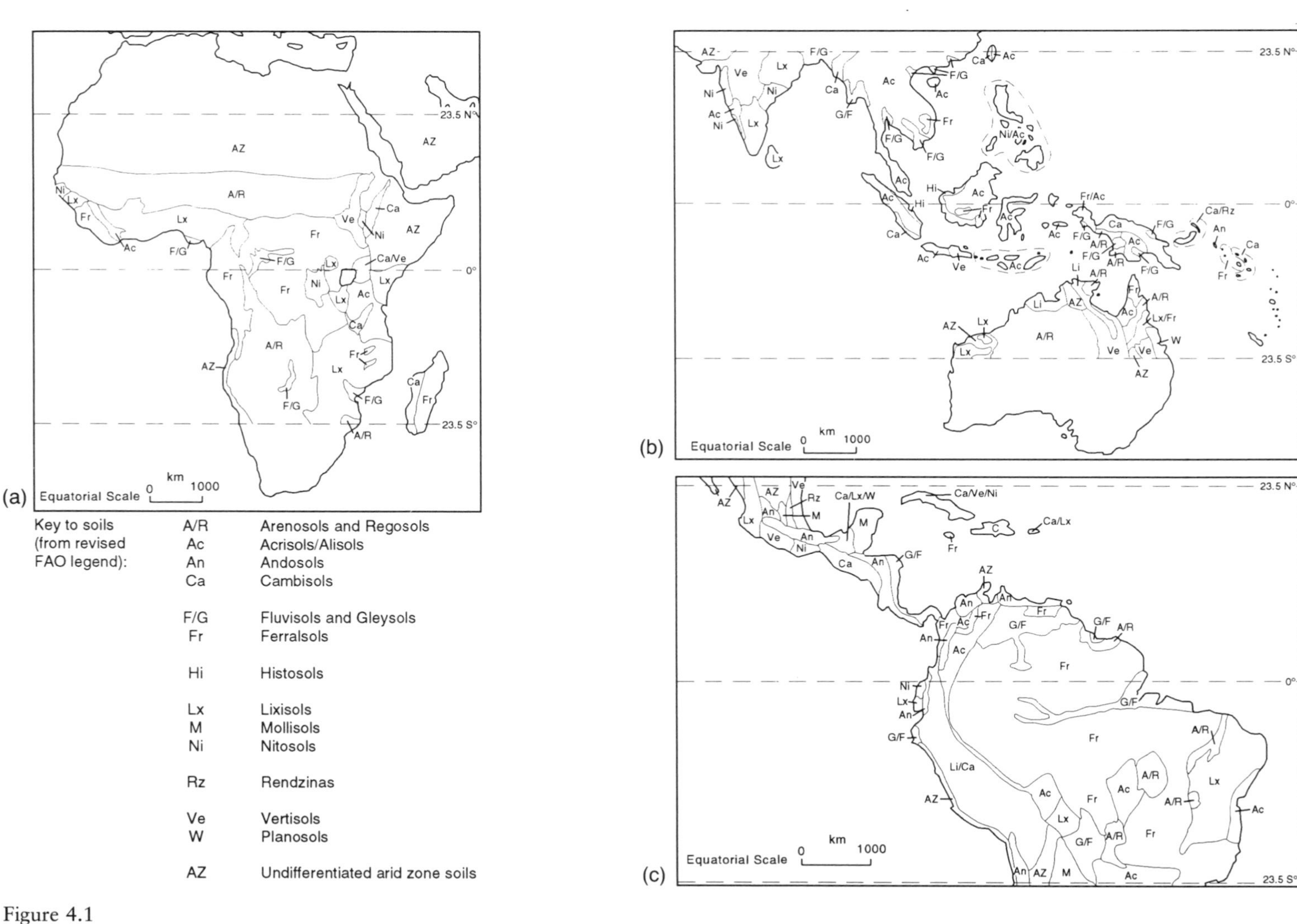

Figure 4.1
Soil distribution in the humid tropics: (a) Africa; (b) Asia and Australasia; (c) Central and South America.

4 There are a number of arid and semi-arid soils whose distributions cover parts of the tropics in all three regions except southeast Asia. A number of soils have been grouped together in the maps (figure 4.1) and table 4.2; these are calcisols, kastanozems, phaerozems, solonchaks, solonetz, xerosols and yermosols.

4.5 Red Soils

A recent reintroduction into the tropical soils literature has been the term *red soils* which is used to describe the many types of soil in the humid tropics which are characterized by red colours (Buol et al., 1980; Cochrane, 1986; Stocking, 1986). Such soils have always held a special interest for foreign scientists and agriculturalists since they were first brought to their attention in India by Buchanan (1807) who also coined the term laterite. The term lateritic soils was first used to describe these soils by Harrassowitz (1926), and the American soil scientist Kellogg (1949) introduced the term latosols. The striking red colours are commonly assumed to be the result of tropical weathering (Buol and Sanchez, 1986) although reddening (rubification) can be attributed to other processes. The degree of redness is often related to the presence of iron as either goethite or

Table 4.2 Distribution of soil associations in the humid tropics

Soil type	*South America*	*Central America and Caribbean*	*Africa*	*South Asia*	*Southeast Asia and China*	*Australasia*	*Total area*	*Proportion of all humid tropics (%)*
Ferralsols	596644	651	377866	272	14848	7704	997985	33.1
Acrisols and alisols	184317	21307	61616	40261	205462	9706	522669	17.3
Nitisols	18169	10791	101698	17104	17470	112	165344	5.5
Lixisols	100499	20244	112302	35486	18891	2428	289850	9.6
Plinthisols	–	807	6514	161	–	–	7482	0.2
Leptosols	67605	17308	50702	10448	11841	11107	169011	5.6
Cambisols	23003	26187	56560	7263	18748	25857	157518	5.2
Arenosols	64983	–	101856	182	8603	2623	178247	5.9
Regosols	5524	2654	29438	3150	6334	5720	52820	1.8
Andosols	4927	9535	3063	–	8876	2695	11571	0.4
Histosols	1383	2492	1654	32	24181	685	30427	1.0
Gleysols	51901	4336	34965	10834	17272	81	119427	4.0
Fluvisols	2538	2568	23885	6551	29317	–	64859	2.2
Planosols	25012	929	3313	–	–	8479	37733	1.3
Vertisols	4381	12972	13852	53712	5078	78228	168223	5.6
Podzols	–	–	–	–	5982	1650	7632	0.3
Arid and semi-arid soils	14471	5943	2458	2786	–	7716	33374	1.1

The areas (× 1000 ha) were calculated from the tables accompanying the FAO Soils Legend and only included the soil extensions corresponding to humid tropical criteria. Adjustments were made for the revised legend because no extensions were available for the revised soil classes. The figures should be used as an indication of the approximate areas of the types of soil.

Source: FAO/UNESCO, 1977a, b, c, d, e, f, 1988

haematite, and if haematite is present soil colours usually exceed 5R hues (i.e. are redder).

Red soils include acrisols/alisols, cambisols, ferralsols, lixisols, nitisols and plinthisols; they are differentiated from each other on the basis of their profile characteristics. These characteristics are a function of the length of time of weathering, geomorphological stability and parent material. The oldest soils which have developed on stable Precambrian (continental) shields are usually ferralsols or ferric acrisols. Soils which have developed on unstable landscapes or in areas where there has been only a short period of weathering are usually termed cambisols. Intermediate between these two types are lixisols and nitisols. Red soils are generally found in three geomorphic environments in the humid tropics: (i) on level or undulating terrain; (ii) on ancient, high elevation surfaces; and (iii) in depressions and on valley floors. The relationships between the main types of red soils, parent material and climate are shown in table 4.3, and an example of the climatic controls on red soil development is provided in table 4.4.

Red soils are an important indicator of humid tropical conditions, and palaeoferralsols found outside the region of contempor-

Table 4.3 Generalized relationships between the main types of red soils, climate and parent materials

	Parent material		
Climate zone	*Basic*	*Felsic to intermediate*	*Felsic, highly weathered*
Humid tropics	Dystric nitisols	Leached ferralsols and ferric acrisols	
Seasonal wet/dry tropics	Eutric nitisols	Ferric and chromic lixisols	Weathered ferralsols and ferric acrisols
High altitude tropics	Humic nitisols	Humic ferralsols	

Source: modified after Young, 1976

Table 4.4 Latitudinal and altitudinal controls on red soils in China

Soil	*Altitudinal range (m a.s.l.)*	*Latitudinal range*
Cambisols (shrubby meadow soils)	>1000	Most northerly
Dystric cambisols (yellow earths)	800–1000	
Haplic acrisols (red earths)	600–800	
Rhodic acrisols (lateritic red earths)	300–600	
Ferralsols	50–300	Most southerly

Source: modified after Zhao Qi-guo and Shi Hua, 1986

ary red soil formation, e.g. in Western Australia, Hawaii, Somalia and Ethiopia, have been used in palaeoenvironmental reconstruction.

Buol et al. (1980) reviewed the processes which lead to the formation of red soils. In high rainfall areas, typically covered by forest, the main processes leading to red soil development are intense, deep weathering and extensive leaching. In seasonal climates, such as the wet savannas, red soils generally develop more slowly, although in these areas more rapid soil development has often taken place in the past under wetter conditions.

4.5.1 Ferralsols and ferric acrisols

One of the biggest difficulties facing soil taxonomists in the humid tropics is the classification of the highly leached and weathered soils which are found in the rain forest and savanna zones. Although there have often been differences in their pedogenic histories, the resulting soils commonly have similar profile characteristics and have often been classified together. The terms oxisol and ultisol are used in the USDA Soil Taxonomy to describe these soils, and are approximately equivalent to ferralsols and ferric acrisols in the legend for the FAO/UNESCO Soils Map of the World. In addition to the similarities between their profiles ferralsols and ferric acrisols are often closely associated with each other in the landscape and it is convenient to consider them together. They have received considerable attention from both soil scientists and geomorphologists (e.g. Eswaran and Tavernier, 1980; El-Swaify, 1980; Paramanathan and Eswaran, 1980; Sanchez and Cochrane, 1980; Eswaran et al., 1986).

Ferralsols and ferric acrisols of the rainforest zone

Ferralsols and ferric acrisols formed by intense leaching are the zonal soils of the equatorial and wet semi-deciduous rainforests. They are best developed where precipitation exceeds 1750 mm, and form over a variety of parent materials including acid and intermediate igneous and metamorphic rocks and various sedimentary rocks, e.g. unconsolidated sands (Tavernier and Sys, 1986) and limestone (Andriesse and Scholten, 1982). They cover large areas of the continental shields of Amazonia and the Zaire Basin, and are also found in the western Ghats, Assam, the Sri Lankan 'wet zone', Indonesia and Malaysia. They occur both on residual (autochthonous) and reworked or transported (allochthonous) parent materials (figure 4.2). They appear on younger landscapes as well, e.g. on basalts and serpentinites on some Pacific Islands and in southeast Asia (Latham, 1983; Van Ranst and Doube, 1986).

The geomorphic controls on ferralsol and ferric acrisol development have been widely studied (Beinroth et al., 1974; Eswaran and Sys, 1976a, b; Guedez and Langohr, 1978; Ojanuga, 1978; Geiger and Nettleton, 1979; Bouckaert et al., 1984; Tavernier and Sys, 1986; Embrechts and De Dapper, 1987). Beinroth et al. (1974)

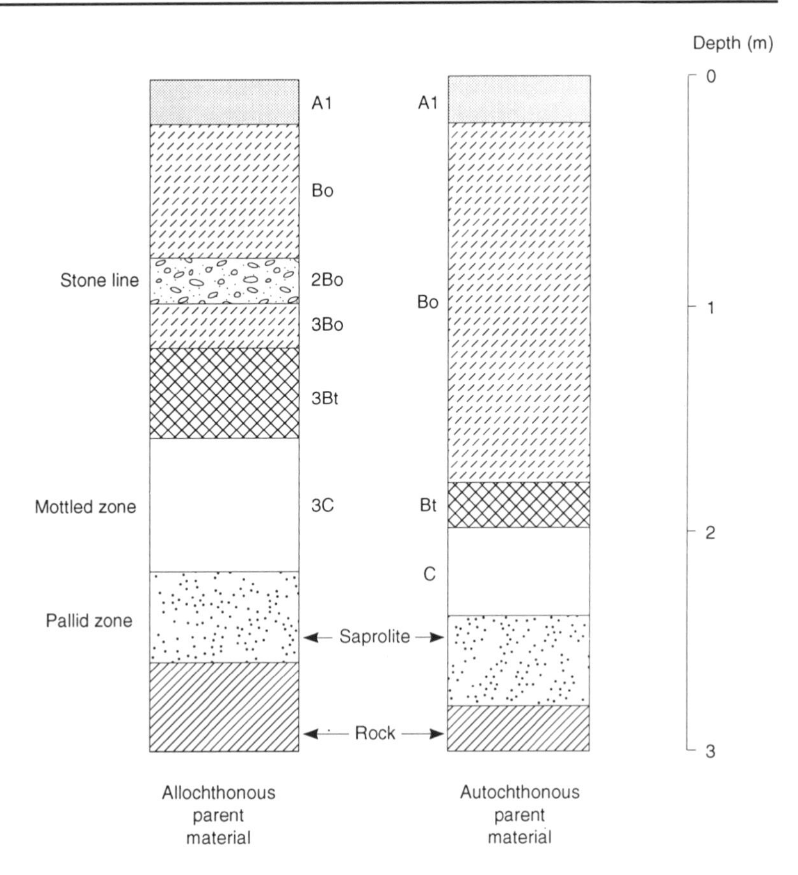

Figure 4.2
Typical ferralsol profile morphologies (after Eswaran et al., 1986, p. 100, figure 4).

have shown that fluvial dissection has a strong influence on soil development on continental shields. In dissected terrain ferralsols often dominate level and gently undulating surfaces, whilst acrisols develop on the rejuvenated slopes adjacent to streams (figure 4.3). Similar distributions of ferralsols have been recorded over other bedrocks such as shales and schists (Bouckaert et al., 1984) (figure 4.4). Moniz and Buol (1982) argue that sloping-ground acrisols are caused by alternating wetting-and-drying cycles, and that they exhibit coarser textures than the ferralsols on adjacent level and undulating terrain due to clay translocation, the removal of fines by erosion and, possibly, the formation of 2:1 clays. These toposequences often show evidence of cycles of landform development and pedogenesis with the formation of inert parent materials (pedisediments) in which some of the grains may have been included in numerous soil profiles since Precambrian!

Weathering in the rain forest zone proceeds rapidly and leads to the breakdown of all minerals, except quartz, to clay-sized particles. The weathering sequence is well illustrated by the Quoin Hill chronosequence from Sabah, Malaysia (Eswaran and Sys, 1976a, b) (figure 4.5). The soils on Quoin Hill have developed on a sequence of lava flows and cinder deposits which date back to the late Pliocene. In the initial stages of the chronosequence primary

Undissected surface

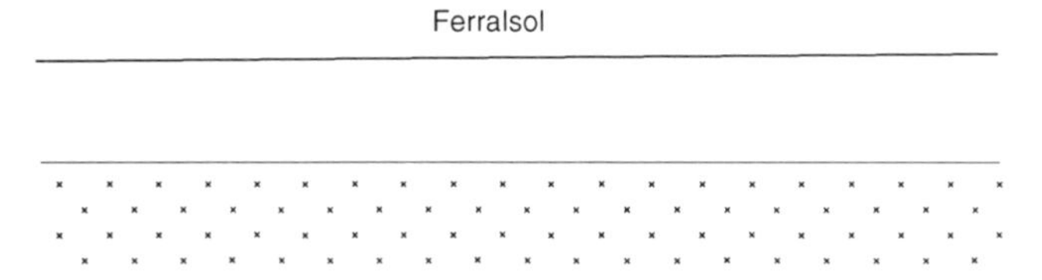

Moderately dissected surface

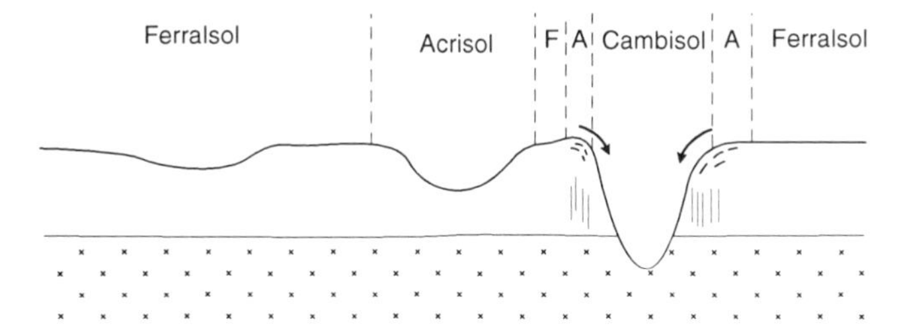

Strongly dissected surface

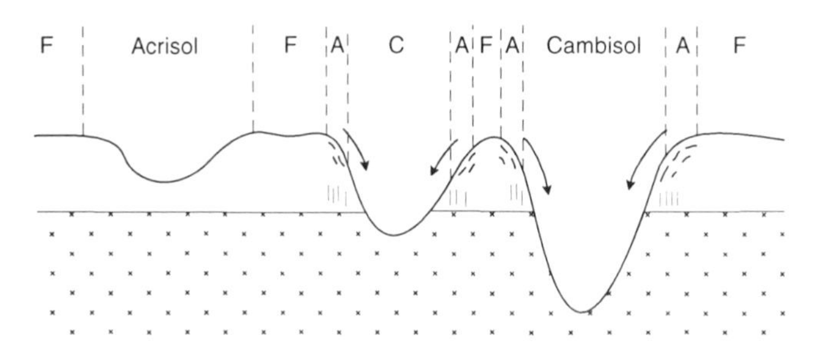

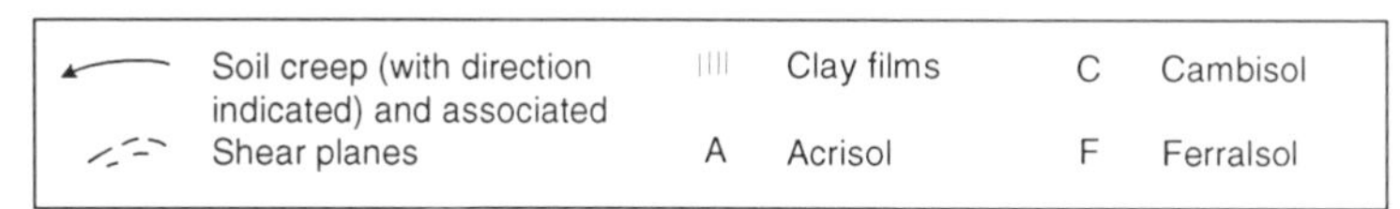

Figure 4.3
Ferralsol–acrisol–cambisol associations (after Beinroth et al., 1974).

minerals weather to bases, silica, and free iron and aluminium. The dominant downward movement of water leads to excessive base leaching and silicic acid is removed in solution. The first soils in the Quoin Hill chronosequence – cambisols (plate 4.1) – show the properties outlined in table 4.5 and figure 4.5, the bases being mainly transported downslope by surface runoff and throughflow because of the steep gradients. The main temporal trends in soil chemistry exhibited in the Quoin Hill soils are a decline in the major cations, cation exchange capacity (CEC) and base saturation, and an increase in the amount of free iron and aluminium. Most of the free iron and aluminium remains in the clay complex. The clay fraction of the cambisol mainly contains allophane, hallyosite, montmorillonite and some feldspar. The amounts of allophane, smectite and hallyosite decline dramatically with time. Kaolinite

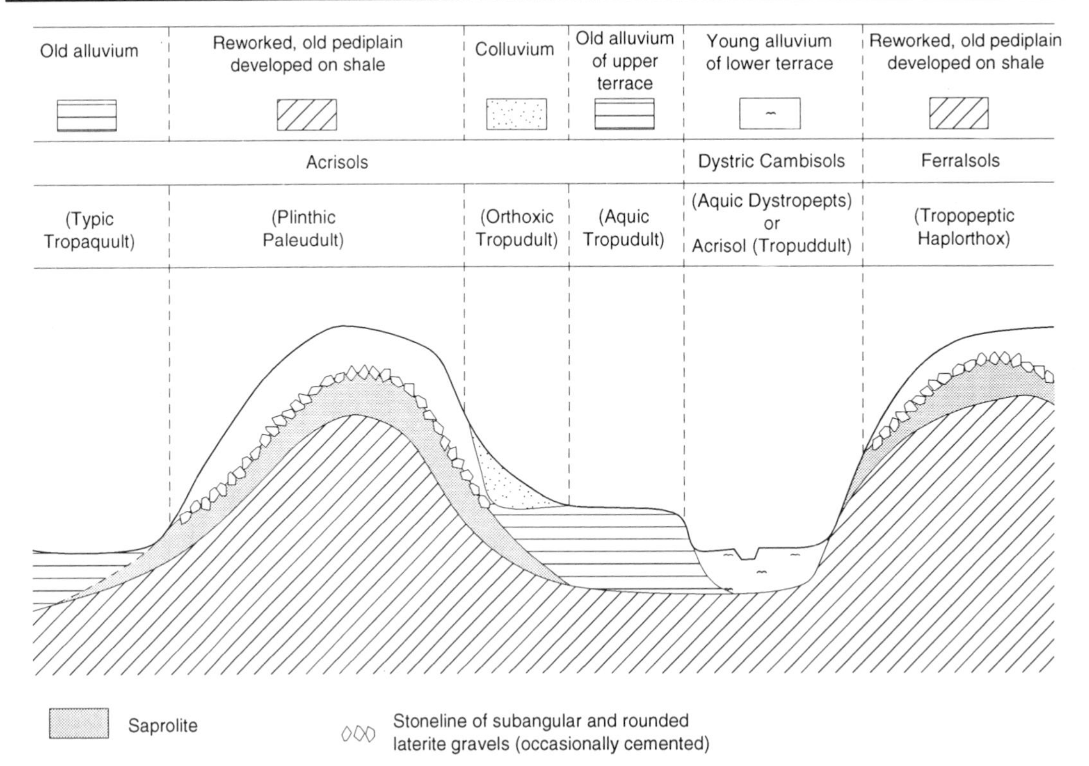

Figure 4.4
Typical toposequence of ferralsols, acrisols and cambisols developed over shale (after Bouckaert et al., 1984).

first appears in the lixisols and the dominant clays in the ferralsols and acrisols are usually kaolinite and gibbsite. Goethite and iron oxyhydrate are also found in the clay fractions of the acrisols and lixisols. Clay translocation first becomes evident in the cambisols and is well developed in lixisols. Illuvial ferri-argillans coat the peds in the lixisols and acrisols, although their occurrence is mainly restricted to ferralsols at Quoin Hill (figure 4.5). The intense base leaching, low exchange capacities of the clays, and free sesquioxides in the clay complex combine to make acrisols and ferralsols strongly acid (pH < 5.0), with low base saturation (< 20 per cent) and a low CEC (< 20 me $(100\,g)^{-1}$ of clay).

The clay mineralogy of the Quoin Hill sequence is slightly different from other red soils because of the basalt parent material. Nevertheless quantitative studies of the clay fraction of a toposequence developed on shales in southern Zaire have recorded similar trends (Tavernier and Sys, 1986). On ultrabasic rocks and limestone, desilication often leads to very intense residual alumina accumulation, usually as gibbsite concentrations or sheets (Eswaran et al., 1977). Desilication on soils developed over basalt in the Leizhou Peninsula, Guangdong, China (Zhao Qi-guo and Shi Hua, 1986) shows a 64.8 per cent decline in silica between the rock and the soil. This is accompanied by augite and olivine decomposition, and kaolinite, gibbsite and iron oxide enrichment. By way of contrast, the residual material on iron-rich rocks is usually dominated by goethite and haematite.

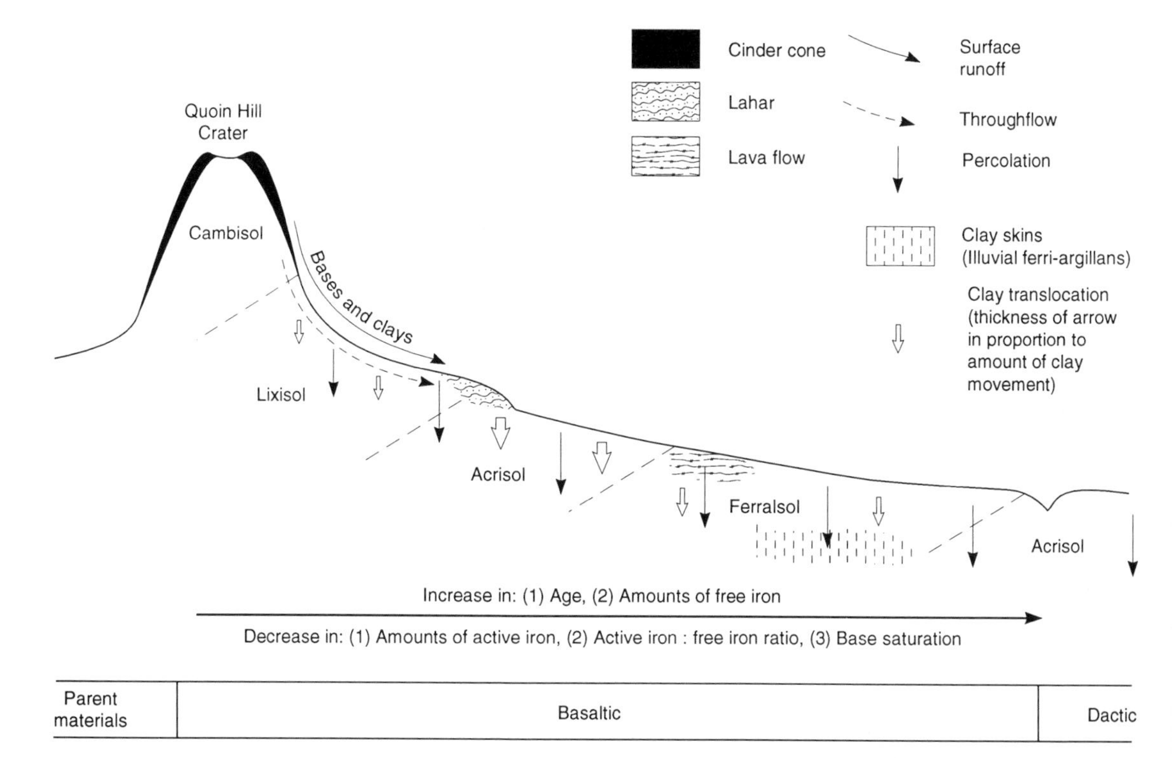

Figure 4.5
Quoin Hill chronosequence, Sabah, Malaysia (topography and soil distribution after Eswaran and Sys, 1976a; data from Eswaran and Sys, 1976b).

The mineralogy of ferralsols and ferric acrisols varies between the different size fractions. The sand-size fractions generally contain quartz, concretionary material, resistant minerals and muscovite if it was present in the parent material. The silt-size fraction contains quartz, muscovite and coarse kaolinite crystals. The clay-size fraction contains kaolinite (if aluminium is present); oxides and oxyhydrates of aluminium, iron, manganese and titanium in either an amorphous form or as goethite, haematite and gibbsite crystals; and, if there was mica in the parent material, illite (Tavernier and Sys, 1986).

Lixisols, acrisols and ferralsols are characterized by the presence of ferricrete. This usually occurs as plinthite nodules (plate 4.2) and occasionally in a massive form. Plinthite is a soft sesquioxide-rich concretion which hardens on exposure to air to form petroplinthite or plinthite gravel (Eswaran and Ragumohan, 1973) (figure 4.6). It can be further cemented to form a massive form of ironstone which pedologists call petroferric contact. Petroferric contact is often impermeable and perched water-tables develop on them which leads to gleying in the soil above. The collective term for these indurated forms is laterite or ferricrete, the latter being preferable because laterite has been defined in other ways. In areas adjacent to ferralsols on plateaux iron comes from the groundwater or from the saprolite zone and is carried upward and laterally resulting in plinthite precipitation (Eswaran et al., 1986). Great thicknesses of plinthite gravel can accumulate in these soils through the selective

Table 4.5 Selected chemical and mineralogical properties of the main soil types on the Quoin Hill chronosequence

	Soil type, and dominant type of B horizon			
	Cambisol, cambic	*Lixisol, argillic*	*Acrisol, argillic*	*Ferralsol, oxic*
Clay (%)	22.1	64.1–71.1	69.7–79.2	3.9–5.0
Sand (%)	15.9	4.9–5.5	2.7–3.2	76.8–81.8
pH (KCl)	4.9	3.8–3.9	4.6	4.7
Organic carbon (%)	0.89	0.32–0.72	0.56–0.99	0.12–0.73
Ca	9.49	5.58–6.25	0.38–0.56	0.27–0.68
CEC	52.77	34.46–39.78	16.04–16.09	1.11–3.17
Base saturation	31.2	30.4–34.0	6.5–10.2	4.6–8.2
Free Fe_2O_3	34.0	5.6–5.9	6.9–7.2	10.0–10.6
Permanent charge (PC)	74.7	17.7	2.5	1.7
pH-dependent charge (PDC)	136.8	48.5	22.7	16.2
Ratio PC:PDC	45.5	49.0	7.7	10.5
Kaolinite and hallyosite (%)	36[a]	50[b]	60[c]	70[d]
Allophane frequency	**	**	(*)	(*)
Montmorillonite frequency	***	**	–	–

Exchangeable cations and CEC values in me $(100\ g)^{-1}$ soil. Charges in me $(100\ g)^{-1}$ clay.
[a] No kaolinite.
[b] Little kaolinite.
[c] Hallyosite and kaolinite.
[d] Kaolinite dominant.

Source: Eswaran and Sys, 1976b

erosion of fines; such gravel-rich soils are common in Africa (figure 4.7) (Smyth and Montgomery, 1962; Fauck, 1963; Stoops, 1968; Collinet, 1969; Lévêque, 1969; Babalola and Lal, 1977; Geiger and Nettleton, 1979; Millington, 1985; Muller and Bocquier, 1986; Dijkermann and Miedema, 1988). In the revised legend to the FAO Soils Maps soils dominated by plinthite are classified separately as plinthisols.

On crystalline parent materials ferralsols and ferric acrisols are often very deep, extending to depths of between 10 and 50 m (plate 4.3), whereas they are usually shallower over sedimentary rocks. Although horizonation is evident most horizons, except the partially decomposed litter- and humic-rich A horizons, merge. Textures are variable and are controlled by the nature of the parent material (figure 4.8). Generally, however, they are dominated by high amounts of clay, pseudo-silt and sand, but little true silt. Clay aggregation leads to the formation of pseudo-silt (Guedez and Langohr, 1978) and in some cases quite large clay balls form (Paramanathan and Eswaran, 1984). Clay-rich textural B horizons may develop in some ferralsols, and in those developed over

Plate 4.1
A cambisol profile from Rokupr, Sierra Leone. This young soil shows evidence of the early downward movement of iron sesquioxides (a slight reddening of the sub-soil is noticeable) and, probably, clays. This profile can be contrasted with that of a mature ferralsol (plate 4.3).

granites the clay content increases down profile. Ferralsols and ferric acrisols generally have a weak to very weak subangular blocky texture which easily becomes granular with a friable consistency; when the soil is picked up it often feels like flour. The subsoil contains nodules of variable internal structure, some argillans, and clay deposition in voids and channels. Nevertheless percolation is good and there is a high infiltration rate giving a linear relationship between clay content and water retained at 15 bar pressure. The upper humic-rich horizons are often dark in colour and the remaining A and B horizons vary from yellow to dark red. The C horizon is speckled red and white; Young (1976) likened its colour and pattern to that of corned beef!

Plate 4.2
A plinthite-rich soil from central Sierra Leone. In this soil the plinthite has taken the form of a pea-sized gravel which can be seen clearly throughout the upper part of the profile.

Although these soils are used extensively for plantation agriculture and shifting cultivation, sustainable cultivation requires careful management to overcome five main problems. First, their low CEC and base saturation is complicated by their zero or negative net charge which leads to high nutrient leaching losses (Uehara and Gillman, 1981; Tessens and Zuayah, 1982; Shamshuddin et al., 1984). This is caused by the presence of iron and aluminium oxide and hydroxide colloids, and the broken kaolinite edges. Second, they have a high phosphorus fixing capacity because the iron and aluminium oxides, especially the commonly occurring amorphous oxides, absorb phosphorus. This, combined with their low pH, creates a phosphorus deficiency. Third, aluminium saturation can exceed 60 per cent, leading to aluminium,

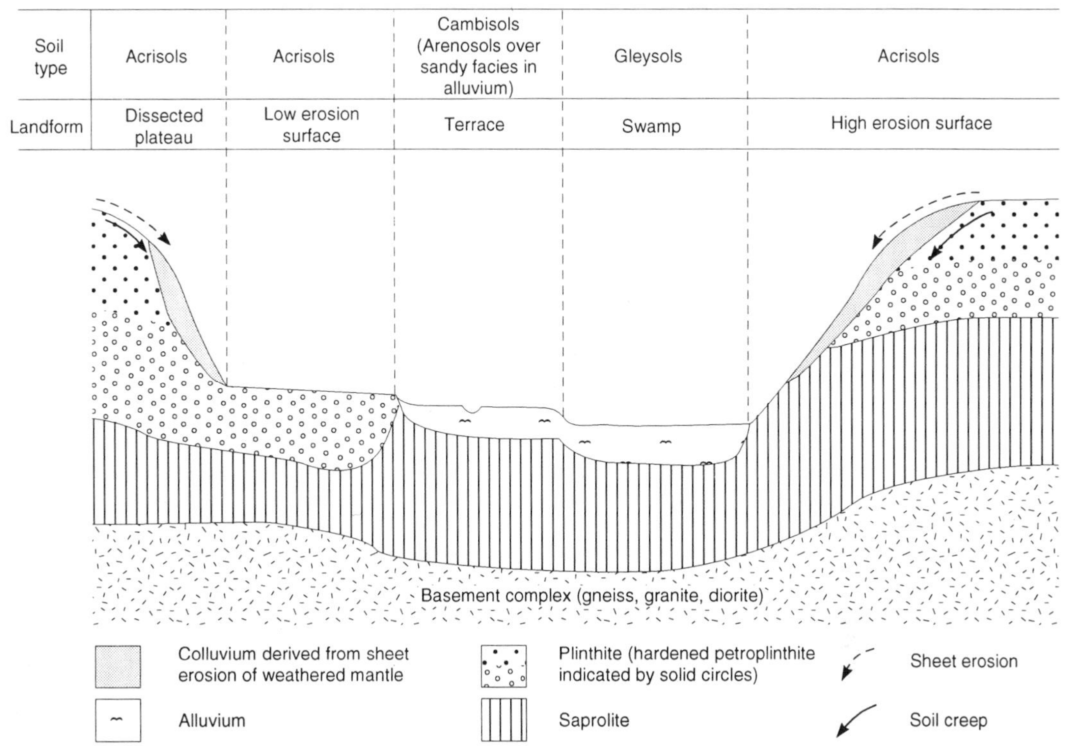

Figure 4.6
Soil–geomorphic relationships from western and central Liberia (information from Geiger and Nettleton, 1979). Particular attention is drawn to the location of sheet erosion and soil creep on steep slopes, and the induration of plinthite in the upper soil horizons.

manganese, nickel, chromium and cobalt mobilization and aluminium toxicity (Bachik et al., 1985). Fourth, trace element deficiencies are common (Sanchez and Cochrane, 1980). Finally, the build up of plinthite gravel can lead to a low available moisture capacity (Tawonas et al., 1984; Humbel, 1988), inhibited root penetration and seedling establishment (Babalola and Lal, 1977) and excessive leaching. However, Anamosa et al. (1990) point out that rapid water movement through gravelly ferralsols and ferric acrisols will not necessarily leach many nutrients from the rooting zone.

Ferralsols and ferric acrisols of the savanna zone
The lower rainfall totals in the savanna zone mean that leaching is less intense than in the equivalent forest zone soils. Therefore intense weathering cannot necessarily be invoked as a primary mechanism for ferralsol and ferric acrisol formation. Here, weathering over extremely long periods on very stable continental shields is preferred to intense leaching over shorter periods as the main mechanism. These soils were first recognized as *weathered* soils, as distinct from *leached* soils, by Botelho da Costa and Cardoso Franco (1965). The simplicity of this argument is complicated by evidence from West Africa which suggests that ferralsols and ferric acrisols have undergone intense leaching when climatic conditions favoured the expansion of humid tropical (rainforest) conditions (Thomas and Thorp, 1985; Thomas et al., 1985; Dijkermann and

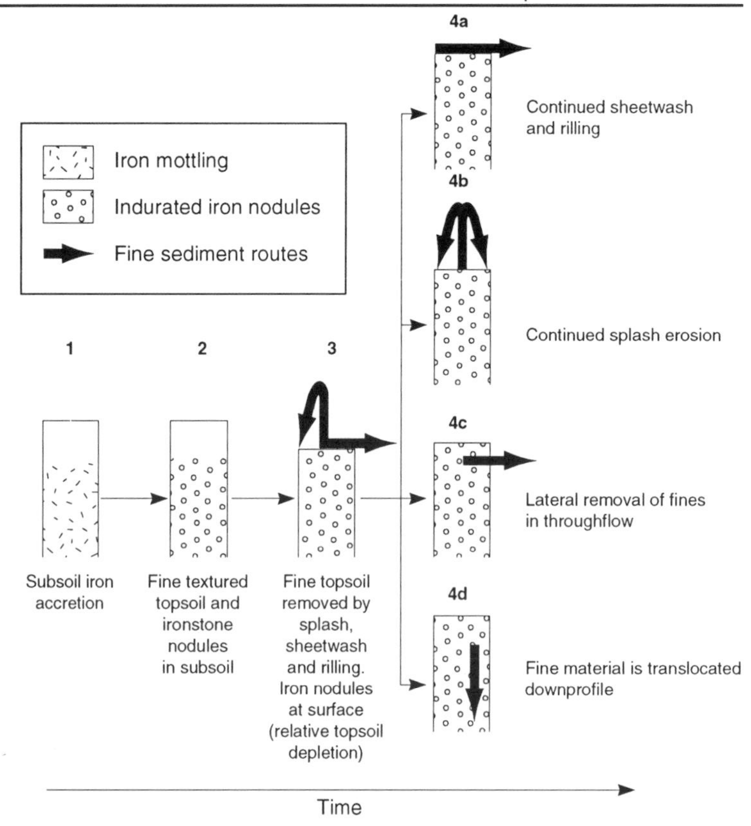

Figure 4.7
Relative accumulation of plinthite gravel in Sierra Leone through the selective erosion of fines (after Millington, 1985). These surface accumulations would be termed lag gravels by geomorphologists.

Plate 4.3
A deep ferralsol profile along a roadcut in central Sierra Leone. The residual core stones and 'weathering aureoles' around them (adjacent to the person standing) provide evidence of extended, deep weathering processes.

Miedema, 1988; Agbu et al., 1989). Whatever the history, whatever the mechanism, the end result is similar: a case of pedological equifinality?

Savanna ferralsols and ferric acrisols occur on stable continental shields where rainfall varies between 500 and 1200 mm, e.g. the

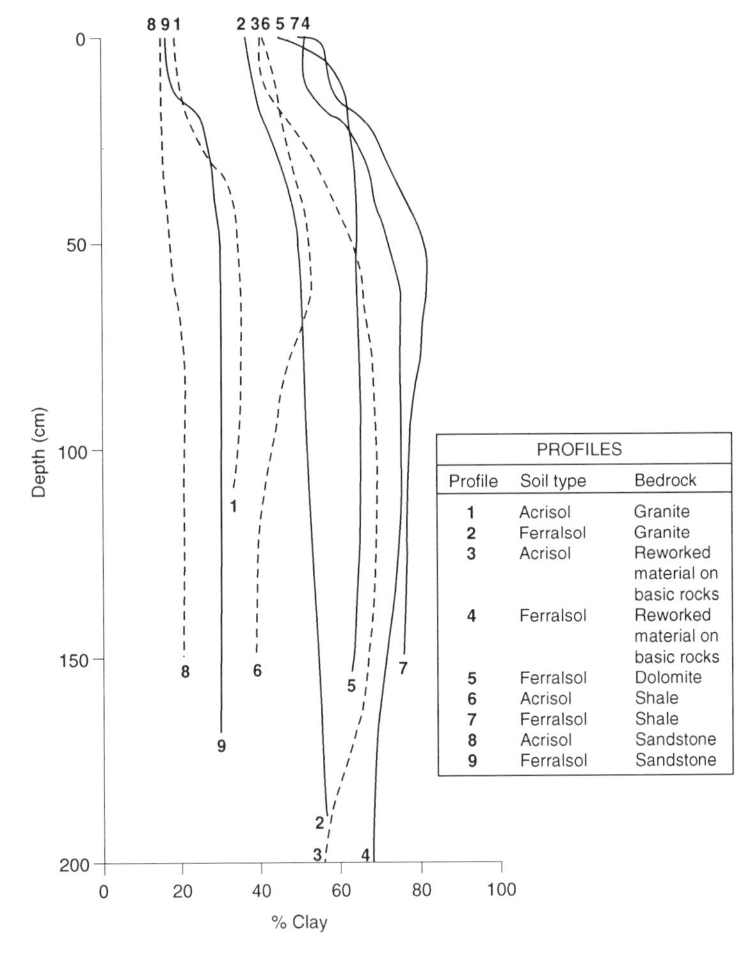

Figure 4.8
Clay texture profiles of acrisols and ferralsols on selected parent materials from southern Zaire (data from Tavernier and Sys, 1986, pp. 172–5, figures 4–8).

Guyanan and Mato Grosso Shields, the Deccan, and on the Gondwana and African surfaces in Africa. They typically form on flat, undulating and gently sloping areas and, so common is this topography–soil relationship, they were once known as *plateau soils* in Africa (Milne, 1936; Trapnell et al., 1948–50). The most common parent materials are acid igneous and metamorphic rocks and siliceous sediments (table 4.3). They are commonly found grading into gleysols in a series of characteristic toposequences (figure 4.9) (Young, 1976). Typical vegetation covers are the *miombo* woodlands of central and east Africa and the South American *cerrado*.

Young (1976) divides savanna zone ferralsols and ferric acrisols into normal and pallid types. Those with normal profiles are dominated by sandy to sandy loam topsoils over a sandy B horizon, although duplex soil profiles with a sandy-clay B horizon also occur. They are weakly fine to medium blocky (plate 4.4) or structureless and there is no evidence of clay skins. The B horizons grade into very thick sequences of highly weathered material in which the structures have been destroyed. The weathered rock lies

Gently undulating surfaces

Crest Upper slope Mid-slope Lower slope Basal concavity Valley

Ferralsols and Ferric Acrisols | Gleysols

Decreasing redness

Sandy & mottled

Surface with inselbergs and pediments

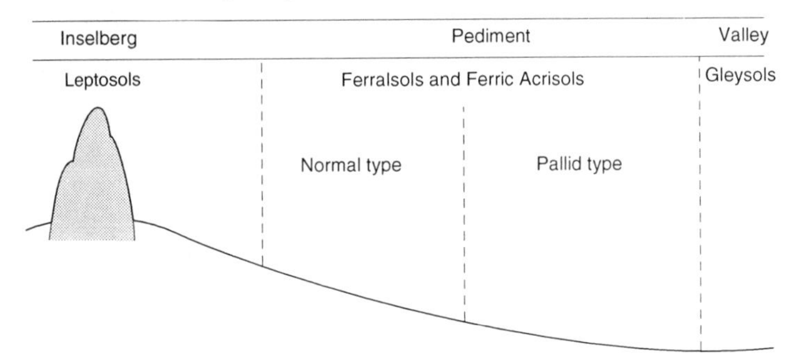

Level plateau surface

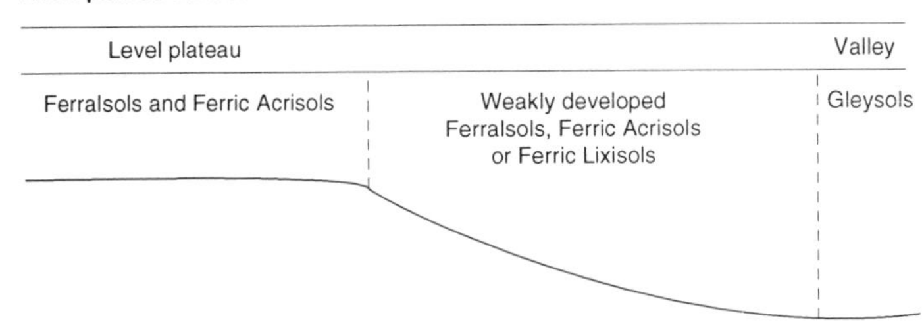

Figure 4.9
Typical toposequences of ferralsols and ferric acrisols under savanna conditions (after Young, 1976, p. 145, figure 7).

at great depth and usually has a very different chemistry to the weathered material. Although the upper A horizon is dark and humus-stained the organic matter content rarely exceeds 2.5 per cent. A very pale and structureless A2 horizon overlies the compact and firm B horizon. Pallid soils have a very pale upper horizon overlying a B horizon which exhibits mottling (plate 4.4) and iron concretions caused by waterlogging. They often occur downslope of freely draining soils in a topocatena where the subsoil drainage is seasonally restricted. Ferricrete is common in both freely drained and waterlogged soils and occurs in concretionary, massive and vesicular forms in horizons of variable thickness. Stone lines are also frequently encountered. Young (1976) considers these *dead soils* because they have very limited interaction with contemporary pedogenic processes, the striking differences between their bedrock chemistry and that of the weathered mantle being indicative of their long evolution. One can theorize that they have formed over very

Plate 4.4
The mottled zone of a ferralsol profile from central Gambia. The white speckling in the dominantly red profile can be seen near the base of the photograph. The soil is clay-rich, and a well-developed columnar to blocky structure is evident.

long time periods or during more active pedogenic phases in the past; either way the parent material that the soils develop from has been highly altered.

The dominant clay mineral in both types of soil is kaolinite, and there are few free sesquioxides. The clay fraction always has a low CEC and there are very few exchangeable bases. Base saturation is variable, ranging from very low to almost 90 per cent, and they are weakly to strongly acid. They have an excellent, very stable structure because the individual clay particles form sand-sized aggregates in which the clays are cemented by amorphous iron and aluminium, although the breakdown products of organic matter can also form important cements (Lugo-López and Juárez, 1959)

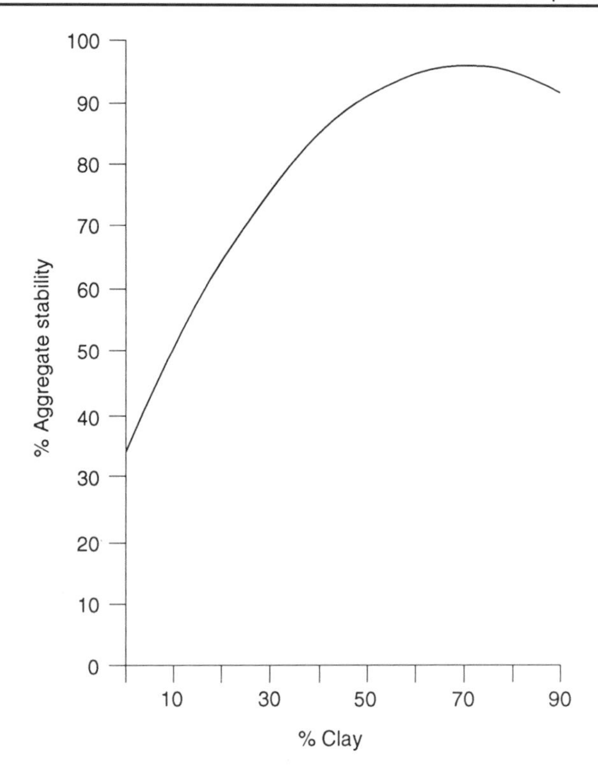

Figure 4.10
The relationship between aggregate stability and clay content of red soils from Luzon, the Philippines (after Briones and Veracion, 1965).

and the orientation of the clay influences aggregate stability (Cagauan and Uehara, 1965). Generally, the greater the clay content the more stable the aggregates (Briones and Veracion, 1965) (figure 4.10) and in soils with a high clay content susceptibility to erosion is high, although sand-rich ferralsols often suffer from compaction and erode.

The management of these soils is often very difficult (table 4.6) (Young, 1976; Stocking, 1986). Intense or continued cultivation often leads to a decline in their structural stability which in turn leads to compaction and accelerated erosion. Nevertheless, with copious fertilizer applications cereals, groundnuts and tobacco are grown. More sustainable long-term cultivation of these soils is possible using shifting cultivation systems, such as the *chiteme* system in east and central Africa (Allan, 1965; Stromgaard, 1985).

Table 4.6 Soil management factors affecting ferralsols and ferric acrisols in savanna environments

Low soil moisture retention
Poor root penetration
Occurrence of ferricrete at shallow depths
Low nutrient status
Low organic matter content
Very low nitrogen levels
Inhibition of denitrification by savanna grasses
Sulphur deficiency

4.5.2 Lixisols

Lixisols are the characteristic soils of the savanna zones and in a humid tropical context are restricted to the seasonal margins. Wet savannas and woodlands are mainly found on ferric lixisols, but in eastern and southern Africa (the *miombo* woodland zone) chromic lixisols dominate. They grade into ferralsols when the precipitation exceeds 1200–1500 mm, and into various types of pedocals when the precipitation drops to around 600 mm. They generally develop over the felsic to intermediate crystalline rocks commonly found on continental shields, but also over various sedimentary rocks. Extensive areas of lixisols are found in west, east and southern Africa and in all account for about 10 per cent of the African tropics (Stocking, 1986). They are also very important in the Deccan and the Sri Lankan 'dry zone'. In the Americas they are mainly restricted to northeast Brazil (figure 4.1).

They are often associated with weathered ferralsols and vertisols in topocatenas (e.g. figure 4.11). They are usually less than 2.5 m deep, and the dominant pedogenic processes are weathering, rubification, plinthite formation, clay translocation, carbonate and base leaching, and silica dissolution and removal. Two variants exist: (i) weakly structured, sand-rich lixisols which develop over sandstones, and (ii) some lixisols, once thought to be podzols, which have a pale leached A2 horizon.

The moderate degree of weathering is indicated by the frequency of partially weathered minerals. Free iron oxides are mobilized under wet conditions and soft plinthite concretions form. During the dry season the iron precipitates and hardens, and iron-rich nodular gravel forms. The amount of gravel and its proximity to the surface increase as drainage conditions worsen. Clay translocation often results in the formation of a textural B horizon with a fine-to-moderate blocky structure. The textural B horizon is a clay plasma which is stained by amorphous iron hydroxides and has well developed illuvial ferri-argillan coatings to the peds. All three of the clay complexes identified by Sanchez (1976) are present, i.e. layer silicates with permanently negative charges, fully oxide-coated silicates with variable charge, and partially oxide-coated silicates; the main clay is kaolinite (although some 2:1 lattice clays also occur) and there are free oxides.

The B horizon has a base saturation of 40–90 per cent, the CEC exceeds 15 me $(100\,g)^{-1}$, and the pH ranges from 6 to 7. Stocking (1986) regards lixisols as one of the more fertile red soils with organic matter contents ranging from 1 to 2 per cent and moderate nitrogen and potassium levels, although phosphorus is easily fixed because of the high levels of free iron. There are important differences here between ferric and chromic lixisols which affect their fertility. Chromic lixisols have less organic matter and a lower pH, but more bases than ferric lixisols. Their moisture-holding capacities are good and root penetration is easy, making them well

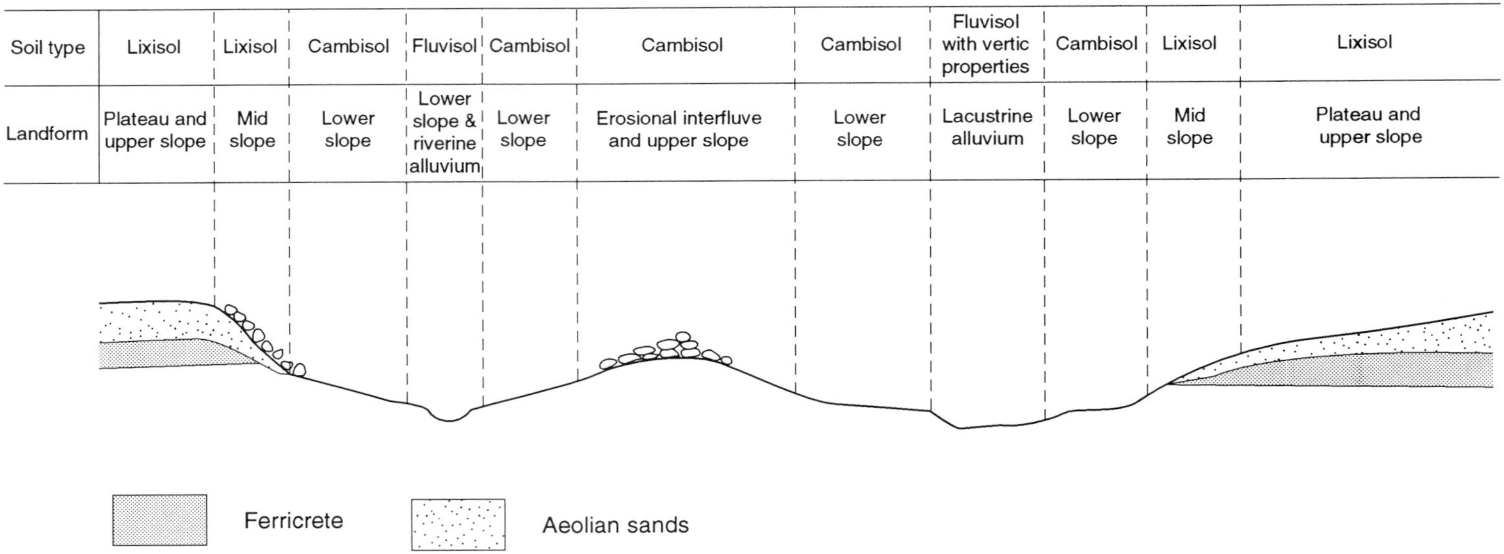

Soil type	Lixisol	Lixisol	Cambisol	Fluvisol	Cambisol	Cambisol	Cambisol	Fluvisol with vertic properties	Cambisol	Lixisol	Lixisol
Landform	Plateau and upper slope	Mid slope	Lower slope	Lower slope & riverine alluvium	Lower slope	Erosional interfluve and upper slope	Lower slope	Lacustrine alluvium	Lower slope	Mid slope	Plateau and upper slope

Ferricrete

Aeolian sands

Figure 4.11
Toposequence of lixisols, cambisols and fluvisols from the Sokoto-Rima basin in northern Nigeria (after Agbu et al., 1989, p. 137, figure 4).

suited to annual crops (e.g. maize, tobacco and cotton). Sand-rich lixisols suffer significant edaphic droughts which restrict their agricultural use to short-season crops such as cereals, cotton and groundnuts. However, the main management problems with lixisols are accelerated erosion and compaction due to their sandy topsoils. Much research has been carried out on lixisol erosion in Nigeria at the International Institute of Tropical Agriculture and is examined in more detail in chapter 8.

4.5.3 Nitisols

Dystric nitisols develop over basic rocks in both the rain forest and wet savanna zones, generally occurring in areas with a mean annual rainfall of 1200–1800 mm spread over one or two wet seasons. In drier savanna areas soils developed over basic rocks are classified as eutric nitisols (table 4.3), but these soils are not considered in this book. Locally important areas of dystric nitisols are found in Congo, Ghana, Malaysia and Samoa; in total they account for between 1 and 2 per cent of the humid tropics.

They typically have a clay to sandy clay loam topsoil with an organic matter content between 2 and 5 per cent. The topsoil is commonly underlain by a deep clay-rich B horizon which exhibits numerous clay skins covering fine to medium blocky peds; the horizon boundaries are usually gradational. Kaolinite and free iron dominate the clay fraction, which occurs as both an iron-rich kaolinitic plasma and illuvial ferri-argillans coating the peds (Stoops, 1968; Eswaran, 1970). Partially weathered minerals, which are common in the upper 2 m, are thought to be due to the continual renewal of iron and bases by weathering. This is exacerbated by their frequent occurrence on steep slopes which leads to much material being removed by erosion.

Variants on this basic profile exist. On ultra-basic rocks they are very dark red to red brown, quartz is rare and trace elements are common (Fox and Hing, 1971). In areas where the rainfall exceeds 2000 mm, e.g. Malaysia and Samoa (Wright, 1963), they become strongly acidic with a base saturation of less than 10 per cent and are classified as rhodic ferralsols.

Dystric nitisols are amongst the most fertile red soils (Young, 1976; Stocking, 1986). They have high organic matter contents and moderately high nitrogen contents, and the continual renewal of other bases maintains fertility levels. In addition they are free draining, roots penetrate easily and they are usually quite resistant to erosion. They are used extensively for perennial crops such as cocoa, coffee and oil palm.

4.5.4 Red soils at high altitudes

At altitudes in excess of 1500 m above sea level humic variants of the red soils found at low elevations are common because of increased organic matter accumulation. Humic nitisols develop on basic rocks, whereas humic ferralsols are found on felsic and intermediate parent materials (table 4.3). Together they account for a small area (<1 per cent) of the humid tropics but are locally important in the Andes, in east and central Africa, on the Ethiopian plateau, in the Western Ghats, and in Indonesia and Malaysia.

Humic ferralsols

Humic ferralsols are shallow soils, often only about a metre in depth, with a dark brown humic-rich A horizon overlying a red–brown B horizon. The latter horizon is sometimes a sandy-clay textural B horizon, but the structure is always weak and there are no visible clay skins. They are highly leached (and as a consequence have few weatherable minerals), are strongly acid and have a low base saturation. Two variants occur: first, a bauxitic humic ferralsol with high aluminium levels; and second, soils with a dark organic-rich B horizon (which have been reported from east Africa), although these may have formed by podzolization.

Humic nitisols

Humic nitisols have deeper profiles than the humic ferralsols, and they are less weathered because they generally occur in areas with slightly lower rainfall. They have well-developed B horizons with fine to moderate blocky structures and clay skins. Nevertheless, they exhibit similar levels of soil acidity and base saturation to humic ferralsols.

Both humic ferralsols and humic nitisols pose problems for agriculture because of their shallowness, acidity and strong leaching. As a consequence they often exhibit potassium and phosphorus deficiencies, although nitrogen levels are high as a result of organic matter accumulation. They are sometimes terraced for tea and coffee cultivation but more often than not are left uncultivated because severe erosion occurs when the forest vegetation found on them is cleared (e.g. Lesotho).

4.6 Hydromorphic Soils

4.6.1 Fluvisols

Fluvisols develop in floodplain and deltaic sediments along all river valleys and the humid tropics are no exception. The definition of a fluvisol in the FAO Soils Legend is a soil developed on recent alluvium with an AC or C profile. On older alluvium B horizons have usually developed, and the soils are classified as either cambisols, lixisols or acrisols (Hassan and Tavernier, 1986). Three types of fluvisol are commonly found in the humid tropics: dystric and eutric fluvisols occur on floodplains, and thionic fluvisols are restricted to coastal estuaries. Fluvisols are closely linked to waterlogged soils – gleysols – and the combined distribution of the two is shown in figure 4.1.

The development of soils on floodplains and deltas is dependent on the type of the alluvium, the time elapsed since deposition, the physiography, the extent of flooding and waterlogging, and environmental and climatic changes. Subtle geomorphological and sedimentological variations lead to significant differences in soil type, which can easily be noted when soils developed on features such as river terraces, levees, sloughs and old channels are compared (figure 4.12) (Stobbs, 1963; Bouckaert et al., 1984; Millington et al., 1985; Hassan and Tavernier, 1986).

There is considerable variation in fluvisol textures arising from the diversity of depositional environments in a floodplain, sedimentation being the main control on floodplain soil texture. In high energy environments, e.g. braid bars, coarse sand is deposited, and the lowest energy environments, e.g. sloughs and backswamps, are characterized by clay deposition. Coastal plains often show a sequence of coarse sand bars with intervening swales

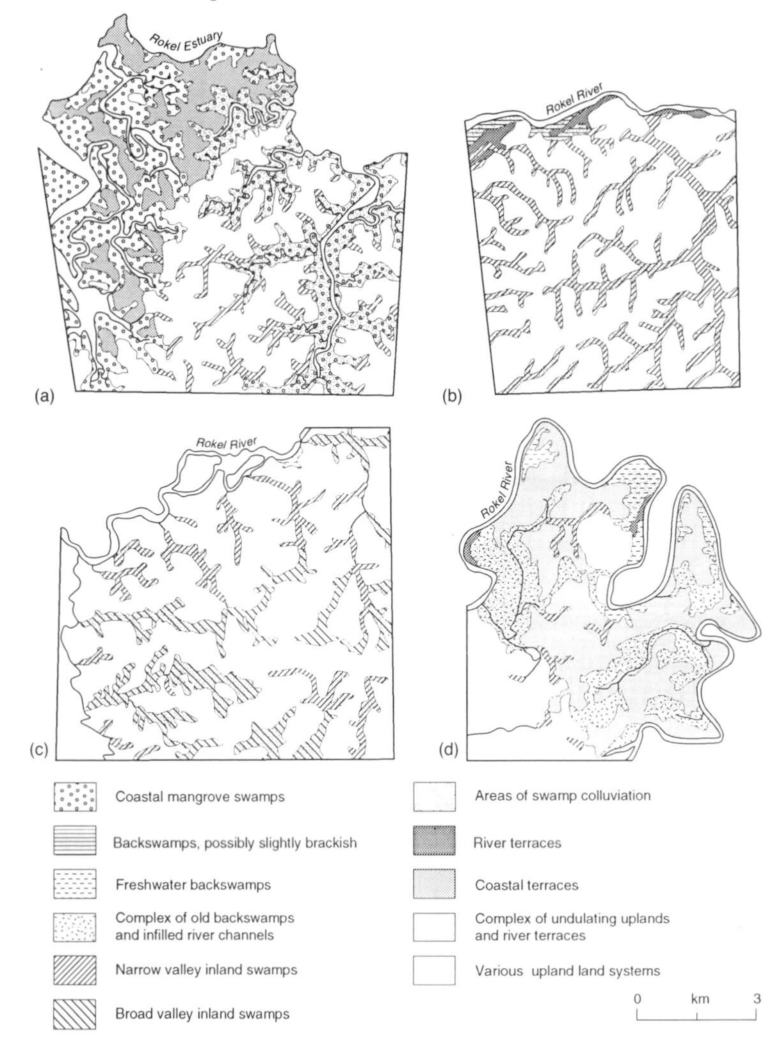

Figure 4.12
Swamp landforms from western Sierra Leone for a sample area (a) within the rocks of Bullom Group; (b) within the rocks of the Kasila Group; (c) within the granite rocks of the basement complex; (d) within the rocks of the Rokel River series. All are interpreted from 1:70,000 infrared aerial photography (after Rhebergen, 1980).

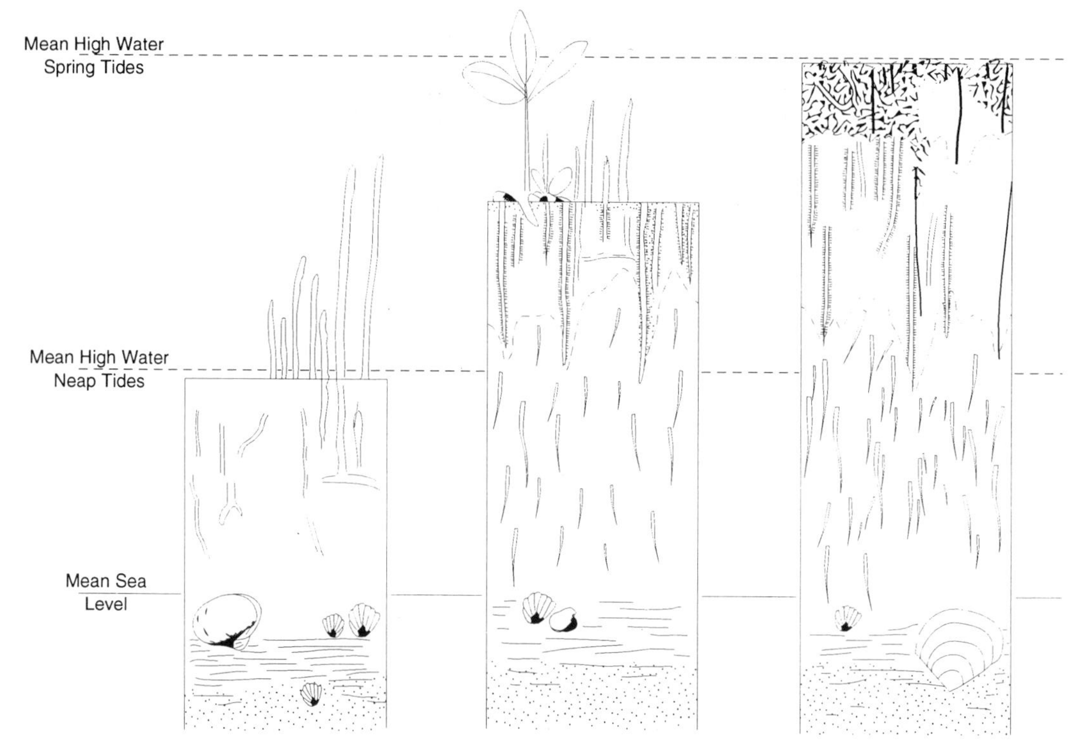

Figure 4.13
Sedimentation and soil development around aerial roots of mangroves (after Dent, 1986, p. 115, figure 4.7).

containing fine-textured fluvisols (Boughey, 1957; Nossin, 1962; Brinkmann and Pons, 1968; Bleeker, 1983; Anthony and Marius, 1984) and in mangrove swamps aerial roots retard water flow causing sedimentation around vegetation (figure 4.13). Consequently there is no such thing as a *typical* fluvisol profile and properties like structure and consistency are also highly variable. Nonetheless some properties do show broad similarities between soils, e.g. organic matter is generally higher than in adjacent soils, as is the pH which usually ranges from 6 to 8.

Fluvisols usually occur on gently sloping or undulating topography and, as the water tables are usually high, they are often waterlogged and gleyed. However, soils which are subject to severe waterlogging are usually classified as gleysols or histosols. Variability between fluvisols is not a function of the past or present sedimentary regimes; the initial processes of soil formation and processes both normal to the climatic zone and specific to alluvial soils are also important.

The initial soil-forming processes in fluvisols are the irreversible processes of soil ripening and homogenization (Pons and Zonneveld, 1965). Initially clay is deposited as a loosely packed slurry of clay minerals and water. During soil ripening the excess water is drained or evaporated, desiccation cracks develop, and the soil becomes aerated. This in turn leads to the oxidation of some of the iron compounds and organic matter. This sequence of processes is accompanied by an increase in soil strength. The elimination of depositional features by bioturbation (i.e. plant root growth, the actions of earthworms, termites and other micro-organisms) is called homogenization. Sediment reworking and homogenization has also been attributed to larger animals (mainly Crustaceans) such as crabs (Bleeker, 1983) and mud lobsters (Andriesse et al., 1973). In the later stages of homogenization clay translocation becomes important, and in floodplains with long periods of seasonal flooding the peds and pores become coated with gleyans. These are grey, reduced, fine-silt and clay skins in which the grains are unoriented (Brammer, 1971). Parallel processes include the reduction of iron, manganese, nitrate and sulphate ions, and changes in other physical properties, particularly structure and permeability (Islam, 1966).

The inherent high fertility of floodplain soils that is typical of temperate regions cannot be assumed in the humid tropics. The alluvium deposited in the floodplains of rivers which drain catchments with much recently exposed sediment (e.g. Ganges, Brahmaputra) or which drain areas with much volcanic ash (e.g. Magdalena) is usually fertile. But rivers draining old erosion surfaces on continental shields (e.g. the Niger, Zaire, Orinoco and Amazon systems) tend to be rich in quartz, kaolinite and iron oxides, and the resulting alluvium is generally infertile (Edelmann and van der Voorde, 1963). In addition, very old river terraces are often highly leached and infertile, although these soils are rarely classified as fluvisols.

One type of fluvisol, the thionic fluvisol (also known as an *acid sulphate soil* or a *cat clay*), merits special consideration because of the severe limitations it poses to cultivation (Dost, 1973; Spaargaren et al., 1981; Dost and van Breemen, 1982). Thionic fluvisols are commonly found in the tidal reaches of rivers and in brackish water, and are usually associated with mangrove swamps, although inland thionic fluvisols are known. Their total extent is about 140,000 km^2 and a further 200,000 km^2 of histosols in Indonesia are underlain by similar soil materials (Spaargaren et al., 1981) (figure 4.1). The location of mangrove swamps adjacent to areas of high population in Africa and Asia and their subdued topography has encouraged the clearance of vegetation for rice cultivation (e.g. Mekong Delta); however, this clearance is usually accompanied by soil acidification which places severe limitations on cropping.

As with all fluvisols the distribution and characteristics of thionic fluvisols are closely related to physiography, rates of sedimentation and type of alluvium (figure 4.14). On the Chao Phraya River floodplain in Thailand (Spaargaren et al., 1981) thionic fluvisols are closely related to topography, being found away from the current river course and in abandoned freshwater channels. Germane to this distribution is the presence of gypsum crystals in the soil. These are closely related to the present river course because gypsum formation requires free lime which is derived from overbank flooding of the Chao Phraya. The thionic fluvisols with gypsum crystals are located on the floodplain in areas where floodwater

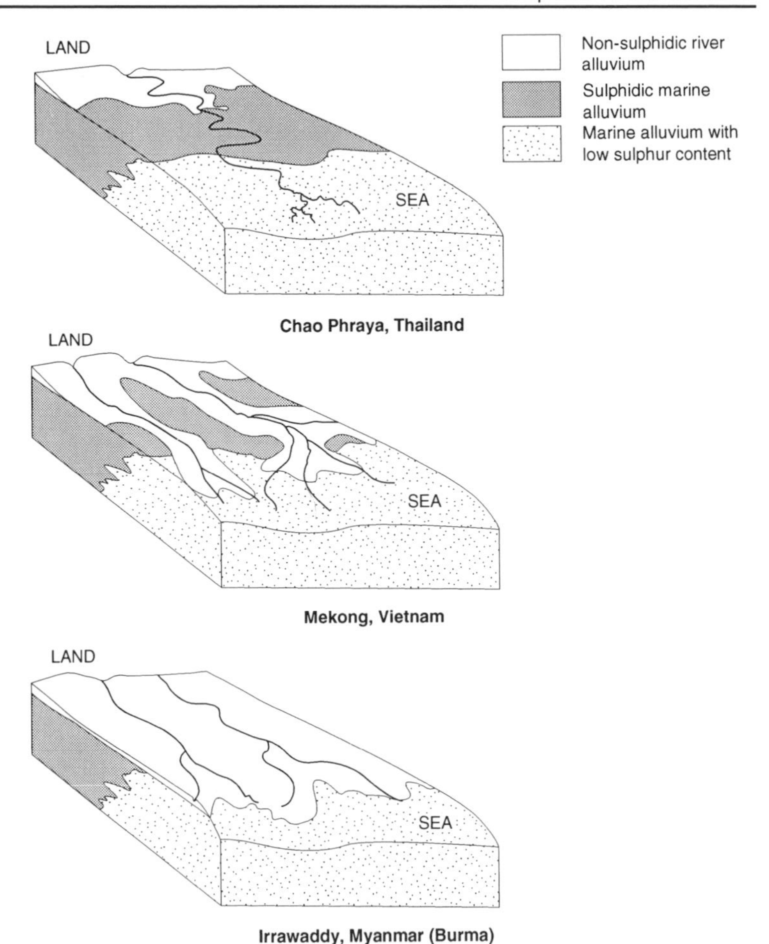

Figure 4.14
The effects of different sedimentation regimes on regional soil patterns in humid tropical deltas (after Pons et al., 1982, p. 29, figure 1.2).

persists and permeability is low, i.e. in low-lying topographic depressions.

Thionic fluvisols occur in mangrove swamps where sediment is deposited around the subsurface and aerial roots of mangroves (figure 4.13). The resulting soil is an organic-rich silty-clay or clay which ranges from dark grey or bluish-black to dark brown in colour depending on the amount of decayed organic material, and normally has a pH of approximately 5.6. Brackish water permeates these soils, leading to waterlogging and gleying. A typical profile from Thailand is described by Spaargaren et al. (1981). Mottling occurs in the Apg horizon, where it is associated with pore and root channels, and in the Bg horizons, where it varies in colour from red through brown to yellow. The gleyed B horizons have different densities of mottles; most pronounced are the yellow jarosite mottles in the Bg3 or *cat clay* horizon and the near absence of mottles in the Bg4 or *soap clay* horizon.

Agricultural colonization of these soils requires both vegetation clearance and drainage. Drainage leads to a number of changes in the soil, the most important being oxidation. Iron sulphates (e.g.

pyrite, FeS_2) occur in thionic fluvisols because of the reducing activity of soil bacteria in the saline pore water environment. Oxidation causes pyrite to change to sulphuric acid and ferric sulphates such as jarosite which are precipitated as yellow mottles in a highly acidic environment. The sulphuric acid in turn reacts with aluminium, magnesium and iron to form sulphates including rare minerals such as tamarujite, pickeringite, halotrichite, bloedite, hexahydrite and rozenite (le Brusq et al., 1987). However, not all thionic fluvisols acidify when drained and vegetation appears to be important in determining the level of acidification. Jordan (1964) noted that *Rhizophora racemosa* swamps always acidified when cleared in Sierra Leone, but swamps colonized by other species of *Rhizophora* and *Avicennia* did not. These differences are probably related to the sediment trapping ability of the vegetation (which is related to the root density) (Hesse, 1961) rather than the biochemistry of the environment.

4.6.2 Gleysols

Gleysols are found in poorly drained sites throughout the humid tropics (see, for example, figure 4.15). They nearly always grade into other soils in which some gleying is present due to seasonal flooding on slightly higher ground. Not surprisingly gleysols and fluvisols are often closely related and the distinction between gleysols and gleyed phases of other soils is often rather subjective. In the FAO Soils Legend gleysols are classified as soils in which hydromorphic properties (e.g. mottling, blue–grey coloration) dominate in the upper 50 cm of the profile, whereas in gleyed phases of other soils these properties are secondary in the upper 50 cm.

Gleysols are common throughout the savanna and rain forest zones of Africa. Characteristic of savannas are broad, gently sloping valleys known variously as *mbuga, dambos* and *fadama*. These landforms exhibit deep, gleyed clay profiles which are often dark grey or black in the upper 2 m. The topsoil has a medium blocky to crumb structure due to the penetration of grass roots but at depth the structure changes to very coarse prismatic with peds approximately 40 cm in length being covered by thick clay skins. An important control on the texture and morphology of the gleysols appears to be the texture of the soils in the upper catena. In the forest zone similar landforms have been identified as *bolis* and inland valley swamps (Stobbs, 1963; Millington et al., 1985) and *bas-fonds*. In these swamps some soils reamin waterlogged for much of the year and a muck horizon often overlies a mottled clay or sandy clay, whilst others are seasonally flooded (figure 4.15). Compared with upper catena soils gleysols exhibit many properties that favour cultivation and many of these soils are used extensively. These properties include large amounts of organic matter, phosphate and potassium, and high CEC and base saturation.

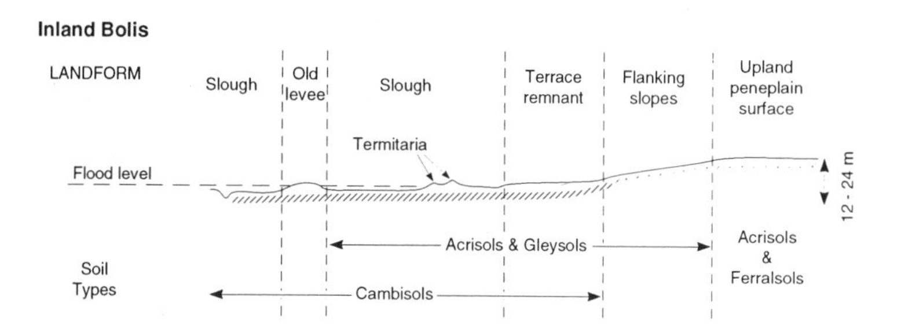

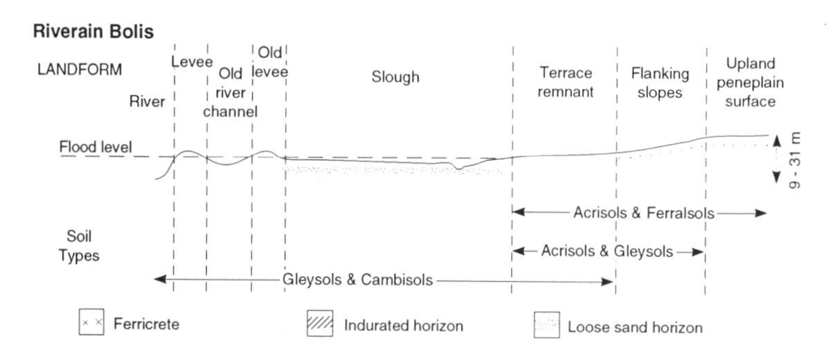

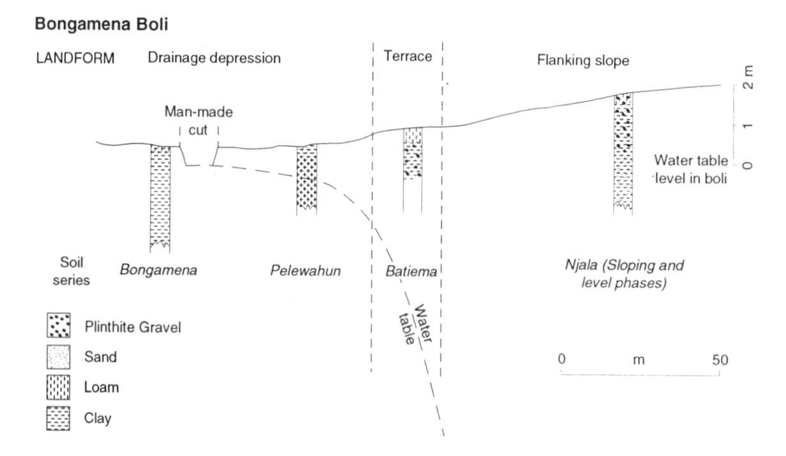

Figure 4.15
Distribution of gleysols in floodplain and swamp landscapes of central Sierra Leone (after Millington et al., 1985, p. 208, figure 5).

4.6.3 The effects of flooding on fluvisols and gleysols

A number of important physical and chemical processes occur when fluvisols and gleysols are flooded. These effects are summarized below, but detailed accounts relating to humid tropical fluvisols and gleysols can be found in Ponnamperuma (1972, 1977a, b), Sanchez (1976) and Young (1976).

When a soil is flooded, water enters the soil and the air in the pores is initially compressed; this leads to aggregate breakdown. For the first ten to twenty days that a soil is flooded the amount of water percolating through the soil increases owing to the gradual loss of trapped air and the production of CO_2. After this period the soil pores become clogged by microbial waste and disaggregated soil particles; this is accompanied by a decrease in percolation (figure 4.16(a)). As flooding proceeds aggregate stability decreases further owing to clay swelling, hydration and the increased solubility of cementing agents. These effects are greatest in pure-layer silicate soils with a high pH or a high sodium content. In particular, smectites swell and this is accompanied by changes in aggregate cohesion (Kita and Kawaguchi, 1960). The complete breakdown of aggregates (puddling) often occurs when flooded soils are cultivated.

In many parts of Asia puddling is an integral part of swamp rice cultivation and is brought about by a series of cultivation measures

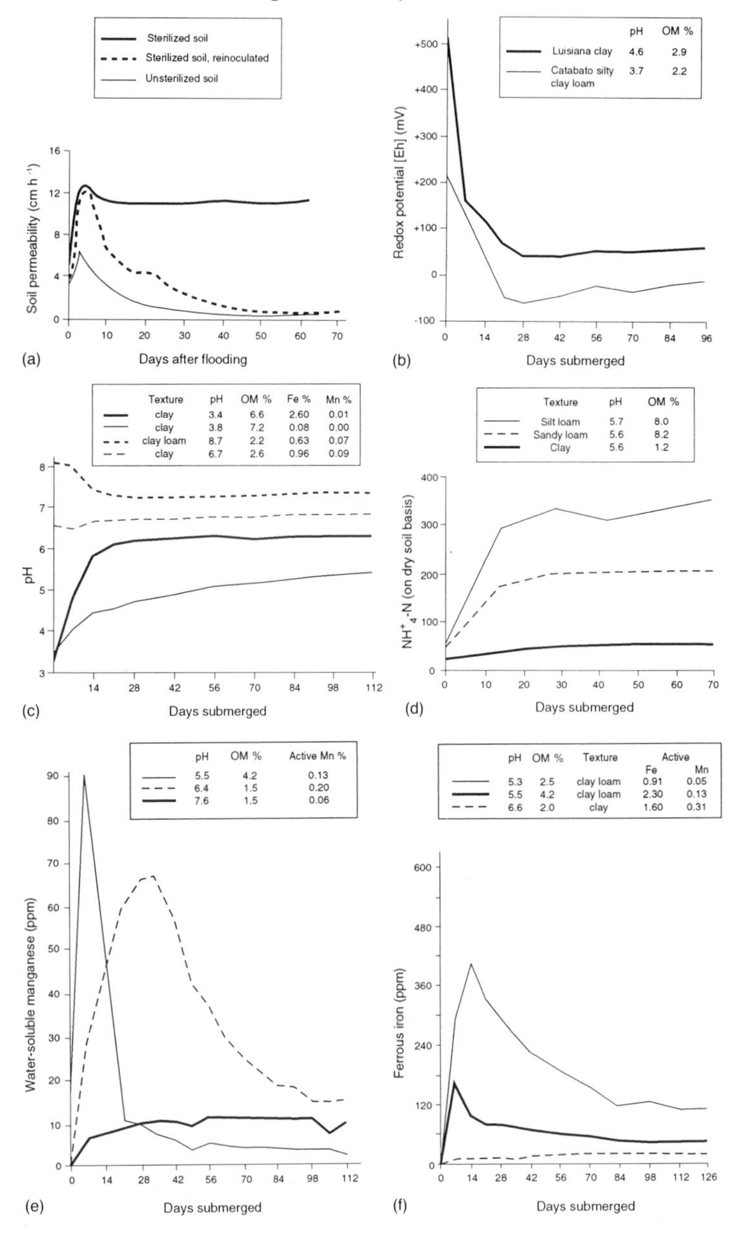

Figure 4.16
Effects of flooding on selected chemical and physical properties of hydromorphic soils: (a) dynamics of soil permeability under flooded, unpuddled conditions (after Allison, 1947); (b) kinetics of solution redox potential (Eh) in submerged soils (after Ponnamperuma, 1977a, p. 12, figure 5); (c) kinetics of the solution pH in submerged soils (after Ponnamperuma, 1977a, p. 4, figure 1); (d) kinetics of ammonification in submerged soils at 25–30 °C (after Ponnamperuma, 1977a, p. 22, figure 9); (e) kinetics of water-soluble Mn^{2+} *in submerged soils (after Ponnamperuma, 1977b, p. 9, figure 5); (f) kinetics of* Fe^{2+} *in the solutions of submerged soils (after Ponnamperuma, 1977b, p. 5 figure 1).*

which change the intra- and inter-aggregate cohesion properties (Koenigs, 1963; Sanchez, 1976). Puddling destroys both the aggregates and the non-capillary pore spaces but increases the number of capillary pore spaces. This leads to increased soil moisture retention, decreased water loss through evaporation and percolation, and soil reduction without flooding due to the absence of soil air (Sanchez, 1976).

The physical changes caused by flooding are accompanied by chemical changes. Within twenty-four hours of flooding the oxygen supply to a soil is effectively cut off because oxygen diffusion is 10,000 times slower through water than through air. This leads to changes in the micro-organism balance, with aerobic micro-organisms (aerobes) either dying or becoming dormant and anaerobes increasing. The anaerobes take over organic matter decomposition from the aerobes and use the oxidized soil components as electron acceptors. The products of decomposition are therefore reduced in a specific thermodynamic sequence (Ponnamperuma, 1965, 1972): first nitrates, followed by manganic compounds, ferric compounds, and finally sulphates and sulphites (table 4.7).

Table 4.7 The thermodynamic sequence and the main reduction reactions which occur when soils are flooded

Stage	*Redox potential Eh_7 (mV)*	*Reduction reactions*
0	+800	$O_2 + 4H^+ + 4e^- \rightarrow 2H_2O$
1	+430	$2NO_3^- + 12H^+ + 10e^- \rightarrow N_2 + 6H_2O$
2	+410	$MnO_2 + 4H^+ + 2e^- \rightarrow Mn^{2+} + 2H_2O$
3	+130	$Fe(OH)_3 + e^- \rightarrow Fe(OH)_2 + OH^-$
4	−180	Organic acids (latic, pyruvic) $\rightarrow 2H^+ + 2e^- +$
5	−200	alcohols
6	−490	$SO_4^{2-} + H_2O + 2e^- \rightarrow SO_3^{2-} + 2OH^-$
		$SO_3^{2-} + 3H_2O + 6e^- \rightarrow S^{2-} + 6OH^-$

Source: adapted from Ponnamperuma, 1965, 1972

An important parameter as soils change from an oxidized to a reduced state is the oxidation–reduction (redox) potential. It is measured in microvolts (corrected to pH 7) and changes from positive to negative as reduction occurs (table 4.6). It can be seen from the table that as each of the stages in the sequence of reduction reactions is attained the redox potential decreases. Consequently the redox potential decreases as the length of time after flooding increases, although the rate of decline varies with the amount of organic matter present (figure 4.16(b)) and temperature. Very low redox potentials, less than −300 mV, can be detrimental to rice growth because sulphites are produced under these conditions (Patrick and Mahapatra, 1968). More typically, though, the high amounts of iron in humid tropical soils give rise to iron reduction processes which enable the redox potential to reach equilibrium at approximately +100 mV.

The soil reaction decreases to between 6.5 and 7.2 within a month of flooding (Ponnamperuma, 1972) (figure 4.16(c)). In soils with a low pH the change to a more neutral pH is caused by the release of OH^- ions as iron hydroxide compounds are reduced. Soils with high pH also approach neutral because of a reduction in the partial pressure of CO_2 and the consequent release of H^+ ions. The only exceptions to these changes in soil reaction are when soils contain few elements that can be reduced (e.g. figure 4.16(d)). Other changes which occur during flooding are as follows.

1 An almost complete denitrification of nitrate occurs within a month of flooding, and nitrite accumulation is absent.
2 Exchangeable and soluble ammonium accumulates, because of its stability under reduced conditions, although this is dependent on the organic matter content and the displacement by exchangeable ferrous iron may promote ammonium leaching.
3 Organic carbon mineralization occurs more slowly as a result of the reduced anaerobe efficiency, and organic nitrogen mineralization occurs more rapidly because more decomposition takes place at higher C:N ratios than would under aerobic conditions (Patrick and Mahapatra, 1968).
4 Large, and sometimes toxic, amounts of CO_2 are released in the first few weeks after submergence.
5 A rapid increase in manganese levels is followed by a gradual decline (figure 4.16(e)). The initial rise is caused by the increased solubility due to the reduction in Mn^{2+} ions, and the later decline is attributable to $MnCO_3$ precipitation.
6 A rapid increase in ferrous iron levels is followed by their gradual decline (figure 4.16(f)). The timing and intensity of this peak is controlled by a number of factors, particularly the pre-flooding pH and the organic matter content. Iron reduction is very important as it raises pH values, increases phosphorus availability and displaces cations. Occasionally iron can reach toxic levels in flooded soils.
7 The amount of phosphorus in solution increases to more than 0.1–0.2 ppm during flooding, and then gradually decreases. Although the reason for this is not clearly known it appears to involve iron and aluminium behaviour, the forms of phosphate present, the original pH, the organic matter content and the soil texture (Patrick and Mahapatra, 1968; Ponnamperuma, 1972; Hossner et al., 1973; Turner and Gilliam, 1976). Phosphorus fixation is also affected by flooding (Patrick and Khalid, 1974).

8 Calcium, magnesium and potassium ions are often displaced from exchange sites into the soil solution by ammonium, iron and manganese.
9 Boron, copper and molybdenum availability is increased (Patrick and Mikkelsen, 1971; Ponnamperuma, 1972) and zinc availability is reduced (Sanchez, 1976).
10 The end products of organic matter decomposition are CO_2, ammonium, methane, amines, mercaptans, hydrogen sulphate, and partially humified residues and resistant humified materials.

Sanchez (1972, 1976) shows that not all of the soil profile is reduced when a soil is flooded (figure 4.17). A surface layer continues to be oxidized owing to the presence of dissolved oxygen in the water, and there is an oxidized rhizosphere due to oxidized compounds being exuded by roots. In addition the subsoil may remain oxidized. In the oxidized zones the chemical and physical changes discussed above do not occur.

Some of the physicochemical effects of flooding can be reversed when the soil dries out. For example, aggregate stability increases because of iron and manganese oxide re-precipitation around clay

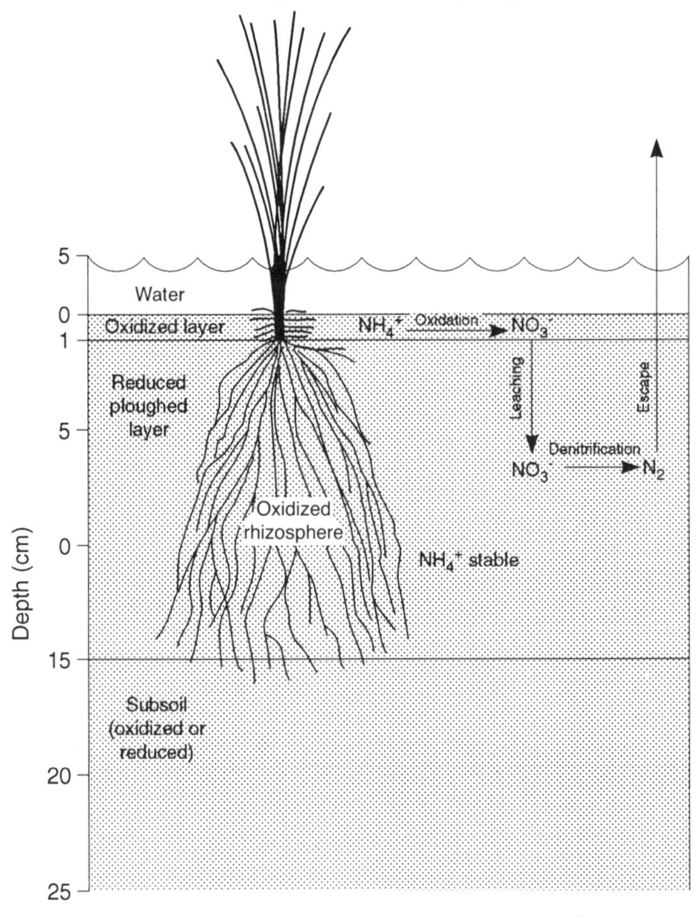

Figure 4.17
Oxidized and reduced zones around a plant in a flooded soil (after Sanchez, 1976, p. 424, figure 11.6).

particles and even puddling is reversible. All of the chemical reactions are reversible although the rate of oxidization is slower in soils that were puddled than in those that were not.

4.7 Arenosols and Regosols

The formation and distribution of arenosols is strongly influenced by the nature of the parent material and topography. Arenosols, as the Latin derivation of the name suggests, are restricted to sandy substrates. Regosols, a separate soil group in the FAO/UNESCO Legend, are also restricted to sandy substrates in the humid tropics and it is convenient to deal with both soil types together. The distribution of arenosols and regosols is azonal and closely related to three types of parent material (figure 4.1):

1 coarse-grained sandstones, e.g. the Kalahari Sands of southern and central Africa, and the Batekes Sands in the Congo and Gabon (Schwartz et al., 1986);
2 unconsolidated sands on old beaches and beach bar sequences, e.g. the Quaternary beach bar sequences of Sherbro Island in Sierra Leone (Anthony and Marius, 1984);
3 beach environments around large lakes, e.g. Lake Malawi (Young, 1976).

4.7.1 Arenosols

A major diagnostic feature of arenosols is their coarse texture, exemplified by very high sand and very low clay contents throughout the profile (plate 4.5). In addition they have significantly less organic material than adjacent soils with higher clay contents. Consequently, arenosols are characterized by a lack of structure, extreme friability (plate 4.5) and a fragmentary consistency. Four types of arenosols can be identified: albic, cambic, ferralic and luvic.

Cambic arenosols are characterized by a cambic B horizon which can only be identified by a slight colour change which is not accompanied by an accumulation of either clays or sesquioxides. Luvic arenosols show a slight clay accumulation in the subsoils indicating that even in clay-poor soils such as these clay translocation is still an active process. Ferralic arenosols are commonly found in areas where the mean annual rainfall exceeds 1000 mm (Young, 1976) and are characterized by a base saturation of <40 me $(100\,g)^{-1}$ in the B horizon.

Albic arenosols have previously been classified as lowland tropical podzols and some researchers still refer to them as podzols. It can be argued that all lowland tropical podzols in the humid tropics are in fact albic arenosols, although the status of those in mountains and on high plateaux is still problematic. Recent

Plate 4.5
An arenosol from Sherbro Island, Sierra Leone. Typical characteristics of arenosols (and regosols) are the lack of structure and extreme friability (here indicated by the core accumulation of loose sand at the base of the profile).

research into such soils has been undertaken in the Congo (Schwartz et al., 1986; Schwartz, 1988) and central Amazonia (Bravard and Righi, 1988, 1989; Righi et al., 1990). The soils studied in the Congo may well be podzols, but those in Brazil are, mostly probably, albic arenosols.

The accumulation of raw humus initially aids the removal of weathered iron oxide, causing rubification, and subsequently intensifies leaching (Andriesse, 1969–70; Young, 1976). An acidophyllous vegetation often develops in response to the strong base leaching in the topsoil. The role of organic acids in the development of these soils has been explored by Bravard and Righi (1989) and Schwartz et al. (1986). They have shown that the complexing of organic acids with metals, particularly the aluminium–fluvic acid complex, is critical to their development. Furthermore, organic material in the B horizon coats the sand grains and fills the pores. Schwartz et al. (1986) were able to subdivide the B horizon on the basis of the morphology of the organo-aluminium gels. In the upper B horizon (B22h) (figure 4.18) the gels form spheres which bind together and in the lower B23h horizon the gels adhere to the clays to give high porosity. The resulting profiles have an acidic, raw humic surface horizon with a high C : N ratio. Below the humic horizon is a humus-stained sandy horizon and underneath this an eluvial horizon (A2) which also has a sandy texture and varies from 20 to 200 cm in depth. The illuvial B horizons vary from thin, irregular and patchy to thick and

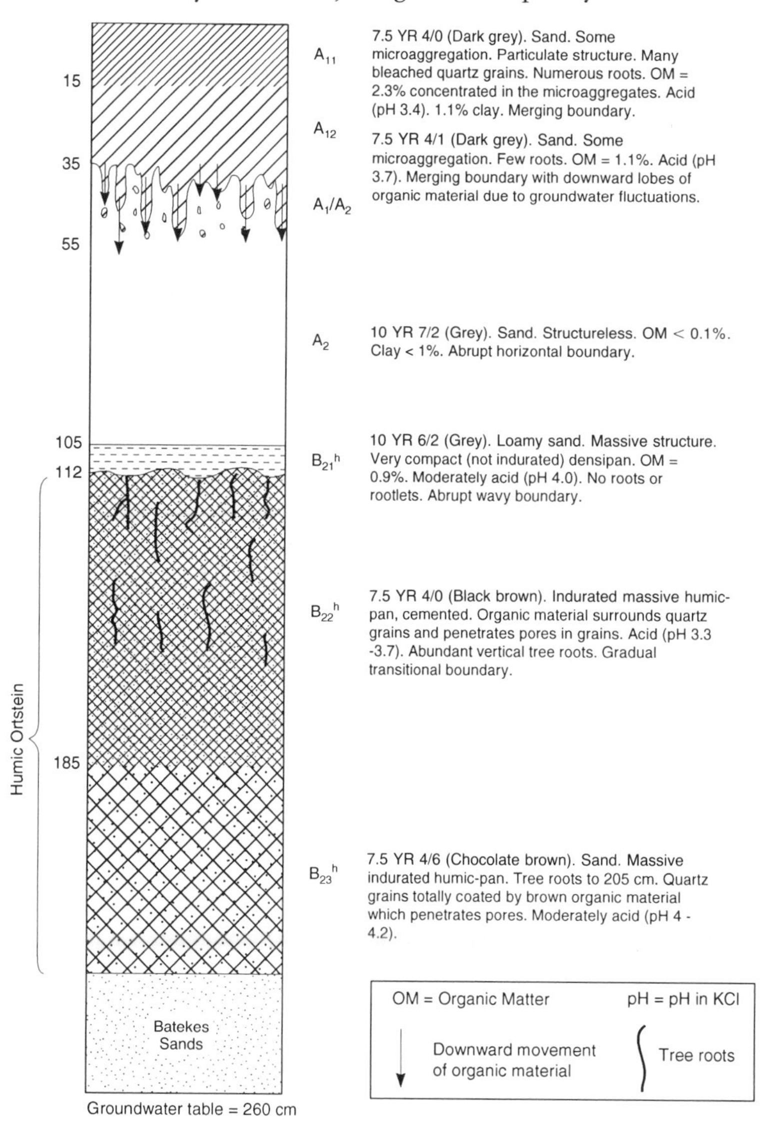

Figure 4.18
Hydromorphic podzol (lousseke) profile from Gangalingolo, Congo (after Schwartz et al., 1986).

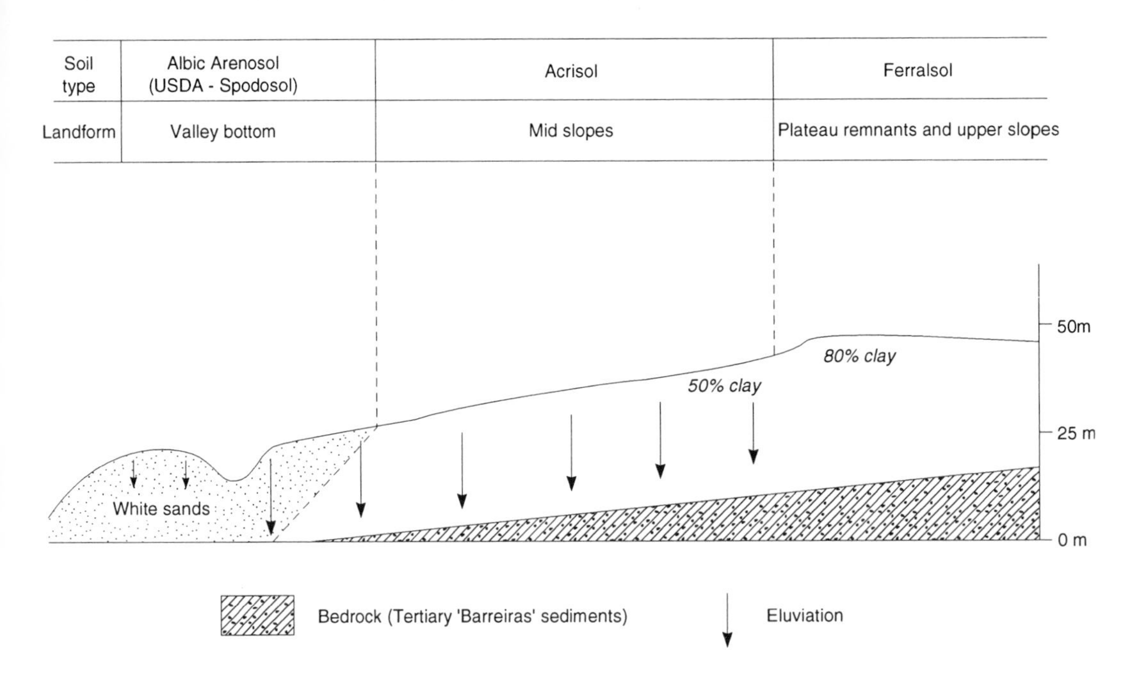

Figure 4.19
Toposequence of albic arenosols, acrisols and ferralsols in central Amazonia (Brazil) (after Bravard and Righi, 1989, p. 280, figure 1).

indurated. Clay and sesquioxides accumulate in these horizons and they often exhibit induration which is thought to result from silica recementation.

In central Amazonia the onset of podzolization only appears to occur when a soil reaches a critical state of impoverishment of all materials except quartz. In the soils studied by Bravard and Righi (1989) relative silica enrichment was caused by the rapid hydrolysis of kaolinite caused by lateral drainage (figure 4.19), although this could as easily be caused by the down-washing of fines which has been noted in many humid tropical soils. The situation is more complicated in the Congo where podzols, known locally as *lousseke*, occur on the Batekes Sand of the undulating, dissected Batekes Plateau. The wet or hydromorphic *lousseke* develops above a perched water-table, whereas the dry *lousseke* has a much deeper water-table. An example of a wet *lousseke* profile is shown in figure 4.18; dry *lousseke* profiles have higher A2 horizons and quasi-horizontal organic lamellae in the A2 horizon, although in other respects they are similar. Schwartz (1988) suggests that these podzols have had a complex history with four periods of pedogenesis (figure 4.20).

On many cays and coastal plateaux in the Seychelles Archipelago, Piggott (1968) noted calcareous sandy soils which he classified as rendzinas. Similar soils in the Solomon Islands have been classified as tropopsamments in the USDA Classification, i.e. arenosols or

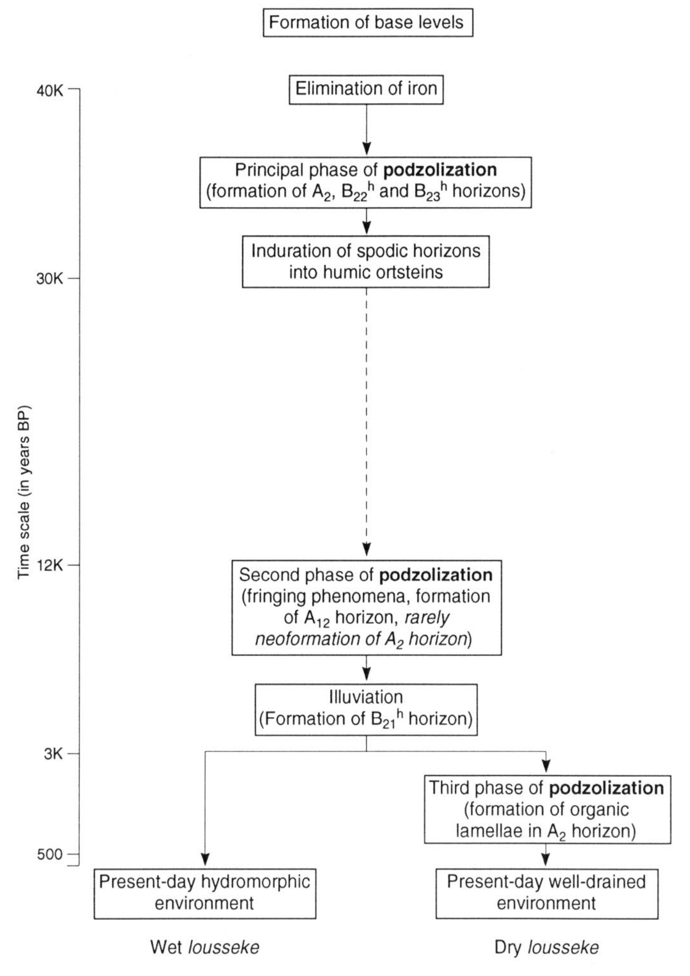

Figure 4.20
Tentative reconstruction of the phases in the development of podzols (loussekes) on the Batekes Sands (adapted from Schwartz, 1988, p. 242, figure 2).

regosols (Wall et al., 1979) and it is appropriate to discuss these soils in this section. The sand-size fraction in these soils is composed of fragments of algae, coral and foraminifera which have been washed onshore by storms. Subsequent calcium carbonate leaching leaves a coarse residue and within some 100 m of the shore the carbonates come into contact with saline groundwater and precipitate out to form beach sandstone.

4.7.2 Regosols

Unlike arenosols, regosols exhibit no B horizon development and typically are restricted to a weak Ah horizon over a sand-rich C horizon. The organic matter contents are very low and even in the Ah horizon do not exceed 1 per cent. A relationship between regosols and arenosols can be inferred (figure 4.21). Newly deposited sand will initially develop a weak organic horizon (Ah) over poorly weathered sands (C) and at this stage a regosol will have developed. Further pedogenesis will lead to a more mature A

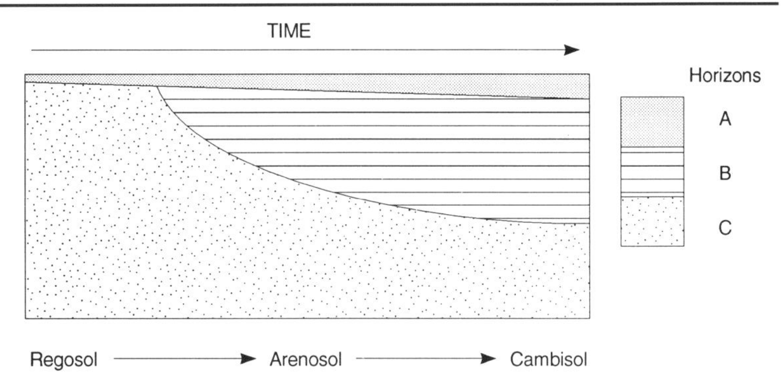

Figure 4.21
Development of arenosols and regosols.

horizon and a weakly developed B horizon; at this point the soil can be classified as an arenosol.

4.8 Andosols

Andosols develop over volcanic ashes, particularly those which contain vitreous material. Their distribution is restricted to areas with active or recent volcanism, e.g. the East Indies, some Pacific Islands, the northern Andes, Central America, and the East and West African highlands.

Volcanic ashes undergo a specific multi-stage weathering sequence in the humid tropics (Sanchez, 1976; Wada, 1985; Delvaux et al., 1989) in which they initially weather to form allophane. Allophane is an amorphous mixture of aluminium and silicates that complexes with organic material. Soils which exhibit this stage in the weathering sequence are called andosols. This part of the sequence has been dated in Papua New Guinea (Haantjens, 1967; Bleeker and Parfitt, 1974; Bleeker, 1983). The high amounts of active aluminium are diagnostic features of andosols (Wada, 1985) but Mizota and Capelle (1988) have shown that the high levels of humus in the A horizons form an alumino-humus complex which retards the formation of allophane and imgobolite by microbial mineralization. Very young andosols that have been studied in Papua New Guinea (Bleeker, 1983), the Solomon Islands (Wall et al., 1979), Costa Rica (Martini and Luzuriaga, 1989) and Cameroon (Delvaux et al., 1989) exhibit very coarse textures, are free-draining, and show little leaching or subsoil hardpan development. During the initial stages of weathering silica is slightly soluble and is leached; but as the soil environment acidifies, owing to calcium and magnesium leaching, silica precipitates in the subsoil and cements coarse particles together (Zijsvelt and Torlach, 1975).

The next stage in the weathering sequence is dependent upon the drainage conditions. If, on the one hand, drainage is free allophane will weather to hallyosite, kaolinite and hydrated and dehydrated phyllosilicates, but on the other hand if it is poor montmorillonite will form (Mohr et al., 1972). Weathering to smectite has also been noted in well-drained East African andosols (Van der Gaast et al., 1986; Mizota and van Reeuwijk, 1987). Drainage rates in these soils can be very high and rates of up to 214 cm day^{-1} have been recorded in the Philippines (Sanchez, 1972). In addition to the clay minerals mentioned above, gibbsite is an important component of andosols in Papua New Guinea and the Solomon Islands and in the latter area gibbsite and allophane also weather to imgobolite (Wall et al., 1979). Mineralogical and other variations in andosols have been attributed to variations in the mineralogy between different ashfalls.

As weathering progresses andosols generally grade into acrisols, lixisols and cambisols. Delvaux et al. (1989) have studied such as sequence in Cameroon and the gradations they found between soils were marked by a series of changes in the A and B horizons (table 4.8). In addition to these changes they found that the amount of exchangeable magnesium was a good indicator of the position of the soil in the weathering sequence and that the evolution of soils had a significant influence on banana yields (figure 4.22).

Andosol formation is not only dependent on the time elapsed since an eruption and parent material. Dudal and Soepraptohardjo (1960), Sanchez (1976) and Tamura et al. (1953) argue that there is also a strong topographic control on their formation (figure 4.23). This is particularly evident on older landscapes where extended weathering has taken place. In Indonesia Dudal and Soepraptohardjo (1960) found an altitudinal zonation in which andosols were only found above 700 m despite the similarity of parent materials at lower altitudes. Similar chronosequences and toposequences have been described from Hawaii (Tamura et al., 1953), and Costa Rican andosols generally occur above 600 m and are best developed between 1000 and 1600 m (Martini and Luzuriaga, 1989). The fact that andosols are restricted to high altitudes on old, ash-rich parent materials is due to the low rates of allophane breakdown because of the low temperatures and high rates of organic matter accumulation that these conditions encourage. The added complexity of a wet windward and dry leeward side to many Pacific Islands reinforces the observations of soil-forming processes and soil distribution on topocatenas developed on volcanic ash substrates (Sanchez, 1976).

Andosols are highly porous with a low bulk density, and tend to leach rapidly in the high rainfall environments in which they are found – hence the rapid rates of weathering and pedogenesis. Very high base losses have been recorded in Colombian andosols – 202–249 kg N ha^{-1}, 0.13–0.23 kg P ha^{-1}, 163–202 kg K ha^{-1}, 776–878 kg Ca ha^{-1} and 232–251 kg M ha^{-1} (Suarez de Castro and Rodriguez, 1958) – and subsoil base enrichment has been reported in Costa Rican andosols (Morelli et al., 1971).

Andosols are generally dark black (their name is derived from the Japanese *an-do* meaning a dark soil) with moderately well-

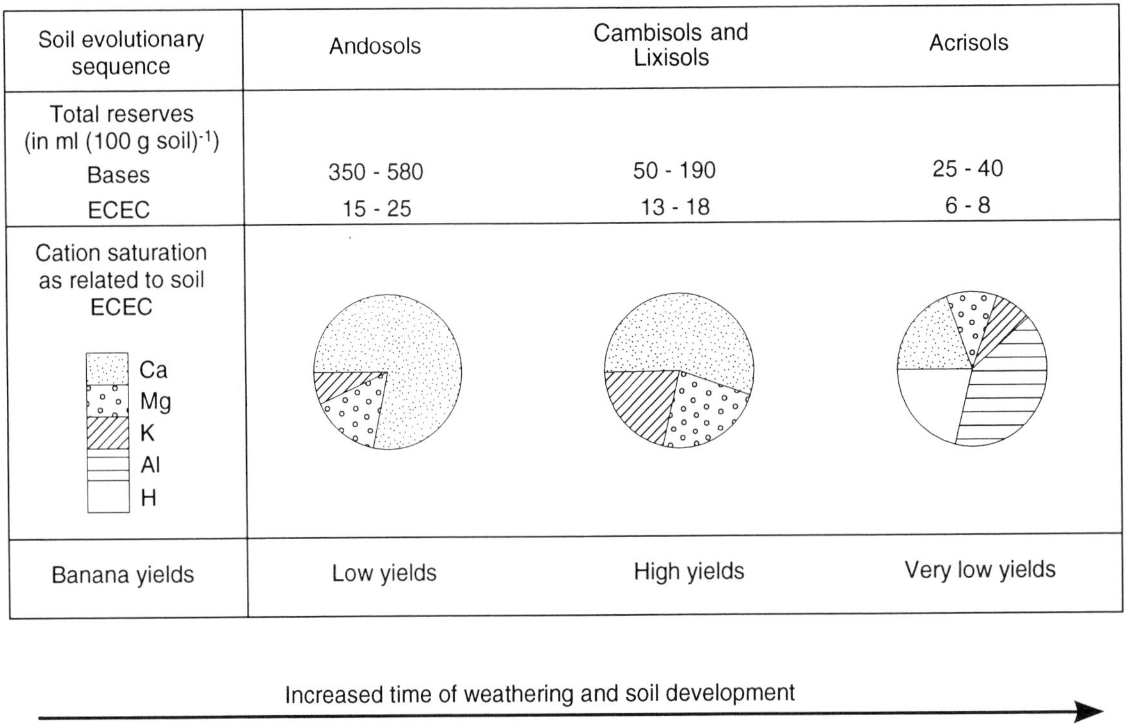

Soil evolutionary sequence	Andosols	Cambisols and Lixisols	Acrisols
Total reserves (in ml (100 g soil)$^{-1}$)			
Bases	350 - 580	50 - 190	25 - 40
ECEC	15 - 25	13 - 18	6 - 8
Cation saturation as related to soil ECEC Ca Mg K Al H			
Banana yields	Low yields	High yields	Very low yields

Increased time of weathering and soil development

Figure 4.22
Relationships between soil type, cation chemistry and banana yields in a chronosequence from southwest Cameroon.

Table 4.8 Principal changes in soil profile characteristics that occur when andosols evolve into acrisols, lixisols and cambisols

Horizon	*Principal changes*
A	Decrease in depth Decrease in carbon content Decrease in cation exchange capacity (CEC) Decrease in phosphorus retention Decrease in oxalate-extractable aluminium, iron and silicon
B	Development of an argillic horizon as evidenced by an increase in bulk density an increase in clay content the development of clay cutans
General	Increase in depth of solum Increase in colour contrast between A and B horizons Decrease in exchangeable calcium and magnesium

developed horizons. A typical profile would include the following characteristics. The A horizon is dark greyish-brown to black with a depth of less than 50 cm. It is usually dominated by silt and fine-sand loamy textures, has an open granular to fine crumb structure, a high porosity, is friable with a greasy consistency and has a very low bulk density (Allbrook and Ratcliffe, 1988). Organic matter is well distributed and ranges from 5 to 20 per cent. The pH ranges from 5 to 7 and the CEC varies between 20 and 100 me $(100\,g)^{-1}$ which reflects the medium to high base saturation. The B horizon is

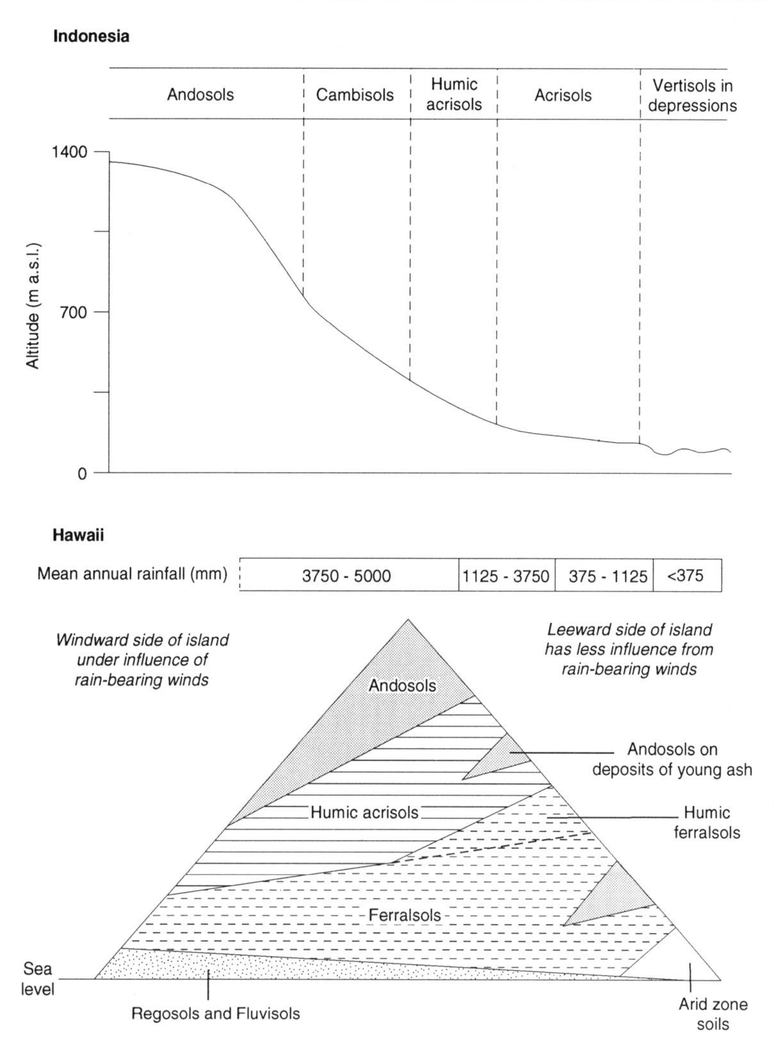

Figure 4.23
Typical toposequences involving andosols from the humid tropics (adapted from Sanchez, 1976).

yellowish to reddish-brown and varies in thickness. It is often very porous but exhibits little clay translocation. These two horizons overlie weathered pumice and ash in which silica hardpans often develop.

Andosols are agriculturally important soils, and those developed over basic ashes have a higher potential than those developed over felsic material or those with silcrete horizons. Basic-ash andosols exhibit several factors which lead to their high agricultural potential. Important physical factors include a high available water capacity, unimpeded free drainage, a fine texture, low bulk density and great aggregate stability due to the strong bonding between allophane and organic matter complexes which reduces soil erodability. Pertinent chemical properties include a high natural fertility marked by a high CEC and a low base saturation deficit, high levels of nutrient retention, soil nutrient renewal by continuous weathering of ash, and high rates of organic matter accumulation in topsoil due to the cool temperatures at high altitudes. However, they can be deficient in nitrogen and phosphorus (Martini and Luzuriaga, 1989).

Andosol soils are also notable for their remarkable geotechnical strength (section 4.6.2). Despite their high percentage clay fraction, high field moisture content and low bulk density they frequently remain stable on slopes of 40° and above in all but the most exceptional conditions. Their strength is attributed largely to the presence of allophane as a highly hydrated gel. The disorganized structure of the allophane mineral also prevents particle orientation along zones of failure (Reading, 1991).

4.9 Leptosols

The strict definition of a leptosol in the FAO classification is a soil of less than 10 cm in depth developed over continuous hard rock, although Young (1976) argues that 25 cm depth is more acceptable when considering the agricultural use of such soils. He further argues that, when discussing leptosols, deeper soils that are dominated by rocks and stones, should be included as well. Consequently he defines three types of soil: leptosols *sensu stricto*, lithic phases of other soil types, and deep stony profiles. In the humid tropics these soils are commonly found in mountainous regions, on inselbergs and over massive ferricrete in lowland areas (table 4.9).

4.10 Histosols

Organic soils are less common in the humid tropics than at high latitudes because the faster rates of organic matter breakdown preclude its accumulation in many circumstances. Andriesse (1988) has recently shown that a little under 10 per cent of the world's organic soils (35.8 million ha) are found in the tropics (table 4.10), although the values for Africa and Latin America are almost certainly underestimated (Beadle, 1960; Suszcynski, 1984). They are most extensive in southeast Asia and Papua New Guinea, where

Table 4.9 Limitations in depth-limited and stony soils

1	Leptosol	Consolidated rock or massive ferricrete at <25 cm
2	Lithic phase	Consolidated rock at 25–50 cm
3	Petric phase	Massive laterite commencing at 25–50 cm
4	Shallow phase	Consolidated rock or massive ferricrete at 50–100 cm
5	Stony	Consolidated rock at depths >50 cm and the upper 50 cm dominated by stones, boulders and gravel
6	Petric phase	Consolidated rock at depths >50 cm and the upper 50 cm dominated by hard ferricrete concretions

Source: after Young, 1976

Table 4.10 Distribution of tropical organic soils

Region	Estimated extents		
	10^6 ha	Proportion of world (%)	Proportion of tropics (%)
Southeast Asia (including Papua)	20.26	4.65	56.6
Caribbean	5.67	1.30	15.8
Amazonia	1.50	0.34	4.19
Africa	4.86	1.11	15.8
South China	1.40	0.32	3.9
Other tropical areas	2.11	0.49	5.9
All tropical and subtropical regions	35.80	8.21	100.0

some 57 per cent of tropical organic soils are found, and are particularly extensive in Indonesia (17 million ha) and Malaysia (2.5 million ha). In the Americas they are found in the Amazon Basin and the basins fringing the Caribbean in Florida, Mexico and Venezuela. They are also common in central Africa and, possibly, on the lowlands bordering the Gulf of Guinea (Andriesse, 1988).

Their occurrence can usually be related to one of eight topographic situations:

1 deltas, mainly in saucer-shaped depressions which are surrounded by levées (figure 4.24(a));
2 coastal basins that have been cut off from the sea and rivers by dunes and levées respectively (figure 4.24(b));
3 coastal lagoons (figure 4.24(d));
4 abandoned meander bends (oxbow lakes) (figure 4.24(c));
5 small isolated peats in bottomlands which are often related to palaeochannels in braided river systems (figure 4.24(c));
6 blocked valleys ranging from large trunk valleys to small tributaries (they are usually blocked by fluvial or colluvial deposits (figure 4.24(c)) but in East Africa peats behind valleys blocked by lava flows have been recorded (Floor and Muyesu, 1986));
7 on coral atolls (figure 4.24(e));
8 high mountain bogs.

All of these peats, except the high mountain peats, are swamp peats which form in aerobic conditions and follow Moore and Bellamy's (1974) model of formation. Most tropical peats, like those at higher latitudes, are less than 10,000 years old and their formation is related to Holocene sea level changes. Significant salt intrusion occurs in coastal peats (Hammond, 1971) and generally tropical peat formation is more sensitive to fluctuations in river and sediment discharge caused by deforestation.

Tropical peats have a number of chemical and physical properties which cause problems to cultivators; in particular the

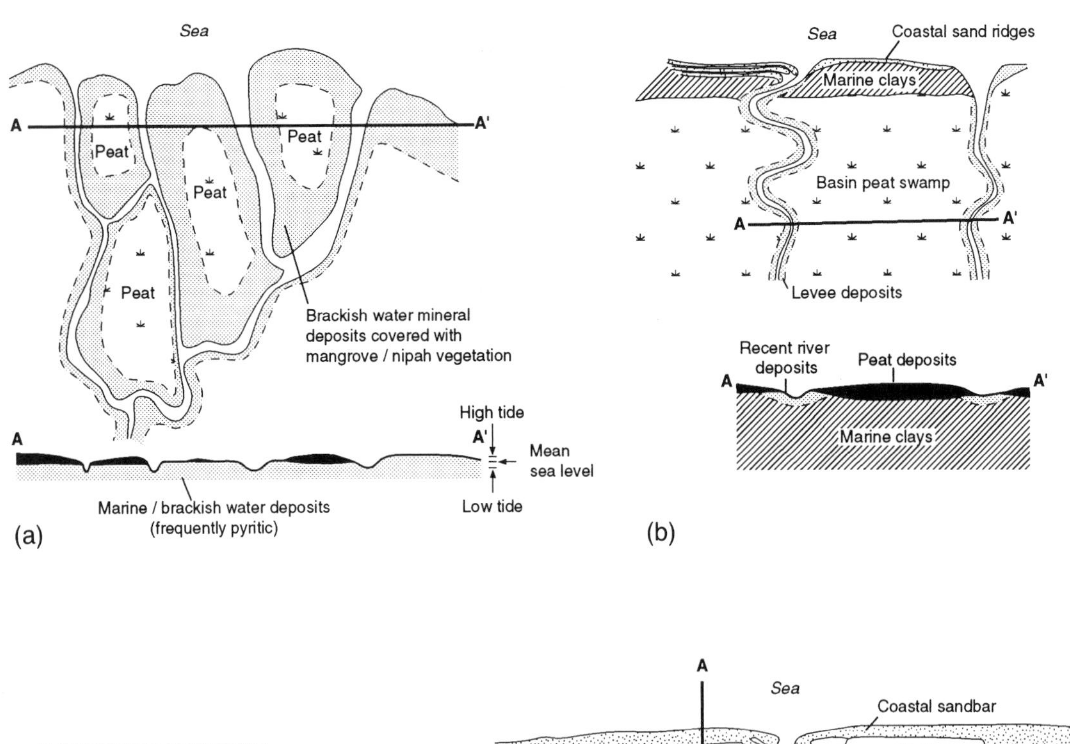

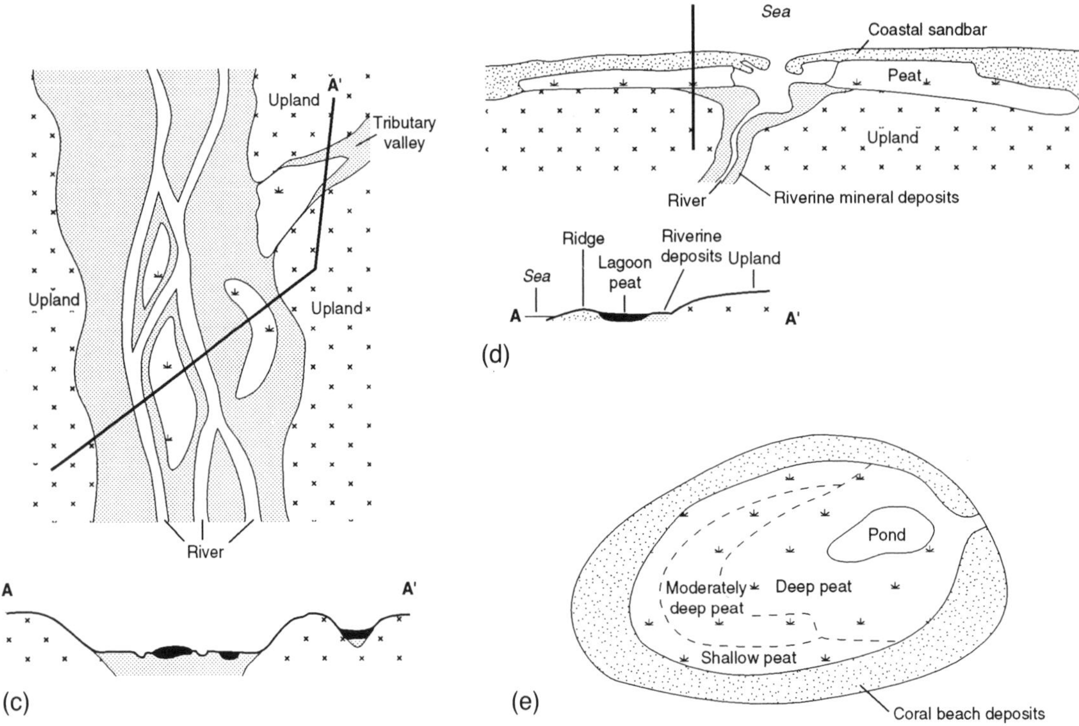

Figure 4.24
Typical situations in which peat forms in the humid tropics: (a) deltaic dome-shaped peat swamps; (b) coastal dome-shaped basin peat swamps; (c) peat swamps in various situations along river valleys; (d) lagoonal peat swamps; (e) peat swamps on coral atolls (modified after Audriesse, 1988).

soil–water relations are very complex. They behave in a similar manner to light-textured soils, i.e. (i) their water retention capacity is low, although it increases with increased decomposition; (ii) their water-holding capacity is variable, though usually high; (iii) they have a high porosity (usually more than 85 per cent); (iv) their hydraulic conductivity is variable; (v) they often need surface irrigation in the dry season to stop irreversible drying; (vi) irreversible drying occurs as a result of their hydrophobic nature which may be due to resinous, iron, carboxyl or phenolic hydroxyl coatings, their high lignin contents or the build up of air films. They are prone to shrinkage on desiccation, and this is most marked where organic matter decomposition is most advanced. Their pH ranges from approximately 3 to 8, and their CEC is strongly pH dependent. Carbon contents range from about 48 to 60 per cent and are greatest in the most decomposed peats, nitrogen contents range from 1 to 3 per cent, free lime is negligible, and the sulphur content may be high which leads to pyrite formation on drying (e.g. in the Orinoco Delta and the papyrus swamps of East Africa).

The chemical and physical limitations, along with the problems of access, have led many agriculturalists to dismiss peats (e.g. Pons and Driessen, 1975), suggesting that either they be removed completely (e.g. Coulter, 1957) or they be left as natural ecosystems. Andriesse (1974, 1988) argued that shallow peats can be cultivated but that deep peats should either be left or be used for forestry. When reclaimed there are potential difficulties related to initial swamp drainage, e.g. access, tree removal, the low load-bearing capacities of peats and the erosion of drainage ditches. Problems of a long-term nature include subsidence, drying out of peats by plant water extraction at depth, lack of support for trees, tree root exposure on shrinkage, oxidation, use of machines and changes in the nature of peat due to fertilizer use. Some crops can be cultivated on peats without significant changes in drainage (e.g. sago, raffia, paddy rice, wetland taro, water spinach, water celery and water chestnuts) and with moderate amounts of drainage it is possible to cultivate vegetables, perennial tree crops such as coconut palms and rubber, and fruit trees table 4.11).

Fosberg (1954) and Piggott (1968) describe island soils in which organic matter accumulation is dominant, but by no stretch of the imagination are they peats. On many oceanic islands organic matter accumulation consists mainly of dead sea birds, feathers, bird droppings and fish waste. This forms a phosphate-rich humus known as *guano* which has been commercially exploited for phosphate, particularly on Nauru, Christmas Island (Indian Ocean), Tuvalu and Kiribati. The resulting soils form by organic accumulation on phosphatic or calcareous sandstones and do not fit easily into any classification. They are inherently fertile but more often than not have been exploited as guano to enrich poorer soils elsewhere!

Table 4.11 Yield of some promising crops on peat in Sarawak and West Selangor, Malaysia

	Yield (t ha^{-1})	
Crop	*Sarawak*	*West Selangor*
Pineapple (*Ananas comosus*)	40.0 (f)	40.0 (f)
Cassava (*Manihot esculenta*)	50.0 (f)	49.0 (t)
Tobacco (*Nicotiana tabacum*)	0.7 (dl)	1.0 (dl)
Groudnut (*Arachis hypogaea*)	1.0 (s)	
Soya Bean (*Glicine max*)	1.5 (s)	
Cowpea (*Vigna unguiculata*)	2.1 (s)	
Bambara groundnut (*Vigna subterranea*)	1.5 (s)	
Sorghum (*Sorghum bicolor*)	1.5 (s)	2.5 (s)
Sweet potato (*Ipomoea batatas*)	14.0 (t)	24.0 (t)
Castor oil (*Ricinus communis*)	2.5 (s)	
Ginger (*Zingiber officinale*)	15.0 (r)	15.0 (r)
Okra (*Hibiscus esculentus*)	6.0 (f)	15.0 (f)
Oil palm (*Elaeis guineensis*)	19.0 (n)	
Sago (*Metroxylon sagus*)	6.0 (ds)	
Coffee (*Coffea liberica*)	1.7 (b)	
Annatto (*Bixa orellana*)	2.0 (s)	
Mulberry (*Morus alba*)	10.3 (fl)	7.5 (dl)

f, Fresh fruit; n, fresh fruit bunch; dl, dry leaf; fl, fresh leaf; s, dry seed; t, fresh tuber; r, fresh rhizome; ds, dry starch; b, fresh berries.

Source: after Tie and Kueh, 1979

FIVE

Humid Tropical Vegetation

5.1 Introduction

To the layman, the commonly held impression of the humid tropics appears to be that of a broad swath of dense, verdant luxuriant forests teeming with exotic plants, animals and indigenous peoples. A further impression is that of rampant destruction which threatens all before it. The logbooks of early explorers in the low latitudes bear testament to the rich forests in the past, the popular literature and film of today, the devastation. Because tropical deforestation is a global issue, the impressions of the past, present and future are reinforced by the media. In reality, the biogeography and ecology of the humid tropics is far more complicated than the impressions may suggest. In particular two points must be made: first, there are many different types of humid tropical forest, and second, not all of the low latitudes were, or are, covered by forest.

The second point begs a set of questions concerning vegetation destruction, disturbance and regrowth and, over a longer time scale, the evolution of humid tropical vegetation. It is almost certain that some low latitude areas have not been covered by forests during the last 2 million years (e.g. high mountains and some of the drier margins) and that humid tropical forests cover a wider area now than at any time in the last 10,000 years. Nevertheless, many areas that are without forest have supported it some time in the last 10 millennia, and the current vegetation of such areas, savanna woodlands and grasslands, has often evolved at the hand of human beings.

In the first part of this chapter we consider the different types of humid tropical vegetation and their present-day distributions. Selected aspects of the ecology of the main vegetation types (e.g. structure, floral and faunal composition, species richness) will be considered next. In the final third of the chapter forest microclimates, forest disturbance and succession, and the evolution of humid tropical forests will be discussed.

5.2 Humid Tropical Vegetation: Classification and Distribution

5.2.1 Classification

Before vegetation distributions can be analysed the main vegetation classification schemes used in the humid tropics must be discussed. Many such classifications exist; the differences between most lie in the nature of the criteria used. Bearing in mind the ecological emphasis of this chapter, classifications based on floristic criteria are less relevant than those based on vegetation structure and environmental parameters.

Although attempts at national and regional vegetation classifications have been made for the humid tropics (e.g. Pires and Prance's (1984) classification of Amazonian vegetation in Brazil, and Cabrera and Willink's (1973) classification for Latin America), global classification schemes for the humid tropics have proved difficult to establish because forests (the major vegetation type) (i) are complex, (ii) are taxonomically difficult, (iii) require very large sample plot areas and (iv) have limited accessibility; these four factors limit the ability to collect the semi-quantitative data needed to apply syntaxonomic procedures.

Longman and Jeník (1987) argued that the only feasible classification is one based on the physiognomy of vegetation structure in combination with environmental parameters, and they produced a slightly revised version of Ellenberg and Mueller-Dombois's (1967) classification (table 5.1). Whitmore (1990) provided a similar classification (table 5.2) although he restricted himself to tropical moist forests.

The most quantitative classification scheme for vegetation classification based on environmental parameters is the life-zone classification (Holdridge, 1967). It is useful to geographers because it relies on climatic data, and it allows vegetation from around the world to be characterized on the basis of similarities in physiognomy, complexity, species diversity, organization and stratification. Put simply, a Vietnamese seasonal monsoonal rain forest will have a different floristic composition to a seasonal rain forest in Ghana with similar climatic conditions, but they can be recognized as the same type of forest. In Holdridge's scheme each bioclimatic zone has a characteristic climax vegetation with a distinct ecology. The differences between the vegetation in each zone can be described thus. For a particular mean annual biotemperature (BT) ecological variations are a function of precipitation, and for any latitude ecological variations are a function of elevation.

Tropical regions are defined as regions with the BT $> 24\,^{\circ}$C and seven temperature–altitudinal zones are recognized (figure 3.1); humid tropical regions are defined on the basis of mean annual

Table 5.1 Classification of tropical forests

A.	Tropical rainforests (Ombrophilous forests *sensu stricto*)
A1.	Tropical lowland rainforest
A2.	Tropical montane rainforest
A2a.	Broad leaved
A2b.	Needle leaved or microphyllous
A3.	Tropical cloud forest
A3a.	Broad leaved
A3b.	Needle leaved or microphyllous
A4.	Tropical alluvial forest
A4a.	Riparian
A4b.	Occasionally flooded
A4c.	Seasonally waterlogged
A5.	Tropical swamp forest
A5a.	Broad leaved, dicotyledon dominated
A5b.	Palm dominated
A6.	Tropical peat forest
A6a.	Broad leaved, dicotyledon dominated
A6b.	Palm dominated
B.	Tropical or subtropical evergreen seasonal forests
B1.	Tropical or subtropical evergreen seasonal lowland forest
B2.	Tropical or subtropical evergreen seasonal montane forest
B3.	Tropical or subtropical evergreen seasonal dry subalpine forest
C.	Tropical or subtropical semi-deciduous forests
C1.	Tropical or subtropical semi-deciduous low forest
C2.	Tropical or subtropical semi-deciduous mountain or cloud forest
D.	Subtropical rainforests
E.	Mangrove forests

Source: data after Longman and Jeník, 1987

precipitation. The mean annual precipitation needed to define a humid tropical class depends on the temperature–altitude zone and is highest (>2000 mm) for lowlands (0–500 m above sea level (a.s.l.)) and lowest (>125 mm) for the nival zone (>4500 m a.s.l.) (figure 3.1). This classification has the advantage of directly relating vegetation communities to climatological parameters. However, its use is limited, first because variations in vegetation patterns due to edaphic, topographic and hydrological factors are not recognized, and second because it represents vegetation types that would exist if there was no human intervention.

Holdridge defined 125 bioclimates of which thirty-eight are tropical and seventeen are humid tropical (Holdridge, 1967). The humid tropical life zones and their dominant vegetation types are listed in table 5.3. The full range of the humid tropical life zones can be seen in the northern Andes, East Africa and some southeast Asian and Pacific islands; but in the lowland humid tropics there is a restricted range.

Using a similar methodology, Walter (1979) used Venezuela to provide examples of the two main climate–vegetation gradients in the humid tropics (figure 5.1): an altitudinal gradient (which in Venezuela extends from sea level to mountain peaks in excess of 5000 m a.s.l.) and a rainfall gradient (which ranges from wet, equatorial regions with more than 3500 mm of rainfall to deserts with less than 200 mm). The main constraint on plant growth along the altitudinal gradient is temperature and this is recognized in the

Table 5.2 Classification of humid tropical forests

Climate	Soil water	Main soil types	Approximate elevation ranges	Forest formation
Seasonally dry	Strong annual shortage			Monsoon forests (various types)
	Slight annual shortage			Rainforests: semi-evergreen rainforest
Ever-wet	Dryland	Zonal acrisols, ferralsols	Lowlands	Lowland evergreen (perhumid) mainly rainforest
			1200–1500 m	Lower montane rainforest
			1500–3000 m	Upper montane rainforest
			3000 m	Subalpine forest to tree-line
		Acrisols, podzols	Mainly lowlands	Heath forest
		Ferralsols	Mainly lowlands	Forest over limestone
		Nitosols	Mainly lowlands	Forest over ultrabasics
Coastal brackish water	Water-table high (at least periodically)	Acrisols, regosols, salt-water histosols		Beach vegetation Mangrove forest
Inland freshwater	Histosols	Histosols		Peat swamp forest
	More or less permanently wet	Gleysols		Freshwater swamp
	Periodically wet	Fluvisols		Freshwater periodic swamp forest

Source: after Whitmore, 1990, p. 13, table 2.1

Spanish names for the altitudinal belts – *tierra calienta* (hot land), *tierra templada* (temperate land), *tierra fria* (cold land) and *tierra helada* (frozen land). The variations in rainfall are caused by a north–south rainfall gradient, seasonal wind reversals which create seasonal wet–dry climates in northern Venezuela, and orographic and rain-shadow effects.

5.2.2 The importance of global vegetation classification schemes

Central to Holdridge's vegetation classification is the broad relationship between macroclimate and vegetation, although it must be recognized that forests alter the local or microclimate (see section 5.4). Climate–vegetation relationships are important because they allow both latitudinal and altitudinal trends in humid tropical vegetation to be examined simultaneously. Such schemes are also useful in that they can be used in conjunction with maps of

Table 5.3 Humid tropical bioclimatic zones

Bioclimatic zone	*Vegetation type(s)*	*Rainfall (mm)*
I Lowland (L) zones (BT > 24 °C)		
L Moist forest	Forest	1500–4000, less than 4 months with less than 200 mm
L Wet forest	Forest	4000–8000, less than 2 months with less than 200 mm
L Rainforest	Forest	>8000, no months with less than 200 mm
II Premontane (P) zones (BT 18–24 °C)		
P Moist forest	Forest	1000–2000, 2–4 months with less than 100 mm
P Wet forest	Forest	2000–4000, less than 2 months with less than 100 mm
P Rainforest	Forest	>4000, no months with less than 100 mm
III Lower montane (LM) zones (BT 12–18 °C)		
LM Moist forest	Forest	1000–2000, 2–4 months with less than 100 mm
LM Wet forest	Forest	2000–4000, less than 2 months with less than 100 mm
LM Rainforest	Forest	>4000, no months with less than 100 m
IV Montane (M) zones (BT 6–12 °C)		
M Moist forest	Forest	500–1000, 2–4 months with less than 50 mm
M Wet forest	Forest	1000–2000, less than 2 months with less than 50 mm
M Rainforest	Forest	>2000, no months with less than 50 mm
V Subalpine or sub-Andean (S-A) zones (BT 3–6 °C)		
S-A Moist with grassland (Puna)	Grassland	250–500, 2–4 months less than 50 mm
S-A Wet with dwarf shrubland (Paramo)	Dwarf shrubland	500–1000, less than 2 months less than 50 mm
S-A Rain with dwarf shrubland (pluvial Paramo)	Dwarf shrubland	1000–2000, no months less than 50 mm
VI Alpine zones: BT 1.5–3 °C with precipitation ranging from 125 to 1000 mm		
VII Nival zones: BT < 1.5 °C, permanent ice and snow		

Source: after Holdridge, 1967

soils (figure 4.1) and climato-geomorphological zones (cf. chapter 6).

However, all global classification schemes are scale constrained and do not allow important intra-zone variations to be considered. For instance, when considering humid tropical vegetation it is important to recognize the deviations from the zonal forest communities which are caused by local factors that lead to ecologically important, edaphic vegetation communities, e.g. swamp forests and grasslands (see section 5.7) and grasslands on thin and/or degraded soils.

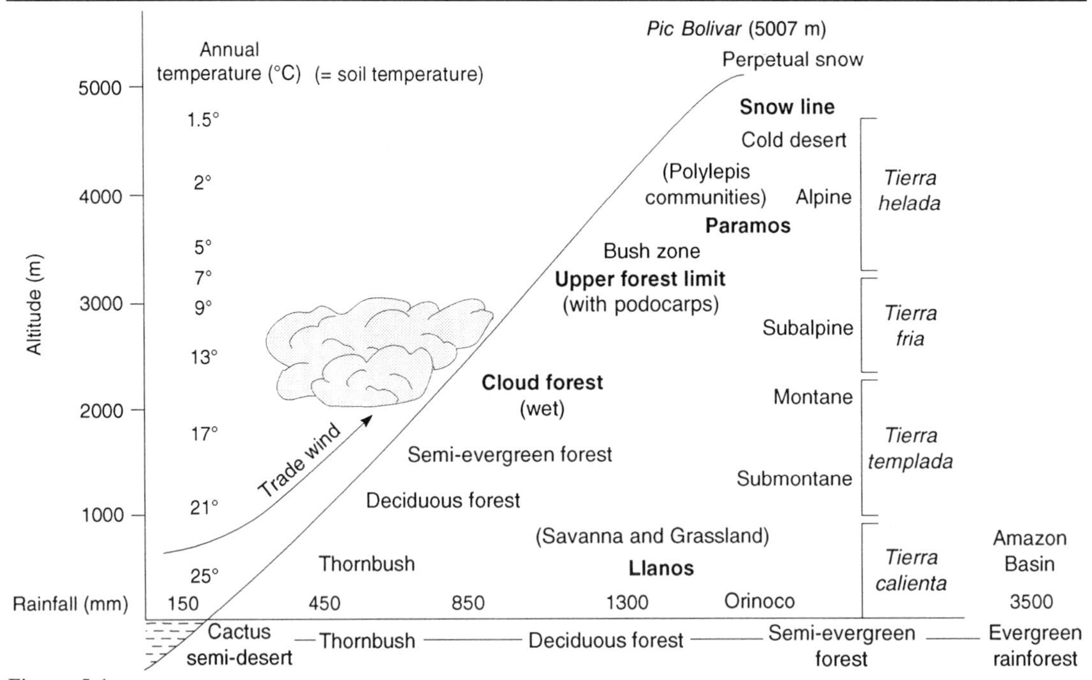

Figure 5.1
Climate–vegetation gradients in Venezuela (after Walter, 1979, p. 34, figure 15).

5.2.3 The main types of humid tropical vegetation

The distribution of the main types of humid tropical vegetation are shown in figure 5.2. Lowland forests are the most extensive vegetation type (table 5.4). This is mainly because large areas of the humid tropics comprise low-lying Precambrian shields and there are many large river floodplains. Humid tropical montane and tropialpine vegetation is mainly found in geologically younger areas such as Central America, the West Indies, the Andes and southeast Asia.

Lowland forest

These generally comprise closed-canopy forests with three or four vegetation strata. The upper stratum, or canopy, reaches a height of 30–60 m with the trees that emerge from the canopy reaching 70–80 m. They have a large proportion of broad-leaved evergreen trees ranging from 60 to 250 species per square kilometre, and an abundance of thick woody climbers (lianas) and epiphytes are

Table 5.4 Proportion of humid tropical vegetation in different bioclimatic zones

Bioclimatic zone	*Vegetation type*	*Proportion (%)*
Lowland	Forest	81
Premontane	Forest	10
Lower montane	Forest	5
Montane	Forest	2
Subalpine	Grassland and shrubland	1
Alpine and nival	Grassland	<1

Source: after Holdridge, 1967

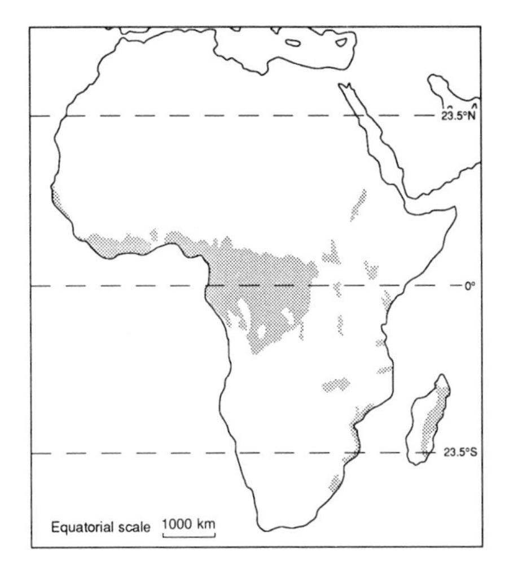

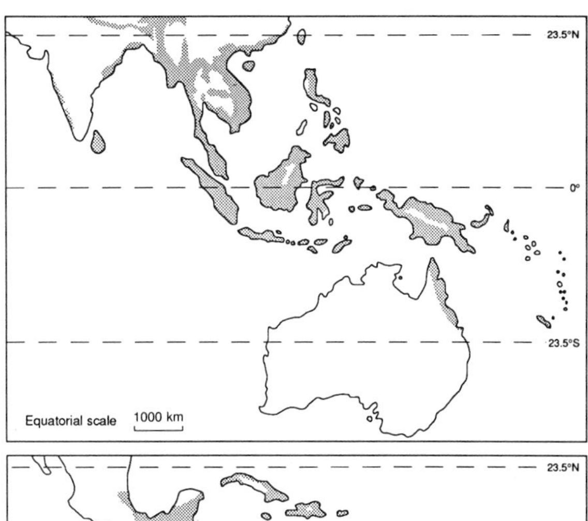

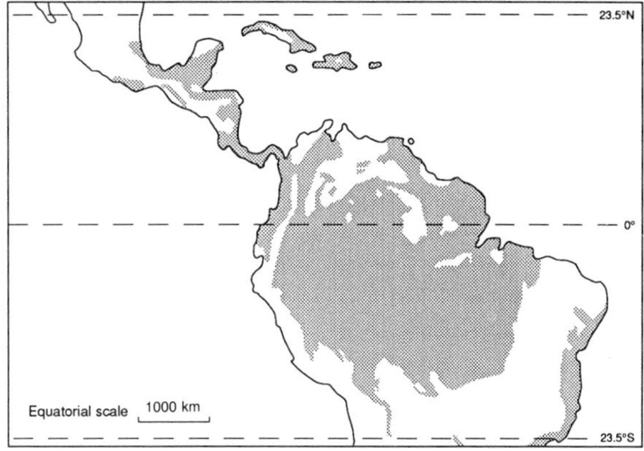

Figure 5.2
Forest cover in the humid tropics.

commonly found growing on the trees. A more detailed description of lowland humid tropical forest ecology is provided in sections 5.3.1–5.3.4. Important hydric forests are found in tidal and freshwater swamps (mangroves and freshwater swamp forests respectively), and heath forests are also found in the lowland forest zone. These are discussed separately in sections 5.7.1–5.7.3.

Premontane forest
Premontane forests, like their lowland counterparts, are closed canopy forests. However, they are of lower stature, the main canopy being found at 20–40 m with emergents reaching 50 m. Species richness is also lower (40–100 species per square kilometre) and, although epiphytes and climbers are present, the latter are rarely woody. These forests are discussed in more detail, along with other montane forest types, in section 5.3.5.

Lower montane forests
These are generally much shorter and poorer forests than those at lower altitudes. They usually have two strata and the highest canopy is found at 10–30 m. There are few emergents and only about twenty-five to fifty species per square kilometre. Moreover,

on poor soils single-species forests are commonly encountered. Epiphytes are common but there are few climbers.

Montane forest
Montane forests have low canopies, always less than 20 m high and often only a few metres high. Emergent trees are usually absent, but when they do occur they grow to heights of about 25 m. There are no climbers, although epiphytes are often present. Single-species forests are again common.

Tropialpine communities
Heathland and grassland communities are typically found in the subalpine and alpine zones of low latitude mountains; these areas are termed tropialpine. Typical of many of these heathlands is a flora consisting of gigantic species of genera commonly found in temperate latitudes.

5.2.4 Tropical forest formations

There are three distinct humid tropical forest formations: the American, African and Indo-Malaysian.

The American or neotropical formation is the largest of the three, covering approximately 40,000 km^2. It is best developed in an area which comprises the Amazon Basin, the Guyanan Plateaux and the western foothills of the Andes. Outliers are found along the southeast Brazilian coast (the *mata atlantica*), in an area extending from Ecuador into Central America, and on many Caribbean islands (figure 5.2). Destruction of the neotropical humid tropical forest formations has been greatest in parts of southern and eastern Amazonia, along the southeast coast of Brazil in the *mata atlantica*, in Central America (especially Costa Rica and Belize) and in the Caribbean (especially in Haiti).

The African humid tropical forest is now almost totally restricted to the Zaire Basin (comprising parts of Zaire, Cameroon, Equatorial Guinea, Gabon and Congo) in central Africa. Only isolated areas remain of the formerly large forested areas found in West Africa (mainly in eastern Sierra Leone, Liberia, southern Ivory Coast and southwest Ghana). Forest destruction has taken a similar toll on the extensive humid tropical forests that once cloaked much of Madagascar, Mauritius and Réunion. In total only about 18,000 km^2 remains.

The Indo-Malaysian formation has a disjointed distribution extending from the western Ghats in India to Fiji, and from Assam to northern New South Wales; in all it covers about 25,000 km^2. In this formation humid tropical forest extends into subtropical regions, as it does in South America. Extensive areas can still be found in Borneo, Indonesia, the Philippines and Papua New Guinea. However, extensive destruction has occurred in Sri Lanka, Java, peninsular Malaysia, Thailand, Vietnam and on many Pacific Islands (e.g. Fiji), and currently large tracts of forest in Sarawak and

Borneo are being destroyed. Longman and Jeník (1987) recognize three divisions within this formation: the Indo-Malaysian (extending from India to southeast Asia), the Australasian (covering Queensland, Papua New Guinea and the western Pacific Islands) and the Hawaiian Islands formation.

5.3 Rain Forest Ecology

5.3.1 Forest structure

Forests are dynamic four-dimensional ecosystems. The vertical (two-dimensional) stratification of tropical forest flora and fauna is well known; but at any one point in time they also vary horizontally, both as a community and in the distribution of individual species (Hubbell and Foster, 1983). Of course, they also vary over time – the fourth dimension. In this section only the vertical organization is considered.

The vertical organization of a hypothetical tropical rainforest was introduced by Richards (1952) (table 5.5) although its existence is now disputed by many ecologists. However, this multilayered structure with three tree strata only achieves its maximum development in mature lowland rainforests. Changes in vertical organization occur both with increasing altitude and during forest regrowth. Despite the clarity attributed to this stratification in textbooks it is often difficult to see in the field. Difficulties are posed by climbers and epiphytes whose main area of growth and leaf development changes over time, e.g. canopy-based climbers often fall to the forest floor and then send up shoots from the soil.

The tree crowns of the upper tree stratum form an interlocking, undulating closed canopy which varies in height from 30 to 50 m. This canopy reaches its greatest height in mature seasonal tropical forest (Longman and Jeník, 1987). A few trees, known as emergents, e.g. in South America the brazil nut (*Bertholletia excelsa*), break through this canopy and can grow to exceptional

Table 5.5 The theoretical vertical organization of mature rainforest and seasonal forest from West Africa

Layer	*Height and main floral elements*
Upper tree	>25 m; emergents, woody climbers and epiphytes
Middle tree	10–25 m; large trees and woody climbers
Lower tree	5–10 m; small trees and saplings
Shrub	1–5 m; tree seedlings, shrubs, pygmy trees and *krummholz* thicket
Herb	<1 m; small tree seedlings, forbs, grasses and sedges, ferns and bryophytes
Upper root	0–5 cm; compact root mass
Middle root	5–50 cm; less abundant tree roots
Lower root	<50 cm; scattered tree roots

Source: after Longman and Jeník, 1987, p. 76, table 4.2

heights. The lower forest strata contain shorter trees, although they are often individuals of the same species that form the canopy.

The shrub layer is poorly developed. Shrubs themselves are rare and many of the components of the shrub layer are either saplings of the taller forest trees or pygmy trees, e.g. *Dracaena* spp. (a common foliage house-plant). The herb layer is less well developed than the shrub layer, sometimes only covering less than 10 per cent of the ground surface.

5.3.2 Flora

The most obvious floristic elements of tropical forests are the trees; but a closer inspection reveals a wide range of plant life whose diversity, luxuriance and, in many cases, strange features have captured the attention of both scientists and plant collectors. Trees over 5 m tall and trunks having a diameter of more than 7 cm at breast height are the dominant life form. The other floral elements include climbers, epiphytes, bamboos, palms, tree ferns, pandans or screw-palms, and the giant herbs of the ground flora.

Trees

The trees range from small, unbranched forms with a single tuft of leaves to the forest giants. The small, unbranched trees mainly grow to between 2 and 4 m in height: Jacobs (1987) refers to them as mini-treelets. However, unbranched trees with tufts of leaves at the top can grow much higher, e.g. the *Cycads* of Malesia, the coconut palm (*Cocos nucifera*) and the sparsely branching screw-palms (*Pandanus* spp.). The forest giants, the type of trees that typify tropical forests to the lay-person, usually reach 35–45 m in height and have crowns ranging from 10 to 20 m in diameter. The forest giants include some of the tallest trees in the world – the tallest tropical forest tree, the Klinki Pine (*Araucaria hunsteinii*) of the Asian-Pacific region, reaches 89 m. Other very tall forest trees include *Eucalyptus deglupta* and *Agathis dammara*. No tree, however, tops the tallest Californian redwoods (*Sequoia sempervirens*) at over 110 m (Richards, 1952; Whitmore, 1990).

Tree boles commonly taper upwards and, although they display a wide variety of bark forms, smooth boles dominate. Many boles exude sap and gum, the most well-known sap being latex from the para rubber (*Hevea brasiliensis*), although in the past a wide range of such products has been collected, e.g. gutta-percha, *damar*, lac and camphor in Malaysia (Dunn, 1975; Andaya and Andaya, 1982). The most interesting parts of the boles are the lower few metres where buttresses and aerial roots are often found. Buttressing of lower trunks occurs frequently in some types of forest, e.g. swamp forests, and provides lop-sided or epiphyte-laden top-heavy trees with a degree of stability. Buttresses act like guy ropes on a tent and, in fact, if struck with a hard metal blade resonate like a steel hawser. Flying buttresses (plate 5.1) are well developed in

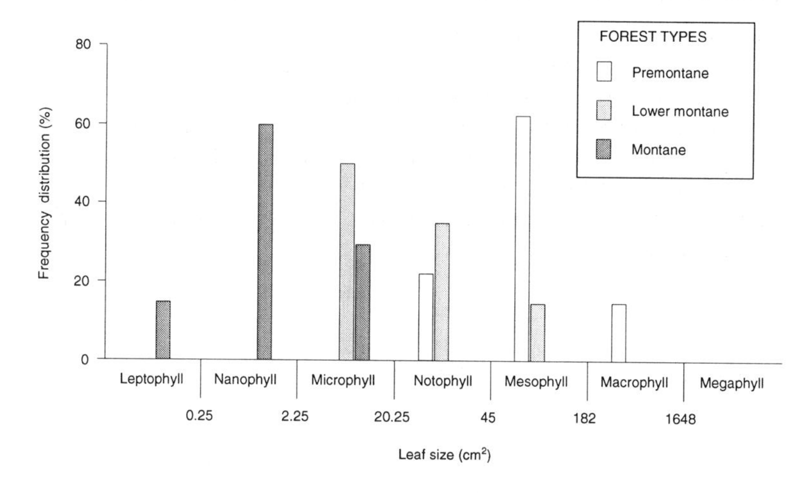

Figure 5.3
Distribution of frequency against leaf size for montane forest types (data from Dolph and Dilcher, 1980).

Ceiba spp. and in the Bombacaceae. Stilt roots, another form of buttressing, are common in pandans and mangroves (plate 5.2).

Twenty-three models of crown form have been identified by Hallé and Oldeman (1970) and Hallé et al. (1978) in tropical forests. They range from those with a single apical shoot which results in a tuft of leaves at the top of the trunk, e.g. palms and many pioneer trees, to the multiple branched crowns of the forest giants. The ecological significance of this wealth of crown forms is unclear, although it is typical of the overall levels of tropical forest diversity.

Mature leaf forms also exhibit great diversity of form (figure 5.3). Nevertheless, most leaves are dark green, leathery and have an oblong-lanceolate entire form (although pinnate, palmate and compound leaves are also common). The leathery nature of leaves is a response to the drought conditions that occur even in tropical forests. Large leaves are common in the lower part of the canopy and on wetter sites (Hall and Swaine, 1976), but leaf size decreases markedly with altitude (Dolph and Dilcher, 1980; Whitmore, 1990). Many leaves display other interesting physiological adaptations to the rain forest environment. Drip tips, elongated ends to the leaves (plate 5.3), facilitate rapid shedding of water from leaves enabling them to transpire freely and reducing the chances of epiphyllous communities establishing themselves. Other leaves, particularly those in the upper canopy, are thickened at their base (articulation) to enable them to adjust position, particularly to sunlight.

The diversity of flowers, fruits and seeds is indicative of the wide range of regenerative organs and the many mechanisms of pollination and fruit and seed dispersal found in tropical forests. In addition to the flowering and fruiting forms commonly found on temperate trees, there are cauliflory and phylloflory. Cauliflory (table 5.6) and phylloflory are the flowering, and subsequent fruiting, directly on leafless woody tissue (i.e. trunks, branches and roots) (plate 5.4) and leaves respectively. Some well-known large

Plate 5.1
A buttressed trunk of Kempasia malacensis (photo: Hugh Vaughan Williams).

tropical fruits such as the durian (*Durio zibethinus*) and the jackfruit (*Artocarpus heterophyllous*) are borne on tree trunks. Jacobs (1987) noted an interesting form of rhizoflory in the Malaysian forests where ground figs flower and fruit on ground runners sent out over 1 m from a small tree which itself bears no flowers or fruit. Other examples of this type of rhizoflory, and geocarpy, have been investigated by van Balgooy and Tantra (1986).

Cauliflory is thought to be an adaptation to aid seed dispersal because trunks and branches are used as pathways by many animals and insects. Clearly this is the case with Malaysian ground figs, which are eaten by pigs, and the argument can be substantiated

Plate 5.2
Stilt roots of Pandanus sp., Bogor Botanic Gardens (photo: Jack Reilly).

further if the main seed dispersal mechanisms are examined (figure 5.4). In the forest understorey the production of fruits on easily located parts of the plants (e.g. trunks) is likely to attract the main agents of seed dispersal – insects, birds, bats and mammals. Some understorey plants do rely on explosive or wind dispersal mechanisms, but these mechanisms are more common in the canopy.

Roots of tropical trees have been extensively studied by Jeník (1978). Root architecture is generally complex and, although some trees have deep tap roots, the majority of the root biomass is found in the upper 30 cm of the soil. This is known as a root mat because

Plate 5.3
The elongated extension at the end of the leaf is known as a drip tip (photo: Paul Wilkin).

Table 5.6 The main types of cauliflory

Ramiflory	Flowers on larger branches and leafless twigs, but absent from the trunk
Trunciflory	Flowers on trunk, but not on branches and twigs (some authors define only this as cauliflory)
Basiflory	Flowers at base of trunk
Flagelliflory	Flowers on pendulous twigs which grow downward from the lower trunk
Rhizoflory	Flowers from the roots

Source: after Longman and Jeník, 1987, p. 97, table 4.5

Plate 5.4
An example of cauliflory: a calabash tree Cresentia cujete, Dominica, West Indies.

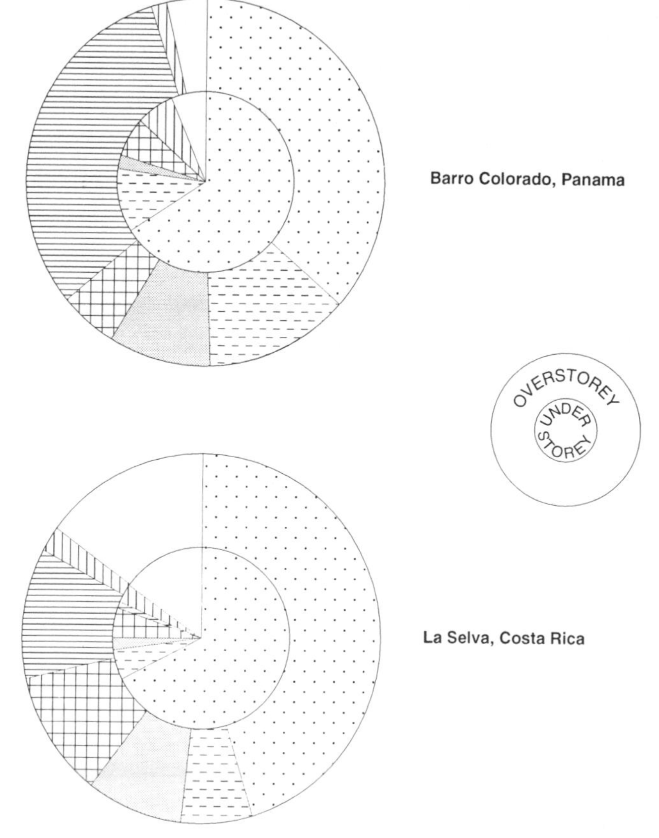

Birds Bats Primates Terrestrial Mammals Wind Mechanical Unknown

Figure 5.4
Fruit and seed dispersal mechanisms in Central American forests (data from Leigh and Windsor, 1982).

it is made up thousands of fine roots, some 20–50 per cent being less than 2 mm in diameter. Mycorrhizae and nitrogen-fixing nodules are also found in the rhizosphere. Also noteworthy are aerial roots, stilt roots and pneumorrhizae. Lower trunks often have adventitious aerial roots dangling down the trunk. They rarely grow very large but stilt roots, which also grow out of the trunk, thicken and grow into the ground (plate 5.2). Stilt roots, like buttresses, promote stability, especially in soft ground, and also facilitate the rapid extraction of nutrients from waterlogged soils. Longman and Jeník (1987) call these two functions the 'snowshoe' and 'shortcircuit' effects respectively. In swamps knee and peg roots occur; these are often called breathing roots (pneumorrhizae) (plate 5.5) (Longman and Jeník, 1987) and may possibly be sites of gaseous exchange (Scholander et al., 1955).

Plate 5.5
Pneumorrhizae.

Climbers and epiphytes

Climbers and epiphytes are common in tropical forests, climbers being found in all successional stages from early regrowth (e.g. the fast-growing Convolvulaceae, Cucurbitaceae and Vitaceae) to mature woody climbers in mature forests (Cremers, 1973, 1974; Vareschi, 1980). The woody climbers often tie tree crowns together and form a pathway for canopy dwelling animals (Montgomery and Sunquist, 1978). The woody climbers fall into two categories – big, woody climbers (lianas) and bole climbers (Whitmore, 1990). Lianas are light-demanding and usually hang freely from tree crowns in the main canopy. Their light-demanding habit means that they are often found in forest gaps and at the edge of forests. They include genera such as *Gnetum, Buahinia* and the rattans (climbing palms that grip the host tree by means of hooked whips at the ends of their leaves). Bole climbers are restricted to the most humid climates and differ from the other type of climbers in that they cling to the tree trunks by specialized roots, often covering the entire trunk. Climbers account for about 8 per cent of tropical forest plant species. In Sabah, 150 genera have been identified (Jacobs, 1987); and on Barro Colorado Island in Panama Montgomery and Sunquist (1978) surveyed thirty-nine trees of which twenty-eight had lianas encompassing fifty-five species.

Epiphytes, plants without a ground contact, are commonly found in humid forests. Their preference for the most humid forests can be illustrated by comparing the number of species present in seasonal tropical forests (nine to twenty-four species) with those in phytogeographically-comparable very humid forests (238–368) (Gentry, 1988). Tixier (1966) distinguished between macro-epiphytes (vascular plants, e.g. orchids, ferns and bromeliads) and micro-epiphytes (mosses, liverworts, algae and lichens). In a mountainous Vietnamese forest he noted 116 species of mosses, 110 liverworts and 36 lichens. Johansson (1974) noted that 40–62 per cent of trees over 10 m in Liberia had epiphytes (mainly orchids and ferns), and that one 14 m high tree had 416 individual epiphytes. Madison (1977) calculated that the eleven main families of vascular epiphytes have over 26,000 species between them.

Bole climbers and epiphytes usually exhibit a vertical zonation in response to the forest microclimate (figure 5.5). They are generally well adapted to moisture stress. Some species can rehydrate after desiccation, whilst others avoid desiccation by having thick leathery leaves that reduce transpiration losses. Orchids often have pseudobulbs and water-absorbent coatings of dead cells. Bromeliads often have leaf bases that are modified to collect and store water, and in some the leaves are fused to form a pitcher that traps water – e.g. *Dischidia rafflesiana* and the endangered *Nepenthes rajah* which is ruthlessly collected from isolated mountains in Sarawak and Sabah by helicopter to supply an international market where the plants are worth up to $1000 each (Briggs, 1985; Aitken and Leigh, 1992). Such plants can also trap litter in their leaf bases which decomposes to humus and provides the plant with a new environment in which to root. Root development around these aerial gardens leads to knots of vegetation that have the appearance of large birds' nests in the canopy.

Some species change form during their life cycle. Some start life as bole climbers and, as the lower parts of the plant die back, they grow upward and become epiphytes. Stranglers start life as crown epiphytes and then send roots downward which completely cover the trunk. Many stranglers belong to the genera *Schefflera* and *Clusia*, but the figs and banyans are stranglers *par excellence* (plate 5.6). In Calcutta Botanical Gardens stands a magnificent banyan tree (*Ficus benegalensis*) which dates back to 1782, measures over 100 m across and has 666 root trunks.

Peculiar epiphytic communities exist on wet, shaded leaves. Initial nitrogen-binding bacterial communities colonize leaf surfaces and are succeeded by algae, fungi, lichens and yeasts. These in turn are followed by unicellular ciliates, flagellates and slime moulds; finally the leaves are colonized by mosses and small arthropods.

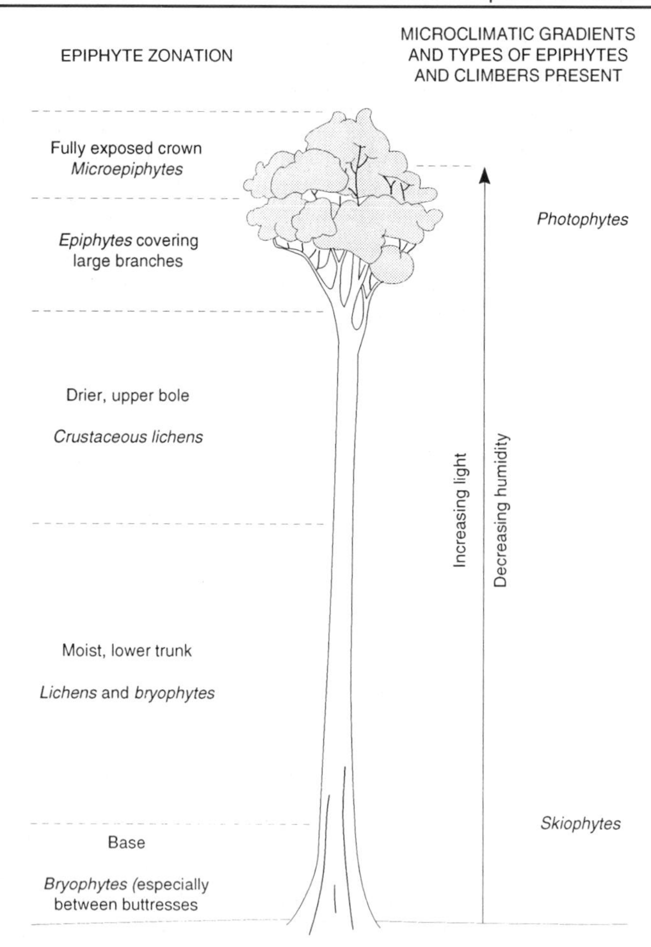

Figure 5.5
Vertical zonation of epiphytes (modified after Longman and Jenik, 1987, p. 102, figure 4.18).

These communities are known as epiphylls; Kiew (1982) noted that lichen communities which appeared after seven months in some Malaysian forests had covered most of the leaf area after five years.

Other plants

Although shrub and herb layers are poorly developed in most tropical forests, the floral elements in these layers are interesting. In the herb layer ferns and their allies are very important and, to many ecologists, the high ratio of ferns to angiosperms is a distinctive tropical forest feature. Common dicotyledons are drawn from the Rubiaceae, Gesneriaceae, Begoniaceae, Melastomaceae and Acanthaceae families and many common foliage house-plants come from these genera, e.g. Begonias, Saintpaulias, Calatheas, Monstera and Philodendrons. The main monocotyledons are bananas, grasses, sedges and the 1300 species of the Zingereraceae family, the most famous of which is ginger.

Other plants such as palms, bamboos and tree-ferns form part of the tree strata (plate 5.7). There are over 2500 species of tropical forest palms most of which are found within forests, although some

Plate 5.6
A strangler fig forming part of the understorey.

form pure stands or take on a shrub-like form. Some have become very rare because of the activities of plant collectors, e.g. *Johannesteijsmannia magnifica*, Malaysia's most endangered plant species is only known from a few sites (Kiew, 1989). Bamboos are light-loving woody grasses that occur frequently in upland forests and often form pure stands (McClure, 1966). Tree ferns (Cycatheaceae) are generally light-demanding and are mainly found in clearings and at forest margins (plates 5.7 and 5.8).

Tropical forest heterotrophs mainly live as hemi-parasites or full parasites on trees. The hemi-parasites are mainly drawn from the

Plate 5.7
Tree fern, typical of tropical and subtropical rainforests, from North Island, New Zealand..

epiphytic mistletoes (Loranthaceae) and the hemi-parasitic trees of the Santalaceae family which includes the sandalwood. Fully parasitic plants include species of *Balanophora* and *Rafflesia* which grow on trees and lianas; of particular note is *R. arnoldi* which bears the world's largest flower at over 1 m diameter.

5.3.3 Fauna

Natural history programmes on television have brought the fascinating world of tropical forest animals into our living rooms. Unfortunately this book has to be more selective in considering the forest fauna. It is estimated that between 2 and 3 million species of

Plate 5.8
Tree ferns and pioneer tree species in a hurricane-disturbed forest, West Indies.

animals live in tropical forests, some two-thirds of all the world's species. Some workers (e.g. Erwin, 1982) put the figure as high as 30 million species. Detailed information on forest fauna is available elsewhere (e.g. UNESCO, 1978; Golley, 1983; Sutton et al., 1983a; Goulding, 1984; Terborgh, 1984) and this section focuses on plant–animal interactions. The vertical stratification evident in many rainforests can also be used as a framework for considering fauna, although diurnal variations in animal activity must also be considered (figure 5.6).

Invertebrates are the most numerous animals in tropical forests, in terms of both species and individuals. Many are decomposers and therefore play a key role in nutrient and energy cycling. Particularly important are micro-arthropods, land snails, millipedes, earthworms and the larger arthropods (e.g. ants and termites). The latter group comprise decomposers, herbivores and carnivores. The invertebrates in turn provide food for many reptiles, birds and mammals such as anteaters and even primates. Their numbers and the amount of plant material they consume are staggering. Madge (1965) found up to 38,000 micro-arthropods in a square metre of soil under a Nigerian rain forest, and termites, which are mainly decomposers, turn over between 1 and 16 per cent of total litterfall in Malayan forests (Collins, 1983). Ants and termites are noted for their high levels of social organization and their success is exemplified by the foraging leaf-cutter ants (*Atta* spp.) which harvest 0.2 per cent of the gross primary production of tropical forests (UNESCO, 1978). Flies, bees, wasps and beetles are equally important, although they tend to be concentrated in the canopy (Sutton, 1983b). The number of species living in the canopy can be very high, e.g. Erwin (1982) found 995 species of beetles in just 19 *Leutiea seemannii* trees in Panama.

Despite their smaller numbers, the vertebrate forest-dwellers are far better known than the invertebrates. The number of vertebrates in tropical forests is probably not much different from that in

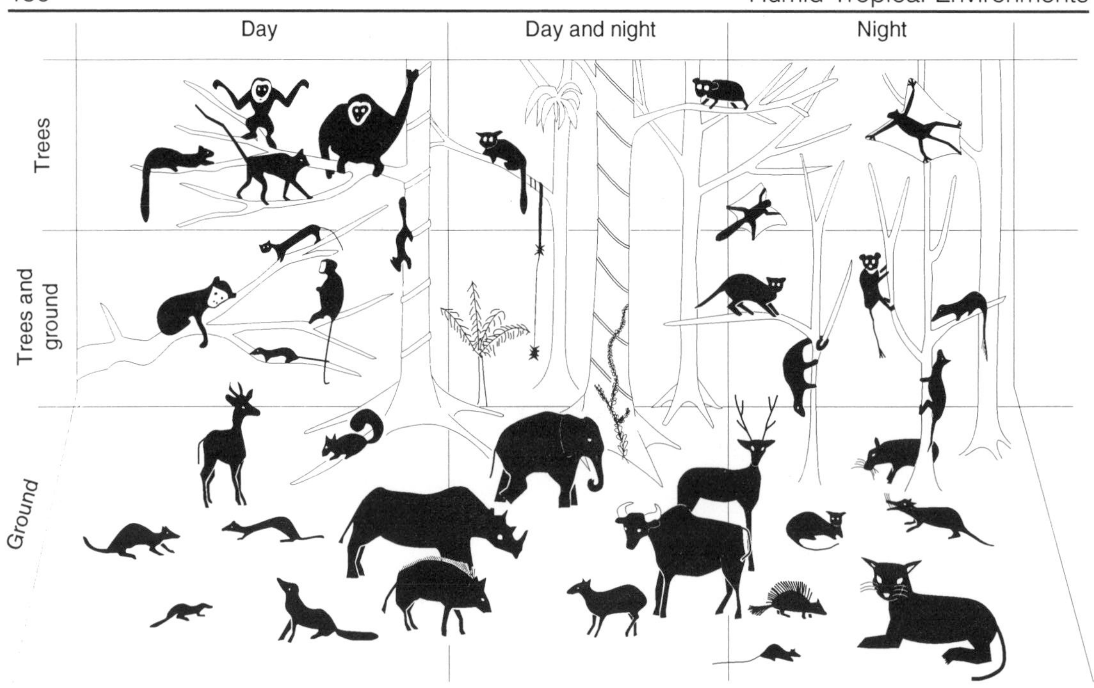

Figure 5.6
Animal activity in forests (after MacKinnon, 1972).

temperate forests. However, the number of species is far greater, and this is particularly true of the avifauna. The mammalian biomass is quite low compared with savanna grasslands and averages about 10–15 kg ha^{-1}, and the avifauna biomass per hectare is similar to that of mammals (Medway, 1978). A high proportion of vertebrates are arboreal; some put the ratio of tree to ground dwellers at about 50 : 50. Davis (1962) noted a higher proportion of arboreal animals in tropical forests (45 per cent of all animals) than in temperate woodlands in Virginia (15 per cent). Most animals show a distinct altitudinal zonation: for instance, species richness of birds is generally felt to decline markedly above 1000 m, although the fact that the distribution of endemic bird species shows an affinity for tropical montane and cloud forests (Bibby et al., in press) casts doubt on this generalization.

Animal–plant interactions often have an important bearing on plant distributions, forest dynamics and nutrient cycling. Many rain forest trees, particularly climax trees, rely on animals for seed dispersal (figure 5.4). Fruit bats are particularly important pollinators and seed dispersers, feeding on a wide variety of species (Ayensu, 1974; Jacobs, 1987). Jacobs (1987) coined the term 'big, odorous fruits' to describe the large, perfumed fleshy fruits borne by many tropical forest trees which attract many animals. Corner (1964) recorded that durians (*Durio zibethinus*) attracted elephants, tigers, tapirs, deer, rhinoceros, monkeys, bears and squirrels – in fact the smell of ripe durians is so strong that they have been banned on public transport in Indonesia! Vivid colouring on other fruits may also be an adaptation to attract birds and mammals.

Some animals interact with many plants, whereas others have far narrower, even obligate, relationships. Orang-utans (*Pongo pygmaeus*) are a good example of a forest-dwelling animal whose diet includes many plants. Rijksen (1978) observed that they ate ninety-one species of fruit and plant material from twenty-two other species. However, some animal–plant relationships, particularly in the field of pollination, are far more specific; e.g. figs can only be pollinated by fig wasps (Wiebes, 1976).

The key herbivores and decomposers in humid tropical forest ecosystems are invertebrates. Ants and termites are important in nutrient cycling, although ants have other ecological roles (Longman and Jeník, 1987). Some species of ants are specific to a plant species, e.g. they may be associated with specific organs, or actually nest in plants and guard them from predation by other invertebrates (Bentley, 1977; Benson, 1984).

5.3.4 Species richness and biodiversity

Arguably the strongest scientific imperative that is being used to try and halt tropical forest destruction is its high biodiversity. Mature tropical forests are only rivalled by coral reefs in species richness and they possibly account for half of the world's gene pool (Longman and Jeník, 1987). Already many plant species are known but, as the rate of discovery of new species proceeds apace, this number is increasing rapidly. In particular many climbers, epiphytes, orchids and bromeliads probably await discovery. Botanical knowledge is patchy, being best in West Africa and parts of southeast Asia and poorest in New Guinea and the neotropical forests. Where details do exist, the statistics can be amazing (see for instance Jacobs, 1987, p. 78). To quote two examples from many: Ashton (1964) calculated that there were over 2000 tree species in Brunei, more than *all* the plant species in The Netherlands (1200 species), despite the fact that Brunei is five times smaller; Myers (1984, p. 53) suggested that, on Mount Kinabalu in Sabah, 'there are five times as many oaks as in the whole of Europe, together with 400 species of ferns and 800 species of orchids'.

The number of species per unit area is an oft-quoted measure of richness (see for example Schultz, 1960; Ashton, 1964). Wyatt-Smith (1953) estimated that at least a hundred tree species could be found in a hectare of southeast Asian forest, the theoretical upper limit being about 400. The upper limit has now been revised downward and the most species-rich plot enumerated so far is in Peru with 283 tree species with diameters greater than 0.1 m and 580 stems in a hectare; it is thought to be near the upper limit of species richness. Richness both within and between regions is variable. Indo-Malaysian forests and those in the Andean foothills are the richest (table 5.7), and the African forests the poorest (Longman and Jeník, 1987). The relative poverty of African forests has been attributed to extensive plant extinctions due to dry periods in the past (see section 5.8). Species richness covers all organisms,

Table 5.7 Species richness in selected Malaysian forests

Location	*Vegetation type*	*Area (ha)*	*Number of species*	*Minimum diameter (cm)*
South Pangkor (Selangor)	Lowland rainforest	62.7	173	120
Sungai Menyala (Negeri Sembilan)	Lowland rainforest	2.0	240	10
Bukit Lagong (Selangor)	Hill dipterocarp forest	2.0	495	10
Gunung Mulu (Sarawak)	Alluvial forest	1.0	223	10
	Dipterocarp forest	1.0	214	10
	Heath forest	1.0	123	10
	Forest over limestone	1.0	73	10
Wanariset (East Kalimantan)	Lowland rainforest	1.6	239	10
Lempake (East Kalimantan)	Lowland rainforest	1.6	205	10
Pasoh (Negeri Sembilan)	Lowland rainforest	11.0	460	–
Pasoh (Negeri Sembilan)	Lowland rainforest	4.0	328	30
Jenka (Pahang)	Lowland rainforest	20.0	375	91
Andalau and Belalong (Brunei)	Lowland rainforest	40.5	760	10
Lambir	Lowland rainforest	0.2	79 (lianas)	1

Source: original data sources are cited in Aitken and Leigh, 1992, p.31, table 2.1

not just trees: total floral species richness can be up to about two times greater than the values quoted for trees alone if other vascular plants are considered (Meijer, 1959; Lawson et al., 1970) (figure 5.7). This is borne out by Whitmore et al. (1986) who found that 57 per cent of all vascular plant species in a tropical rain forest plot were less than 1 m high. Of course, the species richness of animals can also be considered and it is equally rich (see section 5.3.3).

Some major tree families and genera are pantropical, whilst others only occur in one region (see also section 5.8). Moreover, some families just occur in the tropics, whilst others include temperate trees and shrubs. The greatest floristic distinctions between the different tropical forest regions are found amongst single-dominant forests, edaphic forests and secondary forests (Longman and Jeník, 1987). Epiphytes also exhibit significant floristic distinction (Tixier, 1966). Species richness is reduced by human activities such as farming, selective timber extraction and the introduction of plantation trees such as rubber, oil palm, *Eucalyptus* spp. and *Pinus* spp.

At the global scale Rejmanek (1976) described an increase in species richness with decreasing latitude for many groups of organisms, e.g. vascular plants, vertebrates and beetles (e.g. figure 5.8). But the picture is not clear as other groups, e.g. algae, fungi,

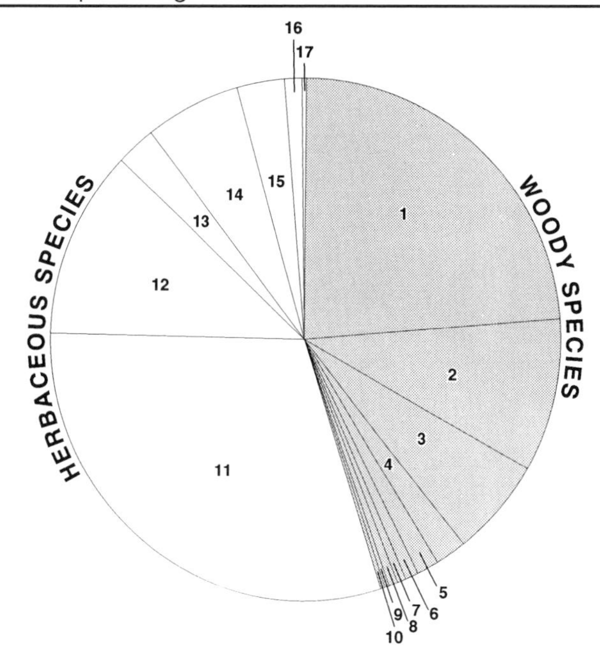

WOODY SPECIES	Trees 1	Shrubs 2	Climbers 3			
	Treelets 4	Creepers 5	Hemi-stranglers 6			
	Tree ferns 7	Palms 8	Pandans 9			
	Mistletoes 10					
HERBACEOUS SPECIES	Epiphytes 11	Low herbs 12	Tall herbs 13			
	Ferns and allies 14	Climbers 15	Creepers 16			
	Bananas 17					

Figure 5.7
An inventory of all plant species except epiphytes (which were estimated at 100 species) (data from West, 1940; quoted in Jacobs, 1987).

lichens and some insects, do not exhibit this trend, and a few, e.g. the parasitic Hymenoptera, even show a decline in richness towards low latitudes (Janzen, 1976b). On a smaller scale, species richness within the humid tropics varies along environmental gradients related to humidity, seasonality and altitude (Ashton, 1964; Janzen, 1976b) (figure 5.9) and, at even smaller scales, with soil and topographic variations.

Why are the humid tropics species-rich? This is a strongly debated issue (Ricklefs, 1973; Connell, 1978; Rosen, 1981; Mabberley, 1983; Deshmukh, 1986) and only the main threads of the arguments can be presented here. The following hypotheses appear to be the most important.

1 *In situ* speciation is more frequent in the humid tropics. This occurs (i) because of the higher frequency of mutations in the humid tropics than in other areas, which may be due to the markedly greater levels of ultraviolet-B radiation which can modify DNA (Caldwell, 1981); (ii) because the

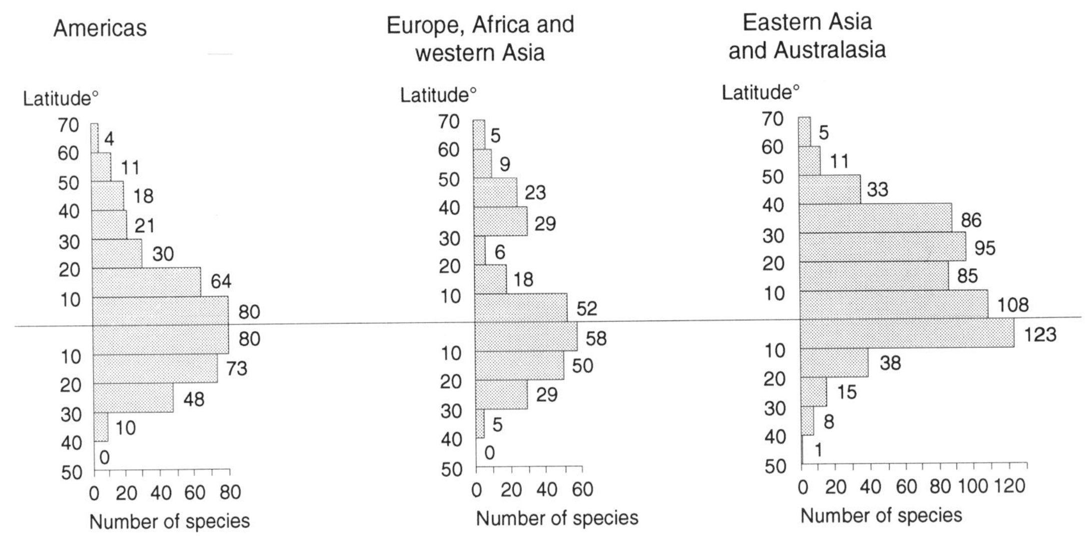

Figure 5.8
*Species richness of swallowtail butterflies (*Papilonidae*) with latitude (from Collins and Morris, 1985).*

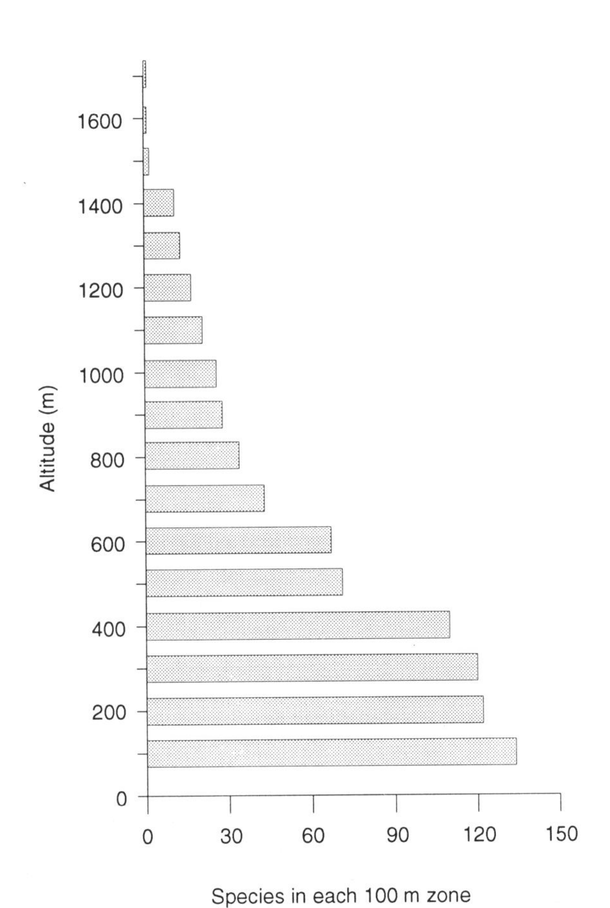

Figure 5.9
Species richness and altitude (from Ashton, 1964).

sedentary nature of tropical populations leads to geographical isolation; (iii) because of faster rates of evolution which can be linked to the greater number of generations over unit time, greater productivity which leads to greater turnover and increased selection, and the marked importance of biological factors which enhances selection; and (iv) because spatial and temporal isolation, which accelerates speciation, is made possible by the forest structure.

2 Rates of extinction are lower in the tropics because (i) competition is less stringent because of increased resource availability and its partitioning over space and time, there are an increased number of ecological niches (Bourière, 1983) and there is control of competitors by predators; and (ii) the constant physical environment and ecosystem stability allow small populations to exist.

The opportunities for *in situ* coexistence (1) and preservation (2) are known as the equilibrium hypotheses.

3 The fact that many parts of the tropics are old and stable has meant that tropical ecosystems have had more time to develop, and that there have been opportunities for species migration along a variety of accessible routes from Pleistocene refugia in the last 2 million years. This argument will be developed further in section 5.8 and is known as the non-equilibrium hypothesis.

The evidence to support these various mechanisms has been summarized by Deshmukh (1986) (table 5.8).

5.3.5 Changes with altitude

Significant changes in forest floristics and structure occur with increasing altitude. Some of these changes are considered elsewhere, and the main variations are listed in table 5.9. The changes are illustrated by two complementary examples from southeast Asia.

Hyndman and Menzies (1990) compared four forest types along a transect in the continuously wet region of Papua New Guinea (figure 5.10, table 5.10). The foothills rain forest, which ranges from 500 to 1000 m a.s.l. in altitude and covers a little more than 40 per cent of the island, is a typical heterogeneous lowland rainforest exhibiting two or three strata. The canopy is formed by the crowns of a number of species which reach about 30 m, epiphytes and lianas. Under the canopy is an irregular tree stratum at 20–25 m and a diverse stratum of young canopy trees at 10–15 m. Pandans (*Pandanus* spp.) and palms are important in the lowest of these two strata. The shrub stratum, at heights of less than 10 m, is dense and below it is a scanty ground flora comprising creepers, ferns, forbs, herbs and orchids.

Table 5.8 An assessment of the evidence in support of the various mechanisms proposed to account for high diversity in the tropics

Mechanisms	*Evidence for high tropical diversity*	*Taxonomic groups*
Time scales		
Seasonal to a few years	Slight	All groups
Primary succession	Poor	All groups
Secondary succession (including disturbance)	Good	All groups
Geological and evolutionary		
Specialization	Fair	Mainly animals
Coevolution	Fair	All groups
Refugia	Fair	All groups
Higher speciation rate	Slight	All groups
Lower extinction rates	Poor	?
Spatial heterogeneity		
More biomes	Good	All groups
More habitats	Good	Mainly animals
Competition and species packing	Fair	Mainly vertebrates
Predation	Fair	Plants and invertebrates
Production		
High production	Poor	–
Year-round production	Good	Vertebrates

Source: after Deshmukh, 1986, p. 217

Table 5.9 Structural and physiognomical differences between lowland and montane forests

Formation	*Tropical lowland evergreen rainforest*	*Tropical lower montane rainforest*	*Tropical upper montane rainforest*
Canopy height (m)	25–45	15–33	1.5–18
Emergent trees with height in (m)	Characteristic to 60–80 m tall	Often absent, to 37 m tall	Usually absent to 26 m tall
Pinnate leaves	Frequent	Rare	Very rare
Principal leaf size class of woody plants[a]	Mesophyll	Mesophyll	Microphyll
Buttresses	Frequent and large	Uncommon, small	Usually absent
Cauliflory	Frequent	Rare	Absent
Big woody climbers	Abundant	Usually none	None
Bole climbers	Often abundant	Frequent to abundant	Very few
Vascular epiphytes	Frequent	Abundant	Frequent
Non-vascular	Occasional epiphytes	Occasional to abundant	Often abundant

[a]Using the Raunkiaer (1934) system.

Source: after Whitmore, 1984, table 18.1

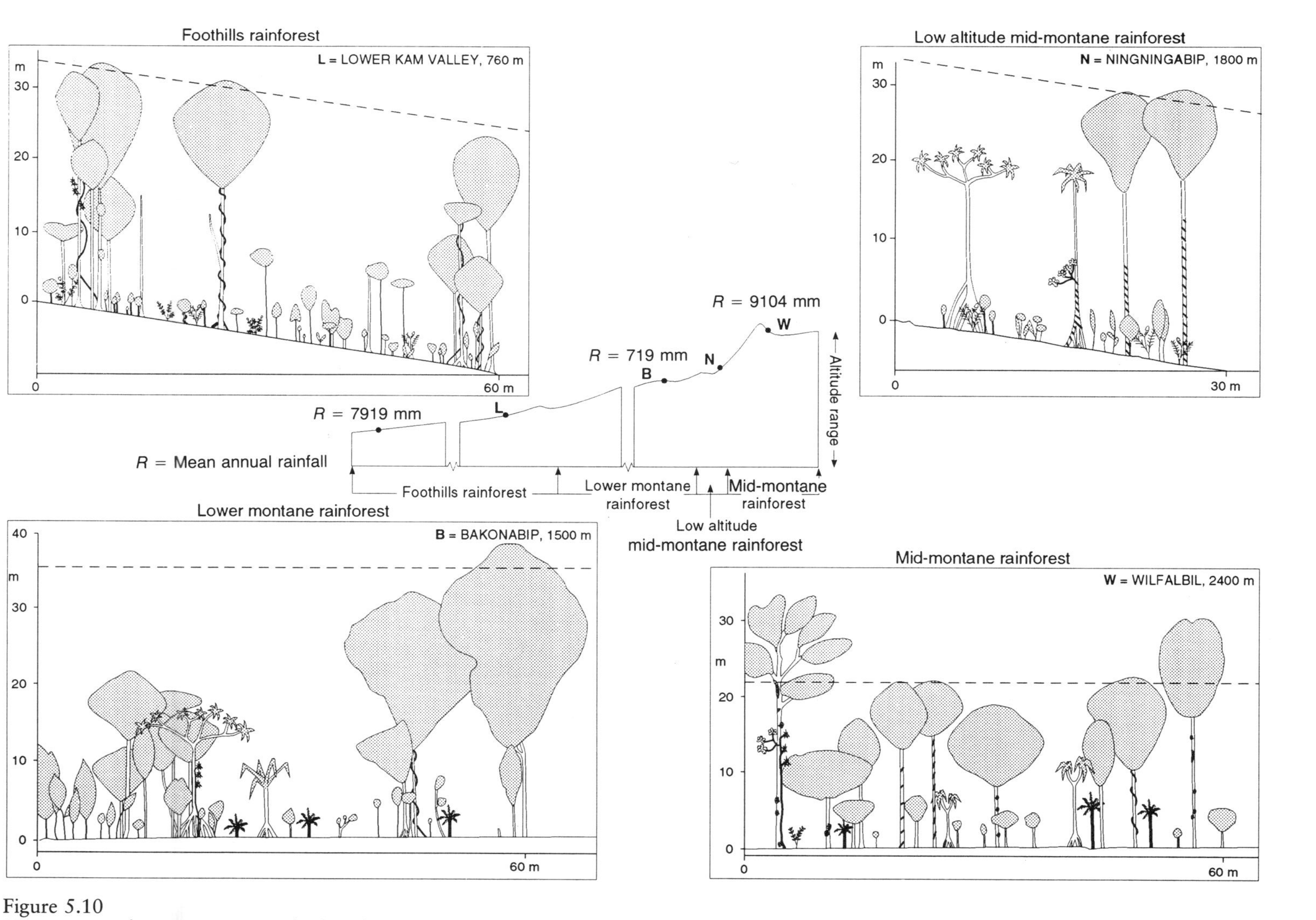

Figure 5.10
Variations in forest structure with altitude in central Papua New Guinea (data from Hyndman and Menzies, 1990).

Table 5.10 Diagnostic structural and floristic elements of the rainforest types of the Ok Teki headwaters, Papua New Guinea

	Foothills (FRF)	Lower montane (LMRP)	Low altitude (LARF)	Mid-montane (MRF)
Altitude	500–1000	1000–1800	1800–2200	2200–3000
Structural elements				
Number of tree strata	3	1–2	2	2
Upper canopy height (m)	30–35	20–25	30	20–25 (emergents to 35)
Shrub layer	Dense	Sparser than FRF	Present	
Ground layer	Scanty cover	Often bare	Very wet and mossy	Sparse cover
Buttressing of large tree trunks	Present	Absent	Absent	Absent
Woody vines	Present	Absent	Absent	Absent
Epiphytic ferns	Absent	Present	Present	Present
Moss on trees	Present	Dominant		
Floristic elements				
Characteristic genera		Oaks (*Castanopsis* spp. and *Lithocarpus* spp.)		Southern beech (*Nothofaqus* spp.)
Orchids	Present	Absent	Absent	Present
Palms	Present	Far fewer than in FRF	Absent	Absent
Tree ferns	Absent	Present	Present	Present
Pandans present (all Pandanus spp)	*P. castaneus* *P. dolichopodus* *P. foveolatus* *P. limbatus*	*P. concavus* *P. conoideus* *P. galorei*	*P. adinobotrys* *P. antaovesensis* *P. galorei* *P. jiulianttii*	*P. brosimos* *P. iwen*

Source: data from Hyndman and Menzies, 1990

The lower montane rainforest is another mixed evergreen forest and covers about 25 per cent of the country. The canopy generally ranges from 20 to 30 m in height with a diverse understorey at 10–20 m (figure 5.10), although it is sometimes reduced to a single canopy between 20 and 25 m. Structural and floristic differences that are typical of other montane forests (table 5.9) become evident as the foothills rainforest grades into lower montane rainforest.

The low altitude, mid-montane rainforest is very distinctive. Structurally it is far simpler than lower altitude forests, having a canopy of *Syzygium* spp. and *Garcinia* spp. reaching 30 m with crowns of *Pandanus antaresensis* and *P. galoeri* almost interlocking at 25 m (figure 5.11). The understorey at about 5–10 m is irregular and comprises tree ferns, young canopy trees and saplings. The shrub layer is composed of bushes of *Rhododendron* spp., gingers and ferns. There are no herbs, the very wet forest floor being covered by mosses and sodden leaf litter.

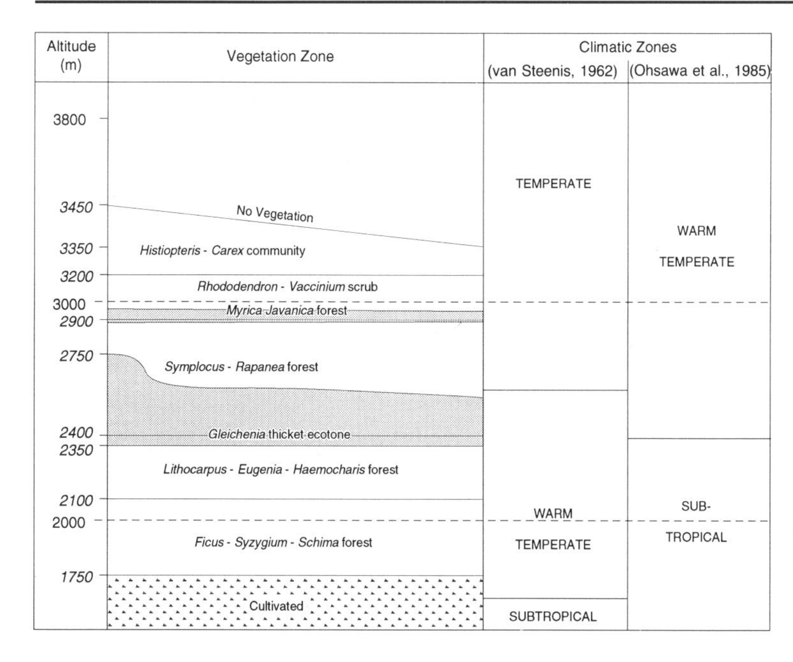

Figure 5.11
Altitudinal zonation of vegetation on Mount Kerinci, Sumatra (adapted from Ohsawa et al., 1985, p. 207, figure 8).

The mid-montane rainforest proper occurs above 2300 m a.s.l. and has a well developed canopy reaching 20–25 m in height with emergents (most notably *Nothofagus pullei* and *Conandrium polyanthum*) reaching 35 m. At about 10 m there is an understorey which includes tree ferns, underneath which is a very sparse ground flora.

At higher altitudes there are further changes in forest structure and flora, and in some places tropialpine heathland with grasses and sedges is found (Hyndman and Menzies, 1990). But to study these communities other complementary studies from southeast Asia can be investigated. Ohsawa et al. (1985) examined the vegetation above 1700 m on Mount Kerinci in Sumatra. They identified five vegetation communities on the basis of dominant species (figure 5.11) and the changes in community structure.

Generally, the forest becomes less complex and shorter with increasing altitude. The *Ficus–Syzygium–Schima* forest is comparable in structure to the lower montane forest of Ok Teki. In the *Lithocarpus–Eugenia–Haemocharis* forest a single canopy at about 18 m is found over shrub and herb layers. At higher altitudes the forests become simpler. There are only tree and herb layers in the *Symplocos–Rapanea* forest, and tree and shrub layers in the *Myrica* forest. The decrease in structural complexity with increasing altitude is mirrored by decreases in the maximum tree height and girth (figure 5.11), although there is little variation in basal area. The number of species and their diversity fall markedly with altitude. Ohsawa et al. (1985) argue that this is due to the fact that altitudinal zonation of vegetation on tropical mountains is brought about by a gradual floral impoverishment of the pool of trees of adjacent lowland forests with no new species being introduced. This hypothesis is supported by evidence from evergreen oak forests from southern China (Wang, 1961).

On mountainous islands the altitudinal limits of vegetation are generally much lower than for the equivalent situations on larger land masses. This is known as the 'telescoping effect' and was first identified by van Steenis (1962). Warm, moist oceanic air (with a high relative humidity) is forced to ascend by the mountainous relief. The steep lapse rates present mean that the resulting cloud base is relatively low (Watts, 1955). Such is its effectiveness that the telescoping effect has been recorded on islands with only a few hundreds of metres of relief in Indonesia (Bush et al., 1986; Hommel, 1990).

5.4 Forest Microclimate

Forests modify the climate considerably, creating a microclimate within the forest that is very different from that outside. Gaining an understanding of this microclimate is more important to ecologists than the macroclimate because it directly affects most plant and animal activity. Hydrologists also require detailed knowledge of the microclimate, particularly with respect to the movement of water above and below the soil surface (sections 3.5 and 7.1).

The pathway of incoming solar radiation (insolation) to the ground is interrupted by the forest canopy, and its subsequent pathways through the forest are of great importance. Both the amount and the spectral properties of solar radiation affect plant growth. The amount of insolation available determines whether light-demanding (heliophilous) plants or shade-bearing (sciophilous) plants will flourish. The wavelengths of the insolation available to plants is also important, in particular the amount of photosynthetically active radiation (PAR) at wavelengths between 400 and 700 nm (figure 5.12). Radiation in other parts of the electromagnetic spectrum controls photomorphogenetic responses to daylength and seed germination (see section 5.6).

5.4.1 Solar radiation

The modification of the insolation pathway by the vertical stratification of forest vegetation creates two zones of exposure to sunlight. The euphotic zone, at the top of the canopy, is fully exposed and receives between 25 and 100 per cent of the insolation. In response to this, a dense vegetation develops, flowering and fruiting is prolific, productivity is high, and heliophilous epiphytes such as orchids, bromeliads and lichens thrive. Below the canopy is the oligotrophic zone which receives less than 3 per cent illumination and in some cases even less (Carter, 1934; Bunning, 1947; Cachan, 1963; Yoda, 1974). Here, vegetation growth is far less prolific and sciophilous epiphytes such as ferns, mosses and liverworts abound. There is fierce competition between plants for

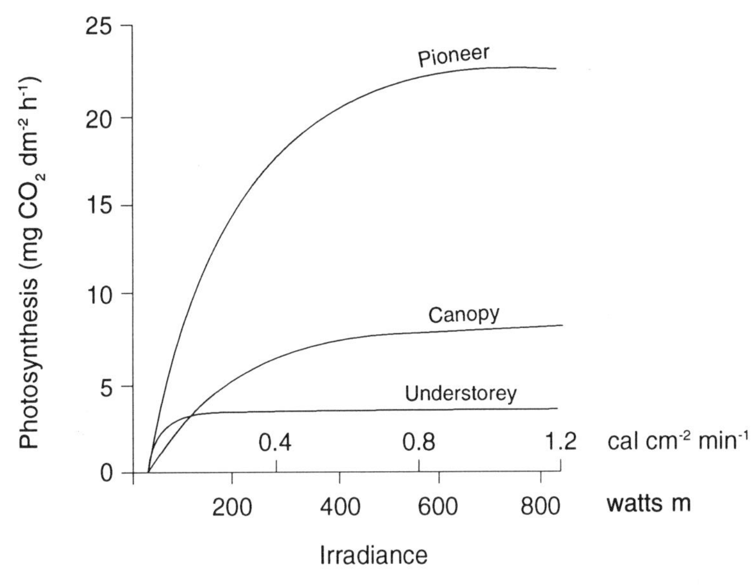

Figure 5.12
Insolation through a forest canopy (adapted from Bazzaz and Pickett, 1980).

light and many plants have leaves specifically adapted to maximize the receipt of insolation (Lee and Lowry, 1974; Lee et al., 1979). In terms of spectral properties the ratio of visible red ($\approx$ 600–700 nm) to infrared (700–1100 nm) radiation is thought to control seed germination. This ratio is high in the canopy but low in the understorey. Horizontal variations in forest illumination have also been identified by Longman and Jeník (1987).

Sunflecks comprise direct and diffuse insolation, and radiation reflected by vegetation (Morgan and Smith, 1981). Generally they do not have a high radiation flux and are prone to many changes in intensity because of the sun's daily cycle and leaf movements. Nonetheless they are important components of understorey illumination. They provide momentary high light fluxes (Kwesiga, 1985) which the leaves of some species, e.g. *Shorea leprosula* and *S. maxwelliana*, respond to almost instantaneously (UNESCO, 1978).

Over longer time scales the amount of insolation received by the forest varies both on a diurnal basis (figure 5.13) and throughout the year (Longman and Jeník, 1987). Seasonality, in particular, has important effects on photosynthesis, and therefore biomass production, as well as on plant phenology.

5.4.2 Temperature

Air temperatures in tropical forests vary markedly, those above the canopy and in clearings being very different from those under the canopy. Understorey maximum and minimum temperatures are lower than those above the canopy and in clearings (Schultz, 1960) and temperature fluctuations are more mixed in the understorey (Cachan and Duval, 1963; Longman and Jeník, 1987) (figure 5.14). Soil temperatures show similar patterns of

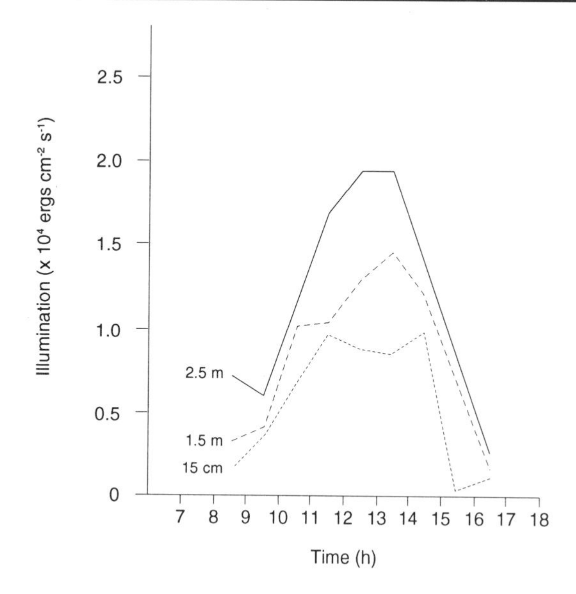

Figure 5.13
Variations in insolation with height above the ground and time in a Surinam rainforest (after Schultz, 1960).

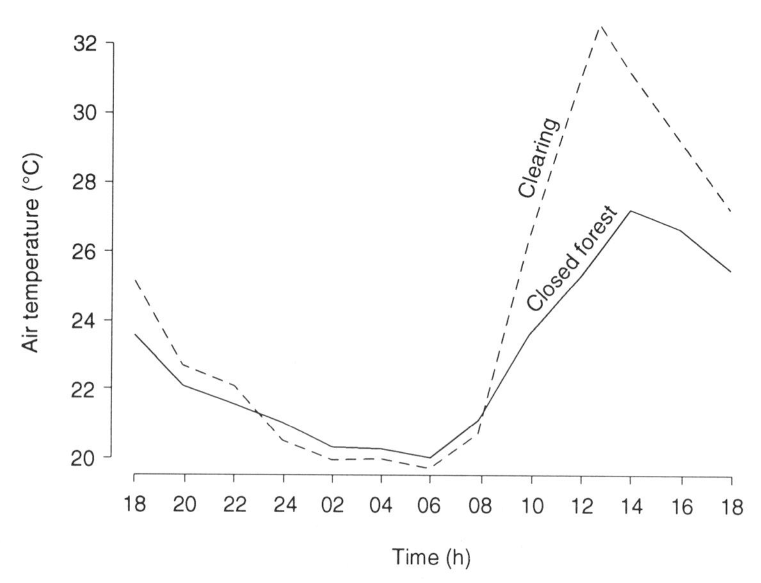

Figure 5.14
Air temperature variations over time in a closed forest and a clearing in southwest Ghana (after Longman and Jeník, 1987, p. 45, figure 3.8).

variation between forested and non-forested land to air temperatures (Schultz, 1960), although such variations are more moderate than those above the ground (figure 5.15). Temperature variations between different types of forest, e.g. evergreen riparian and deciduous forests in Costa Rica (Janzen, 1976a), appear to be minimal.

5.4.3 Moisture

Moisture regimes have been used by many researchers to classify tropical forests (see for example section 5.2.1). However, the *wetness* of tropical forests is often overstressed, and most forests

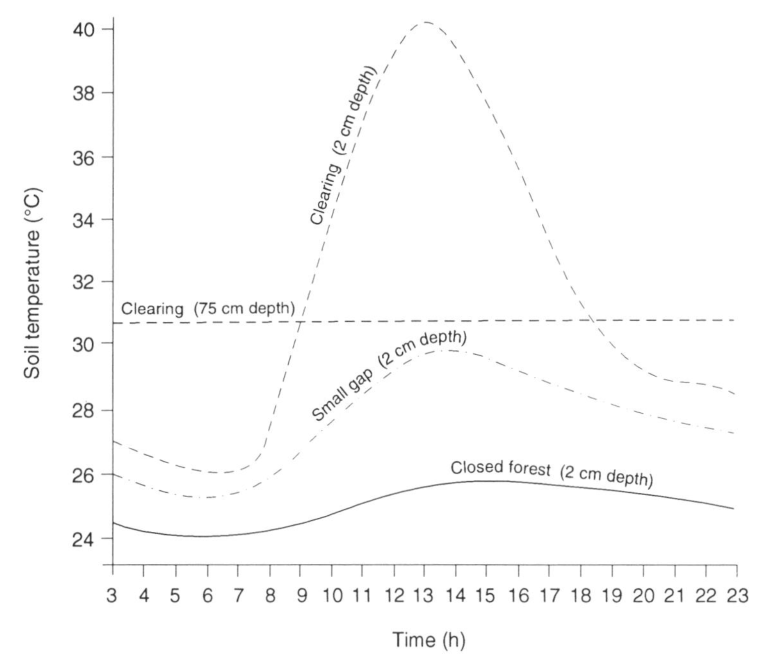

Figure 5.15
Soil temperature variations in closed forest, small gaps and clearings in Surinam. All temperatures were measured at 2 cm depth and are averaged over the dry season (after Schultz, 1960).

are mesic with freely draining soils. This means that forest trees can suffer moisture stresses which, if they continue for a week or so, often promote mass flowering. If they continue longer they can cause ecologically significant changes either directly or through extensive fires such as those recorded in southeast Asia at various times this century (Mackie, 1984; Beaman et al., 1985; Malingreau et al., 1985; Woods, 1989; Aitken and Leigh, 1992) and in the Amazon (Sanford et al., 1985). Diurnal fluctuations in moisture levels also occur, but their ecological significance is unclear.

Rainfall is the most important moisture input into most humid tropical forests, the main exceptions being cloud forests where occult (fog and cloud) precipitation is important. A fair proportion of the rainfall over forested areas probably originates from local evaporation. For example, Salati et al. (1979) found that almost equal proportions of the precipitation in the Amazon Basin originated in the South Atlantic Ocean and by evaporation from the forests themselves. Partitioning of rainfall within the forest is important but more complicated than in temperate forests. Interception is high (table 5.11) owing to the effect of crown epiphytes and climbers. Stemflow, though low, is often harvested by epiphytes and specially adapted leaves, and throughfall is enhanced by the action of drip tips (Williamson et al., 1983). Dew is also an important moisture source and on cloudless nights precipitation equivalents ranging from 100 to 300 ml m^{-2} have been recorded. If dew remains on leaves for more than a few hours significant benefits can accrue to the plant owing to the suppression of transpiration.

Table 5.11 Partitioning of precipitation in some tropical forests (as a percentage of all precipitation)

Forest type	*Location*	*Interception*	*Stemflow*	*Throughfall*	*Reference*
Rainforest	El Verde, Puerto Rico	27–38	0–1	62–73	Sollins and Drewry (1970)
Lowland dipterocarp forest	Jelebu, Malaysia	22	0.5	77.5	Manokaran (1979)
Natural teak forest	Lampang, Thailand	63	0	37	Chunkao et al. (1971)

Source: after Doley, 1981

There are significant variations in both relative humidity and saturation vapour pressure above and below the canopy, between mature forest and clearings, on a diurnal basis and between seasons (Cachan and Duval, 1963; Odum and Pigeon, 1970; Fletcher et al., 1985; Longman and Jeník, 1987) (figure 5.16). Odum and Pigeon (1970) calculated that day-time evaporation rates in the oligotrophic zone were about a quarter of those in the euphotic zone. In fact, the evaporation rates found in the forest canopy are comparable to those found in savanna environments. These data point to an interesting phenomenon: the crowns of tropical forest trees have high water stresses, whilst evapotranspiration rates under their canopy are low.

Soil moisture variations influence the functioning of tropical forest ecosystems. At a medium scale they can be considered as part of the topography–soil–vegetation relationships. However, local variations in soil moisture over time can have profound effects on plant functioning even in tropical forests. Bronchart (1963) found that soil moisture levels reached the wilting point several times over a twelve-month period in humid tropical parts of Zaire.

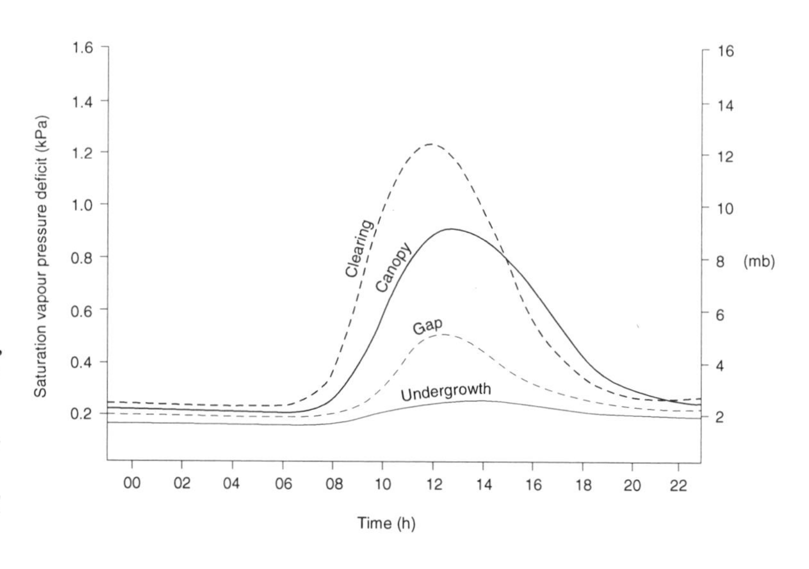

Figure 5.16
Saturation vapour pressure measurements in undergrowth, canopy, 0.4 ha gap and 0.5 ha clearing sites from a lowland rainforest in Costa Rica. The values are geometric means of measurements made between 28 April and 22 October 1981 (after Fletcher et al., 1985).

5.5 Nutrient Cycling

The cycling of plant nutrients is a key element in the ecological functioning of tropical forests, and is of particular relevance for understanding vegetation succession (section 5.6), the soil fertility impacts of deforestation and the global carbon cycle. The main nutrient stores and fluxes of tropical forest ecosystems are illustrated in figure 5.17.

5.5.1 Major components of nutrient cycling

Nutrient inputs to humid tropical forest come from precipitation, by dry (aerosol and dust) deposition, from nitrogen fixation by micro-organisms and from rock weathering. Measurements of bulk precipitation of nutrients in the humid tropics (table 5.12) show considerable variability between sites, much of which can be attributed to experimental error (Proctor, 1987). Nevertheless, variations are also related to trends in precipitation, ashfalls from forest fires and shifting agriculture (cf. chapter 8), volcanic eruptions (Kellman et al., 1982; Proctor, 1987) and regional duststorms (e.g. the harmattan which affects the seasonal rain forests of West Africa). The nutrient load of precipitation varies between rainfall events, and for many forests most of the annual input occurs in a few storms (table 5.13).

Measurements of other types of nutrient inputs to tropical forest ecosystems are rare. No measurements of dry precipitation (dust

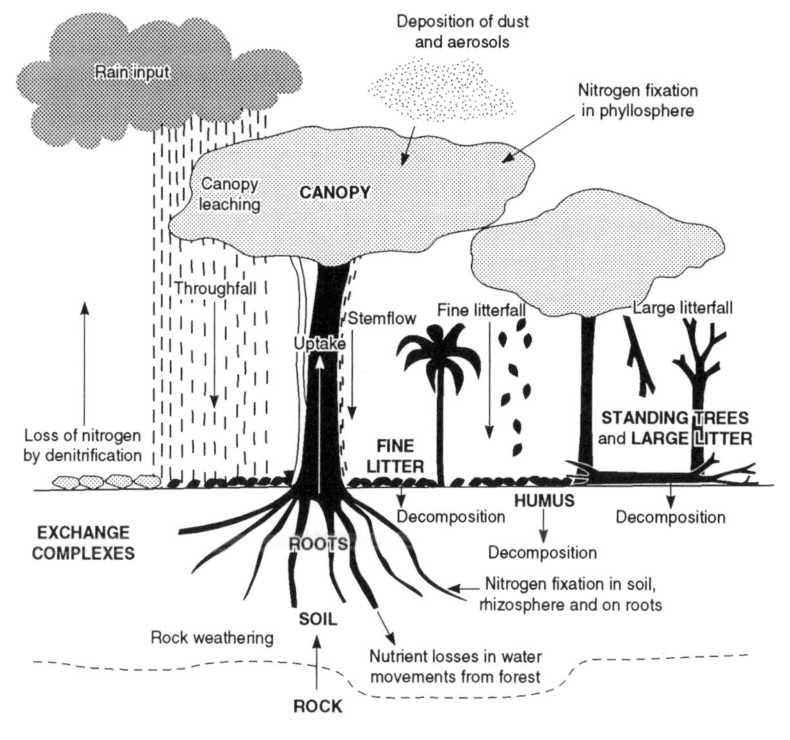

Figure 5.17
Main nutrient fluxes and stores in humid tropical forests (modified after Procter, 1987).

Table 5.12 Annual rates of nutrient precipitation in humid tropical forest environments (kg ha^{-1} yr^{-1})

Country	*N*	*P*	*K*	*Ca*	*Mg*
Lowland forest					
Australia (1)	–	–	4	3	2.5
Brazil (2)	10	0.3	–	3.6	3
Brazil (3)	5	0.4	–	0.3	0.2
Ghana (4)	14	0.4	17.5	12.7	11.3
Ivory Coast (5)	14	–	–	–	–
Ivory Coast (6)	21.2	2.3	5.5	30	7
Malaysia (7)	–	–	12.5	14	3.3
Malaysia (8)	13.5	–	6.4	4.2	0.7
Panama (9)	–	1	9.5	29.3	4.9
Papua New Guinea (10)	–	–	0.8	0	0.3
Puerto Rico (17)			21.8		
Venezuela					
Caatinga (11)	21.2	24.8	23.4	27	3.4
Caatinga (12)	–	16.7	–	16	–
Caatinga (16)				16	
Terra firme (11)	21.7	24.9	24.6	28.4	3.5
Terra firme (12)	–	26.9	11.6	11.6	3
Terra firme (18)	6.1		10.6	8.8	2.4
Lower montane forests					
Papua New Guinea (13)	6.5	1.1	2.6	5.6	5.2
Puerto Rico (14)	–	–	8.9	6.4	19.8
Venezuela (15)	9.9	1.1	2.6	5.6	5.2

Sources: after Proctor, 1972, p. 141, table 2, but with additional data; 1, Brasall and Sinclair, 1983; 2, Anon., 1972; 3, Brinkmann, 1985; 4, Nye and Greenland, 1960; 5, Servant et al., 1984; 6, Bernhard-Reversat, 1975; 7, Kenworthy, 1971; 8, Manokaran, 1980; 9, Golley et al., 1975; 10, Turvey, 1974; 11, Jordan et al., 1980; 12, Jordan, 1982; 13, Edwards, 1982; 14, Clements and Colon, 1975; 15, Grimm and Fassbender, 1981; 16, Herrera, 1979; 17, Jordan et al., 1972; 18, Jordan, 1989

Table 5.13 Nutrient precipitation in Honduras

Nutrient	*Number of rain days in which more than 50 per cent of the annual input fell (total number of rain days 170)*
Calcium	1
Magnesium	1
Nitrogen	3
Phosphorus	9
Potassium	8
Sodium	5

Source: after Kellman et al., 1982

and aerosols) have been made in the humid tropics (Proctor, 1987) although it can be assumed that annual and spatial variability is high, mainly because forest burning produces pulses in dry nutrient precipitation; industrial pollutant levels, however, are generally low. Plant nutrient uptake from rock weathering has been noted in Java (Bruijnzeel, 1983) and Sarawak (Baillie and Ashton, 1983). Nitrogen fixation is very important, and estimates from Amazonia suggest that fixation rates range from 1.5 to 20 kg ha^{-1} on *terra firme* and are over 240 kg ha^{-1} in flooded forests (Sylvester-Bradley et al., 1980).

Nutrients are lost from forests during soil erosion, in solution (throughflow, streamflow and percolation to groundwater), during fires and by denitrification. Quantitative information on nutrient inputs from rainfall and losses measured in streamflow or by percolation in lysimeters (table 5.14) exhibit considerable variability. However, it is apparent that some forests exhibit net nutrient losses whilst others show gains. Measurements of other types of nutrient loss are few and far between. Research into soil erosion has mainly been carried out under disturbed situations but, so far, little emphasis has been placed on nutrient losses. Nitrogen loss by denitrification and volatization to ammonia has been estimated to be around 15 per cent in the Ivory Coast (Servant et al., 1984).

5.5.2 Nutrient fluxes

Proctor (1987) divides nutrient fluxes in forests into two – relatively rapid and relatively slow – although the distinction may be difficult to make in the forest. Rapid fluxes include small litterfall (e.g. the fall of leaves, flowers, fruits) and nutrients transported by throughfall and stemflow. Slow fluxes comprise the decay of large branches and trunks, and transfers through the soil nutrient store.

Relatively rapid fluxes

Small litterfall rates in lowland forests range from $3.4\,t\,ha\,yr^{-1}$ to $15.1\,t\,ha\,yr^{-1}$ (Parker, 1983; Proctor, 1984; Rai and Proctor, 1986). However, it has been difficult to assess their contribution to specific nutrient fluxes because of the spatial and temporal variations of different nutrients (Proctor, 1984) and their strong seasonality.

Leaf-fall has the strongest seasonality of all types of small litterfall (Lugo et al., 1978; Deitrich et al., 1982; Lam and Dudgeon, 1985; Martínez-Yrízar and Sarukhán, 1990) and is greatest in the dry season in seasonal forests. This is because one of the most important controls on leaf shedding is soil moisture

Table 5.14 Comparison of rainfall input, streamflow and percolation losses of some major nutrients in humid tropical ecosystems ($kg\,ha^{-1}\,yr^{-1}$)

Location (and reference)	*N*	*P*	*K*	*Ca*	*Mg*
Lowland forest					
Brazil (1)	29 (−24)	0.3 (0.1)		0.9 (−0.6)	0.5 (−0.3)
Ivory Coast (s) (2)	0.4 (13.6)				
Malaysia (s) (3)			11.3 (1.2)	2.1 (11.9)	1.5 (1.8)
Panama (s) (4)		0.7 (0.3)	9.3 (0.2)	163 (−134)	43.6 (−38.7)
Papua New Guinea (s) (5)			14.9 (−14.1)	24.8 (−24.8)	51 (−50.7)
Venezuela					
Caatinga (6)	16 (7)		2.8 (−13.2)		
Terra firme (6)		30 (−3.1)	4.4 (7.2)	3.9 (7.7)	0.7 (2.3)
Lower montane forests					
Puerto Rico (6)				43.1 (−21.3)	18.2 (8.1)
Venezuela (7)	5.1 (4.8)	0.3 (0.8)	2.2 (0.4)	1.6 (4)	0.6 (4.6)

(s), losses measured in streamflow; all other losses were measured in lysimeters (i.e. percolation losses).
The values in parentheses are the differences between rainfall input and streamflow or percolation loss.

Sources: adapted from Proctor, 1987, p. 143, table 3; 1, Brinkman, 1985; 2, Servant et al., 1984; 3, Kenworthy, 1971; 4, Golley et al., 1975; 5, Turvey, 1974; 6, Jordan, 1982; 7, Grimm and Fassbender, 1981 deficit (a delayed response to declining rainfall) although photosynthetic effects cannot be discounted. Martínez-Yrízar and Sarukhán (1990) analysed a five year record of litterfall in the tropical deciduous forests of western Mexico and found significant variations in the timing and amount of leaf-fall in two contrasting forests (figure 5.18(a)). Seasonal variations are more muted in moist forests (Dantas and Phillipson, 1989) (figure 5.18(b)). Leaf-fall can occur during wet periods when it can be attributed to the heliophytic nature of some trees, continuous leaf renewal, wind-blow, grazing (which is most pronounced after leaf flushes) and short dry spells.

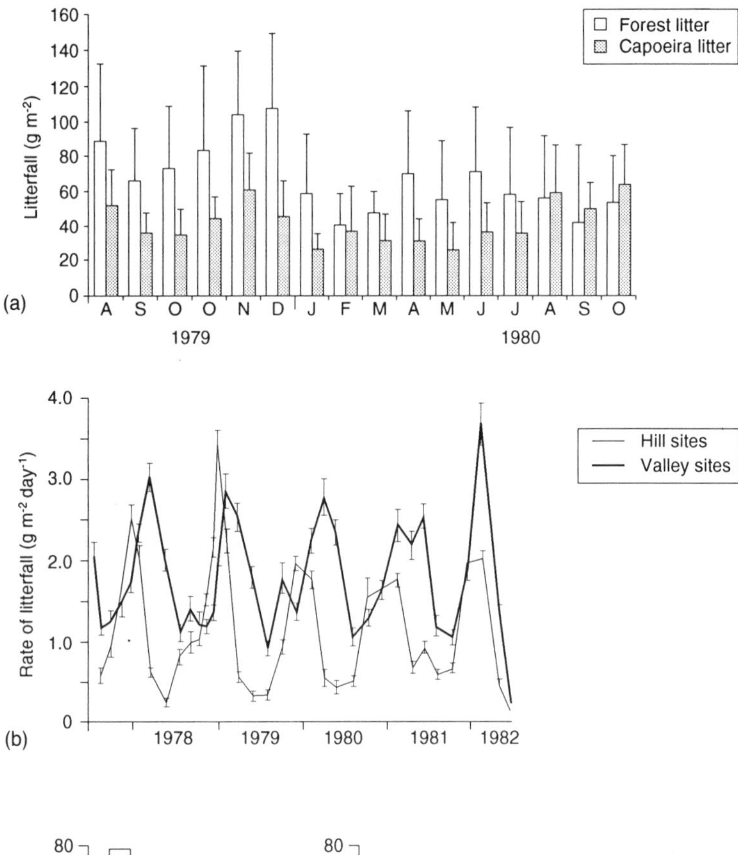

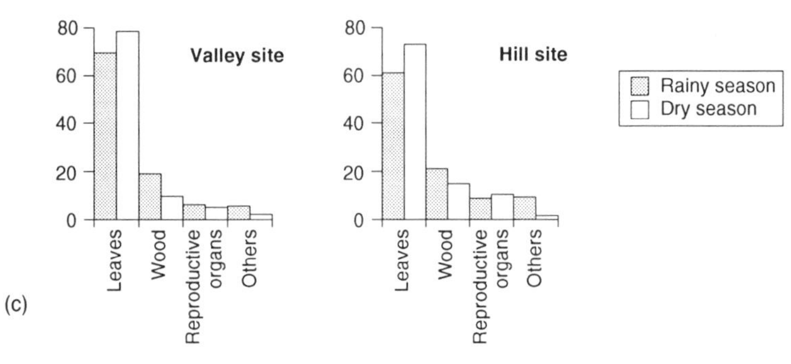

Figure 5.18
Seasonal variations in leaf fall (a) in a humid tropical forest (after Dantas and Phillipson, 1989) and (b) in tropical deciduous forests, Jalisco, Mexico (data from Martinez-Yrizar and Sarukhan, 1990); (c) vartiations in components for small litterfall between wet and dry seasons (data from Martinez-Yrizar and Sarukhan, 1990).

Other types of small litterfall (small branchfall, twigfall, the loss of flowers and fruits) show less seasonality than leaf-fall (figure 5.18(c)). Small branches show a slight increase in frequency of fall during the dry season, but there is a noticeable time lag which is due to the differences in leaf and branch abscission. The reproductive organs often do not show marked seasonal effects as these are masked by animal foraging. Other variations in litterfall may be related to forest age.

Parker (1983) reviewed throughfall and stemflow in forests. Both processes are efficient at leaching nutrients from plant surfaces. The rate of leaching from leaves, for instance, is dependent on temperature, rainfall, the leaf area index and the residence time of water on the leaves. Throughfall and stemflow also collect nutrients by washing off frass (insect residues and remains). Parker also claimed that nitrogen fixed by leaves is leached, but this is disputed by Edwards (1982). What is clear, however, is that water passing through the canopy becomes enriched with certain nutrients (table 5.15) and reduced in others, most notably phosphorus, calcium and sulphur (Jordan et al., 1980). This is probably due to direct absorption by leaves, aerial roots and other plant surfaces, as well as by epiphytes (Benzing, 1983). It is also probable that some nutrients move preferentially along particular pathways, e.g. potassium is most concentrated in throughfall.

The partitioning of nutrients between small litterfall, throughflow and stemflow is important in terms of vegetation growth. This is because nutrients in solution are immediately available to plants, whereas litter has to decompose before the nutrients are released. Generally, rates of litter decomposition (turnover) in tropical forests are thought to be very rapid but this is by no means common (Anderson and Swift, 1983). Proctor (1987) calculated turnover

Table 5.15 Canopy nutrient enrichment (throughfall – rainfall) for some lowland and lower montane forests (all values in kg ha^{-1} yr^{-1})

Forest types (and reference)	*N*	*P*	*K*	*Ca*	*Mg*
Lowland forest					
Australia (site A) (1)	–	–	121	47	23
Australia (site B) (1)	–	–	93	53	18
Brazil (2)	20	0.5	–	10.2	6.6
Ghana (3)	12.3	3.7	220	29	18
Ivory Coast (Banco Plateau) (4)	58.8	−0.1	59.5	9	34
Ivory Coast (Banco Valley) (4)	59.8	7.5	169.5	17	41
Ivory Coast (Yapo) (4)	11.8	3.7	82.5	5	16
Panama (5)	–	0.6	50	37	10
Venezuela (6)	–	–	6.5	−5.2	−0.9
Lower montane forest					
New Guinea (7)	23.1	2.0	63.8	15.4	9.6
Puerto Rico (8)	–	–	79	25.8	47.5
Venezuela (9, 10)	8.5	1.4	69.7	6.9	3.3

Sources: adapted from Proctor, 1987, pp. 144–5, table 4, but with additional data; 1, Brasell and Sinclair, 198?; 2, Brinkmann, 1985; 3, Nye, 1961; 4, Bernhard-Reversat, 1975; 5, Golley et al., 1975; 6, Jordan, 1989; 7, Edwards, 1982; 8, Clements and Colon, 1975; 9, Fassbender and Grimm, 198?; 10, Grimm and Fassbender, 1981 coefficients for a variety of forests (table 5.16). All lowland tropical forests have rapid turnover rates, but those in cooler, montane forests are slower. Factors other than temperature are also important and it is likely that variations in the chemical and physical composition of litter, the decomposer community and the chemistry of the micro-environment are influential.

Relatively slow fluxes

Much less is known about slow fluxes than fast fluxes. Brasell and Sinclair (1983) suggest that large branchfall accounts for a *significant* amount of recycled nutrients. Measurements of the nutrient contents of tree trunks from Jamaican montane forests (Tanner, 1980) and the Ivory Coast (Bernhard-Reversat, 1975) suggest that relatively few nutrients are immobilized in wood each year although the actual pool of nutrients can be very high.

Table 5.16 Litterfall and turnover coefficients for tropical forests

Country	*Turnover coefficient[a] Leaves*	*Small litter[b]*	*Reference*
Lowland rainforest			
Brazil	1.5	1.1	Klinge (1973)
Ghana	2.5	2.0	John (1973)
Ivory Coast (Banco)	3.3		Bernhard (1970)
Ivory Coast (Yapo)	3.6		Bernhard (1970)
Malaya (Pasoh)		3.3	Ogawa (1978)
Malaya (Pasoh)	3.6		Yoda (1978)
Malaya (Penang)	1.1	1.0	Gong and Ong (1983)
Malaya (freshwater swamp, Tasak Bera)		1.9	Whitmore (1984)
Panama	2.6	1.2	Healey and Swift, quoted in Proctor (1987)
Sarawak (alluvial)	1.8	1.7	Anderson et al. (1983)
Sarawak (dipterocarp)	1.7	1.3	Anderson et al. (1983)
Sarawak (heath forest)	1.4	1.3	Anderson et al. (1983)
Sarawak (over limestone)	1.7	1.5	Anderson et al. (1983)
Montane forests			
New Guinea (lower)		1–1.4	Edwards (1977)
Puerto Rico (lower)	0.9		Wiegert (1970)
Jamaica (upper)	0.5–07		Tanner (1981)
Colombia (1630 m a.s.l.)		0.6	Whitmore (1984)
Costa Rica (1000 m a.s.l.)		1.6	Heaney and Proctor (1989)
Costa Rica (2000 m a.s.l.)		1.1	Heaney and Proctor (1989)
Costa Rica (2600 m a.s.l.)		0.8	Heaney and Proctor (1989)

[a]The turnover coefficient is litterfall (t ha^{-1} yr^{-1}) divided by forest floor litter (t ha^{-1}). A coefficient of 1 indicates that all the litter is decomposed in a year, greater than 1 indicates a faster decomposition rate, and less than 1 a slower.
[b]Small litter is defined as fruits, small branches and twigs of less than 2.5–50 mm diameter, and leaves if not specified separately.

Source: adapted from Proctor, 1987, p. 147, table 5, but with additional data.

Food chains and webs involving large numbers of herbivores, carnivores and decomposers can be included in the group of relatively slow fluxes; examples of complete food chains only exist for a very few, very simple chains.

5.5.3 Nutrient stores

Nutrients in tropical forests are stored in the living vegetation above and below the ground, in above-ground plant litter and in the soil. From measurements made of the dry weight of biomass (which were reviewed by Proctor, 1987) it is clear that the majority of biomass is stored above ground (figure 5.17). Below-ground biomass and litter rarely contribute more than 20–25 per cent of the total. It has already been noted that soil nutrient stores are very low in many types of soil, e.g. arenosols, regosols, ferralsols and acrisols.

However, these data must be treated with caution because of sampling errors associated with their collection (Proctor, 1987). With the exception of nutrients stored in trunks and large branches, measurements of the other parts of the ecosystem are probably very variable and poorly sampled.

5.6 Forest Dynamics

Humid tropical forests are dynamic, but succession is a far more complex phenomenon than in temperate forests. It is rarely predictable (Bazzaz and Pickett, 1980) and there are many pathways leading from a pioneer community to a mature forest (Ewel, 1980).

The destruction of forest communities (disturbance) can be either exodynamic or endodynamic (Longman and Jeník, 1987). Exodynamic disturbance is brought about by external forces such as cyclones, volcanic eruptions, landslides, the action of bulldozer herbivores and humans; it usually occurs on a large scale. It is often assumed that the resulting secondary forest will eventually mature to form a climax community. However, evidence from the Yucatan Peninsula, where forests have been developing since the decline of the Mayan culture, does not bear this out: succession is clearly different from that in species-poor temperate forests. Succession in large, disturbed areas in tropical forests has been studied by Budowski (1965), who identified pioneer, early and late-successional, and climax trees growing in the clearings. Examples of succession from different exodynamic disturbance regimes are rare but three situations are noteworthy: first, vegetation establishment on volcanic ashes, e.g. the long-term monitoring of vegetation on the Krakatau Islands from 1883 (Whitaker et al., 1989); second, vegetation establishment on the alluvial deposited by migrating river channels, a situation exemplified by Salo et al. (1986) (figure 5.19); third, vegetation succession on abandoned cultivated areas

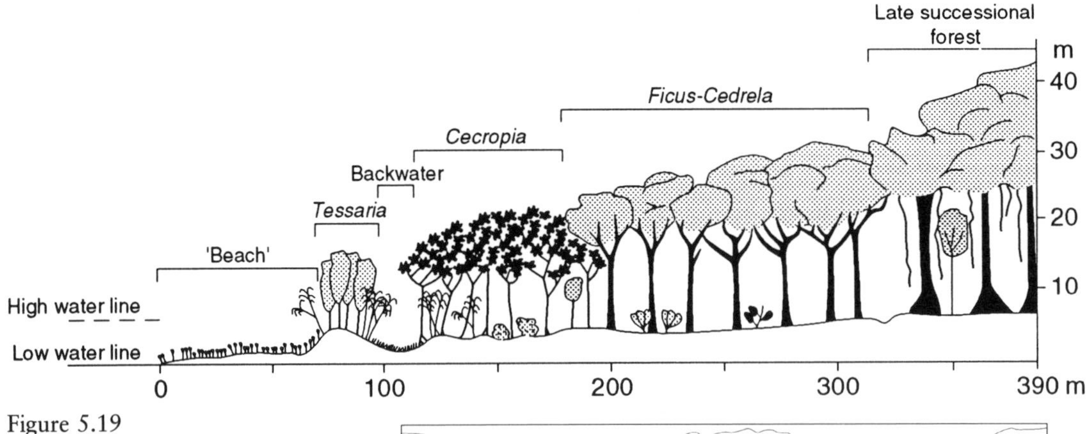

Figure 5.19
Primary succession on newly deposited riverine alluvium, at Cocha Cashu on the Rio Manu in Peruvian Amazonia (Salo et al., 1986, figure 3).

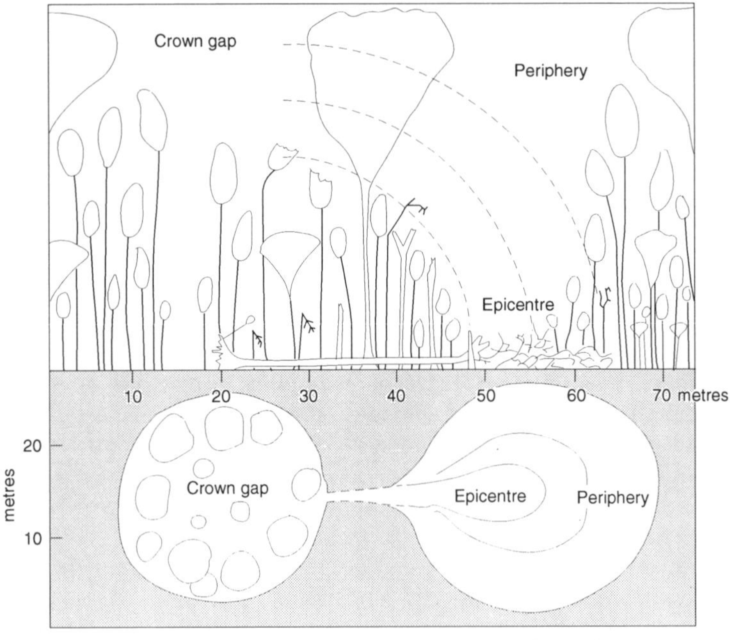

Figure 5.20
Vertical and horizontal projections of a chablis (after Tomlinson and Zimmermann, 1978).

(e.g. Purata, 1986), although the spatial scale of the latter type of disturbance is variable and has an influence on the type of succession. Gómez-Pompa et al. (1991) provide a series of examples of regeneration after exodynamic and endodynamic disturbances.

Small forest clearings are created by endodynamic disturbances (e.g. dead tree falls, wind throw, lightning strikes and lianas bringing down crowns), and the study of forest succession in such clearings is known as gap dynamics. Although such clearings are small, ranging from 1000 to 2000 m^2 in lower montane rain forests to 200 m^2 in lowland rain forest (Kramer, 1933; Whitmore, 1990), they are areally significant. Poore (1968) found that recently formed gaps accounted for up to 33 per cent of Malaysian forests. Oldeman (1978) introduced the concept of a chablis into tropical forestry, a term that embraces tree fall, the resulting canopy gaps and soil disturbance (which leaves gaps in the rhizosphere), and the accumulation of biomass (figure 5.20). A chablis comprises many ecological niches that can be defined in terms of microclimate, soils, plant litter (table 5.17). The term is a rehabilitation of a medieval French word and has no obvious connection with the excellent wines of the same name! A chablis then is the endodynamic process that initiates succession. Many chablis successions have been studied (e.g. Wyatt-Smith, 1966; Swaine and Hall, 1983) and three types of successional trees can be identified:

1 small, short-lived pioneers which need light for their seeds to germinate;
2 large pioneer trees which are found in the mature forest but that also need light for their seeds to germinate;
3 primary or late-successional species that germinate in shady conditions.

Initially a chablis is colonized by short-lived pioneer trees and shrubs. The seeds of pioneer species are produced in abundance, widely dispersed (often by wind) and can remain dormant for long periods. Not surprisingly large banks of pioneer species seeds have been found in most tropical forest soils. Germination of these seeds is stimulated by an increase in insolation levels (photoblasty) or, as is the case with balsa (*Ochroma lagopus*), changes in the soil temperature (thermoblasty). Pioneer trees are opportunistic and aggressive. They grow very fast to fill the canopy gaps because they are light demanding and need to preempt competition from slower growing trees, which they subsequently suppress. Swaine and Hall (1983) measured growth rates of $4\,m\,yr^{-1}$ in some secondary species in Ghana and specimens of *Trema orientalis* averaged around 10 m growth in five years, with one tree growing 17 m over the five years. However, such trees are short-lived, rarely surviving longer than twenty years, e.g. *Cecropia* spp. (Hallé et al., 1978), and put much of their effort into rapid upward growth and reproduction. Leaf growth and trunk thickening are less important. They usually form a simple canopy ranging in height from 2 to

Table 5.17 The main environmental characteristics of a chablis

Zone	*Lower tree and shrub layers*	*Soils*	*Microclimate*[a]
Crown gap	Intact	Soil spread as root plate is ripped out of ground	Increased insolation
Epicentre	Destroyed	Increase in leaf litter	Great increase in insolation
Periphery	Transition from partial destruction to intact	Little effect	Partial increase in insolation

[a]Additionally, in all zones the spectral properties of insolation change.

more than 30 m. Most pioneer species are drawn from seven families (Euphorbiaceae, Malvaceae, Moraceae, Stericuliaceae, Tiliaceae, Ulmaceae and Urticaceae) and in some genera all of the species are pioneers, e.g. *Trema* (Ulmaceae), *Macranga* (Euphorbiaceae) and *Cecropia* (Urticaceae). Most genera and species are widely distributed, which bears testament to their efficient dispersal mechanisms.

Pioneer and climax species are not, however, mutually exclusive. They often grow together, especially in the later stages of a succession. The climax trees grow upward more slowly, ultimately breaking through the canopy of pioneer trees. In fact, the situation is more complex than this as a series of canopies are formed during succession which are broken through by increasingly taller trees – this is termed relay succession. Tree densities in pioneer stands increase rapidly in the first year but then decrease. This trend is paralleled by a decline in the number of pioneer species and an increase in late-successional and climax trees (figure 5.21).

Late-successional and climax trees have markedly different properties from pioneer species. For example, they grow more slowly and many do not start reproducing until they are fifty or sixty years old. They generally produce fewer seeds than pioneers, although the seeds are often very large (the biggest being that of the ironwood (*Eusideroxylon zwageri*) at 230 g). Unlike the seeds of pioneer trees, they usually cannot stand the extreme conditions of soils in forest clearings such as moisture stresses and low temperatures although the seeds of Leguminosae are an exception to this generalization. Moreover, they are large because they

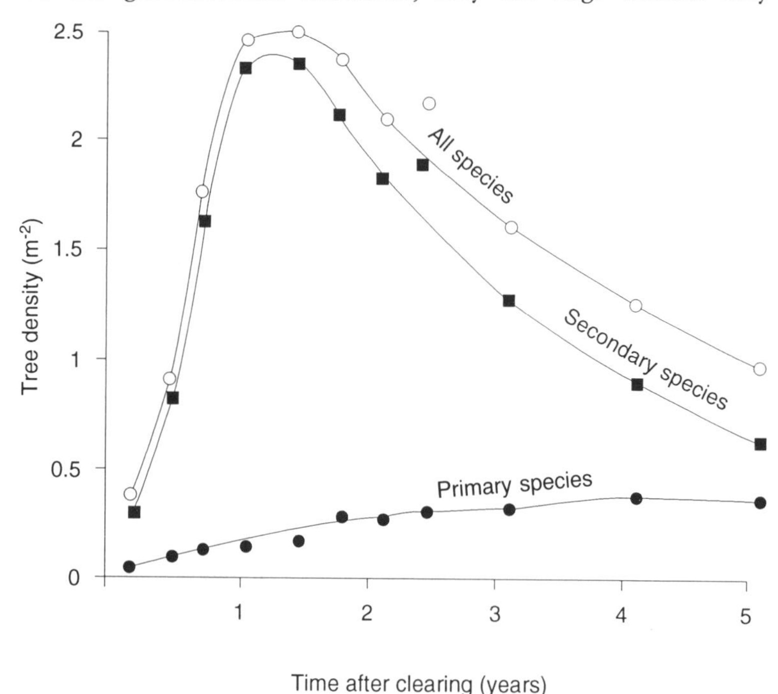

Figure 5.21
Density of trees in a clearing in Ghana during the first five years of succession (after Swaine and Hall, 1983).

contain much food and have to germinate rapidly before they are attacked or decay. Consequently few last more than three to four weeks, most germinating in three to four days. The majority of seeds are dropped close to the mother trees, although survival through to maturity is most likely to be achieved by the seeds that travel furthest. The few seeds that germinate and produce trees that grow to maturity can survive a long time: Dipterocarps in Indo-Malaysian forests often attain ages of 200–250 years; specimens of *Lophira alata* and *Guarea cedratia* in Nigeria up to 300 and 350 years old have been recorded (Jones, 1956); and a survey of trees at La Selva (Costa Rica) gave ages of 52 to 422 years, with a mean of 190 years (Leiberman et al., 1985). The essential differences between pioneer and climax trees are highlighted in table 5.18.

One of the main problems in the study of tropical forest dynamics is the fact that in many gaps there are few seeds of the climax forest trees. This poses problems in modelling and predicting forest succession. Two hypotheses, the mosaic or cyclical theory of regeneration (Aubreville, 1938) and the spot-wise regeneration theory (van Steenis, 1958), have been used to overcome this difficulty. Both hypotheses promote the idea that tropical forests are a mosaic of small patches at different stages of succession (from recent gaps to mature forest) with different species composition (Riswan et al., 1985; Swaine and Hall, 1988). The constantly occurring gaps will be filled by trees recruited from the mosaic and will therefore not necessarily have the same composition as the gap prior to disturbance. Swaine et al., (1987) describe natural and (relatively) undisturbed forests as 'self maintaining; in accordance with local processes of mortality, growth and regeneration, lost trees are continually replaced by new recruits; the vegetation thus continues, in dynamic equilibrium or steady state, to be forest'. Put in more simple, geographical terms – forest composition changes in both space and time over a small scale, but remains the same over the broader spatial and longer time scales unless there is an exodynamic disturbance.

5.7 Other Types of Humid Tropical Vegetation

This chapter has so far focused on humid tropical forests. In this section reference is made to some other important humid tropical vegetation communities.

5.7.1 Wet coastal ecosystems

Two vegetation communities – mangroves and salt-marshes – are found in the estuaries and tidal reaches of low-lying tropical rivers (figure 5.22) (Chapman, 1977). The most important and well known of these are mangrove forests which reach their optimum development in the tropics (West, 1977). Their distribution is controlled by a wide range of environmental factors, and in turn

Table 5.18 The main characteristics of pioneer and climax trees in the humid tropics

Characteristic	Pioneer trees	Climax trees
Germination	Only in canopy gaps allow sunlight to reach forest floor	Usually below canopy in shade
Seedlings	Cannot survive below canopy in shade; never found there	Can survive below canopy; forms a 'seedling bank'
Seeds	Small, produced copiously and more or less continuously and from early in life	Often large, not copious; produced annually or less frequently and only on trees that have (almost) reached full height
Soil seed bank	Many species	Few species
Dispersal	By wind or animals, often over considerable distances	By diverse means, including gravity, sometimes over a short distance
Dormancy	Capable of dormancy (orthodox); commonly abundant in forest soil as a seed bank	Often with no capacity for dormancy (recalcitrant); seldom found in soil seed bank
Growth rate	Carbon fixation rate, unit leaf rate, and relative growth rate	Equivalent rates lower
Compensation point	High	Low
Height growth	Fast	Often slow
Branching	Sparse, few orders	Often copious with several orders
Growth periodicity	Indeterminate (sylleptic); no resting buds	Determinate (proleptic), with resting buds
Leaf life	Short, one generation present, i.e. high turnover rate	Long, sometimes several generations present so slow turnover rate
Herbivory	Leaves susceptible, soft, little chemical defence	Leaves sometimes less susceptible due to mechanical roughness or toxic chemicals
Wood	Usually pale, low density, not siliceous	Variable, pale to very dark, low to high density, sometimes siliceous
Ecological range	Wide	Sometimes narrow
Stand table	Negative	Positive
Longevity	Often short	Sometimes very long

Source: modified after Whitmore, 1990, p. 107, table 7.4

they exert a powerful influence on coastal landform development in the humid tropics (section 6.6.3).

At a macroscale, mangroves are restricted to areas where the air temperature in the coldest month exceeds 20 °C and the annual range is of the order of 10 °C. In addition they are only found in low wave-energy, shallow water areas such as deltas, lagoons, estuaries, and behind bars and spits. The water has to be brackish to saline. At the microscale, factors such as the texture and salinity of the

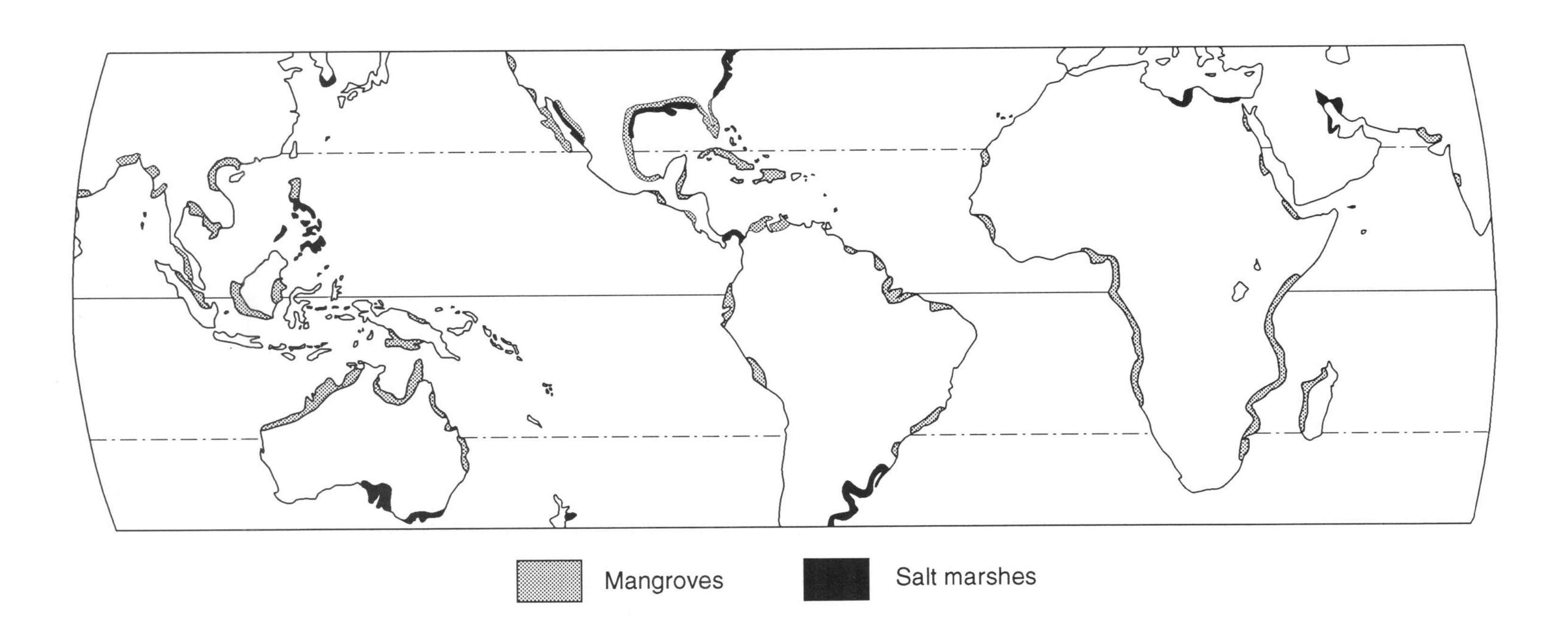

Figure 5.22
Distribution of salt marshes and mangroves in the tropics and subtropics (modified after Chapman, 1977).

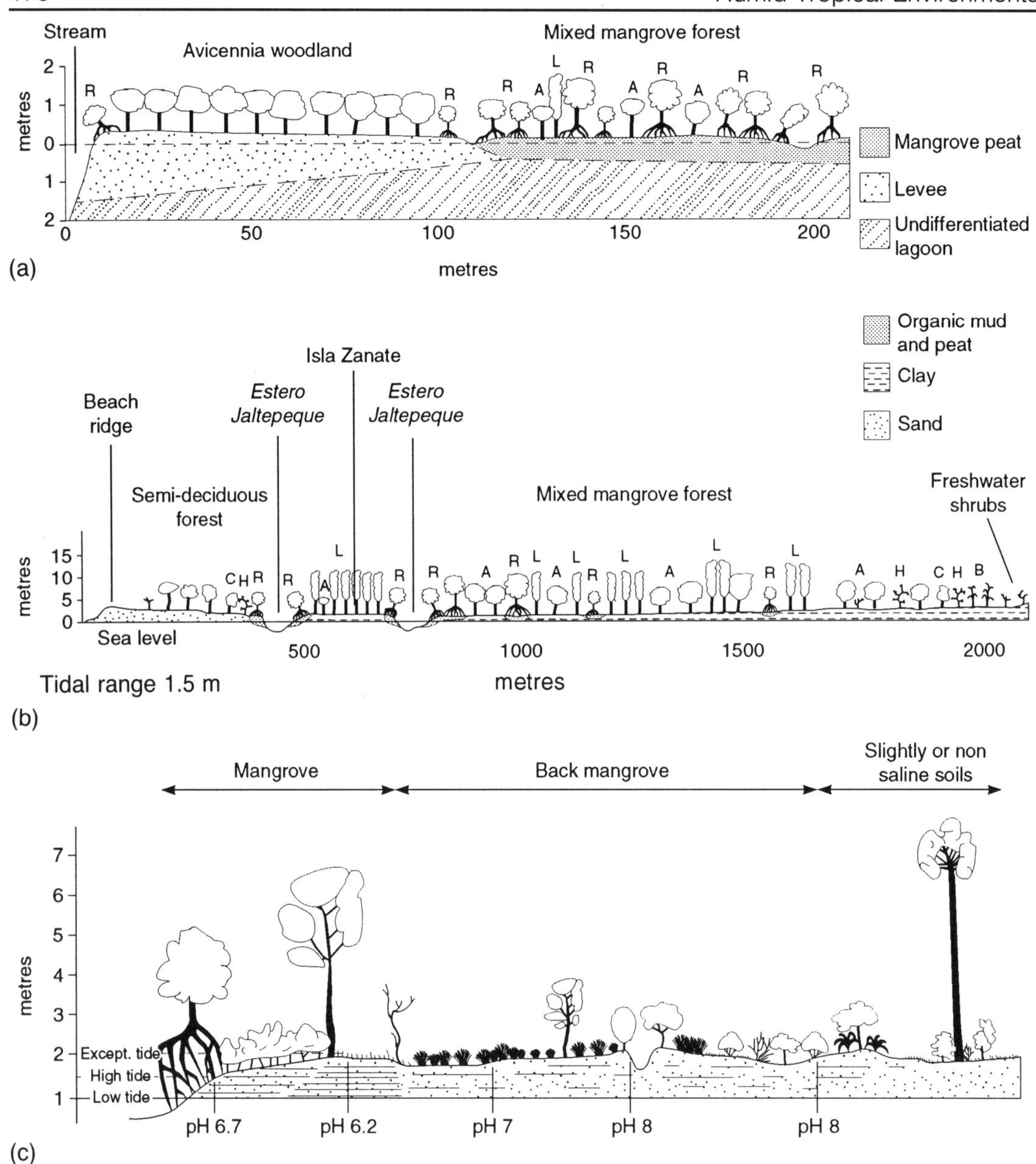

Figure 5.23
Typical humid tropical mangrove formations (A, Avicennia*; C,* Conocarpus*; H,* Hibiscus Hiliaceous*; R,* Rhizophora*; B,* Bactris subglobosa*; L,* Laguhcularia*): (a) inactive levee and interdistributary habitats, Grijalva Delta, Tabasco State, Mexico (after Thom, 1967); (b) Jaltepeque Lagoon, El Salvador (from West, 1977); (c) Cuavery Delta, India (from Blasco, 1977).*

substrate and the tidal range determine their species composition and ecological complexity.

Mangroves occur inland of bare mudflats and sandflats (figure 5.23); on the landward side there are usually transitional freshwater swamps or salt marshes which are dissected by drainage channels.

The distribution of mangrove species at a particular site is closely related to topography, substrate and salinity. Such distributions are clearest in complex deltaic and lagoonal landscapes, and it is possible that edaphic factors are more important than succession in such areas (Thom, 1967). The ecological niches produced by these environments give rise to bands of mangroves parallel to the drainage channels. One of the most comprehensive studies of the spatial patterns of mangrove communities was carried out as early as 1928 in Malaysia (Watson, 1928), but since then there have been many studies most of which are summarized by Chapman (1977). Mangrove communities are dynamic, and the spatial patterns outlined above relate closely to a serial succession.

Humid tropical mangroves are arborescent communities and trees up to 46 m high have been recorded in the Niger Delta (Rosevar, 1947). The undergrowth is severely limited, usually being restricted to mangrove seedlings, and the absence of an understorey makes mangroves unique amongst humid tropical forest formations (Janzen, 1985; Corlett, 1986). Pioneer species establish themselves in semi-liquid mud and as the substrate stabilizes, and other environmental parameters change, the pioneers are replaced. The species exploiting the range of niches provided in these complex environments varies along distinct phytogeographical lines (Chapman, 1977). Nevertheless, important distinctions have been made between genera. The rooting patterns of the trees are distinctive and complex, and include prop roots (*Rhizophora* spp.), aerial pneumatophores (*Avicennia* and *Sonneratia* spp.) and knee roots (*Bruguiera* spp.). *Avicennia* spp. appears to favour finer-textured substrate, whereas *Rhizophora* spp. seems to prefer organic mucks and peats.

5.7.2 Freshwater swamps and flooded forests

Many types of freshwater swamps and flooded forests are found throughout the humid tropics. In Malaysia peat-swamp and freshwater-swamp forests occur frequently (Whitmore, 1984). Freshwater-swamp forests are well known from West Africa, and in Sierra Leone Cole (1968) identified two main types of swamp forest – Raphia Palm Forest and Fringing Swamp Forest – which are found throughout the region (plate 5.9). Flooded forests are perhaps best known from Amazonia. Irmler (1977), Prance (1979) and Pires and Prance (1984) recognize two main types: varzéa (areas flooded by muddy or white waters) and igapo (areas flooded by black or clear waters) and four subtypes of varzéa. Goulding (1984) estimates that flooded forests cover about 100,000 km^2 of the Amazon Basin.

Specific adaptations of these forests to seasonal inundations are a rich understorey, trunk buttressing, low density wood which enables trees to float when banks are eroded (Ducke and Black, 1945) and floating seeds (e.g. those of para rubber, *Hevea brasiliensis*), some of which are collected for oil extraction

Plate 5.9
A raphia-dominated freshwater swamp forest from eastern Sierra Leone.

(e.g. andiroba (*Carapa guianensis*) and ucuuba (*Iryanthera surinamensis*)).

5.7.3 Heath forests

Heath forests are best known from southeast Asia where they replace lowland humid tropical forest on poor soils – usually podzols or albic arenosols related to sandstone outcrops and raised beaches (Ashton, 1971; Whitmore, 1984). They are very extensive in Sarawak where they are called *kerangas*. In terms of floristic composition they are similar to the surrounding forests, the main difference being in the structure. The canopy is low, there are few emergents, they take on a thicket-like form at ground level and the foliage colours are dominated by reds and browns (Ashton, 1971). Although they have no endemic species, species richness is low (860 species versus 1800–2300 species in lowland forests in Sarawak) (Brünig, 1973).

Brünig (1969, 1970, 1971) has argued that their occurrence is the result of periodic water deficiencies related to droughts, and that this is borne out in their structure and physiognomy. Other workers (e.g. Richards, 1952; Proctor et al., 1983a, b) argue that they occur because of nutrient deficiencies in the soil, and that the plants are adapted to reduce the intake of toxic hydrogen ions and phenols from the acidic soils.

5.8 The Palaeogeography of Humid Tropical Vegetation

Research into the palaeogeography of humid tropics has produced a plethora of important biogeographical contributions (e.g. Flenley, 1979). More recently, it has received renewed stimulus as scientists attempt to identify areas of species richness, endemism and forest refugia to lend weight to the fight to preserve humid tropical forests as biosphere reserves, national parks and various other types of reserve. It is only possible to illustrate the main dimensions of this research in the remainder of this chapter and the reader is directed to the extensive biogeographical literature covering the area.

5.8.1 Floristics and plate tectonics

The three tropical forest formations lie on parts of the great southern continent of Gondwanaland which broke away from Pangaea about 180 Ma ago, and achieved its present configuration at about 10 Ma during the late Tertiary. This break up has exerted a significant influence on floristic patterns and a number of phytogeographical patterns have been distinguished.

Pantropical distributions are those which have representatives in all three forest formations. Fifty-nine plant families and 339 genera are pantropical. The large number of pantropical genera is probably due to the fact that the major steps in the evolution of flowering plants had been achieved before Gondwanaland broke up. Disjunct distributions are those with representatives in two of the three areas. They indicate either extinction in the third region or evolution after one of the 'continents' had split off. Finally, there are those families and genera which are restricted to one region; this is thought to be indicative of post-splitting evolution. The palaeogeography of humid tropical vegetation is dealt with in some detail by Whitmore (1990) and excellent accounts of the evolution of the phytogeography of the southeast Asian forests are provided by Whitmore (1981) and Flenley (1979).

5.8.2 Humid tropical vegetation during the late Tertiary

Little is known about humid forest vegetation between *c.* 10 Ma BP and the beginning of the Quaternary. During the late Tertiary palynological evidence from a core dating back to 3.5 Ma BP (van der Hammen, 1974) yielded twenty-seven oscillations suggesting that vegetation fluctuations similar to those that have been described for the Quaternary (see section 5.8.3) occurred in the late Tertiary. Whitmore (1990) argues that these fluctuations are the cause of the amphi-Pacific disjunct plate distributions described by van Steenis (1962). Moreover, he suggests that drier phases during the Tertiary may have led to the poverty of the African humid tropical flora, particularly of bamboos, ferns and palms. In addition, the floristic differences that are found between Australia and the adjacent islands may also be a result of Tertiary desiccation.

5.8.3 Humid tropical vegetation during the Quaternary

Despite the paucity of studies compared with temperate regions, it is clear that climatic fluctuations during the Quaternary have influenced humid tropical vegetation. Humid tropical climates during the glacial phases were cooler, with a lower annual but more seasonal rainfall than at present. During such phases the forest areas shrank. Interglacial periods were warm with an increased annual rainfall that had a less seasonal rainfall distribution in the low latitudes. Forests expanded in response to these conditions. Furthermore, during the glacial maxima sea levels were up to 180 m lower than at present and areas that are now shallow seas were exposed. For example, the Sunda and Sahul Shelves were exposed enabling many islands of the southeast Asian Archipelago to join up and allowing species migration between 'islands' (figure 5.24).

Most palaeoecological studies of humid tropical vegetation have been carried out on cores from high altitude lakes and bogs in the montane forest zones. Studies of material from lowland peat deposits are less common. Walker and Chen (1987) provide a comprehensive review of studies made up until the mid-1980s. Two types of vegetation fluctuations appear to have occurred: first, changes in the altitudinal boundaries between forest types; second, fluctuations between lowland rainforest and savanna woodland or grassland.

The longest records of montane humid tropical vegetation fluctuations during the Quaternary come from the northern Andes. The evidence from these cores indicates that there were oscillations in the relative abundance of tree and shrub pollen compared with herb and grass pollen. This is taken to indicate that altitudinal

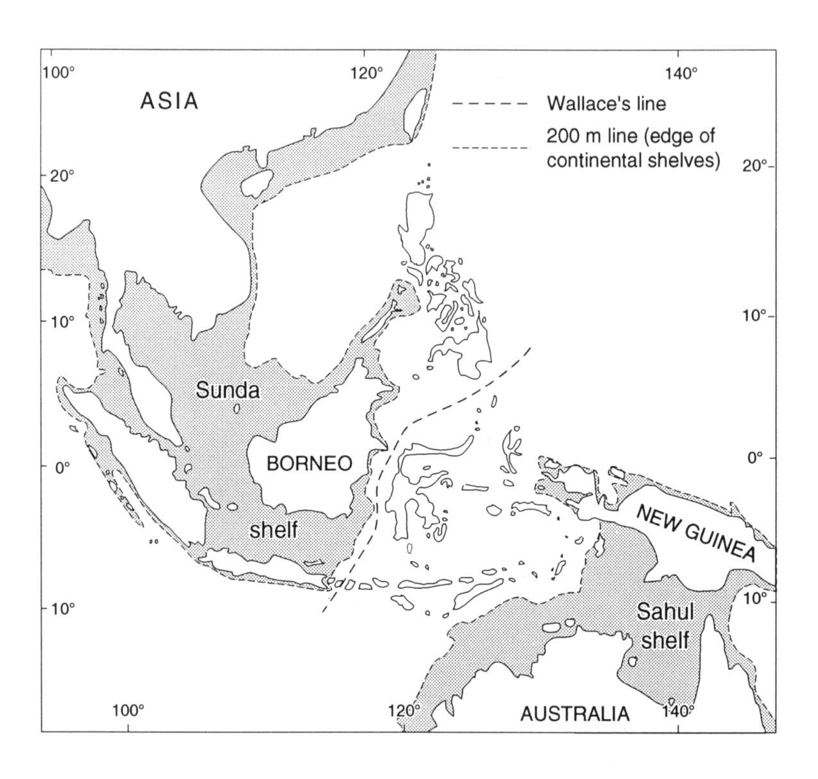

Figure 5.24
Exposure of the Sunda and Sahul Shelves during the Quaternary (after Whitmore, 1984, figure 1.9).

vegetation belts were depressed and compressed during cold periods, and that during warmer, interglacial periods the tree-line rose and the vegetation belts expanded (figure 5.25). Other studies in East Africa, Indonesia and Papua New Guinea have shown comparable fluctuations of altitudinal vegetation belts over the last 30,000 years (Flenley, 1979). The forest limit in all three regions was depressed prior to 30 KA BP and again between 15 and 16 Ka BP; it rose rapidly after this time reaching its maximum elevation around 8 Ka BP. Walker and Chen (1987) argue that lowland tropical vegetation patterns showed substantial changes at the times when montane forest fluctuations were depressed. In lowland areas, the vegetation fluctuations indicated by palynological studies were generally between forest and savanna communities. This again indicates fluctuations between wetter and drier conditions, and this is supported by a variety of pedological and geomorphological studies.

The key date in the late Quaternary history of the humid tropics appears to be around 10 Ka BP. Walker and Chen (1987) describe conditions prior to this as cryomeric, i.e. cooler and generally drier. Such conditions may have existed for up to 2 million years before 10 Ka BP, although such conditions would have been punctuated by warmer periods or thermomeres. Under cryomeric conditions lowland forests would have occupied a much smaller area than at present, altitudinal forest belts would have been depressed, and there would probably have been floristic and ecological adjustments to the smaller range of niches available. For example, it has been calculated that forests in Africa and New Guinea contracted dramatically, and that upland forests were as low as 700 m a.s.l. on Mount Kenya and 1500 m a.s.l. in New Guinea. Furthermore, some researchers argue that during the last glacial maxima (*c.* 18 Ka BP) lowland forests contracted to a few isolated refugia, which were often located in wet, lower montane regions. These researchers argue that the refugia can be identified at the present time by areas with the highest levels of endemism and hybridization (figure 5.26). However, palaeoecological evidence to support the refugia hypothesis is lacking (Walker, 1982; Colinvaux, 1987). Moreover, the theory has been criticized on evolutionary grounds and the calculated rates of forest evolution in Africa (Hamilton, 1976) also fail to lend support.

The changes in vegetation around 10 Ka BP occurred in response to a change from cryomeric to thermomeric conditions. Palynological evidence suggests that vegetation responded rapidly. In New Guinea the tree-line rose 6 m every century for 2000 years (Hope, 1976), and on the Atherton Tablelands in northeast Australia dry, fire-controlled seasonal *Eucalyptus* woodland was replaced by seasonal rain forest over a period of 1000–1500 years (Chen, 1986). However, it is likely that changes were slower in areas where forests already existed because of competition from forest species.

To summarize, it appears that humid tropical forests are more widely distributed at present than at any time between 2 Ma BP and

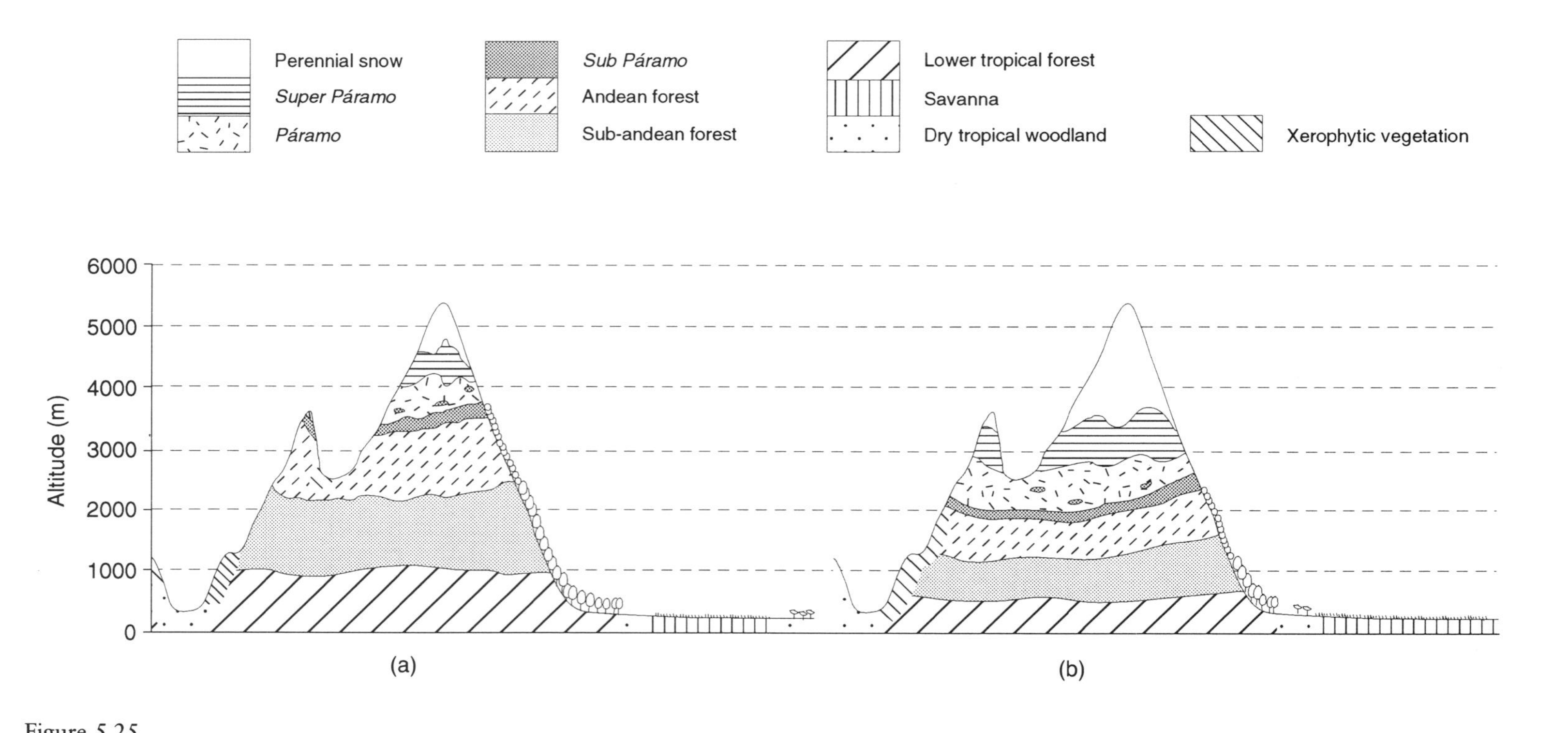

Figure 5.25
Quaternary vegetation dynamics on the northern Andes, Colombia: (a) present day; (b) 14,000–20,000 BP *(after van der Hammen, in Flenley, 1979, figure 4.27).*

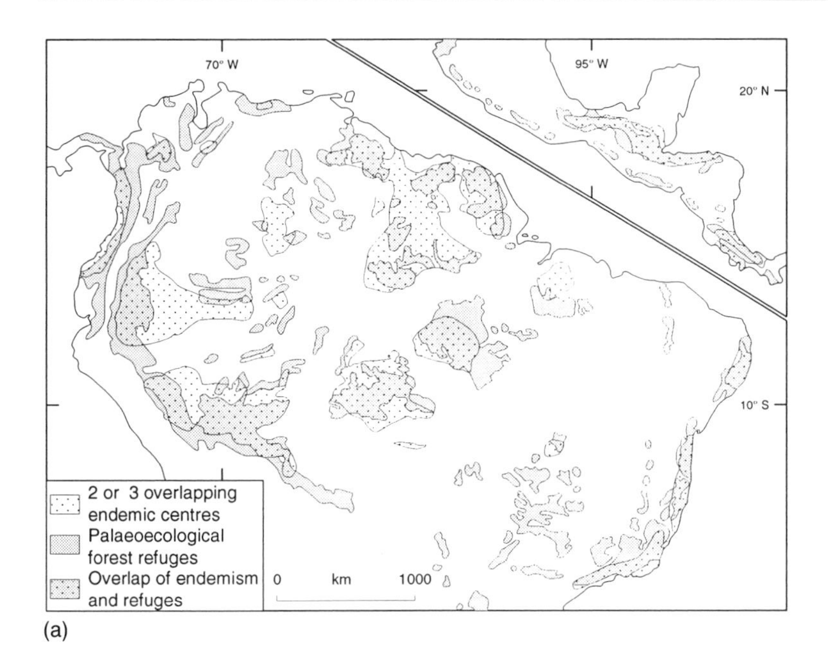

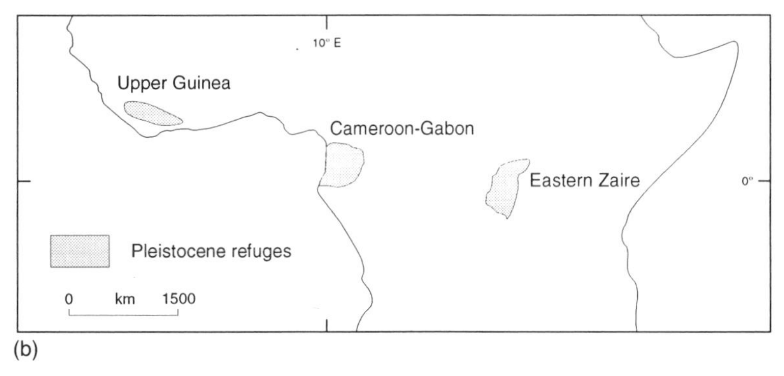

Figure 5.26
Postulated Pleistocene refugia in the American and African humid tropics: (a) overlap of centres of endemism with palaeoecological forest refuges in Central and South America (after Whitmore, 1990); (b) centres of high species richness and endemism in Africa (after Mayr and O'Hara, 1986).

10 Ka BP. Walker and Chen (1987) conclude by arguing that humid forests are relaxed at the present time; i.e. their spatial distribution is near some theoretical maximum and they occupy a range of niches that is near the potential maximum. The important implications that this has for tropical forests in the future will be addressed in chapter 9.

SIX

Humid Tropical Landscapes

6.1 Introduction

Humid tropical landscapes are extremely diverse and often complex and are the result of the interaction between factors such as climate, lithology, structure and tectonics, scale, age (the time factor), vegetation and increasingly the impact of human activity (figure 6.1).

It was the European explorers and naturalists of the eighteenth and nineteenth centuries who first formulated ideas about the nature of landscape development at low latitudes and European scientists have continued to dominate geomorphological research in

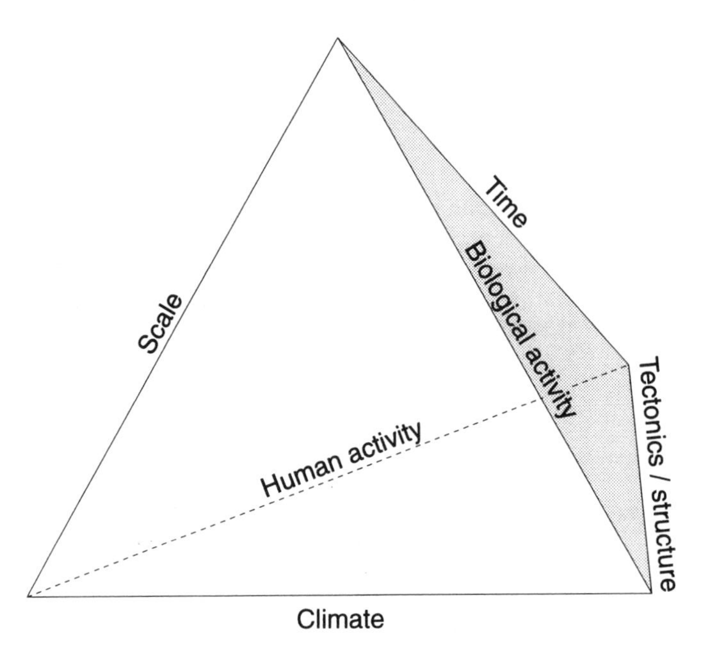

Figure 6.1
Controls on the humid tropical denudation system.

the humid tropics (figure 1.5). Many carried with them measurement techniques and conceptual assumptions about the operation of the humid tropical denudation system based upon their experiences in higher latitudes. Whether or not this has detracted from a true understanding of the tropical physical landscape is a moot point (Thorn, 1988). Nonetheless, the history of geomorphology at low latitudes provides an important context within which current ideas on the subject should be appraised.

In the early part of this century, the development of ideas in tropical geomorphology were influenced greatly by the concept of climatic geomorphology (e.g. Davis, 1899; Cotton, 1942; De Martonne, 1946; Büdel, 1948; Peltier, 1950). This concept postulates that a particular set of climatic variables, primarily rainfall and temperature regimes, are responsible for the operation and intensity of various geomorphological processes and therefore for the development of a distinctive set of landforms. The ideas of climatic geomorphology are incorporated in several geomorphology texts including three important texts on humid tropical geomorphology (i.e. Tricart, 1972; Thomas, 1974; Faniran and Jeje, 1983) and have had a significant effect on the development of geomorphological ideas by British, French and German scientists. Criticism of the approach has been severe (e.g. Stoddart, 1969; Derbyshire, 1976). Briefly, limitations of the approach increase as the scale of investigation decreases and as site-specific conditions or randomness become more relevant.

Our understanding of the humid landscape has also been influenced by the fact that geomorphological investigations have tended to concentrate on 'unusual' landscapes (compared with temperate latitudes) in areas visited and colonized by Europeans. Douglas and Spencer (1985) review the major contributions to the development of geomorphological ideas made by British, French and Dutch scientists working in their respective colonies and the important contributions made by German geomorphologists working throughout the Pacific and Africa. The majority of the areas studied lie within the large and tectonically stable Precambrian cratons (e.g. tropical Africa, India and South America). Geomorphological investigations cited in mountainous and tectonically active regions can be found in major journals (e.g. Imray, 1848; Daly, 1882; Behrmann, 1921). However, it is the investigations of, for example, the 'extraordinary isolated hills' (Thomson, 1882, p. 69) and the great depths of weathered material (e.g. Darwin, 1890) that characterize the great plain-lands which have dominated the subject. Tricart's (1972) classic text *Landforms of the Humid Tropics and Savannas*, for example, deals almost exclusively with the tectonically inactive parts of the tropics and includes large sections discussing the landforms and the relevance of landscape models which relate exclusively to the oldest parts of the tropics. Later texts by Thomas (1974, 1994) and Faniran and Jeje (1983) continue in this style.

An alternative approach to the study of geomorphology at low latitudes takes account of the importance of tectonics and structure

Table 6.1 Machatschek's (1955) classification of tropical landforms into six major relief zones

1	The south and southeast Asian mountain chain
2	The Old World Gondwanaland area
3	Oceania
4	Central America (mainly Tertiary fold mountains)
5	Andean South America (mainly Tertiary fold mountains)
6	South America outside the Andes (essentially Gondwanaland)

in producing variations in landscapes. For example Machatschek's (1955) study of tropical landforms is carried out on the basis of a sixfold division into major relief zones (table 6.1). Douglas (1978) has also argued that large-scale tectonic factors probably constitute a major control over the geomorphological diversity of the humid tropics. Tectonic classifications of humid tropical landscapes have recently been re-evaluated (Douglas and Spencer, 1985). Summerfield (1985), for example, uses one as a framework to discuss landform development in tropical Africa.

The importance of tectonics and structure is emphasized throughout this chapter by differentiating the denudational systems and landforms of the tectonically inactive shields and plateaux from the denudational systems of higher relief and tectonically active volcanic areas and fold mountains. Karst (i.e. solutional) landscapes and coastal environments are considered separately as distinctive process domains. Weathering is considered first since it affects all regions and is, in most cases, the first stage of denudation.

6.2 Weathering

Chemical alteration and physical disintegration processes affect all surface materials to some extent. In the humid tropics, the availability of water and consistently high temperatures maximize the efficiency of chemical reactions and in the oldest parts of the tropics these processes have been in operation for an extremely long time. In many of these regions rock weathering is complete, weathered products are abundant and the significance of weathering to landscape development most dramatic. Weathered products are not only important in denudational terms, i.e. as precursors to removal and redistribution, they are also directly and indirectly important in landform development. Weathered products often have a very different resistance to erosion than parent material. For example, weathered soils containing metal oxides of iron, aluminium, manganese etc. can become hardened and form areas resistant to erosion (section 4.2). Also, as weathering progresses and the depth of weathered regolith increases, slopes may become progressively less stable. In this case rapid mass movements are likely to take place in a cyclical pattern, once a critical amount of weathering has taken place (figure 6.2). Some weathered products, e.g. kaolinite (china clay) and the ores of

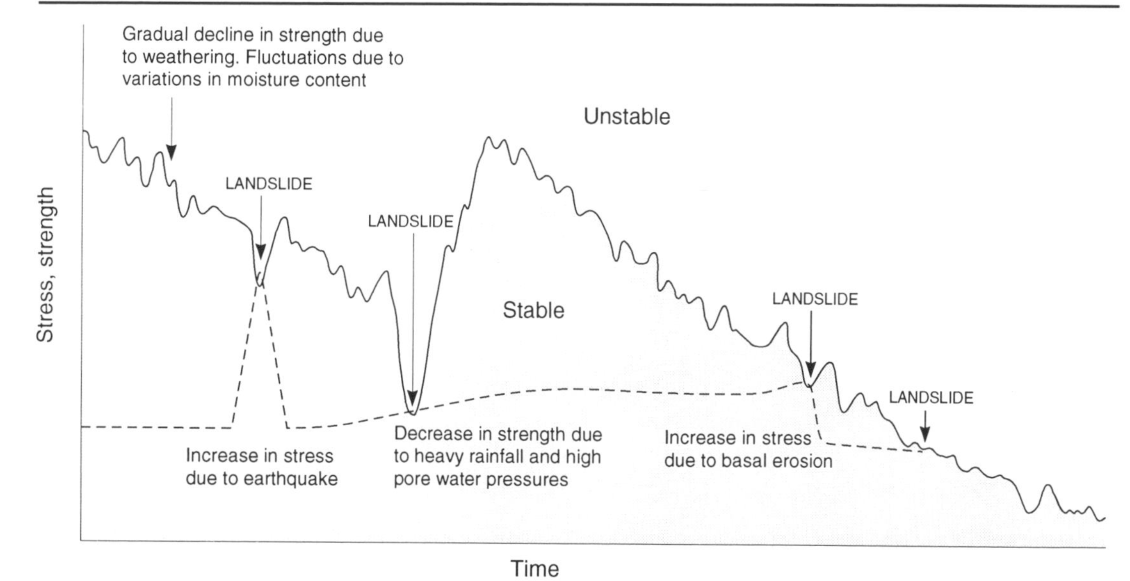

Figure 6.2
Cycles of landsliding due to changes in hillslope strengths and stresses.

aluminium and iron, form important natural resources (section 8.3). Aluminium ore (bauxite) represents a very advanced stage of weathering and is therefore only found in areas which have been subject to prolonged and intense weathering (section 8.2.1).

Weathering profiles vary enormously. On a regional scale, the variations are largely a response to differences in rates of weathering and erosion caused by climatic factors (in particular rainfall) and also variations in rock type, structure and permeability. At smaller scales, topographical conditions such as slope gradient, shape (straight, convex, concave), position (top, middle, bottom) and aspect (ubac, adret) become increasingly important. The literature contains numerous descriptions of weathered profiles few of which use consistent or comparable terminology (table 6.2). The idealized profile described below and illustrated schematically in figure 6.3 is taken largely from Deere and Patton (1971). It is particularly applicable in terms of the engineering characteristics of the material. Three main divisions are recognized:

Zone 1 Residual soil } Regolith
Zone 2 Weathered rock }
Zone 3 Relatively unweathered (fresh) bedrock

Zone 1: Residual soil

This uppermost layer, which is often highly organic, is a zone of eluviation, depleted of minerals by the movement through the profile of infiltrated water. It is this part of the weathering profile which is the subject matter of chapter 4. Upper parts of this layer (corresponding loosely to the pedologists' B horizon) are most altered from the parent material. Weathering may have progressed to a stage where even quartzites are fragmented and corroded and the original structure and texture of the parent

Table 6.2 Classifications of weathered profiles for igneous and metamorphic rocks

Vargas (1953)	*Kiersch andd Treasher (1954)*	*Moye (1955)*	*Ruxton and Berry (1957)*	*Sowers (1954, 1963)*	*Knill and Jones (1965)*	*Vargas et al. (1965)*
Gneiss, basalt, sandstone	Quartz, diorite	Granite	Granite	Igneous and metamorphic rock	Gneiss	Granite, gneiss, schists, basalt, sandstone
Mature residual soil		Granitic soil	ZONE Soil A horizon Soil B horizon I Residual debris	The upper zone	GRADE Gneissic soils	Upper zone
Young residual soil	Highly weathered	Completely weathered granite	IIA Residual debris with corestones	The intermediate zone	IV Completely weathered gneiss	Intermediate zone
Disintegrated rock layer	Moderately weathered	Completely weathered granite	IIB Residual debris with corestones	The partially weathered zone 1	IIIa Completely weathered gneiss	Lower zone
		Moderately weathered granite	III Corestones with residual debris (grussi)		IIIb Moderately weathered gneiss	
	Slightly weathered	Slightly weathered granite	IV Partially weathered		II Slightly weathered gneiss	Partially weathered or fissured rock
Sound rock	Essentially fresh	Fresh granite	Bedrock	Unweathered rock	I Fresh gneiss	

<table>
<tr><th>Korzhenko and Shwets (1965)</th><th>Sowers (1967)</th><th>Little (1967, 1970), Saunders and Fookes (1970), Fookes and Horswill (1970)</th><th>Vargas (1969)</th><th>Barata (1969)</th><th colspan="2">Deere and Patton (1971)</th></tr>
<tr><td>on-clayey
ocks</td><td>Igneous,
metamorphic
and others</td><td>A variety of
rocks</td><td>Gneiss,
schist,
granite,
clay,
sandstone</td><td>Gneiss</td><td colspan="2">Igneous and metamorphic</td></tr>
<tr><td rowspan="2">layey soil</td><td>A horizon</td><td rowspan="2">GRADE VI
Soil or true
residual soil</td><td>ZONE I
'Porous'
clay or
sand layer</td><td rowspan="2">ZONE I
Mature
residual
soil</td><td rowspan="3">1
Residual soil</td><td>1A
A horizon</td></tr>
<tr><td>B horizon</td><td>II
Stiff clay or
clay sand</td><td>IB
B horizon</td></tr>
<tr><td>aprolite</td><td>Saprolite</td><td>V
Completely
weathered</td><td>III
Young
residual
soil</td><td>IIA
Young
residual
soil</td><td>IC
C horizon,
Isaprolite</td></tr>
<tr><td rowspan="3">otten
naterial</td><td rowspan="3">Transition
zone</td><td>IV
Highly
weathered</td><td rowspan="3">IV
Weathered
rock 2</td><td>IIB
Young
residual
soil</td><td rowspan="3">Weathered
rock</td><td rowspan="2">IIA
Transition from
residual soil or
saprolite to
partly
weathered
rock</td></tr>
<tr><td>III
Moderately
weathered</td><td>III
Very
altered
rock 3</td></tr>
<tr><td>II
Slightly
weathered</td><td>IV
Fractured
or
fissured
rock</td><td>IIB
Partially
weathered
rock</td></tr>
<tr><td>ock</td><td>Solid rock</td><td>I
Fresh rock</td><td>V
Fissured or
sound rock
strata</td><td>V
Mother
rock</td><td>III
Unweathered
rock</td><td>III
Unweathered
rock</td></tr>
</table>

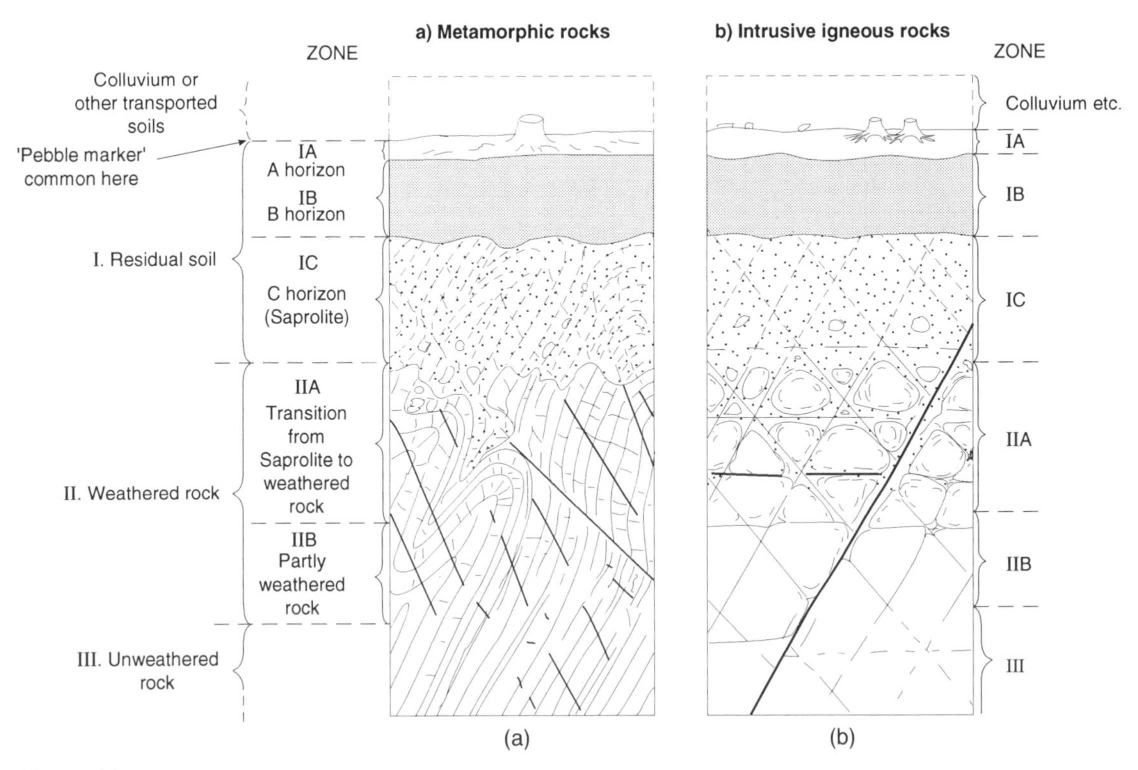

Figure 6.3
Idealized weathering profiles for (a) metamorphic rocks and (b) intrusive igneous rocks (after Deere and Patten, 1971).

material is lost. Although leached of its soluble constituents, clay-sized minerals remain, particularly those containing silicon, iron and aluminium. Desiccation of this layer may cause induration and the formation of a duricrust (section 6.3.1). In lower parts of the residual soil (C horizon), features of the original rock structure, joints, faults and mineral orientations etc. are recognizable but clay minerals such as the kaolinites will have replaced the rock-forming feldspars and micas. Other primary minerals (with the exception of quartz) will also be altered. The material may appear to have the same consistency as the parent rock but rock mass strength is invariably reduced (it is frequently soft enough to be penetrated with a knife). In engineering terms its properties more closely resemble soil than rock. This material is termed saprolite.

Zone 2: Weathered rock
This zone is usually arbitrarily distinguished from true saprolite by the presence of at least 10 per cent by volume of unweathered material in the form of corestones (plate 6.1). This zone is typically highly permeable, especially in its upper sections, and contains minerals in a wide range of weathering stages. Its engineering properties are difficult to predict. In its lower sections, discoloration and alteration of minerals is visibly more pronounced along joints and other planes of weakness.

Plate 6.1
A corestone in weathered pyroclastic material, Dominica, West Indies.

Zone 3: Unweathered bedrock
This zone is reached at a depth where there is little alteration of feldspars and micas and little staining, even along joints, which can be attributed to weathering processes. The greatest depths at which unweathered rock is reached occur where humid tropical conditions have occurred for a prolonged period of time, i.e. in the ancient cratons at the centre of the stable shields. Thomas (1965) has measured weathered rock to a depth of 90 m in Nigeria and Ollier (1960) has recorded similar depths of weathering in Uganda. In Australia, a record depth of 275 m has been measured in New South Wales, 45 m in Queensland, 35 m in Western Australia and 80 m in Victoria (Faniran and Jeje, 1983).

6.3 Processes and Landforms Characteristic of Old and Tectonically Stable Parts of the Tropics

6.3.1 Duricrusts

Early European explorers commented that mountains and hills in the humid tropics were much more rounded and less angular than elsewhere. This can be attributed to the soft texture of the weathered regolith which almost everywhere mantles the unweathered rock below. Under certain conditions, such as desiccation, however, induration of the surface layers of weathered material results in the formation of a tough and impenetrable surface – a duricrust. The term duricrust is a generic term used to describe all types of surficial crusts formed from the processes of rock weathering and soil formation; their chemical composition allows further classification into types (Goudie and Pye, 1983). The form and composition of duricrusts relates to the nature of the weathering regime (in particular the movement of water through the weathering profile), rock type, topography, vegetation and biological activity, climate and time. In humid tropical environments deep, intense and prolonged chemical weathering removes alkalies, alkali earths and silica leaving only iron and alumina in upper parts of the weathered profile. Iron- and alumina-rich laterites and bauxites form under these conditions. Under more seasonal leaching regimes such as are found in savanna landscapes, silica may remain, allowing the development of silcrete crusts. Duricrusts occur across large areas of the present-day tropics and subtropics (Petit, 1985).

Duricrusts are geomorphologically significant in several respects. The importance of duricrusts in landform evolution has attracted considerable attention in the literature (e.g. Goudie, 1973). Their indurated character means that they are often relatively resistant to erosion and thus encourage the formation of positive relief. Where they occur as a capping on residual hills and plateaux they encourage parallel retreat of hillslopes and hence the maintenance of relief through time. Undercutting and collapse of duricrust-capped relief may result in the formation of boulder fields and pseudo-karst topography. Where they overlie geotechnically weak sediments overburden pressure can result in valley cambering. In areas of low relief their resistance to erosion can, in time, result in an inversion of drainage and relief.

In their control on relief, duricrusts also have an effect on slope hydrology and sediment transfer processes. When wet, duricrusts are often soft and sticky, massive and continuous in form and highly impermeable. In this state runoff is encouraged. When desiccated, they often harden and crack and their permeability increases. This encourages the rapid infiltration of surface water. In the wettest parts of the tropics, where the movement of soil water is almost always downwards, the hydrological and engineering changes associated with desiccation and induration of mineral-enriched soils are often irreversible. Changes in the moisture regime of these materials are often indicative of environmental degradation.

The formation of duricrusts, and in particular the importance of topography in their development, has also received considerable attention by geomorphologists. Faniran and Jeje (1983) provide a review of major ideas. The accumulation of a chemical residuum requires that products of chemical weathering are allowed to accumulate more rapidly than they are removed by mechanical erosion. High relief and steep slopes therefore discourage duricrust formation; however, relief changes due to uplifting, down-faulting etc. may mean that duricrusts are found on hilltops, plateaux and other interfluves. Laterites and bauxites are essentially residual accumulations of chemically resistant materials (McFarlane, 1983). Accumulation and enrichment processes are discussed by McFarlane (1983) and considered in sections 8.2.2 and 8.2.3 in relation to the formation of mineral ores.

The presence of duricrusts can also be used to provide evidence of palaeoenvironmental conditions. For example, the accumulation of thick deposits of weathered material requires substantial periods of landscape stability when tectonic and erosional activity was subdued. The composition of the duricrust can be used as an indicator of broad-scale climatic or local hydrological conditions at the time of formation (Goudie and Pye, 1983).

6.3.2 The residual landscape

Over a large proportion of the ancient shields, tectonic stability, a preponderance of particular lithologies (i.e. basement complex) and (in comparison with higher latitudes) relative climatic stability have contributed to the development of a landscape dominated by wide undulating plains at various altitudes. The terrain, which extends well beyond the boundaries of the humid tropics, can be conveniently differentiated into wide valley floors, wide interfluvial areas and footslopes, and isolated and resistant residual hills.

Debate over the origin, development and denudation of this landform appears inexhaustible and contributions made by German geomorphologists (e.g. Sapper, 1935; Büdel, 1970; Wirthmann, 1977; Bremer, 1979, 1980) deserve particular mention. It is one of the few remaining strongholds of research into landscape development on a regional scale, a theme which has been abandoned by the majority of modern geomorphological research in favour of systems modelling (Thorn, 1988). Numerous geomorphological concepts have been developed to explain the residual landscape; those of peneplanation (Davis, 1899), pediplanation (King, 1953, 1957) and etchplanation (Weyland, 1934) are the best known and have had the greatest impact on the subject and subsequent research. The theory of slope evolution through slope replacement (Penck, 1924) should also be mentioned. All these ideas and their subsequent refinements are described in detail in many general geomorphological texts (e.g. Twidale, 1964; Chorley et al., 1984) and also in Thomas's (1994) and Faniran and Jeje's (1983) books on tropical geomorphology. Only their major points are summarized below.

Peneplanation (figure 6.4(a)) begins with the rapid uplift of a region to produce an initial surface of irregular relief. Weathering and fluvial erosion emphasize this relief and rivers develop consequent on regional slopes. Along river channels, vertical incision occurs more rapidly than interfluve lowering and relative relief is increased (youthful stage). Vertical incision continues until base level (grade) is reached at the river mouth (mature stage). Rates of incision are then exceeded by interfluve lowering, valley widening occurs and relative relief declines. As grade progresses up river valleys valley widening becomes more and more extensive, eventually producing large areas of low relief (peneplains) separated by low erosional residuals (monadnocks) in the old age or senile stage. Although this concept, which connotes slope lowering, may have some credence in explaining landscape development in geologically simple volcanic terrains, it cannot explain the

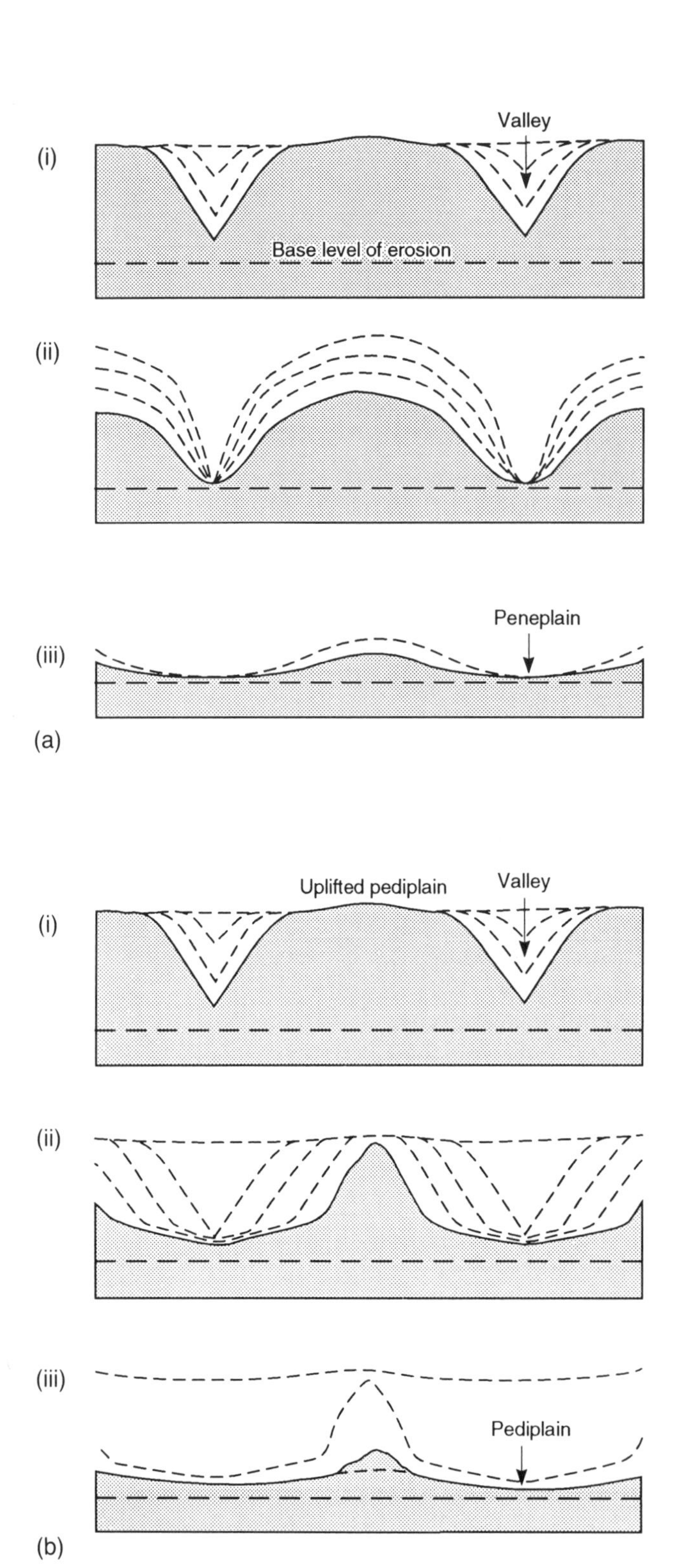

Figure 6.4
Stages of (a) peneplanation and (b) pediplanation (based on Thomas, 1974).

'staircase' pattern of terrain (surfaces separated from one another by scarps) found throughout large parts of the older tropics.

The concept of pediplanation differs fundamentally from that of peneplanation in that it proposes the extension of plains at the expense of interfluves through the retreat of slopes with a constant declivity. Basic to this concept is that hillsides have four distinct slope elements; a top-slope (convex), freeface, debris slope and base-slope (concave). The scarp retreats parallel to itself, at a rate controlled by weathering and the removal of weathered material by overland flow or mass movements, to reveal an erosional pediment thinly covered by weathered debris. Parallel retreat has been shown to occur most readily in landscapes with hard resistant rocks and a strong structural control (Savigear, 1960; Thomas, 1965). Within the humid tropics, parallel retreat is also frequently observed at gully heads where it occurs as a consequence of undermining and basal sapping which initiate slumping (Faniran and Jeje, 1983).

Pediplanation (figure 6.4(b)) is initiated by tectonic uplift which results in accelerated stream incision and the formation of knickpoints, falls rapids and gorges along river valleys (youthful stage). With the establishment of new base levels (mature stage), rivers cease to deepen their valleys and lateral erosion becomes dominant. Hillslopes begin to retreat parallel to themselves at constant angles, leaving a low-angled pediment. Pediments widen and those at lower levels may regrade across those at higher elevations eventually producing a landscape of low-angled plains (pediplains) separating rocky hills called *kopjes*. The concept of pediplanation has been used to explain landscape evolution in humid and dry parts of Africa (e.g. Ruxton, 1958; Pugh, 1966). It cannot account, however, for the fact that all erosion surfaces are not flat or for the existence, in the ancient shield areas, of extensive land surfaces with deep weathering and several rock outcrops. In the latter areas the concept of etchplanation (figure 6.5) has attracted considerable attention (e.g. Erhart, 1956; Büdel, 1965, 1982; Thomas, 1965).

Central to the concept of etchplanation is that the shield areas have been subjected to continual chemical weathering during the Cainozoic era. During this time, periodic tectonically induced sea-level and climatic changes have resulted in 'swings' in the relative efficiency of weathering and removal processes. Thus periods when weathering has outpaced erosion, resulting in the accumulation of great depths of weathered material, have been punctuated by phases when erosional processes have outpaced weathering, stripping the surface of its weathered mantle. Erhart (1956) expressed these ideas in terms of 'biostacy' (periods of stability lasting up to 10^9 years and favouring the development of deep and leached soil profiles and duricrusts) and 'rhexistacy' (phases of instability when erosional processes cut down and removed the soil and saprolite). Büdel's (1965) expression of the concept of etchplanation in terms of a 'double surface of levelling' has been widely accepted in general terms. His hypothesis evokes the idea of a double surface of

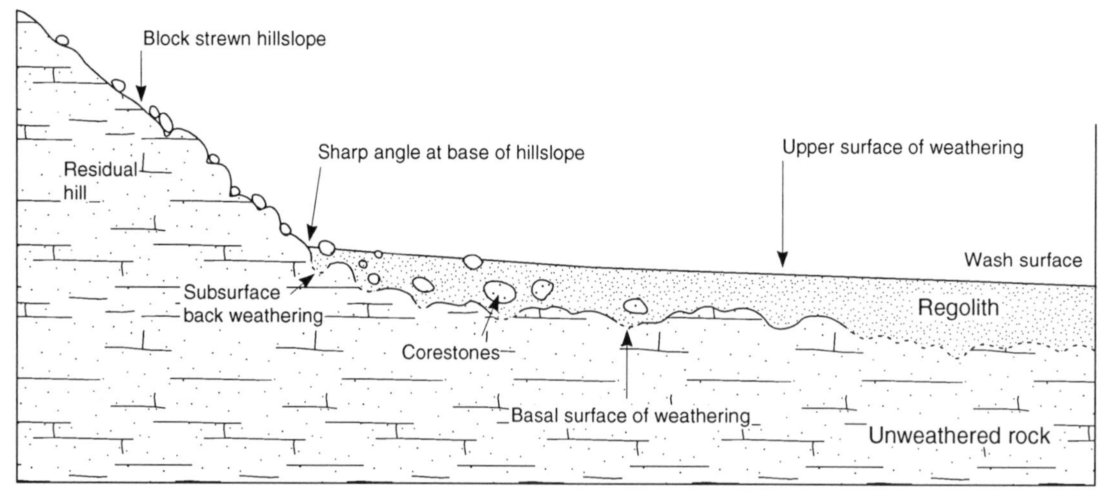

Figure 6.5
Etchplanation (after Büdel, 1957).

weathering, the lower surface (the interface between weathered and unweathered rock) will bear little resemblance to surface topography; instead it will follow variations in the resistance to weathering of the parent material due to structural and lithological controls (Pye et al., 1986). 'Etching' of the upper ground surface may occur through incision by rivers or by slope retreat involving sheetwash and basal sapping, and slumping on duricrust surfaces (plate 6.2).

The extent to which the unweathered bedrock has been exposed and incised has been used to distinguish up to five types of etchplain (figure 6.6). Mabbutt (1961), Bishop (1966) and Ollier (1960) restrict the term 'etchplain' to areas of stripped and exposed unweathered bedrock (figure 6.6, types (d) and (e)). Thomas and Thorp (1985) distinguish three categories of etchplain in their study of the West African craton:

1 plains resulting from one or more distinct episodes of stripping of a relict saprolite mantle;
2 plains continuing to evolve by surface erosion acting on pre-weathered materials;
3 plains within which emergent rock forms coexist with more rapidly evolving forms controlled by the shifting balance between weathering penetration and surface lowering within a system evolving active or dynamic etchplanation.

All three categories can exist within a single landscape, a fact which helps explain the complexity of landforms which exist within the plains. Most etchplanes are thought to be polycyclic. For example in Guyana and Surinam, five storeys have been identified, while in Brazil, Nigeria and Malaysia, three storeys are clearly visible. Each storey represents a cycle of etchplanation and is frequently separated by a steep scarp. The plains may also contain structural landforms such as depressions. The Chad basin and

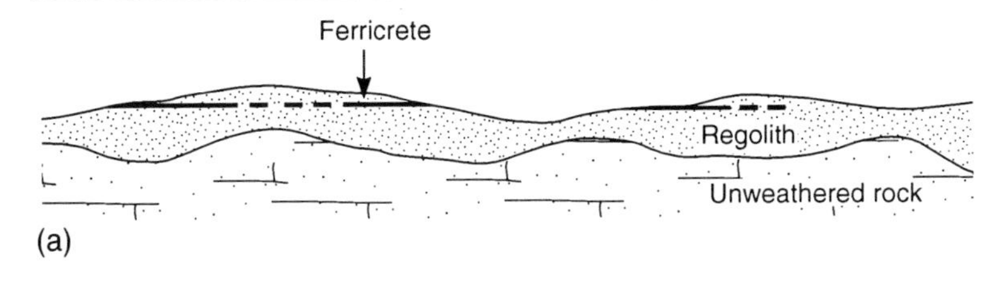

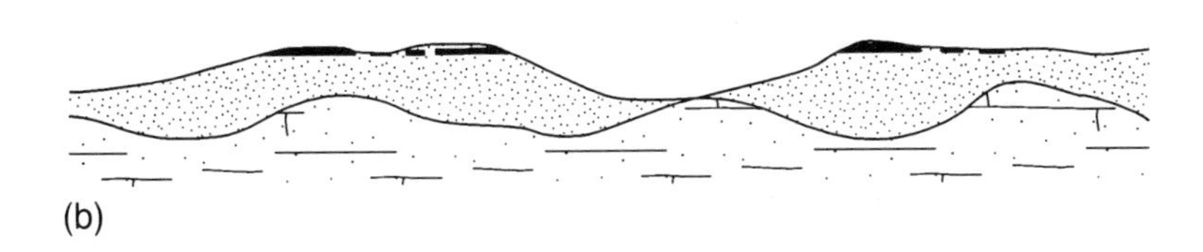

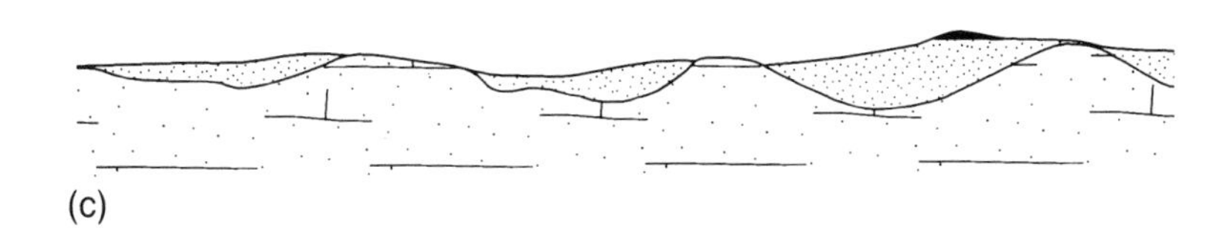

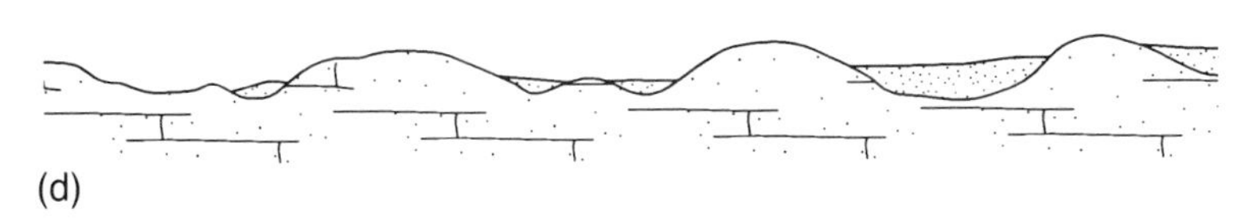

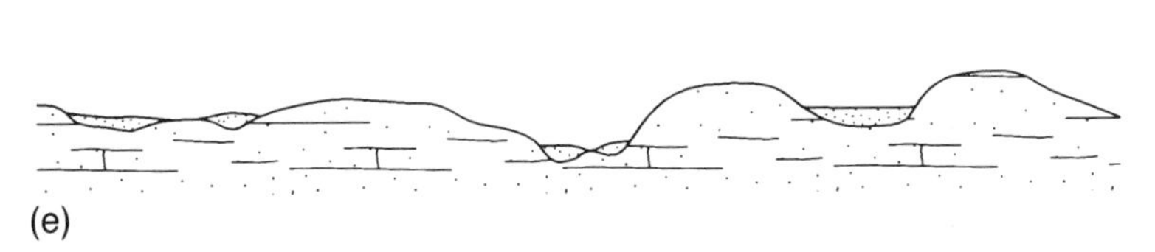

Figure 6.6
Types of etchplain (based on Thomas, 1974; Faniran and Jeje, 1983): (a) laterized etchplain; (b) dissected etchplain; (c) partially stripped etchplain; (d) dominantly stripped etchplain; (e) incised etchplain.

Congo depression represent two of the largest structural depressions in Africa. Sediments deposited in these have also been eroded into a series of erosional landforms.

The concept of etchplanation has an empirical foundation and is able to account for the complexity of landform assemblages in the humid tropics. However, it requires further validation, particularly within a geological and time context. The time taken for etchplanes to develop and the timing and placement of rhexistacy phases and

Plate 6.2
Slumping of duricrust, West Africa.

their correlation with periods of saprolite removal have been addressed (Schumm, 1963; Arnhert, 1970; Kronbert et al., 1986), but conclusions remain uncertain. Also needed are many more empirical measurements within specific areas of supposed etchplanation and information on the current balance of geomorphological processes.

It is perhaps unwise to look for a single theory to explain the development of all residual landscapes adequately since it is possible that residual landscapes develop by various routes. Much research remains to be carried out before we can fully understand and model the development of individual landforms or the entire humid tropical landscape.

6.3.3 Residual hills

The other very striking feature of the tropical plains is the existence of rock hills and mountains which rise abruptly from their bases. The origin and maintenance of these rock remnants, termed inselbergs by the explorer Bornhardt (1900) in Tanzania, continue to evoke a great deal of discussion and debate.

It is possible to identify residual hills in almost all climates (Kessel, 1973). However, many geomorphologists now accept that, while residual hills occur in many different climates, conditions are particularly favourable for their formation in the seasonal tropics. Residual hills are best developed on volcanic material, particularly granites and gneisses with widely spaced joints and a high potassium content. However, they may also develop in sedimentary material, in particular within areas of sandstones, conglomerates and sedimentary quartzites.

Residual hills occur in a range of sizes and shapes. Their diversity of form and spatial occurrence has resulted in a preoccupation with description and classification and the establishment of terminology which is confusing and often ambiguous. The term inselberg is frequently used generically to describe almost all abrupt rises from the plains. Duricrust topped hills are the major exception. Thomas (1974) has suggested that the term inselberg should be restricted to exposures of granite and that sedimentary residuals (e.g. Ayers Rock) should be classified as mesas or buttes. Faniran and Jeje (1971), however, differentiate residual hills according to morphology. The terms tor or boulder inselberg are used to described spheroidally weathered boulders rooted in bedrock and exposed as a basal rock surface or concealed by a waste mantle (plate 6.3). Bornhardts or domed inselbergs are described as monolithic dome-shaped or elongated whaleback forms occurring in cupola shape or as sugar loaves (plate 6.4). The third category, castle kopjes, are distinguished by their castellated forms. They are elongated slabs of rock showing little signs of spheroidal modification (plate 6.5). Other classifications differentiate between inselbergs of position

Plate 6.3
Tor or boulder inselberg, Zimbabwe.

Plate 6.4
Smooth, dome-shaped bornhardt, Zimbabwe.

Plate 6.5
Castle Kopje, Zimbabwe.

(hills remaining because they are on interfluves and most distant from areas of active erosion) and inselbergs of resistance (those remaining because of their superior resistance to erosion) (Goudie, 1989).

Central to most current ideas on the evolution of residual hills is that they are the result of the stripping of weathered regolith from a differentially weathered surface (Faniran and Jeje, 1983). The major debate lies in the necessity of deep weathering as a precursor to hill formation and exhaustive arguments and counterarguments can be found in the literature. Models such as those by Linton (1955) and Thomas (1978) propose a two-stage model of formation (figure 6.7) requiring the development of a mass of unweathered rock below ground by differential weathering of variably jointed rocks before exhumation. Opponents of this idea cite examples of very tall residual hills rising above the plains as evidence that weathering and erosion must take place simultaneously. The diversity of residual hills present in the landscape indicates that both schools of thought may have some validity. Another contentious issue among present-day geomorphologists relates to the status of current geomorphological processes in the residual environment.

6.4 Processes and Landforms Characteristic of Young and Tectonically Active Regions

The majority of geographical literature concerning the humid tropical environment deals with weathering processes or erosional landforms. Removal processes are often only briefly discussed, using generalized and inexact statements. By comparison, in the dry tropics the examination of removal processes forms an area of

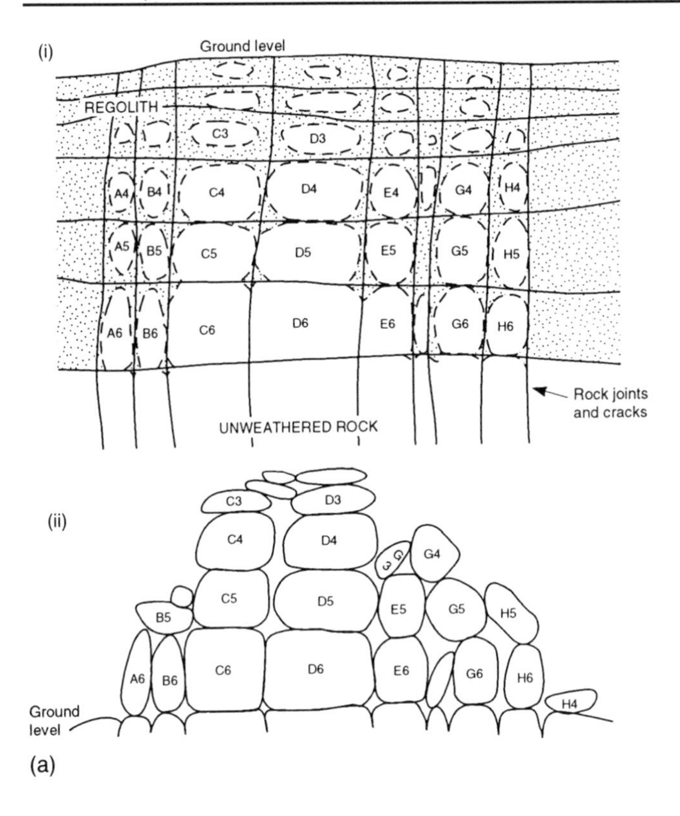

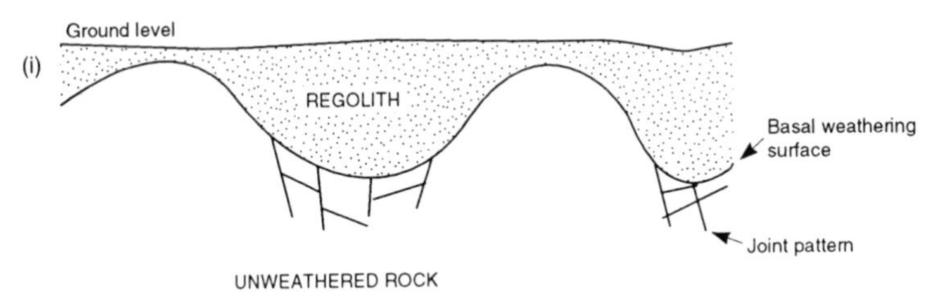

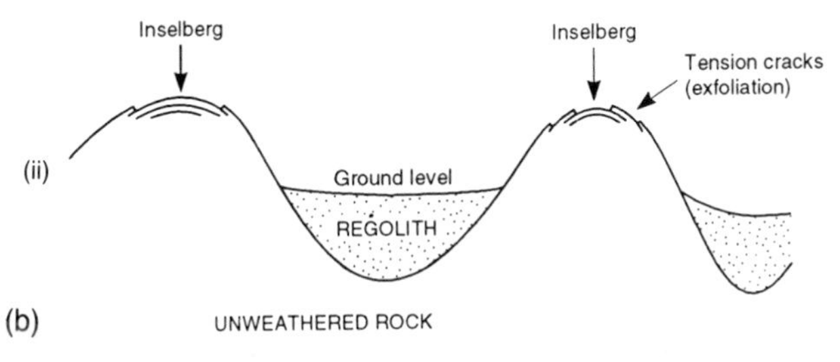

Figure 6.7
Two stage models of residual hill development: (a) Linton's (1955) model of tor development; (b) Thomas's (1978) model of inselberg development.

major study, particularly in relation to slope denudation and pediment development. One reason for the brevity of discussion of humid tropical removal processes has been, at least until recently, a lack of empirical information on the subject. This is somewhat surprising since many of the processes are particularly active in a humid tropical environment, particularly a young or tectonically unstable environment. These processes should therefore be relatively conducive to observation and monitoring.

Two categories of removal processes can be distinguished, and those which remove material under the influence of gravity (generically known as mass movements) are considered below. An examination of removal processes involving fluvial activity can be found in chapter 7.

Faniran and Jeje (1983) describe the humid tropics as an area prone to mass movements – in mountainous and rugged parts of the humid tropics it is often the dominant denudational process (Spencer and Douglas, 1985; Reading, 1986). Various methods have been suggested to classify mass movements on the basis of the type of material (e.g. soil, weathered regolith, rock), the mechanism of movement (i.e. heave, slide, flow fall) or the rate of downhill movement (i.e. fast, slow, imperceptible). Most classifications are based upon several criteria (cf. Sharpe, 1938; Varnes, 1958; Hutchingson, 1968; Carson and Kirkby, 1972). However, in the field, landslides are frequently complex, falling into several categories (Reading, 1986).

6.4.1 Soil creep

Soil creep is defined as the slow or imperceptible movement of soil and rock debris downslope under the influence of gravity. It involves the frequent rearrangement of the constituent particles at a rate slow enough to prevent the formation of a surface rupture, instead leading to deformation of the microrelief. Creep is caused by the downslope migration of material due to volume changes which induce heave normal to the ground surface and settlement approximately vertically downwards (figure 6.8).

The geomorphological significance of creep in any environment has proved difficult to assess since empirical investigations have to be undertaken with the utmost accuracy over long periods of time, often in difficult terrain. In thickly vegetated areas, problems of identifying creep and relocating underground monitoring markers are magnified. Frequently the existence of creep is implied directly, from evidence of deformed trees and leaning posts, displaced roads, cracked buildings or the accumulation of gravelly material at the base of slopes. Most published literature on creep refers to research carried out in temperate and periglacial regions, where heave is

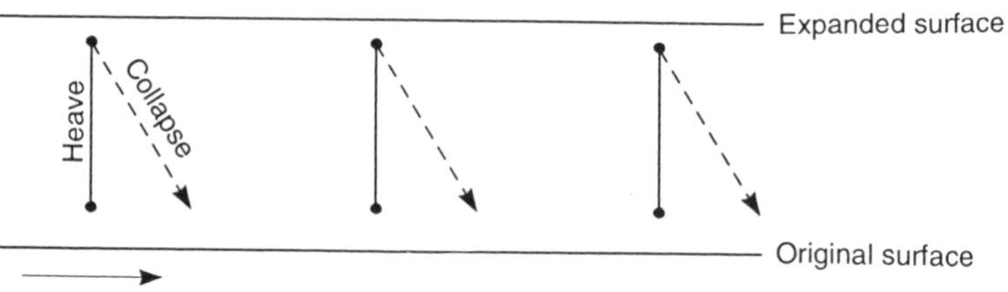

Figure 6.8
The process of soil creep.

associated with volume changes due to the seasonal freezing of water (Davison, 1889; Kirkby, 1967; Young, 1978). In the tropics, the major mechanism of heave is volume changes associated with the swelling of clay minerals as they become wet. A number of articles, including Schumm (1956), Carson and Kirkby (1972) and Kirkby (1967), describe creep in the drier parts of the tropics but little information exists for the truly wet tropics. However, the presence of a thick zone of highly weathered material, low in strength and rich in secondary clay minerals, which undergo considerable volume changes under varying moisture regimes, are conditions likely to favour creep. The presence of montmorillonite, a layer lattice mineral which has a massive capacity to expand when wet, is likely to be very conducive to creep. Rougerie (1960) was probably the first to appreciate the importance of creep in a humid tropical environment following his experiments in the per-humid forests of the Ivory Coast. Detailed work has since been carried out in other parts of the humid tropics, including Puerto Rico (Lewis, 1974, 1976), Australia and Malaysia (Thomas, 1974) and Rwanda (Moeyersons, 1981, 1988).

There are major differences in creep in temperate and tropical regions, principally in the depth, pattern and speed of movement through the profile. In temperate environments, decreasing moisture contents with depth tend to result in a progressive reduction in the rate of creep down the soil profile, in line with Davison's (1889) and Kirkby's (1967) model (figures 6.9(a) and 6.9(b)). In the forested humid tropics, throughflow often results in a high moisture zone below the surface (chapter 7), giving rise to an S-shaped creep profile (figure 6.9(c)). It has also been suggested that tree root systems retard movement in the upper layers of the soil. However, Moeyersons (1988) suggests that creep may be sufficiently deep-seated to allow even very large trees to 'float' on the layer of creeping mantle.

Creep has frequently been overlooked as a denudational element in the humid tropics and yet, from the limited information available, it appears that creep may be very active in frost-free areas with cyclical changes in soil moisture. Creep is also likely to be important in terms of human activity and, for example, should be taken into account in modern building and infrastructure design and in soil conservation strategies. Rapid mass movements are much more widely recognized as a major geomorphological agent and, because of their unpredictability and rapid speed, are disruptive and a potential environmental hazard.

6.4.2 Rapid mass movements

Rapid mass movements are shear failures which result in surface rupture rather than deformation. They are known also as landslides or landslips but this terminology should be discouraged as it suggests movement in a specific manner. Rapid mass movements may travel in free air above an undisturbed surface

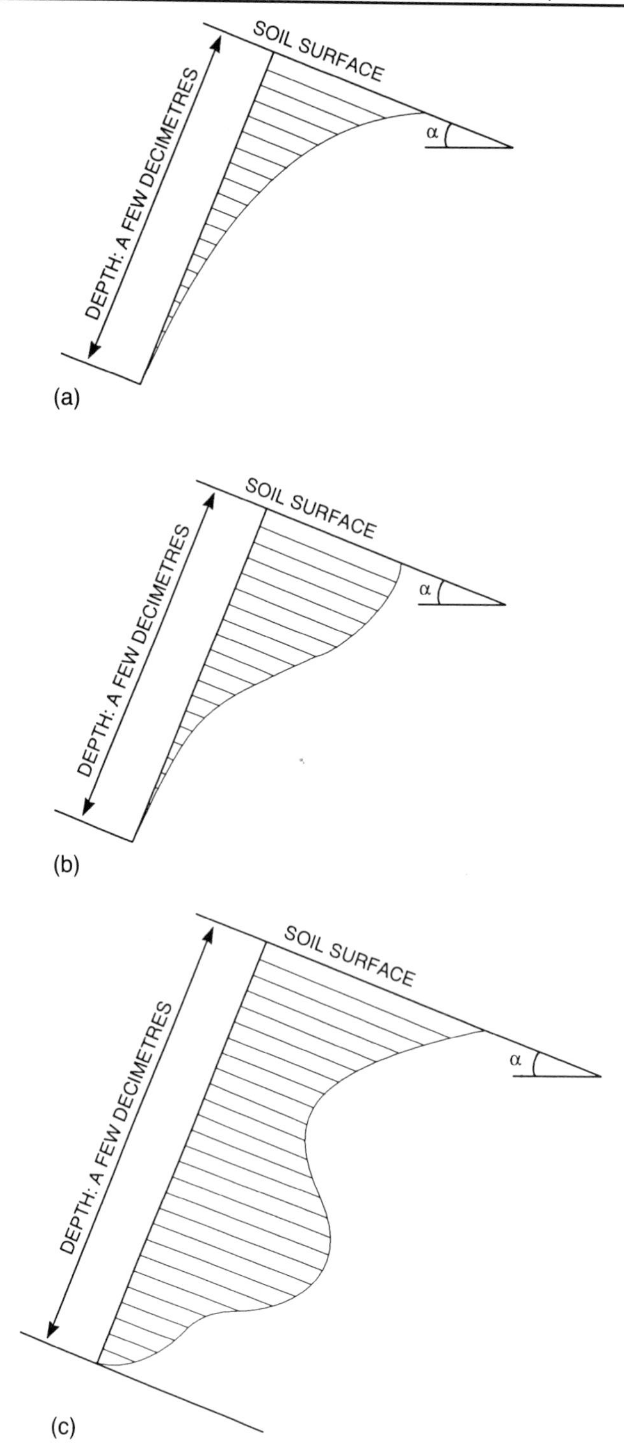

Figure 6.9
Theoretical velocity profiles of soil creep: (a) temperate soils (Davison, 1889); (b) temperate soils (Kirkby, 1967); (c) tropical soils (Lewis, 1976; Moeyersons, 1988).

(e.g. in the case of a fall or certain types of flowslides) (figures 6.10(a) and 6.10(b)), as a more or less intact mass above a discrete failure zone (e.g. as a glide or slide) (figures 6.10(c) and 6.10(d)) or

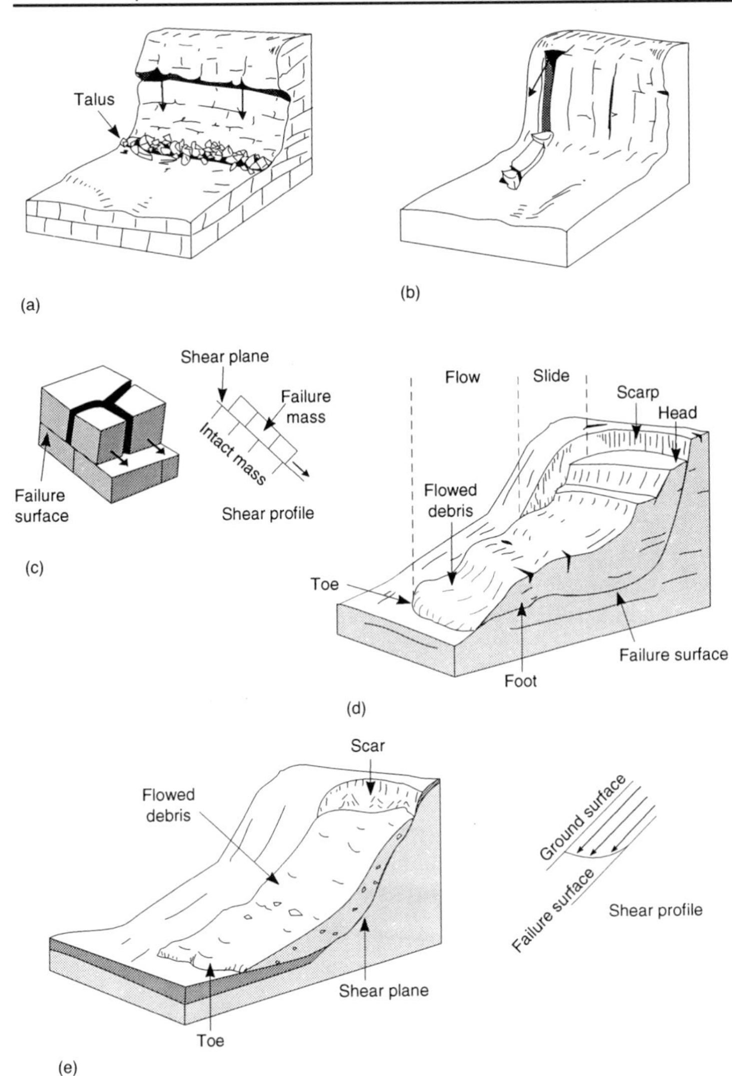

Figure 6.10
Types of rapid mass movement: (a) fall; (b) topple; (c) slide (planar); (d) slide (rotational, with flow); (e) flow.

as a jumbled mass with the characteristics of a fluid (i.e. as a flow) (figure 6.10(e)). The failure surface frequently represents a zone of weakness or a permeability unconformity above which pore water pressure can build. Rapid mass movements are most common in mountainous volcanic parts of the humid tropics where they represent a major denudational element. In the residual soils which mantle volcanic slopes, the failure zone is often located within the soil's B horizon or at the soil–weathered rock interface. At such shallow depths (rarely more than 2 m), the failure surface usually develops parallel or translational to the ground surface. However, in areas where the soil is much deeper or where the weathered material below is of homogeneous strength, an arcuate failure surface may develop, giving rise to a rotational failure (e.g. figure 6.10(d)).

While it is unlikely that creep involves shear at its base (Lewis, 1976) most rapid mass movements are associated with shear and abrasion of subsurface material. In many cases rapid mass movements change in status as they move downslope, e.g. from a slide to a flow, as shear is distributed through the failing mass from the failure surface (figure 6.10(d)). High moisture contents encourage the distribution of shear. Flows are therefore particularly common in the humid tropics. The nature of the movement is geomorphologically significant since flows travel further distances (plate 6.6) and at greater speeds and result in a different scar and toe topography than slides (Cooke and Doornkamp, 1990).

Rapid mass movements are often associated with material of low shear strength, usually weathered material rich in clays. Heavy and protracted rainfall, which promotes saturation of the soil and regolith and the build up of positive pore water pressures, is also conducive to failure. The relationship between rainfall and mass movement in the humid tropics is well known (e.g. So, 1971; Crozier and Eyles, 1980; Fukuoka, 1980) although the mechanism by which rainfall triggers failure in residual soils is not well understood (Brand, 1989). The frequency of rapid mass movements is often attributed to the copious rainfall, highly weathered surface materials and long steep slopes which characterize many areas in the young and active parts of the humid tropics. Vibrations caused by volcanic activity, earthquakes, falling trees or even the movement of heavy vehicles may disturb the structure of the soil and weathered rock and may initiate failure (Simonett, 1967). In more seasonal parts of the tropics, the wetting of the soil may be aided by the presence of a cracked and jointed lateritic crust and the build up of positive pore water pressures promoted by a major structural and permeability unconformity at the weathered–unweathered rock boundary.

Plate 6.6
A landslide running the length of the slope, Dominica, West Indies.

While rapid mass movements are common in the active humid tropics (Pitts, 1983; Brand, 1985; Kuruppuarachchi et al., 1987), detailed studies of the regolith and particularly of the soils developed on volcanic extrusive rocks (ash, pumice and basalt) have revealed surprisingly high shear strengths. Volcanic clay soils (andosols – section 4.8) frequently have field moisture contents of well over 100 per cent and have a very low unit weight and high void ratio (table 6.3). However, it is not uncommon to find long stable slopes, under forest or even short rooted annual crops, at angles of more than 40 °. Laboratory measurements of stability angles consistently underestimate true values owing to the unrepresentative nature of standard shear testing equipment (Rouse et al., 1986). In temperate conditions such slopes would be expected to fail at angles as low as 10 ° (Skempton and Delory, 1957).

The key to understanding the remarkable stability of tropical volcanic clay slopes lies in an appreciation of the structure and water-holding properties of the soil and, in particular, in the nature of the clay minerals which form readily under humid tropical conditions. In most humid tropical environments there is a net loss of minerals due to leaching. The intensity and consistency of the downward loss affects the type of clay minerals which form. For example, silica-rich smectoid clays (e.g. montmorillonite and illite) form in areas with a seasonal leaching regime, while kaolinitic clays (e.g. halloysite, kaolinite) are produced by a regime of continual leaching. Alumina-rich allophanic clays (e.g. allophane and gibbsite) are produced and persist due to a regime of continual and intense leaching.

Table 6.3 Typical *in situ* soil properties for Dominican samples

Soil type	*Sample name*	In situ *moisture (% dry weight)*	In situ *unit weight ($kN\ m^{-2}$)*	*Dry unit weight ($kN\ m^{-2}$)*	*Saturated unit weight ($kN\ m^{-2}$)*	*Void ratio*	*Degree of saturation*	*Porosity*	*Saturated water content (% dry weight)*
Allophane	Grandfond	72.0	13.26	7.71	14.8	2.62	78.2	0.72	90
latosolic	Pichelin	69.5	13.39	7.90	14.93	1.96	78.1	0.66	69
	Carhome	90.0	10.42	5.47	13.35	4.09	62.8	0.81	144
	Bells	48.0	11.35	7.67	14.78	2.63	51.8	0.72	92
	Rosalie	70.5	11.89	6.98	14.33	2.99	66.8	0.75	105
	Attley	73.5	11.62	6.71	14.16	3.15	66.0	0.76	111
	Fresh Water Lake	88.0	11.98	6.37	13.94	3.37	74.1	0.77	119
Allophane	Pont Cassé	180.0	11.23						
podsolic						2.22	49.3	0.69	78
Kandoid	Vielle Case	38.5	12.00	8.66	15.42	1.93	70.8	0.66	68
latosolic	Calibishie	48.0	14.09	9.52	15.98				
Smectoid	Mero	13.0	13.67	12.09	17.65	1.30	28.3	0.57	46
	Soufriere	29.5	14.23	10.98	16.93	1.54	54.6	0.61	54
	Galion	18.5	21.07	17.81	21.35	0.56	92.0	0.36	20

In many old and intensely weathered regoliths kaolin is the final weathering product (sections 4.2, 6.2). The unusual stability of kaolinite-rich regoliths was first documented by engineers working on dam construction in Africa (Terzaghi, 1958; Dixon and Robertson, 1971). There are several reasons for these unusual properties. First, their low shrinkage coefficient ensures that water passes down the soil profile by percolation through pores, allowing the dissipation of pore water pressures. It cannot (as is the case with swelling clays) pass directly and at great speed to lower parts of the soil profile via cracks. The low shrinkage coefficient of kaolinites also reduces the potential for heave and soil creep. Second, kaolinites have a high liquid and plastic limit (figure 6.11), meaning that they require a very high moisture content before they will flow.

In the younger parts of the humid tropics, where the regolith contains minerals at a transient stage of weathering (e.g. halloysite), other factors affect the stability. For example, hydrated halloysite exists as cylindrical tubes, rather than flat plates, which tend to ride over each other less easily during shear. When desiccated, the cylinders unfurl and further increase interparticle friction. The breakage of halloysite tubes may explain an improvement in shear strength frequently associated with desiccation following, for example, the creation of road cuttings. Allophanic minerals have no regular crystalline structure or composition and thus sliding between sheets of minerals is not appropriate. Allophanes occur

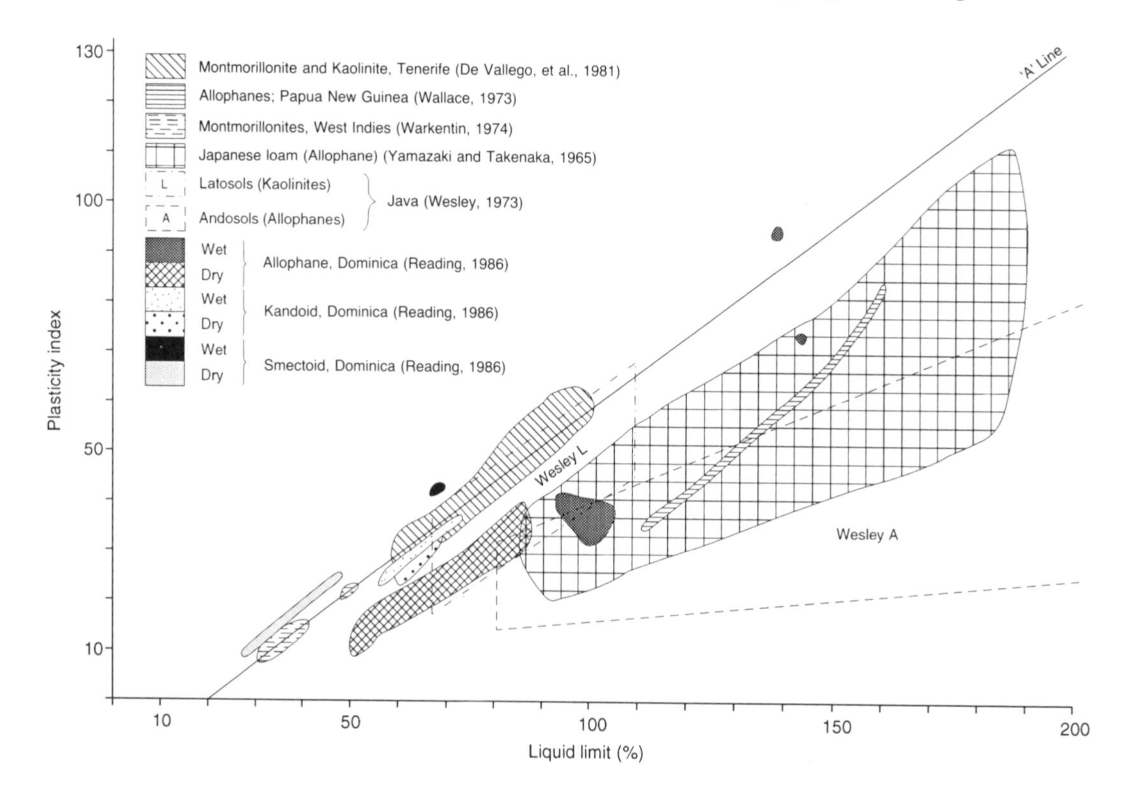

Figure 6.11
Plasticity chart: a selection of tropical volcanic soils (after Reading, 1986).

only in perennially wet environments, and desiccation results in the contraction and destruction of their highly hydrated 'gel' structure and their ability to disperse shear stresses imposed upon them. Thus in a wet state allophanic soils can be very strong but they may weaken dramatically if drying occurs (table 6.4).

Although humid tropical residual volcanic soils often have very high water contents their high void ratios (table 6.3) and rapid vertical and lateral permeabilities mean they are difficult to saturate. They also have very high liquid limits and are therefore able to retain the properties of a solid at very high water contents (figure 6.11). In summary, tropical residual clay material is able to remain stable at very high moisture contents and it is only during the most extreme conditions (e.g. during a tropical cyclone or earthquake) that stability criteria are exceeded (Reading, 1986). For example, over 200 shallow landslides occurred in the Luquillo experimental forest of Puerto Rico as a result of Hurricane Hugo in September 1990 (Larsen and Torres-Sanchez, 1990).

The depth of weathered material on a slope appears to affect the frequency of failure and the position on the slope is important with respect to the actual location of the scar head. For example, mass movement is often initiated approximately one-third of the way down the slope, at the head of first-order valleys, at a point where the underground flow net converges. It appears that failure only takes place when a critical depth of weathered material allows the build up of critical stresses on the failure plain. However, as this depth progressively increases, the magnitude of event required to initiate such critical stresses decreases. Thus rapid mass movements occur with a cyclical frequency (figure 6.2). Any changes in the frequency of extreme events, or in the speed of regolith development, will therefore affect the frequency of failures. Natural factors (e.g. climatic change) and human factors (e.g. land use and hydrological changes) have the potential to do this. For example, many failures are initiated by poorly positioned road or land drains. Rapid mass movements are sporadic and unpredictable and can occur at speeds estimated at several hundred miles an hour (Brunsden and Prior, 1984), often in association with extreme rainfall and flooding. In parts of Indonesia, Hong Kong and tropical South America they are one of the most important environmental hazards (Brand, 1985) (section 8.5).

Table 6.4 Changes in shear strength due to desiccation: examples from Dominica, West Indies

Sample	*Soil type*	Φ_r' *undried sample*	Φ_r' *air-dried sample*
Calibishie	Kandoid	30°	38°
Vielle Case	Kandoid	30°	35°
Grand Fond	Allophane	39°	38°
Mero	Smectoid	34°	34°

6.4.3 Subsidence

There is one other type of mass movement which is usually classified separately since it is not dependent on slope. Subsidence describes movement in a predominantly vertical direction. It may affect an area of a few square centimetres or several tens or hundreds of square metres. Its most common causes are subsurface erosion, e.g. due to the removal of material in solution or by the collapse of soil pipes. Alternatively it may occur as a result of surface loading or prolonged desiccation and compaction. The role of human activity in initiating or exaggerating subsidence is enormous; examples include the mining of rock, ores, minerals and liquids (oil, gas, water) and the draining of wetlands.

6.4.4 Depositional landforms

The mountainous and rugged landscapes of young and tectonically active parts of the tropics are high energy environments by virtue of their relative relief and the availability of readily weatherable material. For example, rapid mass movements frequently continue to the valley bottom where the failed debris is easily removed by fluvial processes (plate 6.6). Fluvial depositional landforms are also poorly developed since flow velocities are generally sufficient to remove material rather than redistribute it. Erosional processes and lanforms are therefore much more important than depositional processes and landforms. There are local exceptions, however. For example debris fans and colluvial deposits may be well developed in structural basins and hollows.

6.4.5 Drainage networks

Drainage networks develop according to the interplay of climatic, bioclimatic and lithological factors within the framework of time (Walsh, 1985). Unfortunately, much of the work on rates of drainage network development, drainage densities and their response to changing environmental conditions has focused upon areas outside the tropics (see Gregory, 1976). Drainage density and relief development in the volcanic Windward Islands, West Indies, has been studied in some detail by Walsh (1980, 1985). He notes that the influence of climate is both direct, in particular through variations in the magnitude/frequency of heavy rainfalls, and indirect via the zonal influence on soils and vegetation. The dimension of time is an important factor in determining the rate and nature of soil development, in particular the soil's hydrological properties (e.g. infiltration, percolation and permeability rates) and erodibility. In the youngest volcanic landscapes contrasts in drainage density and hence also drainage network development were principally related to rock permeability, with the lowest drainage densities and slowest drainage network development found in areas with the most permeable deposits.

Differences in the methodology and scale of other studies make direct comparisons with other parts of the humid tropics difficult. However, drainage densities in the older parts of the Windwards appear high compared with humid temperate regions but lower, for a similar mean annual rainfall, compared with values from other humid tropical areas (table 6.5). This may be explained by differences in the permeability of the lithology and the overlying soils, the volcanic deposits of the Windwards and the residual soils which develop upon them having high to very high permeability (sections 4.8 and 6.2).

6.5 Tropical Karst

The term karst is used here to describe landscapes which are topographically distinct due to the fact that chemical solution has been the dominant or critical factor in landscape denudation. Pseudokarst is a term used to describe karst-like landforms produced by weathering processes other than solution (Watson and Pye, 1985). Development of a karst landscape will fundamentally depend upon the presence of large areas of relatively soluble rocks at or close to the surface. The thickness and areal extent of the rocks, their purity and hydrological connectivity, directly affect the degree of development of karst features, in particular their

Table 6.5 Drainage density in some tropical areas

Territory	*Lithology*	*Vegetation/land use*	*Annual rainfall (mm)*	*Drainage density (km km^{-2})*	*Source*
Windward	Volcanics	Seasonal/rainforest	1000–8000	3–11	Walsh (1985)
Islands	Granite	Tea plantations	1346–5436	5–15	Madduma Bandara
Sri Lanka	Various	Semi-arid to			(1974a, b)
Queensland	Granite	rainforest	889–3988	4–14[a]	Abrahams (1972)
Malaysia	Arenaceous	Rainforest/cultivated	*c.*2250–2540	7.5	Eyles (1966)
Malaysia	Calcareous	Rainforest/cultivated	*c.*2250–2540	9.3	Eyles (1966)
Malaysia	Various	Rainforest/cultivated	*c.*2250–2540	9.0	Eyles (1966)
Malaysia	Granite	Rainforest/cultivated	*c.*2250–2540	7.5–12.4	Morgan (1976)
Fiji	Granite	Grass and scrub	1778–2540	15.5–18.6	Wright (1973)
Brazil	Sandstones/clays	ND	1270–1524	6.5	Cunha et al. (1975)
Brazil	Gneiss, micaschists	ND	1270–1524	4.0	Cunha et al. (1975)
Uganda	and granite	Savanna	1016–1270	1.2–2.2	Doornkamp and King (1971)
Uganda	Quartzites and conglomerates	Savanna	1016–1270	3.0–4.0	Doornkamp and King (1971)
India	Rhyolite	Sparse	356	1.9	Ghose et al. (1967)
India	Granite	Sparse	356	1.3	Ghose et al. (1967)
India	Sandy alluvium	Sparse	356	0.8	Ghose et al. (1967)

[a]Highest drainage densities in Abrahams' study were in semi-arid areas, which were the only areas with maps of adequately large scale for accuracy.

Source: Walsh, 1985

microform. Limestone most commonly satisfies the prerequisites for karst development and the term karst is often used as synonymous with limestone terrain. Under certain conditions, however, karst may also be associated with siliceous rocks, in particular with types of granite and sandstone (Otvos, 1976; Jennings, 1983; Young, 1986).

During the last two decades there has been increasing recognition of the existence of solutional landforms on quartzose rocks. Most of the major examples reported so far (e.g. huge solutional features in the Roriama quartzites of the Guyana Shield, and large caves, dolines and towers in Venezuela, Australia and South Africa) are in the humid or seasonally humid tropics (e.g. Sczerban and Urbani, 1974; Martini, 1981; Jennings, 1983; Pouyllau and Seurin, 1985; Young, 1988). Young (1986) has emphasized that the general neglect of quartzose karst in tropical geomorphological texts has resulted in a lithological bias in the theory of tropical landscape evolution, since conventional theory is drawn almost exclusively from studies of limestone and igneous rocks (especially granite). Much more needs to be known about the distribution and form of quartzose karst; our knowledge of the chemistry of quartz solution, the manner in which the constituents of quartz rocks are dissolved, also needs to be improved. By comparison, details of carbonate solution processes are relatively well understood (Ford and Williams, 1989), yet our understanding of the morphology and evolution of limestone karst terrains remains incomplete. One reason for this is the complexity of the relationship between process and landform. It is a function of the interplay of geological, climatic and topographic factors, the detailed nature of which is a subject of much debate.

Limestone solution occurs through a complex series of chemical reactions (comprehensively described by Loughnan, 1969; Wilson, 1983; Ollier, 1984; Ford and Williams, 1989; and others). It can be described simply as the reaction of calcium carbonate or calcite ($CaCO_3$) or sometimes magnesium carbonate or dolomite ($MgCO_3$) with dilute carbonic acid (H_2CO_3) to form calcium bicarbonate ($Ca(HCO_3)_2$). Calcium bicarbonate forms an aqueous solution with water (figure 6.12). Like all chemical reactions, the potential rate of

1

$$CO_2 + H_2O \rightleftharpoons H_2CO_3$$

Carbon dioxide + water ⇌ carbonic acid

2

$$CaCO_3 + H_2CO_3 + H_2O \rightleftharpoons Ca(HCO_3)_2$$

Calcium carbonate + carbonic acid + water ⇌ calcium bicarbonate as an aqueous solution

Figure 6.12
Limestone solution.

conversion is controlled by temperature (van't Hoff's rule). In the tropics, the potential rate of solution may be four times as great as in cold climates (Ford and Williams, 1989). However, there are a number of complicating factors in carbonate dissolution chemistry (White, 1984) which serve to enhance (boost) or inhibit (depress) the solubility of carbonate minerals (figure 6.13). One of the most important factors in controlling the actual rate of solution is the availability of dissolved carbon dioxide for the formation of carbonic acid.

The solubility of carbon dioxide depends upon its concentration in air, measured as a partial pressure, and the temperature of the solution (Smith and Atkinson, 1976). At a constant partial pressure, increasing temperatures reduce the solubility of carbon dioxide. Hence, in the tropics, water is able to dissolve less carbon dioxide per unit volume than in colder areas before becoming saturated. This situation is complicated, however, by the production of biogenic carbon dioxide as a by-product of respiration of soil organisms. This can result in high levels of carbon dioxide in soil pores and hence higher rates of solution than would otherwise be expected (Viles, 1988). Rates of carbon dioxide production by roots and soil bacteria increase with temperature and are also sensitive to soil moisture conditions, which are at an optimum in soils at 50–80 per cent of field capacity. In the humid tropics, conditions for the production of biogenic carbon dioxide are therefore particularly favourable (table 6.6).

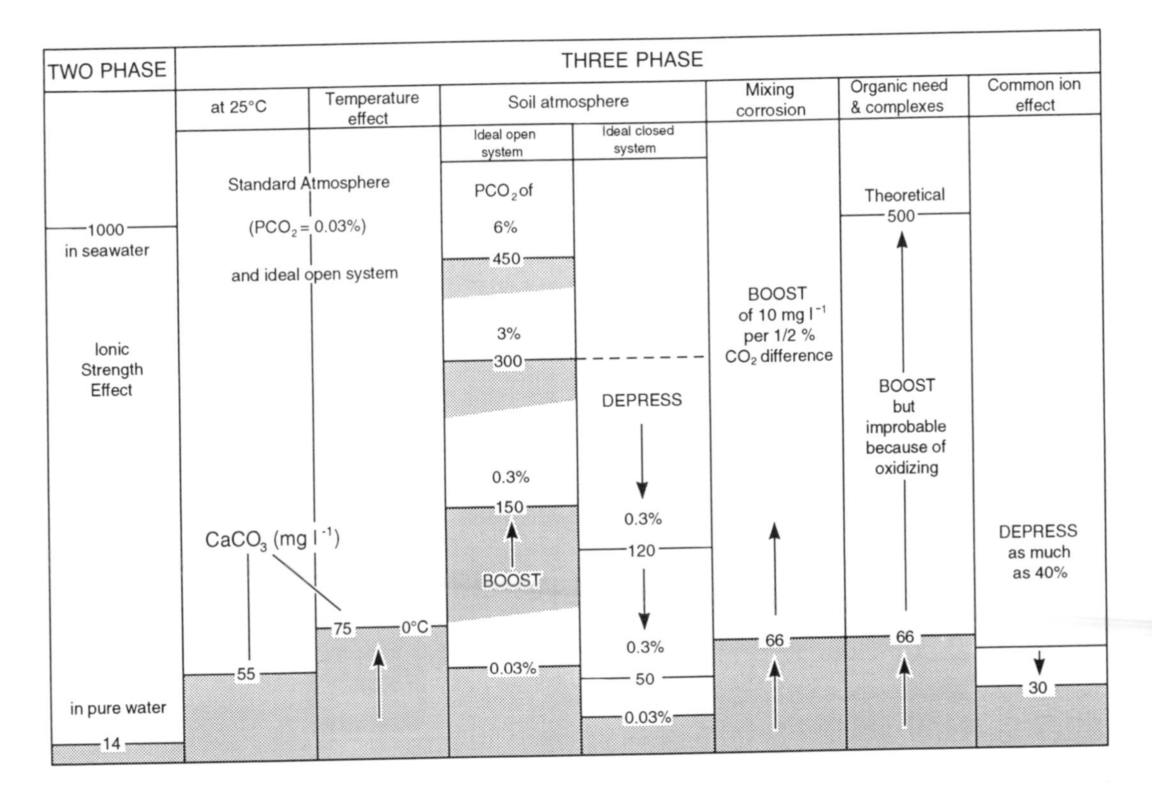

Figure 6.13
Principal complicating effects in carbonate dissolution chemistry (after Ford and Williams, 1989, figure 3.7). Values indicated refer to $CaCO_3$ mg l^{-1}.

Table 6.6 Carbon dioxide concentration in soils (measured as PCO_2)

Tropics	0.2%–11%
Temperate	0.1%–3.5%
Arctic tundra	0.2%–1%

Source: Ford and Williams, 1989

These facts form the basis of an argument concerning the rates of limestone denudation in different climates. Corbel (1957) and his followers argue that denudation rates are highest in cold climates while Lehmann (1970), Sweeting (1966) and others argue that the greatest rates occur in the humid tropics. A comprehensive review of the debate may be found in Smith and Atkinson (1976, pp. 367–409). They conclude that mean annual runoff is the major determinant of denudation rate and acknowledge the importance of temperature largely in its effect on the development of a continuous soil and vegetation cover. They stress that traditional divisions of climate do not necessarily correspond with geomorphological divisions. A multitude of other interacting geological, tectonic and inheritance variables also play a part. Ford and Williams (1989) question the conceptual and empirical accuracy of published data on limestone solution. First, they note the ambiguity of the term 'solution rate' widely used in geomorphic literature as a synonym for 'solutional denudation rate'. They also point out that 'karst (limestone) denudation rate' should describe the sum of chemical and mechanical denudational processes. The fact that it frequently describes only solutional rates is a 'major deficiency in karst (limestone) process research' (Ford and Williams, 1989, p. 97). Errors in empirically measuring solutional denudation (Williams and Downing, 1979; Gunn, 1981) are seen as the second great failing and the reason why Smith and Atkinson's (1976) equations have proved inadequate for predicting solutional denudation. They argue that published solutional denudation values derived by the Corbel method may be over 100 per cent in error.

The causal relationship between environmental variables and landform is poorly understood. However, it is generally accepted that, while some karst landforms occur throughout the world (e.g. solution hollows, karren, speleothems), some are exclusive to or at least more highly developed in tropical areas. Polygonal karst, a landscape pitted with smooth-sided and soil-covered depressions (plate 6.7), and tower karst, a landscape of upstanding monoliths scattered across a surface of low relief (plate 6.8), are the most frequently cited examples of an exclusively tropical karst landscape. The vast majority of these landforms occur in limestones and it is this majority which is considered in sections 6.5.1 and 6.5.2.

It has been suggested either that polygonal forms represent an advanced stage of limestone denudation and are therefore only visible in regions which have received a long and uninterrupted

Plate 6.7
Aerial view of polygonal karst, Jamaica, West Indies.

period during which solution has been the dominant process, or alternatively that differences in the operation of solutional processes between temperate and tropical regions act to produce a dry valley and doline landscape on one hand and a polygonal landscape on the other (Ford and Williams, 1989).

6.5.1 Polygonal karst landscapes

Carbonate rocks outcrop throughout the tropics (figure 6.14), and limestone karst landscapes demonstrate a wide variety of large- and small-scale solutional features. However, in the tropics it is the polygonal (or cockpit) landform that has attracted most interest. Descriptive accounts and morphometric analyses make up the bulk of the literature (Williams, 1972; Jennings, 1985) whereas the dynamics of these landscapes has received less attention. Sweeting (1958) describes cockpit karst as a succession of cone-like hills with alternating enclosed depressions or cockpits. The hills are usually remarkably uniform in height, ranging from 30–70 m in Papua New Guinea to 100–160 m in Jamaica. Base diameters vary from around 50 m in Zaire (e.g. Kissenga plateau) to around 300 m in Jamaica (Faniran and Jeje, 1983).

A feature which distinguishes carbonate from siliceous rocks is that they have a greater tendency to fracture and fault, rather than fold and buckle, under stress. The initiation of solutional hollows is thought to result from preferential solution along these fractures. Williams's (1985) hypothesis for the initiation and development of polygonal karst (figure 6.15) stresses the importance of the three-dimensional pattern of permeability created by cracks and fractures. He postulates that the interconnecting pores, bedding planes and fissures provide a hydrological network which is easily corroded and widened by solution, especially close to the surface where compressive pressures are less. During heavy rain, this subcuta-

Plate 6.8
Tower karst, Papua New Guinea (Photo: Erlet Cater).

neous zone of high hydrological connectivity, which lies below the soil but above the phreatic zone or layer, is analogous to a leaky aquifer (the epikarstic aquifer). Williams (1985) suggests that the spatial variability in fissure permeability will allow the concentration of flow along preferred flow paths at the base of the subcutaneous zone. Also, if flow rates are sufficiently great, it will result in the local depression of the suspended water-table in the same manner as a cone of depression forms around a pumped well. The extra flow along the preferred flow line (i.e. the epikarstic stream line) will result in accelerated corrosion and the further enhancement of vertical permeability in a positive feedback loop.

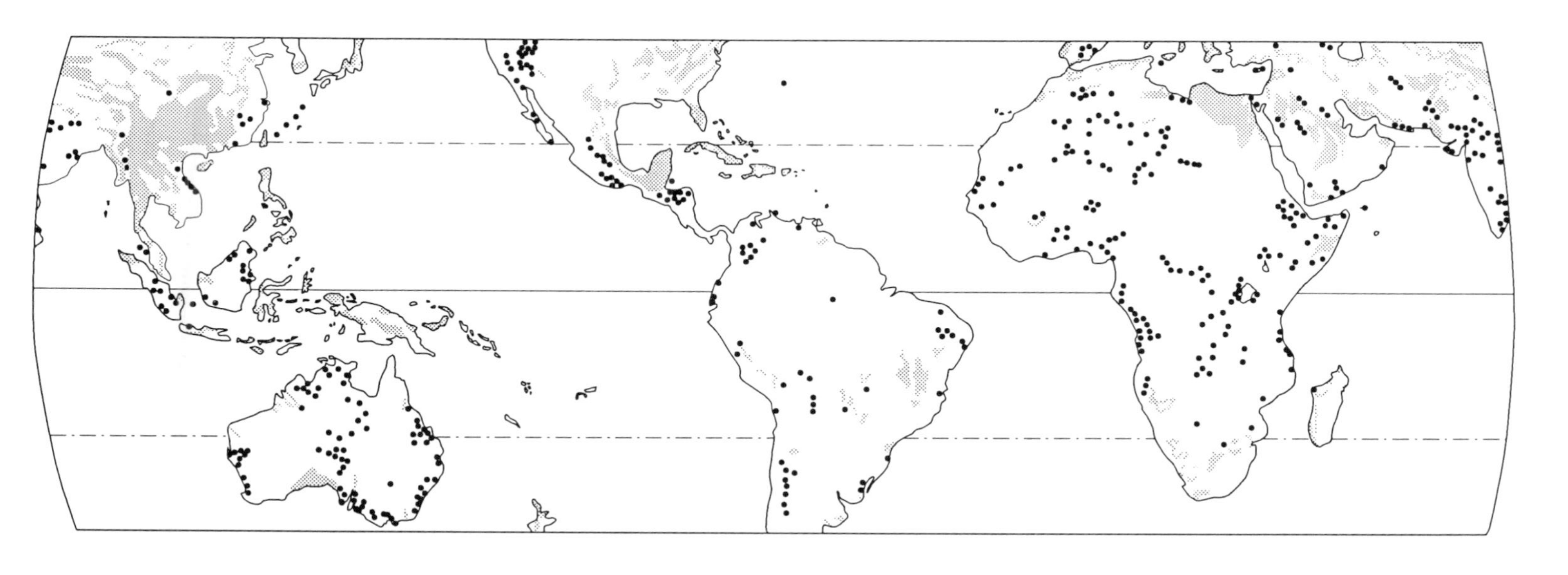

Figure 6.14
Major outcrops of carbonate rocks in the tropics (after Ford and Wiliams, 1989).

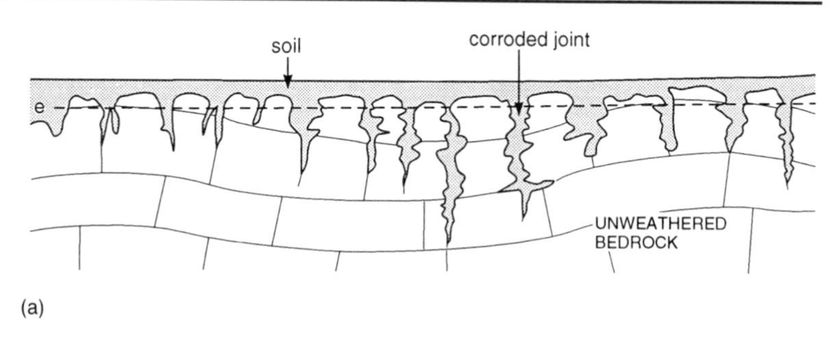

(a)

soil

s

NEAR SURFACE ZONE OF INTENSIVE WEATHERING

epikarstic water-table

r

RELATIVELY
UNWEATHERED
BEDROCK

rapid percolation
route

(b)

Figure 6.15
The initiation of doline development (after Williams, 1985): (a) initiation of karstification (e, epikarstic water-table near the surface); (b) initiation of hydrological conditions favouring solutional doline development (r, radius of epikarstic cone of depression; s, subcutaneous zone).

The radius of the cone of depression will then initially determine the size of the surface hollow which develops as a result of the enhanced solution. Once the solution hollows have become established the centripetal focusing of flow and hence corrosion will enhance their development. Slope processes will also encourage the maintenance and enlargement of a hollow in a series of positive feedback loops. For example, the downslope centripetal focusing of flow, and hence corrosion, and the accumulation of soil which is preferentially moist will encourage the biogenic production of carbon dioxide and thus further aid solution (Williams, 1985). However, soil which is thick may also have the effect of reducing the permeability of hollow bottoms and may then retard corrosion (Crowther, 1984). These processes are essentially azonal. Williams (1985) argues that the contrast between the shallow dolines of temperate regions and the deep cockpits of tropical regions is probably a consequence of the strength of the feedback mechanisms. The enlargement of hollows in humid tropical regions may also take place by the capturing of minor depressions.

According to Williams (1985), vertical development will eventually extend to the permanent or regional water table, forcing water to escape laterally rather than vertically at the base of the hollows. This will inhibit further vertical deepening and lead instead to a widening of the hollow bottoms and the development of a corrosion plane (figure 6.16). Further denudation of the slopes of the cockpits and the enlargement of the plain will lead to a residual landscape described as tower karst.

6.5.2 Tower karst landscapes

Tower karst scenery is a landscape of isolated hills or towers which rise abruptly from flat alluvial plains. It is well developed in a number of locations, including the Caribbean, Indonesia and Malaysia. Southern China is often cited as the 'type area' (plate 6.8). Tower karst landforms may be found in siliceous material (Young, 1986) but it most frequently occurs in limestone, often adjacent to polygonal karst. In any area the towers are much more variable in size than the conical hills of cockpit karst. For example, in Malaysia they range from a few metres to over 300 m in height.

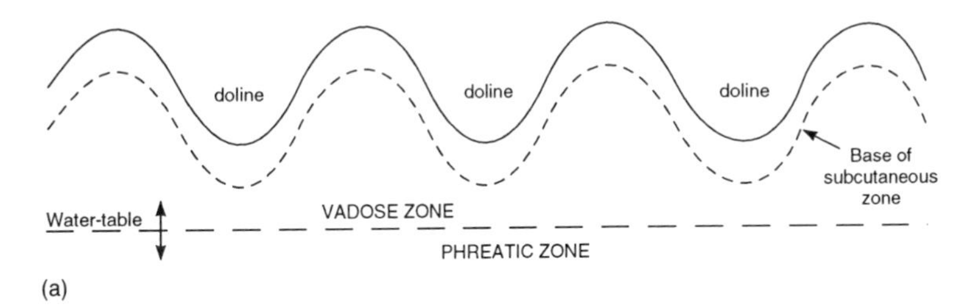

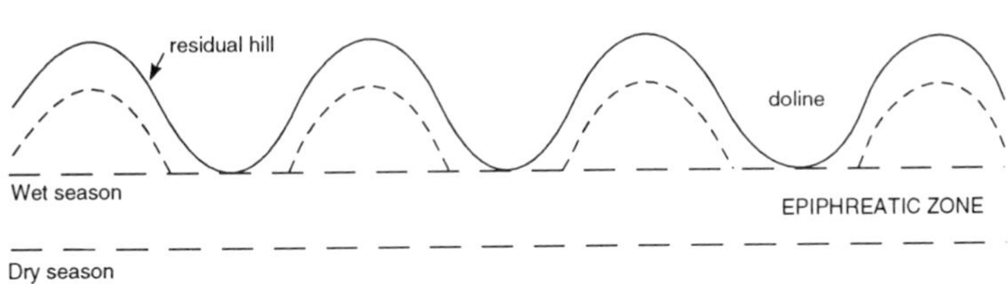

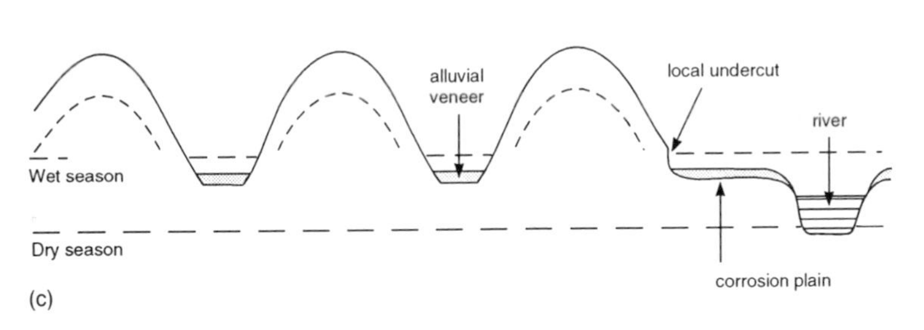

Figure 6.16
Development of a residual karst landscape from a polygonal form (after Williams, 1985): (a) dolines incise vertically through the vadose zone; (b) incising depressions reach the epiphreatic zone; (c) hills reduced by parallel slope retreat and undercutting; depression floors coalesce, corrosion plain develops.

The largest tower, the Api tower in Sarawak, rises some 1510 m and covers an area of around 38 km^2. Tower sides are typically steep (>60 °), often with cliffs and overhangs. Caves, furrows and solution notches are also common at their bases. There is no consensus of opinion regarding the formation of tower karst but it is unlikely that any single hypothesis can adequately explain all towers. Ford and Williams (1989), for example, classify tower karst into four generic types (figure 6.17).

Panos and Stelcl (1968), and Corbel and Muxart (1970), among others, believe that towers develop in a manner similar to other residual hills, i.e. by the differential erosion of rock with varying resistance. While this may be true for Ford and Williams's (1989) type (b) or type (d) tower karst (figure 6.17), it does not explain the development of towers which emerge from a common bedrock base. Differential corrosion of massive high density limestones along structural weaknesses has been suggested as the mode of

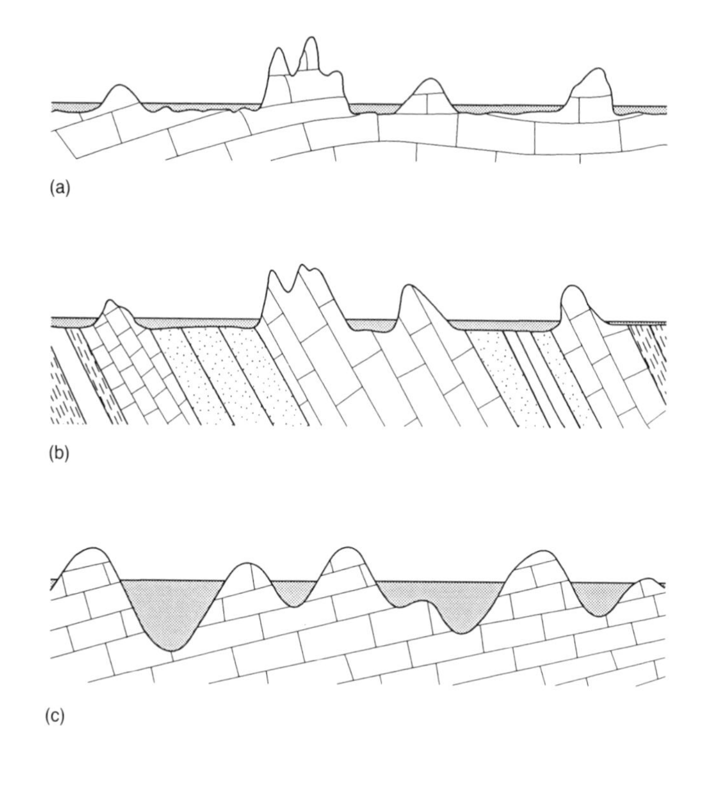

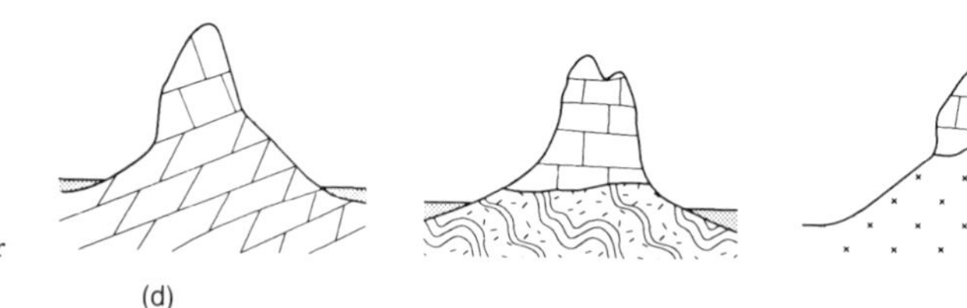

Figure 6.17
Genetic types of tower karst (after Williams, 1987): (a) residual hills on a planed limestone surface; (b) residual hills emerging from limestone inliers; (c) residual hills protruding through an alluviated surface; (d) towers rising from sloping pedestals of various lithologies.

development of tower karst in northwest Australia and northern Canada (Jennings and Sweeting, 1963; Jennings, 1969; Brook and Ford, 1978, 1980). Other theories relate the development of towers to the formation of cockpit karst. From work in Indonesia and Java, Balazs (1968) claims that the formation of karstic cones and towers represents an instantaneous state of development under the particularly favourable lithological, relief and climatic conditions provided by the humid tropics. The difference in form is explained by the hardness of the limestone and their height above base level.

Williams (1987, p. 457) argues that the frequent juxtaposition of polygonal and tower karst 'points more compellingly to a sequential development', i.e. with tower karst developing from cone karst. He suggests that when polygonal karst has dissolved vertically to the water table (figure 6.16(c)), incision is replaced by *in situ* corrosion of the cockpit hills which retreat across a developing plain to become progressively isolated residual towers. The process by which the hills are reduced is yet another area of contention. Lateral solutional undermining of the tower slopes by sheetwash has been widely reported (e.g. Lehman, 1954; Gerstenhauer, 1960) but more recent work (e.g. McDonald, 1985) argues that river action is responsible for the development of caves and the retreat of towers. Williams (1987) emphasizes the significance of the morphological and structural characteristics of the cockpit cones in the control of tower slope angles.

Reduction of cockpit karst to the water table is most likely to occur first around the edges of a karst region. This may explain Verstappen's (1960) observations that in Java there is an orderly progression of morphology from the base to the top of plateaux, from towers to steep and then shallower cones. The relative frequency of cone and tower karst will ultimately be largely controlled by the depth of the water table. Polygonal karst will persist if the water table is sufficiently deep to allow continual vertical solution. Once the water table has been reached, tower karst will begin to develop by lateral erosion (Williams, 1987).

6.6 Tropical Coasts

The coastal zone extends landwards from the sea to the limit of penetration by marine processes (figure 6.18). The action of waves and tides, the movement of sediment and the consequence of sea-level changes are to a great extent azonal. Yet the world's 500,000 km of coastline is tremendously diverse, the result (at regional and local scales) of variations in lithology and the relative activity of a number of geomorphological and biological processes, many of which are climatically sensitive.

Directly or indirectly the tropical climate, and in particular the constant high temperatures and, in the humid tropics, the abundant rainfall, play a dominant role in the formation of a number of

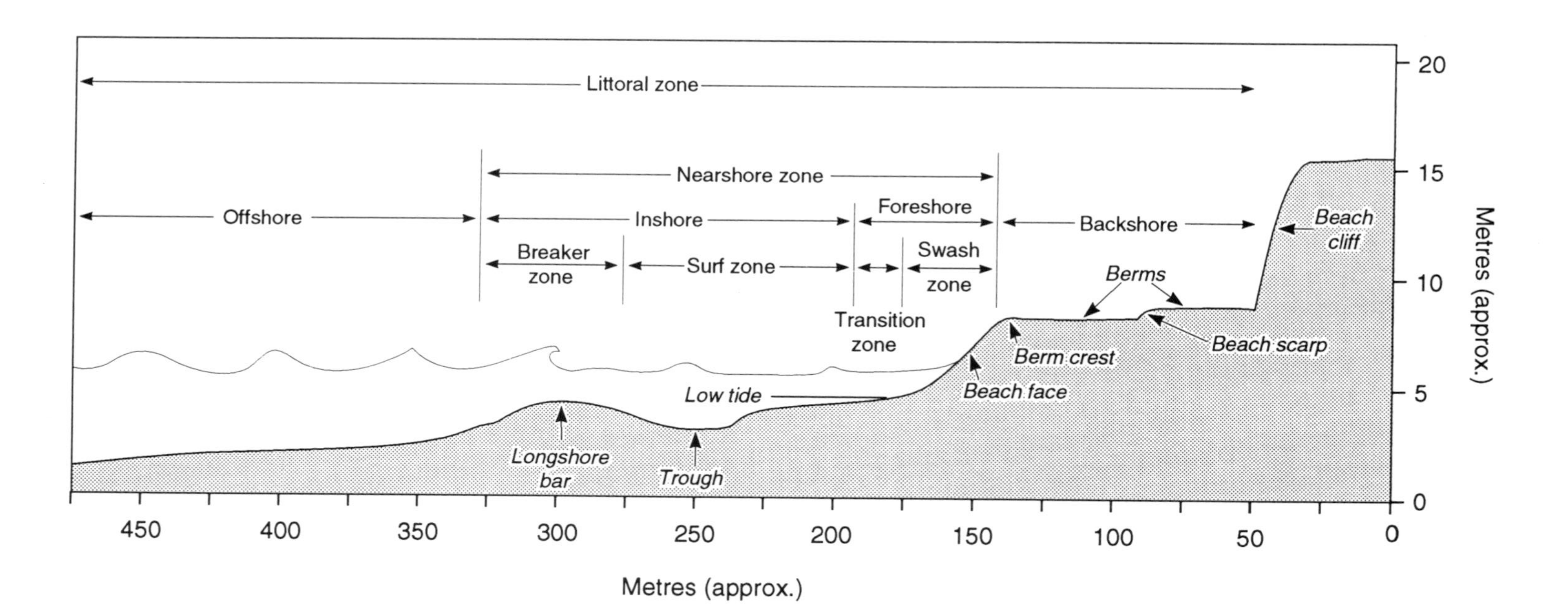

Figure 6.18
Major divisions of the littoral zone.

distinctively tropical coastal landforms. Tropical coasts differ from land margins at higher latitudes in several respects. One of these is the abundance of sediment, largely sand-sized or smaller, which is the result of the weathering of exposed continental shelves during periods of low sea level and the consequence of fluvial deposition by the numerous rivers which drain tropical lands. Tropical coasts are frequently fringed by extensive fluvio-marine sediment plains and by large and well developed sand-bars, spits and other depositional features. Large dunes tend to occur poleward of the tropics (Tinley, 1985). Some fine examples of depositional landforms can be found along the West African coast between Senegal and Nigeria and in the state of Bahia in Brazil. In contrast, coastal areas in temperate latitudes are dominated by erosional landforms such as cliffs and wave cut platforms. The coastal geomorphology of small oceanic islands, of which there are many in the tropics, are a special case and are considered in detail by Nunn (1994).

In any environment the balance between forces of erosion and deposition is a function of the energy available. Along coasts, it is the amount of wave energy which is of fundamental importance. Tropical waves are generally low energy waves; they often have a very long reach (frequently several thousands of kilometres) and approach the coast with a long wavelength and low amplitude. Rather than eroding sediment, tropical waves tend to redistribute it. Erosional waves occur infrequently, during tropical storms, but are capable of causing massive erosion during the few hours they are present. Tropical coasts are primarily low energy coasts; for the majority of the time they receive low energy waves. The existence of a low energy environment is a prerequisite for the proliferation of a number of tropical organisms, such as algae, coral polyps and mangroves, which are geomorphologically significant.

Regions prone to tropical storms (cyclones) sporadically and temporarily become high energy environments, a fact which is also geomorphologically significant. One feature which all coastal environments share is a high sensitivity. Disturbance to the balance of physical and biological processes in operation, due to 'natural' forces or human intervention, is likely to result in rapid and substantial geomorphological change. This is especially true of island landscapes (Nunn, 1994).

In common with many other branches of geomorphology, before the upsurge of interest in process and systems studies a great deal of attention was focused upon classification. Bird (1969), for example, devoted a whole chapter to a review of this subject. King's (1972) classification of coasts, based on the relative movement of sea level, the effect of marine processes and the form of the land–sea contact zone, is the most widely accepted today (table 6.7). Of the seven categories proposed by King, coral coasts and mangrove coasts are exclusive to the tropics while delta coasts are best developed at low latitudes. The remainder of this chapter will concentrate on these.

Table 6.7 King's (1972) classification of coasts

1	Barrier coasts
2	Mangrove coasts
3	Delta coasts
4	Submerged coasts
5	Emerged coasts
6	Rocky coasts
7	Coral coasts

6.6.1 Coral coasts

Corals are bottom-dwelling sessile marine organisms which live singly or communally in colonies. Many secrete external skeletons of calcium carbonate. Coral reefs are a complex organogenic framework of calcium carbonate, composed primarily, today, of the branching skeletons of Madreporian colonial corals, infilled with fragments of reef material, mollusca, echinoid debris and foraminifera. In many cases, calocareous algae cover and bind the coral colonies. Calcium carbonate, precipitated from the seawater in the form of a 'chalky mud', may also help cement together the reef components.

The geomorphological significance of reefs is enormous. They form an underwater barrier forcing waves to break and dissipate their energy seaward of the coast, thus protecting the shore (plate 6.9). They also create a very low energy environment between the reef and the foreshore which favours the accumulation of large amounts of fine fluvio-marine sediments and creates conditions suitable for mangrove development. The protective role of reefs is especially important during tropical storms (Stoddart, 1971).

Reef corals cannot survive prolonged periods of subaerial exposure. As a result, the vertical development of a reef is limited by the level of low water. Falls in sea level will lead to the death of exposed coral but may result in the emergence of a reef above high

Plate 6.9
Coral islands, Maldives. The extent of the coral reef is indicated by the change in water depth.

water, creating areas of low-lying land such as the keys or cays of Florida. These islands become colonized by terrestrial vegetation which encourages further deposition of sediment, increasing the land surface. Eventually lithogenesis of the corals' calcite skeletons will convert the reef to coralline limestone.

The distribution of reefs is governed by the strict life requirements of the corals, in particular by the phototropic needs of a zooxanthellae algae which forms a symbiotic relationship with the reef forming polyps (table 6.8). Reefs are largely absent along the western side of continents affected by cold ocean currents and are poorly developed seaward of land areas receiving large amounts of rainfall, or at river mouths where fresh water reduces salinity and brings choking sediment. Coral growth is upwards and outwards wherever ecological conditions are suitable but reef formation is the result of two antagonistic actions. The growth of polyps and their skeletons enlarge the reef and encourage the accumulation of debris, while light conditions and the low tide level limit vertical development. Littoral action tends to fragment the reef and lead to the production of a detrital slope (Bird, 1969).

Sea conditions during an intense tropical storm frequently result in massive erosion of the reef and a reduction in fish and invertebrate life. However, in an area prone to cyclones, reefs will have suffered such disturbances on a number of occasions. While the effects of a storm may be destructive in the short term, the form of the reef will have developed taking account of these infrequent extreme events. Cyclone damage to reefs will only result in significant long-term changes if the frequency or intensity of the cyclones changes or if the sensitivity of the reef to the storms is in some way altered. For example, a reef may become more sensitive to break up or death if it is already under stress. Most living reefs

Table 6.8 Life requirements of reef corals

Warmth	Minimum 18 °C, optimum 25–30 °C. Distribution of reefs correlates with 20 °C isotherm for coldest month
Light	For photosynthesis by symbiotic algae. Restricts most corals to around 25 m depth; some grow to depth of 90 m
Salinity	Tolerances 27%–40%, optimum 34%–36%
Turbulence	For oxygen, nutrient supply and to stabilize temperature
Substrate	Preference for firm base but can develop on rubble and fine sediments
Sediment	Reduces light and may bury polyps. Clear water required
Exposure	Desiccation results in polyp death but some protection at low tide afforded by mucus coating
Currents	Larvae are free-floating plankton and will be dispersed by offshore winds
Food	Zooplankton; in turn these require inorganic mineral salts
Sea level	Can accommodate moderate rises by vertical growth. Rapid change causes death

Source: adapted from Viles, 1988

in the tropics today are affected by human activity (e.g. section 8.2.5). Accelerated sedimentation due to poor land management and dredging is 'responsible for more damage to reef communities than all other forms of human insult combined' (Johannes, 1977, p. 593) and also lasts longer than most other forms of stress. Thermal and chemical pollution, sewage and accelerated surface runoff are also important reef stressors. In the Pacific over-fishing and the collection of reef organisms are also important. The sensitivity of reefs and their geomorphological and economic importance is well known, yet research on the tolerances of reefs to certain types of stress and the implications of reef destruction is sorely lacking.

Sea-level fluctuations are also likely to be significant to reef development in the long term. However, to a certain extent, reefs are able to contend with sea-level rises. This growth during periods of rising sea may explain the existence of very thick reefs (e.g. 1450 m in Tuamoto islands, over 1000 m in the Bahamas and 4488 m on Andros island). Most coral reefs respond comparatively promptly to changes in the relative sea level. This has led to the use of reef accumulation rates, calculated by dating sections through the reef limestone, as surrogates for the rate of past relative sea-level rise (Nunn, 1994). This information is also useful when considering the likely reef response to predicted future sea-level changes (section 9.10.2). Upward growth is often cited as evidence for Darwin's (1842) subsidence theory of reef development (figure 6.19). Darwin's (1842) morphological classification of reefs is still widely used. He identified three types of reef: fringing reefs bordering a coast, barrier reefs separated from the coast by a deep channel and atolls, encircling a lagoon. The major amendments have been the addition of platform and patch reefs which, it is suggested, evolve from former shorelines (Fairbridge, 1967).

6.6.2 Algal mats

Biological influences on landform development along tropical coasts are not limited to reefs. Micro-algal communities and macrophytes trap and bind sediments, often with mucus, and so encourage stabilization. Large marine plants form efficient energy buffers protecting the shore. Biological processes are also implicated in the precipitation of carbonate which, along with aragonite, form the cementing agents which bind together calcitic beach material on the intertidal slope to form beachrock (Stoddart and Cann, 1965; Viles, 1988). The formation of this seaward dipping deposit (plate 6.10) can be very rapid, e.g. Second World War artifacts and even divers' aqualung equipment have been found in the hardened beach material of several South Pacific Islands. Once again, where it occurs, beachrock affords protection to the tropical foreshore.

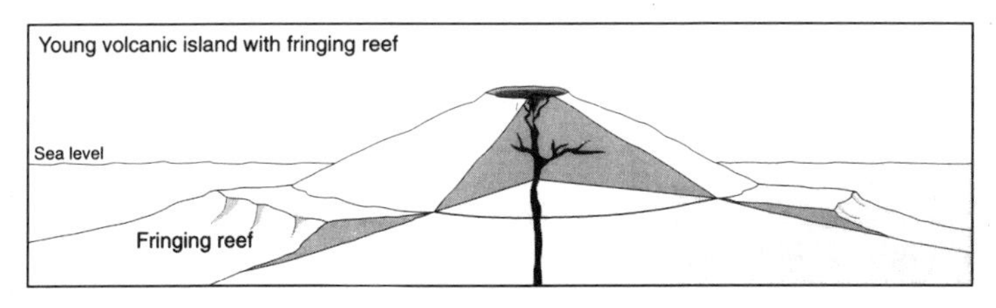

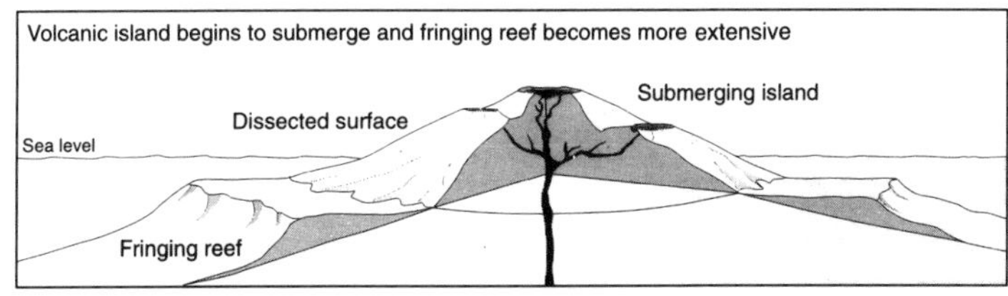

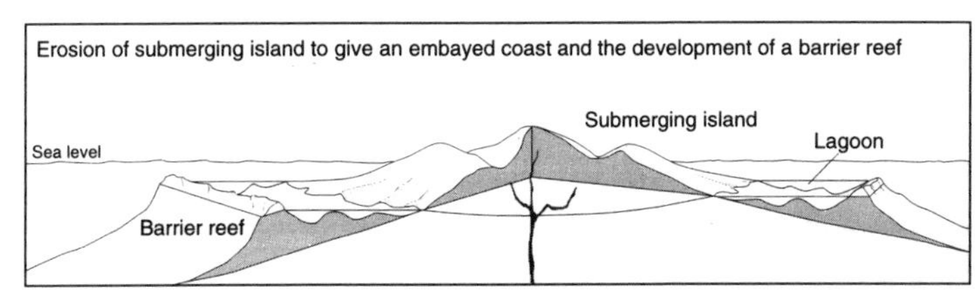

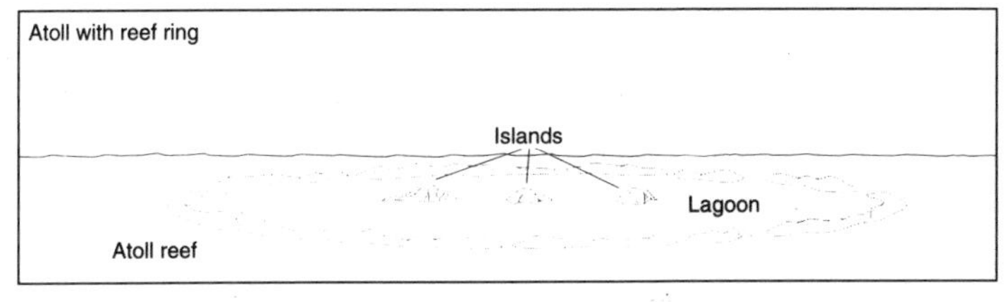

Figure 6.19
Darwin's model of reef development by subsidence of a volcanic island.

6.6.3 Mangroves

Mangroves, which are exclusive to tropical and subtropical regions, occupy approximately 75 per cent of the coastal fringe, protecting it from major erosion during tropical storms and helping to stabilize and accumulate fine coastal sediments (figure 5.22). Mangroves also provide a rich and diverse habitat for a wide variety of other plants and animals and are a valuable timber and fuelwood resource.

Extensive mudflats are a feature of mangrove coasts (plate 6.11); lying just above low tide they are inundated as the tide advances through an intricate network of anastomosing creeks and gullies. Mangroves are frequently cited as important and competent 'land builders', helping to trap sediment around their specialized roof structures and stabilizing sediments by compaction and agglutination. Studies by Thom et al. (1975) and Thom (1967, 1982, 1984), however, stress the complex interrelationships between biological activity and geomorphology. Mangroves are most effective at 'land building' through the trapping of floating and suspended stream load along deteriorating or abandoned channels within the already colonized coastal sediments. The accreting mudflats, however, may often remain bare until they have obtained particular conditions of

Plate 6.10
Beachrock running parallel to the shoreline, West Indies.

elevation and consistency. This suggests that mangroves follow silting rather than cause it. The electrolytic precipitation (flocculation) of suspended materials in river water on contact with salt water is another cause of deposition of fine particles along the coast.

In parts of Hawaii, Florida and along the coast of South Africa, mangroves have been purposefully introduced to help stabilize disturbed or inconveniently mobile coastal material (Bird, 1969). Elsewhere, however, mangrove swamps are being deliberately and inadvertently destroyed (section 9.5.4). In Asia, for example, many thousands of hectares of mangroves have been cleared for timber, mariculture, agriculture and a variety of industrial and commercial

Plate 6.11
Mangroves and associated mudflats, West Africa.

activities. Many more hectares have been disturbed or destroyed by placer mining and the deposition of waste material (section 8.2.5) and by changes in salinity levels caused by the diversion of freshwater. In the future they may also come under threat from increased sea levels associated with global warming (section 9.14.1). Whatever the cause, the effects on land are increased sea incursions and accelerated erosion.

SEVEN

The Humid Tropical Hydrosphere

7.1 The Hydrological Cycle

The circulation of water within the earth–atmosphere system is essentially the same throughout the globe (figure 7.1). However, the total and relative amounts of water which travel along the various pathways vary enormously from catchment to catchment and on a regional basis, depending upon a host of environmental (e.g. lithological, pedological and topographical) and atmospheric variables.

In most humid tropical catchments, precipitation is by far the most important hydrological input. In a small number of catchments water derived from aquifers whose areas of recharge lie beyond catchment boundaries, or water imported from adjacent catchments as part of water resource management schemes, may also contribute significantly. Except for a small number of catchments whose headwaters lie at very high altitude or in areas of persistent low cloud, rainfall is the most important type of precipitation.

The characteristics of rainfall in the humid tropics are discussed in section 3.6. Those aspects of rainfall which are especially significant to the humid tropical denudation system and to water availability in ecological terms, include (Jackson, 1989)

1 rainfall seasonality – its distribution on an annual basis
2 rainfall variability – variations in annual totals from year to year
3 rainfall reliability – the consistency in the timing and duration of the rains
4 rainfall amount – the amount of rainfall which falls per unit time
5 rainfall duration – how long the rain lasts
6 rainfall frequency – how often it rains

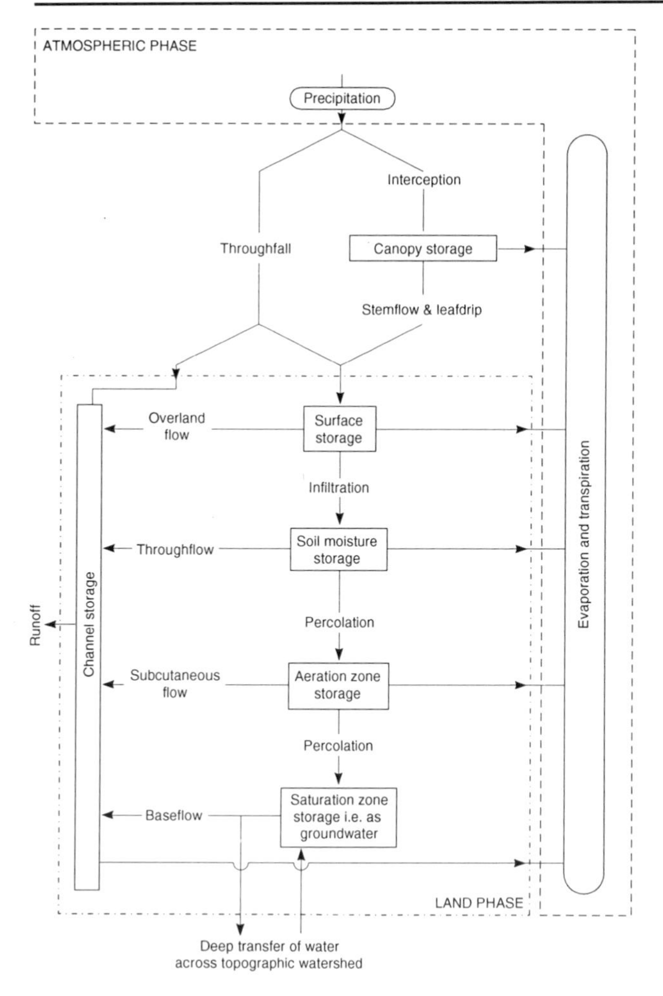

Figure 7.1
Water flows in a vegetated catchment.

The seasonality of rainfall exerts an overall control on water availability while the intensity, duration and frequency of individual rainstorm events affect the pattern of hydrological routeways through the catchment. A key point is that, in the tropics, a high proportion of rainfall occurs during high magnitude, high intensity storms (section 3.6.3). Experiments undertaken in East Africa (Hudson, 1971) suggest that rainfall becomes (potentially) erosive at intensities above 25 mm h^{-1}. The degree to which this potential is realized will depend upon the soil's characteristics and surface cover. A figure of around 40 per cent is often cited as representative of the proportion of rainfall which falls at intensities above 25 mm h^{-1} in tropical regions. A further 12 per cent falls at intensities of 20–25 mm h^{-1}.

7.1.1 Interception

In humid tropical forests, the process of interception by vegetation (figure 7.2) is particularly complex and results in important local

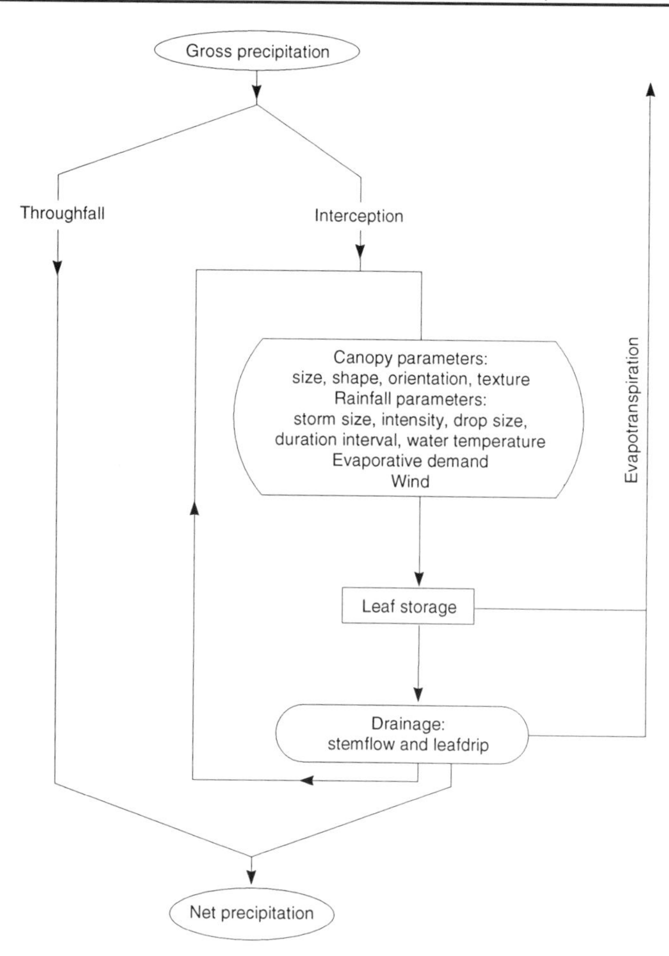

Figure 7.2
The interception process.

spatial variations in the amount, timing and energy of water reaching the ground surface. In general, vegetation is effective at reducing the energy of impact of rainfall reaching the surface. Wiersum (1985), however, notes that erosion can be locally enhanced where accumulated raindrops fall from a forest canopy onto a ground surface with a thin vegetation and litter cover. Herwitz (1988) describes the role of tree buttresses in concentrating stemflow and causing local erosion. Localized erosion caused by concentration of water by single canopy crops such as maize and bananas has also been the subject of study (cf. Lewis, 1976).

Interception also reduces the amount of water reaching the ground surface due to direct evaporation from the canopy; however, this should not necessarily be regarded as a loss in water input since it may, at least in part, be compensated for by a reduction in evapotranspiration from the canopy because of the utilization of energy (Jackson, 1989). The effectiveness of interception, as measured by the proportion of gross rainfall which is intercepted, varies according to a number of controlling variables. These include plant type (e.g. leaf type and arrangement), plant

Table 7.1 Interception data for tropical forests

Source	*Location*	*Interception*
Vaughan and Wiehe (1947)	Mauritius	34% gross rainfall
Hopkins (1960)	Uganda	35% gross rainfall
Wiersum (1985)	Java	12% gross rainfall but variable between storms
Pathak et al. (1985)	Himalayas	8%–25% according to species
Mohr and Van Baren (1954)	Surinam	20% of 40 mm storm, 80% of 1 mm storm
Jackson (1975)	Tanzania	12% of 20 mm storm, 85% of 1 mm storm

density, wind conditions, rainfall intensity, duration, amount of evaporation during the storm and the interval between storms. Interception data (table 7.1), whilst indicating conditions during the monitoring period, will therefore not necessarily be representative of other periods or other locations. Jackson (1989) stresses the requirement to relate interception to various precipitation and vegetation characteristics in order that comparisons can be made.

In some upland forested areas, where cloud belts develop in response to the forced ascent and cooling of moist air across topographic barriers, a significant water input may originate from water condensed or 'captured' from the cloud onto vegetation surfaces. There is also an indirect effect, since evaporation losses in the almost saturated air will be very low. These so-called 'cloud forests' are a distinctive ecosystem (section 5.2.4) and, in certain tropical regions, they extend across large areas. For example, in central America between 7000 and 15,000 km^2 is classified as cloud forest (figure 7.3). In Costa Rica, around 8.7 per cent of the total land surface is covered by montane forest and cloud moisture interception makes a considerable contribution to water yield in a number of catchments, providing potable water for lowland urban areas (Zadroga, 1981). The consequences of deforestation in these areas are likely to be particularly severe (sections 9.3 and 9.4).

Measurement of cloud moisture interception (also known as mist precipitation, horizontal precipitation, condensation or fog drip) is extremely complex. It requires the estimation of the rate of moisture collection per unit area of vegetation, as controlled by the effectiveness of interception of liquid water droplets by the vegetation and the rate of vapour condensation (itself affected by the presence of vegetation, which can act as condensation nuclei). A further complication arises from the fact that the interception efficiency of any part of the vegetation will vary with condensation drop size. Empirical data are not directly comparable. Ekern (1964, 1982) has demonstrated that the additional moisture gained varies spatially and temporally and according to cloud conditions. He cites an increase of almost a third for the Norfolk Island pine forests (mean annual rainfall 2600 mm) of Hawaii but only a 10 per cent increase in moisture for the leeward forests of

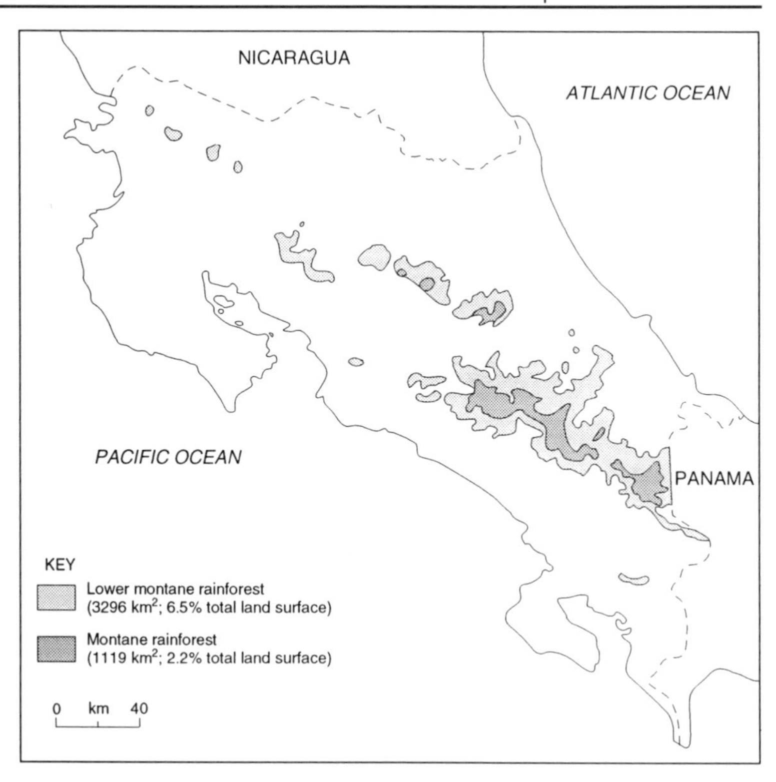

Figure 7.3
Distribution of montane cloud forest in Costa Rica (from Zadroga, 1981).

the Koolau range, Oahu (mean annual rainfall 2835 mm). In another study in Hawaii, Pereira (1973) showed that the relative contribution of cloud moisture interception increased with height, to a maximum of 65 per cent at 2500 m.

7.1.2 Evaporation and transpiration

In the humid tropics, a large proportion of the surface is covered by vegetation (chapter 5), giving rise to two important upward fluxes of water into the atmosphere: first from the soil (evaporation) and second via the stomata of leaf tissue (transpiration). Since they are difficult to separate, they are usually considered together as evapotranspiration (section 3.5). Knowledge of rates of transpiration (as opposed to evapotranspiration) is important for ecological studies and provides an indication of the importance of vegetation to water cycling. Unfortunately, however, it is exceedingly difficult to measure, requiring sophisticated equipment or complex mathematical calculations. Data for the tropics are rare; however, Jordan and Kline (1977) have recorded transpiration losses of between 2.71 litres day^{-1} for small understorey trees and 1180 litres day^{-1} for large canopy trees in the heterogeneous Amazon forest. These results were obtained by injecting a tritiated water tracer into the base of the tree and measuring the radioactivity of sampled twigs over time. Jordan and Kline (1977) also describe a strong correlation ($r = 0.96$)

between transpiration and sapwood area, a relationship which may be useful for similar comparative studies.

7.1.3 Infiltration and percolation

Infiltration describes the process by which water enters the soil surface and percolation describes its progress through the unsaturated soil zone. Their rate and capacity control the relative proportion of precipitation which travels to the streams by surface and subsurface routes. Infiltration rates vary enormously both spatially and temporally according to factors such as soil type (its structure and moisture content), land cover and land use (table 7.2).

Table 7.2 Typical steady state infiltration rates

Soil type	*Infiltration rate*
Sandy and aggregated silts	8–12 mm h^{-1}
Loess and sandy loams	4–8 mm h^{-1}
Clays	1–4 mm h^{-1}
Swelling clays	<1 mm h^{-1}

Humid tropical soils, particularly those derived from deeply weathered regoliths, are generally regarded as having very high (and often variable) infiltration rates, despite having initially high moisture contents (table 7.3). Norman et al. (1984), however, argue that initially high rates steadily decline during protracted infiltration and final steady rates are often low, a function of low terminal hydraulic conductivity. Bonell et al. (1982) found that it took between five and eight hours to achieve steady state infiltration rates in the initially very moist catchment soils of South Creek, northeast Queensland. Roy and Ghosh (1984) admit that they were unable to achieve a constant rate for the alluvial Ganges soils of India and Reading (1986) has also described the problems of obtaining meaningful infiltration values for some West Indian soils. The failure of field experiments to achieve a steady state or minimum infiltration rate may at least in part explain some of the large differences in infiltration rates quoted for very similar soils.

Minimum infiltration values are conventionally used for comparative purposes. However, if the time taken to achieve steady state infiltration exceeds that of the most prolonged rainstorm this rate will never be reached in the field. For hydrological purposes it is therefore more relevant to relate infiltration rates to storm duration. In seasonal parts of the tropics soils often contain the swelling clay montmorillonite. When dry, these soils shrink and crack and, at the onset of rain, water can infiltrate the surface via the cracks at very high rates. However, wetting of the soils causes expansion of the clay molecules, sealing the cracks and reducing infiltration to very low

Table 7.3 Typical infiltration rates for tropical soils

Source	*Location*	*Soil*	*Land use*	*Rate (mm h^{-1})*
Reading (1986)	Dominica	Allophane podzolic	Forest	60–960
		Allophane latosolic	Forest	12–360
		Kandoid	Forest	30–60
		Smectoid	Forest	60–360
Walsh (1980)	Dominica	Allophane latosolic	Forest	46–6096
Gilmour et al. (1987)	Nepal		Grass-forest	39–524
Bonell et al. (1983c)	Queensland	Podzolic (ultisol-inceptisol)	Forest	18–107[a]
		Kranozem (oxisol)	Forest	40–103[a]
Lundgren (1980)	Tanzania	Humic nitosol	Forest	830–850
		Humic nitosol	Cultivated	635
		Humic nitosol	Denuded (after 90 minutes)	270
Lugo-Lopez et al. (1968)[b]	Puerto Rico	Ultisols	Cultivated	75–241 (after 8 hours)
Van der Weert (1972)[b]	Surinam	Sandy loams	Forest	720
Wilkinson and Aina (1976)[b]	Nigeria	Alfisols		
	(Iwo)		Bush fallow	1189 ±171
				1st crop 447 ± 66
				2nd crop 495 ± 33
	(Oba)		Bush fallow	1077 ± 125
				1st crop 488 ± 65
				2nd crop 216 ± 38 (after 120–150 minutes)
Lal (1976)[b]	Nigeria	Alfisols	Bare fallow	2100–120[c]
			Maize[d]	2100–60[c]

[a] Metres per day.
[b] 2 hour infiltration figures. In Lundgren (1980).
[c] 0–2 years after forest clearing.
[d] Under various rotation and cultivation practices.

rates. This 'surface capping' may occur before lower levels of the soil are wetted, leading to a theoretical increase in permeability with depth. It is theoretical since if water was to reach the lower soil levels this would also initiate swelling and a reduction in permeability.

Estimates of the rate of movement of water through the soil can be made in the field using an infiltrometer or permeameter apparatus to calculate hydraulic conductivity, described by the parameter K in Darcy's law (Freeze and Cherry, 1979). Methods and techniques used in tropical situations are reviewed by Bonell et al. (1983a). They advise caution in the comparison of results because of differences in determination techniques but acknowledge that figures provide a useful indicator of subsurface conditions. A rapid decline in K with depth is frequently cited. This in the main reflects a decrease in the number and size of pores with increasing depth due to increasing pressures and the reduction in the effectiveness of biological activity in 'opening up' the soil structure. It is also likely to reflect a lack of parity in the results from surface infiltration and subsurface permeameter apparatus (Talsma and Hallam, 1980).

7.2 Runoff Processes

Numerous models have attempted to quantify the relationship between water inputs and runoff for humid temperate regions. In recent years, evaluation of these models has become a major theme in hydrology (Takasao and Takara, 1988) and, increasingly, empirical evidence is overturning theory (Kirkby, 1988). Unfortunately, the mechanisms through which runoff is generated and their consequence to hillslope hydrology in the humid tropics has received scant attention. Frequently, runoff processes are monitored and discussed in terms of techniques and models developed from experience in temperate environments.

In the humid tropics the proportion of rainfall which eventually runs off (i.e. the runoff coefficient) varies enormously, ranging from almost zero (0.1 per cent) for experimental plots under small-scale agriculture in Tanzania (Lundgren, 1980) to 65 per cent for forested catchments in monsoonal northeast Queensland (e.g. Bonell et al., 1983a) (table 7.4). Precipitation totals and seasonal distribution are obviously important variables and differences in plot size and monitoring technique will also affect these figures. Runoff response is also influenced by geology, soil characteristics, hydraulic geometry, relief and vegetation (Ledger, 1964). Ogunkoya et al. (1984) investigated the significance of these 'environmental' factors on runoff response in fifteen third-order catchments on the Precambrian basement complex of southwest Nigeria. They demonstrated that, while the runoff coefficient was statistically related to basin geology, relief and land use, there was no significant relationship between rainfall parameters and runoff. They concluded that (p. 34) 'the runoff coefficient is overwhelmingly influenced by the nature of the saprolite and rocks underlying the basin'. Another important observation was that runoff was not only persistently low but also highly variable, ranging from 1 per cent to 40 per cent, reflecting the heterogeneous nature of hydrological processes even within a relatively small area.

7.2.1 Overland flow generation

Detailed investigations of the mechanisms of overland flow generation in the humid tropics include studies undertaken in the perennially wet forests of Malaysia (Leigh, 1978), central Java (Bruijnzeel, 1983), Dominica (Walsh, 1980a, b), Amazonia (Nortcliff and Thornes, 1981), monsoonal northeast Queensland

Table 7.4 Runoff from experimental plots

Source	*Location*	*Soil*	*Land use*	*Slope (degrees)*	*Annual rainfall (mm)*	*Runoff (% rainfall)*
Mitchell (1965)	Tanzania	Nitosol	Coffee	9.5	1660	3–11
			Coffee	18	1300	5
			Maize	18	1300	2–3
			Bananas	18	1300	2
			Grass	18	1300	1
Othieno (1975)	Kenya	Nitosol	Tea	6	2160	2–7
Lundgren (1980)	Tanzania	Nitosol	Forest	12–24	1115	0.5–1
			Forest	14–22	820	0.8–1.3
			Cultivated	11–19	630	0.1
			Bare	1	1410	34
Hutchinson et al. (1958)	Uganda	Ferrasol	Mulch	1	1410	5–10
			Grass	1	1410	4
			Cultivated	1	1300	21
Roose (1967)	Senegal	Ferruginous	Forest	1	1300	1–1.2
Roose (1970)	Ivory Coast	Ferralitic	Semi-deciduous forest	5	1750	1.5
Lal (1976)	Nigeria	Alfisols	Bare fallow	0.5–8.5	1010	37–42
			Maize	0.5–8.5	1010	<0.1–17

These data are not comparable with catchment scale runoff coefficients calculated as a proportion of the annual precipitation which runs off past a river gauging station.

Source: adapted from Lundgren, 1980

(Bonell and Gilmour, 1978; Bonell et al., 1983b), the middle hills of Nepal (Gilmour et al., 1987) and the seasonal forests of the Ivory Coast (Roose, 1982). In addition there are a number of other studies which have addressed the process of overland flow generation with respect to its relationship with sediment production and erosion. A considerable amount of this work has been carried out in seasonal Tanzania (e.g. Rapp et al., 1972; Temple and Sundborg, 1972; Lundgren, 1980), Nigeria (e.g. Lal, 1981) and the Ivory Coast (e.g. Roose, 1970). Similar studies in wetter parts of the tropics include those by Wiersum (1985) in Java. Unfortunately most of the research was conducted over a short period and is therefore not necessarily representative of long-term conditions. Only Bonell and Gilmour (1978), Bonell (1988) and Bruijnzeel (1983) attempt to describe the relative contribution of the various runoff components in a quantitative manner.

There is a commonly held belief that surface flow of any type in the humid tropics is highly restricted in its occurrence owing to the rapid infiltration and efficient percolation of the deep forest covered soils and highly weathered mantle. For example, Thomas (1973, p. 145) states that 'rapid infiltration of heavy rain into the deep tropical soils prevents much surface flow taking place except on steep slopes'. This he attributes to the presence of highly permeable kaolin clay soils and the effectiveness of the rainforest at intercepting rainfall and thereby reducing its intensity.

In vegetated humid areas, infiltration capacities and topsoil percolation rates frequently exceed even the highest rainfall intensities and infiltration excess overland flow (Hortonian overland flow) is indeed rare. For example, Bruijnzeel (1983, p. 170), referring to the results of a monitoring exercise in the forested Kali Mondo catchment in central Java, suggests that 'this type of overland flow has never been observed on the forest floor. Even rainfall intensities of 200 mm hr^{-1} ... were not sufficient to produce Hortonian overland flow on non-compacted surfaces.' However, in localized areas of compacted soils (covering 0.9 per cent of catchment) Hortonian overland flow was reported during intense storms (figure 7.4).

Lundgren (1980) describes an interesting example of widespread but temporally localized Hortonian overland flow in the seasonal forested Usumbara Mountain catchments, Tanzania. At the end of the dry seasons (February and October), the first rainwater to reach the soils does not infiltrate but runs off across the surface. It is only after the soil has been wetted that it can accept the rainwater. The percentage of surface runoff is therefore initially very high but, as wetting progresses, it rapidly diminishes to very low levels. The soils on which this phenomenon occurs are described as kaolin-rich humic nitisols (section 4.5.3) and the effect is seen to be especially significant in those areas where organic contents are greatest. This pattern is opposite to that more usually found in swelling montmorillonite soils.

A highly localized form of Hortonian overland flow, derived from stemflow, has been implicated by Douglas (1973) as making an important contribution to stream runoff in areas of undisturbed rainforest with little ground cover. His findings agree with earlier observations by Ruxton (1967) concerning the role of stemflow in the generation of Hortonian overland flow in the rainforests of Papua New Guinea.

The total absence of overland flow, even during the most intense downpours, has been reported for rainforested parts of French Guyana (Cailleux, 1959), Puerto Rico (Lewis, 1976), in the wettest areas of Dominica (Walsh, 1980a,b), in the more seasonal forests of Bahia, Brazil (Tricart, 1972), and in the Ivory Coast (Dabin, 1957). However, the existence of saturation overland flow in other humid tropical regions is widely acknowledged. Several articles dealing with runoff processes in seasonal areas report on the existence of widespread saturation overland flow, especially where rocks are impermeable and soils are thin. Again, much of the work has been carried out in Africa, but results and conclusions differ. For example Rougerie (1956, 1960) and Tricart (1972) describe widespread saturation overland flow under semi-evergreen forests in the Ivory Coast. In contrast, Roose and Lelong (1976) compared cultivated and bare plots with natural vegetation in both the Ivory Coast and Upper Volta (Bukina Faso). They conclude that while saturation overland flow is significant in the bare and cultivated plots, it is of little

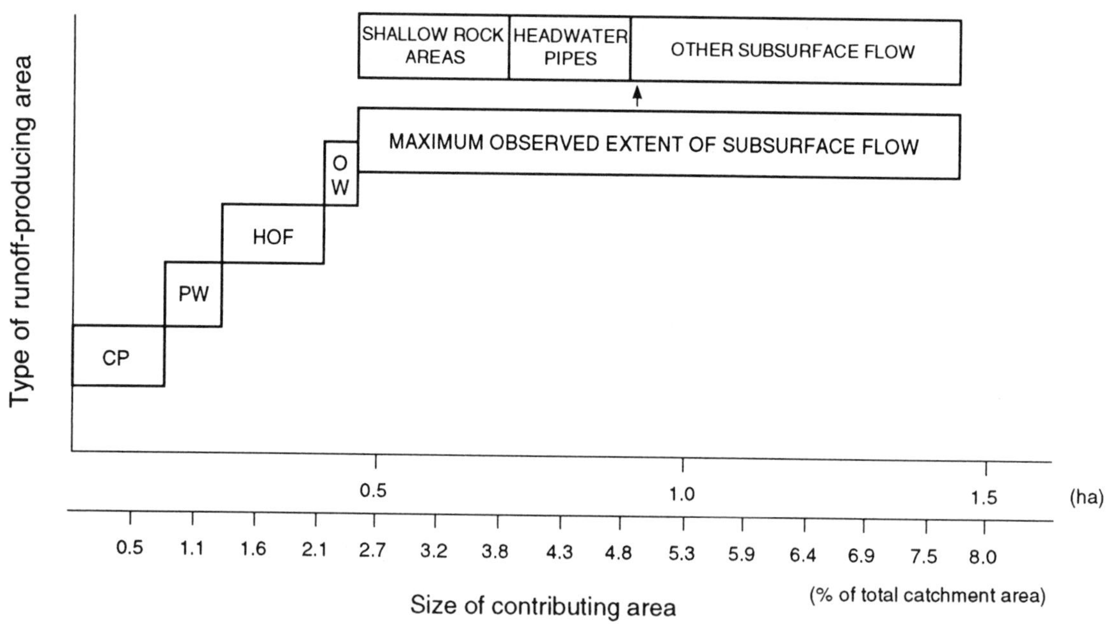

Figure 7.4
Contributing area and runoff type in the Mondo basin, central Java (from Bruijnzeel, 1983): CP, channel precipitation; PW, permanently wet; HOF, Hortonian overland flow; OW, occasionally wet.

significance except during the heaviest storms in the more dense savanna and rainforest.

Many of the soils in the seasonal tropics have lateritic crusts which are hydrologically significant (section 6.3.1). Kesel (1977) has described the hydrological importance of lateritic crusts on various slope segments in Guyana. On ridge tops, the presence of a highly impermeable primary laterite, at a shallow depth below the surface, restricts the soil water storage capacity and hence facilitates saturation and saturation overland flow. Secondary laterite close to the surface in the valley bottoms, where there is also a convergence of the flow net, promotes a second area of saturation overland flow. However, on the slopes where the laterite is poorly developed and water drains easily, saturation overland flow is uncommon.

In wetter parts of the tropics most of the available evidence points to the existence of saturation overland flow only in highly localized locations, such as in valley bottoms, along channel margins or below areas of concentrated canopy drip (Bonell, 1988). Despite the fact that it may account for a relatively small proportion of storm runoff, it remains largely responsible for the quickflow element of stream discharge. Nortcliff and Thornes (1981) describe a situation of very occasional overland flow, restricted to valley bottoms, in the equatorial Manaus catchment in Amazonas. More frequent saturation overland flow, principally in channel head locations, is described by Morgan (1972) for the Malayan peninsula and Walsh (1980a, 1982) for Mt Mulu, Sarawak and parts of Dominica.

In contrast, work carried out along the wet coastal strip of northeast Queensland (Bonell and Gilmour, 1978) reports substantial overland flow at all sites during prolonged, high intensity monsoonal rainfall. Explanations of these differences require considerations of the climatological parameters (principally rainfall intensity and duration characteristics) and soil hydraulic properties peculiar to the site. Once again, they demonstrate the great heterogeneity of humid tropical environments and the difficulties involved in making generalizations.

In Amazonas, the oxisol soils monitored by Nortcliff and Thornes (1981) retained high saturation conductivity levels to considerable depth (3.8 m day^{-1} to 60 cm and 1.47 m day^{-1} to 90 cm). Hence, even the large amounts of water accepted into the soil would flow down towards the floodplain rather than ponding up above the subsoil. In this case, most wet season river flow would be generated by a rapid rise of the water beneath the floodplain and adjacent lower slopes as a direct result of rainfall infiltration. Throughflow would be unimportant. The estimates of 'deep' saturated conductivity made in Amazonas lie within a broad band of values quoted for other wet tropical soils (although data vary significantly, even for the same soils). For example, rates of 0.36–1.44 m day^{-1} have been cited for coarse textured ultisols in Venezuela (Lal, 1980) and 1.2–2.2 m day^{-1} for the 'undisturbed' cores of Hawaiian oxisol samples (El-Swaify, 1980). Foote et al. (1972) have recorded larger but wider ranging K values of between 0.38 and 3.84 m day^{-1} using a different technique.

The explanation provided by Bonell and his colleagues (e.g. Bonell and Gilmour, 1978; Bonell et al., 1981; Bonell et al., 1983a, b) for the Queensland situation is complex. Here mean surface K values (0–10 cm depth) range from almost 10 to over 100 m day^{-1} and are partly explained by the strong crumb structure and high organic content of the surface soil. K values decrease in the 10–20 cm soil layer but are still sufficient to accommodate the highest rainfall intensities (up to 150 mm h^{-1}). The possibility of Hortonian overland flow is therefore negated. Critically, however, below 20 cm, hydraulic conductivities fall sharply, ranging from 0.2–1.47 m day^{-1} at 20–50 cm depth to 0.01–0.66 m day^{-1} at 0.5–1m depth. Thus, during heavy and prolonged storms, precipitation intensities exceed deep saturated conductivity values. This allows the build up of water in the soil and the possibility of widespread saturation overland flow above a 'perched' water table.

During the post-monsoonal 'transition season', saturation overland flow may persist throughout storms on the upper slopes but is confined to the precipitation intensity peaks in the lower incised areas. In these circumstances, the 'variable source area' concept (Hewlett, 1961; Hewlett and Hibbert, 1967) is applicable. This model, widely accepted for humid temperate regions, describes how an area of saturation expands upslope during the course of a storm. However, during very heavy and prolonged monsoonal storms, the upper soil store is persistently full allowing widespread saturation overland flow to occur for the most part of these storms. In these conditions, the variable source area concept clearly does not apply.

A further variation in runoff generation is described by Bruijnzeel (1983) for the forested Kali Mondo basin in monsoonal Java. Field mapping in this area has indicated that the contribution to stormflow by channel precipitation, saturation overland flow and Hortonian overland flow originates in constant and well-defined areas in a quite predictable manner (figure 7.4), wherein the lumped hillslope model (Takasao and Takara, 1988) is considered to be more applicable. Unfortunately, the mechanisms responsible for this runoff pattern are not well understood.

7.2.2 Throughflow

Although the occurrence of throughflow has long been recognized in both temperate and tropical environments, relatively few attempts have been made to quantify rates and volumes of flow or to assess its contribution to stream flow under various soil moisture and rainfall conditions. Empirical measurements are usually made directly, using vertical soil pits, or calculated in terms of a (saturated) soil moisture flux according to Darcy's law (Freeze and Cherry, 1979). Results from the two methods, however, are not entirely comparable.

Several of the studies which have taken place comment upon the great variability of rates and amounts within as well as between catchments and over time. It is therefore perhaps not surprising to find that, whereas, for example, Weyman (1970, 1973) and Kirkby and Chorley (1967) conclude that throughflow accounts for a considerable part of storm discharge, Dunne and Black (1970) and Freeze (1972) suggest that its contribution is minimal.

In the humid tropics, especially where overland flow is absent or highly localized, throughflow is widely thought to be an important hydrological routeway. However, comparative data are lacking. Much of the work carried out during the 1960s and 1970s (Iwatsubo and Tsutsumi, 1968; Godefroy et al., 1970; Jordan, 1970; Roose, 1970; Swan, 1970; Morgan, 1972) was directed towards an appreciation of the role of throughflow in lateral mineral eluviation, and hence slope development. Leigh (1978) and Walsh (1980) review this work as part of their case studies of throughflow in Malaysia and the West Indies respectively. More recent information, such as that provided by Nortcliff and Thornes (1981), is discussed in Bonell's (1988) account of hydrological processes in northeast Queensland.

Despite the problems of comparing data derived from different monitoring periods and procedures, a number of consistent patterns between sites are apparent (table 7.5). For example, although the Malaysian (Pasoh) and the Ivory Coast (Adiopodoume) sites differ in their vegetation and soil characteristics, their rates of throughflow are considered to be in the same general range (Leigh, 1978). The fact that throughflow is relatively more important at Pasoh is attributed to compaction (and thereore low surface infiltration rates) on the slopes at Adiopodoume, cultivated with rubber trees. Surface compaction would also explain why rates of throughflow could increase with depth at the Adiopodoume site, a trend

Table 7.5 Surface wash and throughflow (litres)

Source	*Location*		*Depth (cm)*				
			Surface	*0–8*	*8–70*	*70–125*	
Leigh	Pasoh	pit 1	86.9	37.9	118.0	26.4	
(1978)[a]	Malaysia	pit 2	313.9	91.3	150.7	35.6	
			Surface	*0–30*	*30–60*	*60–100*	*100–150*
Roose	Adiopodoume	1966	467.0	5.4	13.4	24.4	25.9
(1970)[a]	(Ivory Coast)	1967	277.0	1.3	46.9	60.5	95.2
		1968	184.5	52.7	74.6	34.8	59.8
		1969	255.0	28.2	43.5	78.5	35.0
		Average	283.5	21.9	44.6	49.9	54.0
			A horizon	45–60	90–105		
Walsh	WAES	pit 1	15	2	1		
(1980)[b]	(Dominica)	pit 2	4	1	0		
		pit 3	13	2	0		
		pit 4	2	0	–		
		pit 5	5	2	0		
		pit 6	3	3	–		
		pit 7	4	2	5		
		pit 8	8	2	0		
		pit 9	0	3	9		

[a] Figures refer to volumes of flow in litres.
[b] Figures refer to the frequency of throughflow during a 40-day monitoring period.

opposite to that observed at Pasoh, Malaysia (Leigh, 1978), and in the very wet Wet Area Experimental Station (WAES) site in Dominica (Walsh, 1980a).

Walsh (1980a) has demonstrated the existence of two major throughflow areas using throughflow pits in the humic latosolic soil at WAES. The first is at the base of the A horizon and the second at the junction between soil and weathered bedrock (B–C horizon). Substantial (i.e. > 1 litre day^{-1}) throughflow was found to be two to four times as frequent in the upper (A horizon) throughflow zone than in the lower throughflow zone. A general relationship between precipitation and flow volumes was found at all levels. Walsh (1980a) describes a situation of rapid and short duration throughflow following a storm and concludes that topsoil throughflow is a relatively efficient mechanism by which water is transported downslope at this site.

Leigh (1978) was able to quantify the amounts of throughflow collected at various depths at Pasoh and to establish a relationship between precipitation and profile throughflow (figure 7.5). The intercept on the x axis represents the average amount of rainfall required to produce throughflow at any depth and is greatest for the uppermost collection tray. From this he concludes that only during the larger storms is throughflow in the upper soil horizons important. Both Leigh (1978) and Walsh (1980a) note the importance of antecedent moisture conditions in the rainfall–throughflow relationship, with a faster response of throughflow in wetter antecedent conditions. Leigh (1978) found that the majority of the residuals of the regression lines (figure 7.5) represented falls of less than 20 mm during the wet season.

Available evidence therefore points to the fact that throughflow, in certain humid tropical catchments and under certain soil moisture and rainfall conditions, makes an important contribution to catchment drainage. However, much more quantitative and comparable data must be accumulated before valid generalizations on the role of throughflow in the humid tropics can be made.

7.2.3 Runoff models in the humid tropics

From the limited field evidence available it is clear that rates of infiltration and percolation and surface and subsurface flow patterns vary enormously within the humid tropics in response to differences in rainfall (i.e. amount, intensity–duration characteristics and antecedent conditions), soil properties (e.g. depth, water storage capacity and permeability), vegetation (e.g. species and density) and other factors such as topography, lithology and land use. It is therefore highly unlikely that a single runoff model can adequately describe runoff processes within this zone. Walsh (1980b) has proposed a series of models as a framework within which to view and understand some of the contrasts which occur (table 7.6), but acknowledges that not all parts of the humid tropics will fit neatly into single categories. For example, in

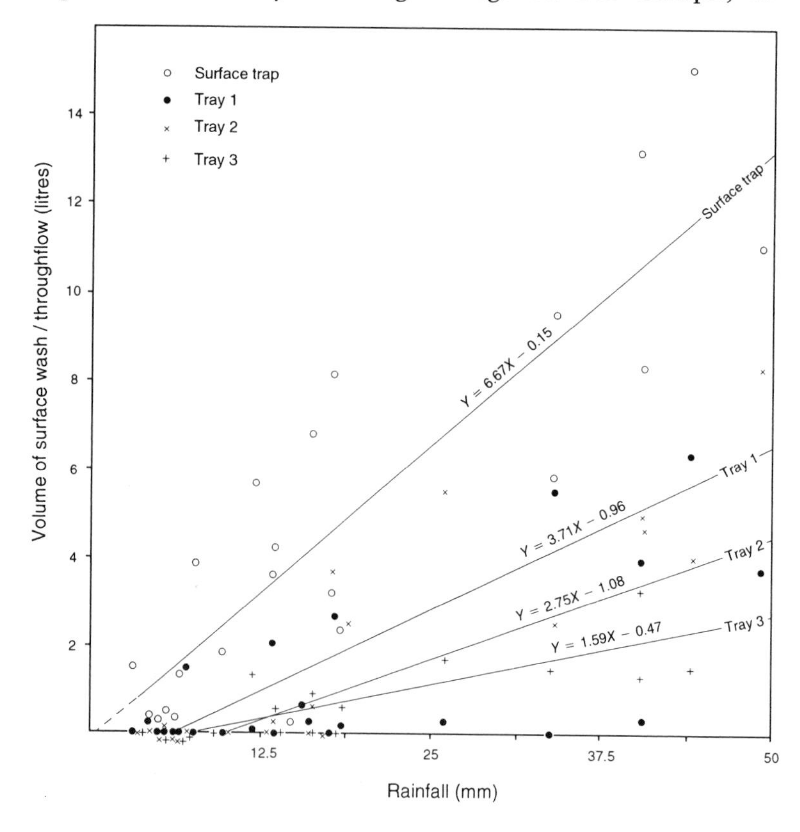

Figure 7.5
Regression of throughflow and surface wash on rainfall for single storm collections (from Leigh, 1978).

Table 7.6 Storm runoff models in the humid tropics

Storm runoff model	*Principal climatic tropical zone*	*Soil factors (controlled by climate, lithology etc.)*			*Other major factors influencing type of runoff model*			
		Topsoil permeability	*Topsoil depth*	*Subsoil permeability*	*Topography*	*Type of geology*	*Rainfall intensity*	*Interference by man*
Hortonian overland flow	Semi-arid and arid	Low	–	–	–	↑ Impermeable	↑ High	↑ Intense
Widespread saturation overland flow	Seasonal	Moderate	Low	Low to very low	↑			
Localized saturation overland flow dominant to	Ever-wet to very wet	High +	High	Moderate to high	Extensive catchment flats			
Throughflow dominant	Ever-wet to very wet	High +	Very high	High	Steep slopes, narrow ridges and valley bottoms ↓			
Rapid throughflow	Extremely wet	Very high	Very high	Low		Permeable	Low	Slight
Groundwater	Azonal	Very high	–	Very high	–	↓	↓	↓
Pipeflow	Azonal?	High +	–	Low?	–			

Source: adapted from Walsh, 1981

Amazonas, both throughflow and overland flow are thought to contribute little to wet season river flow. Also, in Queensland, the influence of exceptionally long and intense rainstorms during the monsoon period outweighs the bioclimatic and lithological influences which would otherwise suggest that a saturation overland flow–throughflow model would be applicable throughout the year.

Land use changes inevitably result in a significant alteration of the hydrological relationships within a catchment. Monitoring and prediction of these changes and the development of ways in which they may be minimized or managed is a priority in applied hydrology. The effects of deforestation and the establishment of agriculture have been investigated at a number of locations (e.g. Roose, 1970, in the Ivory Coast) and research is ongoing (section 9.3). However, less is known about the potential and likely consequences of urbanization in the humid tropics, although both Douglas (1978) and Gupta (1984) emphasize that its effect is likely to outweigh all other influences on runoff generation. Furthermore, rates of urbanization in the humid tropics are greater than those in temperate regions and the nature of the humid tropical physical environment tends to accentuate the changes in geomorphological processes that urbanization brings about. The gaps in our understanding of the basic workings of humid tropical hydrological systems are all too evident and these must be filled before it is possible to evaluate the impact of human activity effectively.

7.2.4 Pipeflow

The existence of concentrated throughflow in natural tunnels or pipes in the soil or weathered mantle has been reported in a number of humid tropical environments. These include the alluvial lowlands of Johore and Singapore (Burton, 1964); monsoonal West Bengal (Banajee, 1972) and Vietnam (Fontaine, 1965); seasonal Natal (Downing, 1968); weakly seasonal parts of Dominica (Walsh and Howells, 1988) and the perennially wet forests of Colombia (Feininger, 1969) and Sarawak (Baille, 1975; Walsh, 1982). Unfortunately, these studies are all largely descriptive and little is known about the contribution of pipeflow to streamflow generation or the transfer of sediment. In some seasonal areas, the episodic development of desiccation cracks has been implicated as important for pipe development but elsewhere it is assumed that conditions conducive to pipe development in temperate areas occur for at least part of the year. These include a permeable surface underlain by a markedly less permeable subsurface; material which is sufficiently elastic to allow tunnel development but cohesive enough to prevent tunnel roof collapse; and a gradient which is sufficient to encourage lateral flow and prevent the ponding up of water *in situ*. Lewis (1976) and Jones (1981) discuss the nature of soil piping in much greater detail.

7.3 Sediment Transport

7.3.1 Sediment transport by surface wash

Estimates of soil loss based upon artificial or experimental plot data range from almost zero to around 1200 t ha yr^{-1} (Lal, 1981), according to the interplay of a number of environmental parameters and the technique of assessment. Data usually include material moved by rainsplash and stemflow as well as surface wash.

Results from plot, trough and erosion pin studies are not entirely comparable. For example, Coster (1938) found that rates of surface lowering, based on erosion pin measurements in Java, were consistently higher than rates calculated from fractional-acre plots in the same area. Leigh (1978) also notes that while average rates of ground lowering at Pasoh, Malaysia, are of similar magnitude to those reported by Rougerie (1960) using similar pin techniques in the Ivory Coast (2.6 mm yr^{-1}), both are high in comparison with results obtained by others using different techniques of measurement. Reasons advanced for the overestimation of soil loss by pins include their interruption of water flow and the creation of turbulent (and hence erosive) eddies around the base of pins. In seasonal areas there is also a possibility that the intermittent wetting and drying of clay-rich soils may promote an upward displacement of pins. Erosion pin plots are therefore best suited for monitoring the relative amounts and direction of soil movement on individual hillslopes. Here they have the advantage of being cheap and easy to establish and are well suited to monitoring over periods of several years.

Notwithstanding the problems of comparing data in a quantitative manner, larger scale spatial patterns of soil loss differ widely. In general, the loss of sediment by surface wash under dense tropical forest is considered low (Fournier, 1962; Douglas, 1969; Young, 1974); Brunig (1975) measured surface loss under virgin forest at around 0.2 t ha yr^{-1} and UNESCO/UNEP/FAO (1978) quote a figure of less than 1 t ha yr^{-1} for the humid tropics in general. In contrast, however, Birot (1968) argues that, under tropical forests, erosion can be relatively high in comparison with temperate forests, owing to the existence of openings and gaps in the forest canopy. In the absence of ground flora this allows large raindrops from frequent intense tropical downpours to reach the ground surface. The rapid decomposition of organic matter will also tend to reduce the protection afforded by litter. Birot's (1968) ideas may be representative of areas of high relief, where tree fall and mass movements more easily produce gaps. However, in most other areas high erosion rates are likely to be very localized, and due to concentrated stemflow or leaf drip.

The importance of soil depth, permeability, mineralogy (and thus ultimately parent material) is noted in a number of studies but again there is no consensus as to their effects. For example, Peh (1978) describes higher rates of runoff and particle detachment for soils developed on shales than those developed on granites in peninsular

Malaysia. This he attributes to the lower porosity and clay content and the higher silt content of the shale soils. Studies by Leigh (1978) in the same locality are in general agreement, although the highest rates of runoff and sediment erosion were restricted to the upper-slope segments.

Peh (1978) describes a positive correlation between the amount of sediment collected and distance from the crest of the slope. A similar pattern is noted by Swan (1970) and Morgan (1973) in lowland parts of Malaysia and is explained by the fact that the clay content of the upper soil horizons decreases downslope and hence volumes of surface wash increase downslope. These results conflict with the findings of Leigh (1978) who could find no significant correlation between soil texture and distance downslope and who cites a negative correlation between rates of sediment transport and distance from the crest. In a later study Leigh (1982) makes reference to a relationship between increased rates of soil loss and the development of litter and humus layers above overland flow traps.

Notwithstanding the considerable problems of soil loss measurement, a complicated pattern of soil loss is perhaps to be expected under a forest canopy with a mosaic structure of regeneration, building and mature stages and hence a heterogeneous pattern of litter, humus and soil development.

Hudson and Jackson (1959) describe a clear relationship between slope angle, even over low gradients, for maize-covered slopes in Zimbabwe (table 7.7) and the importance of land cover and land use is discussed in a number of studies. For example, the significance of rainfall intensity–duration characteristics is likely to be higher in agricultural areas which do not have a continuous ground cover. Hutchinson et al. (1958) undertook a six year investigation of the control of rainfall intensity on runoff and hence soil loss in Numulongo, Uganda. They conclude that while 'it is usual in studies of the effects of cropping patterns to measure susceptibility to erosion in terms of runoff ... in view of the dependence of runoff on precipitation intensity, the use of percolation rate (i.e. infiltration capacity) is to be preferred as an index of surface treatment' (p. 258).

Table 7.7 Effect of slope on soil erosion: soil loss

	Slope in degrees		
Season	*3.5*	*2.5*	*1.5*
1953–4	6.6	4.1	4.6
1954–5	2.8	1.0	1.8
1955–6	6.9	2.5	1.7
1956–7	11.0	7.1	3.6
1957–8	1.2	0.5	3.3
1858–9	11.5	6.9	3.5
Average	6.7	3.7	3.1

Data from three clay slopes with continuous maize in Zimbabwe (soil loss in m^3 ha^{-1}).

Source: adapted from Hudson and Jackson, 1959

The highest rates of soil loss recorded in the humid tropics occur in areas of disturbance. Any method of deforestation is likely to increase runoff and soil loss and the indiscriminate use of heavy machinery is often cited as a major cause of excessive rates of soil erosion (Lal, 1981) (section 9.3). Investigations of methods of deforestation and post-development soil management which maintain rates of runoff and surface wash at 'forested' levels form the basis of many soil conservation and catchment management strategies (Oyebande, 1981; Gladwell and Bonell, 1988). There are fortunately many examples of cases where agricultural practices have been successfully combined with very low rates of soil loss (e.g. Lal, 1976; Lundgren, 1980).

7.3.2 Sediment transport by rivers

Geomorphologists have debated, at great length, the efficacy of tropical rivers in the mechanical process of erosion. Workers, notably Büdel (1957), Birot (1958, 1968), Tricart (1972) and Louis (1964), claim that since humid tropical rivers flow over deeply weathered material they lack the abrasive tools for erosion. They cite as evidence the fineness of river bed material and the persistence of resistant layers of rock, resulting in rapids and waterfalls rather than incised gorges (figure 7.6). It is certainly true that many of the world's most spectacular waterfalls lie within the humid tropics, e.g. Angel Falls in Venezuela (800 m high), Victoria Falls in Zimbabwe (120 m high), Iguazu Falls in Brazil (100 m high) and others in Surinam, Guyana and tropical Africa. Many tropical waterfalls occur in ancient surfaces of low relief and cannot be explained by recent tectonics.

Bakker (1957) and Michel (1973) suggest a number of reasons why humid tropical streams lack the tools for erosion. Their studies demonstrate that fragments of rock, where they occur, are transported over relatively short distances before being completely worn away. Bakker (1957) quotes a figure of 6 km for a waterfall complex in Surinam.

In some tectonically active and Tertiary mountain areas (e.g. Puerto Rico and the Amazon headwaters), Birot (1970) concedes that large pebbles up to 50 cm in diameter do frequently line the bed of rivers. However, this material is considered to be reworked bed

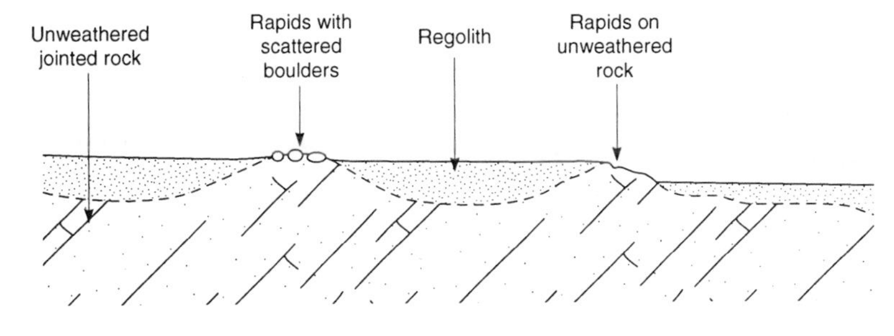

Figure 7.6
The typical long profile of tropical rivers (adapted from Tricart, 1972).

or old river terrace material formed when the climate was drier. The river is described as having a 'passive baselevel', being unable to erode vertically into the weathered material below it. These generalizations have been widely accepted by geomorphologists and their students but, as Löffler (1977) and Douglas and Spencer (1985) emphasize, they can be misleading. Löffler (1977) provides examples from mountainous Papua New Guinea where, owing to the great tectonic relief and copious rainfall, fluvial erosion is undoubtedly the most important process operating. Rivers in Papua New Guinea erode deeply into the bedrock and flow in narrow valleys with steep and straight sides. Most of the debris transported to the foot of the slope is removed quickly by the rivers whose bed contains large amounts of coarse gravel and boulders which travel intact considerable distances downstream. A similar situation occurs in parts of volcanic Dominica where landslides are the most important mechanism by which material is supplied to the rivers (Reading, 1986).

In these rugged environments, bedload obviously forms an important contribution to rivers' total sediment load. Unfortunately, quantitative data for bedload in the humid tropics is rare. Löffler (1977) provides a case study of the Leron River in Papua New Guinea having its bed raised by 50 cm in a three month period and describes partially buried trees in several locations indicating the deposition of up to 2.7 m of gravel during a two year period. While not intended to be representative of the humid tropics in general, these results provide an indication of the potential magnitude of bedload transport in extreme cases.

On a global scale, Walling and Webb (1983) suggest that the average contribution of bedload to total yield is around 10 per cent and the NEDECCO (1959) figures (5–6.5 per cent) for the Lower Benue and Lower Niger suggest that these may be reasonable for the less dynamic parts of the humid tropics. Such values are conveniently considered as insignificant and most estimates of net erosion and denudation rely wholly on measurements of suspended load.

A number of attempts have been made to compare yield (more precisely suspended sediment yield) on a regional basis. Global maps produced, for example, by Fournier (1960), Milliman and Meade (1983), Walling and Webb (1983) (figure 7.7), Dedkov and Mozzherin (1964) and Jansson (1988) show dramatic variations between and within different climates. It should be noted, however, that there are considerable differences between the maps which stem from inconsistencies in the methods of data generation and differences in data treatment and incorporation. Catchment size is also likely to be important. Oyebande (1981), for example, reports that in Nigeria catchments below 10,000 km^2 can yield up to eight times more sediment per unit area than catchments whose size is greater than 100,000 km^2, the reason being the relationship between catchment size and relief (smaller catchments tend to occur in mountainous areas).

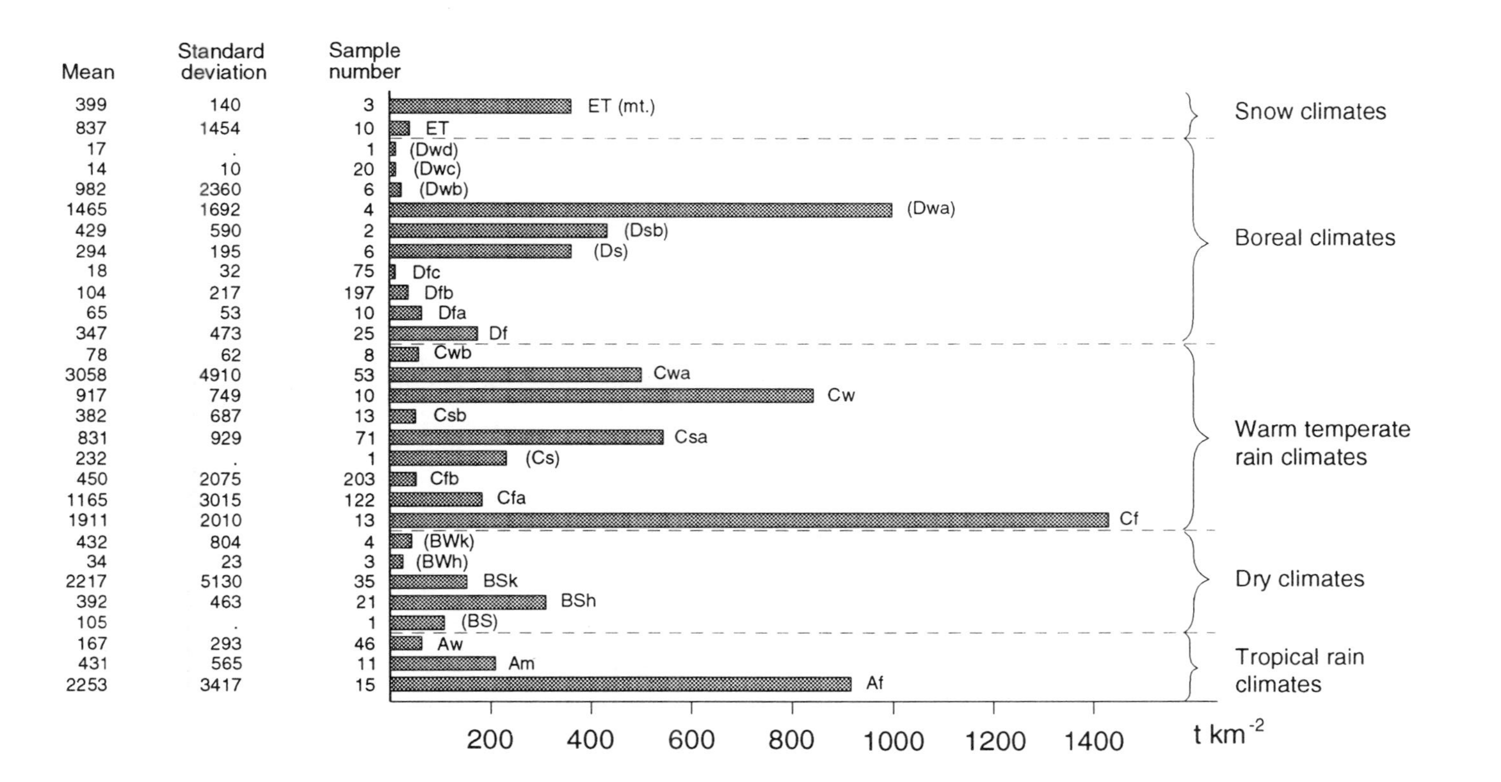

Figure 7.7
Median sediment yields in different climates (from Jansson, 1988). Data from rivers going through more than one climatic type are excluded. Symbols follow Köppen's classification.

Jansson (1988) only includes data for catchments between 350 and 100,000 km^2 and uses a modified version of Köppen's classification, which takes account of the effects of altitude, to categorize catchments according to climate. The perennially wet tropics (denoted as Af in Köppen's classification) are shown to have 'medium' sediment yields with over half of the forty-two catchments sampled yielding in excess of 1000 $t\,km^2\,yr^{-1}$ and only one river yielding less than 100 $t\,km^2\,yr^{-1}$. Seasonal parts of the tropics are described as having variable but relatively low median yields (figure 7.8). This pattern is explained by the greater runoff of the wet tropics. Median, rather than mean, values are considered the most representative as the standard deviations show that the latter will be strongly influenced by a few exceptional values.

Contrasts within the tropics illustrate the importance of age and relief. For example, sediment yields in Papua New Guinea and Indonesia are some of the world's highest, frequently in excess of 1000 $t\,km^2\,yr^{-1}$. Southeast Asia and the Himalayas also have high yields but, with the exception of the Indus valley, the Indian lowlands sediment yield values are roughly halved. Other ancient shield areas also consistently produce low sediment yields, except where human activity has exaggerated soil loss (e.g. the Tana River catchment in Africa) or where rainfall is particularly intense and prolonged (e.g. monsoonal northeast Queensland). Reiger and Olive (1988) explain the low yields of the rest of tropical Australia (50 $t\,km^2\,yr^{-1}$) by the fact that, despite the considerable supply of fine, deeply weathered material available for transport, the low gradients and large temporal variability in river discharge combine to form an inefficient sediment delivery system. Sediment yields often differ enormously, even within a catchment. Gibbs (1967) has demonstrated the spatial variations which occur within the Amazon basin (figure 7.9). His sediment yield data probably underestimate true values by up to 50 per cent (Meade et al., 1985) but nevertheless illustrate the importance of the mountainous Andean headwaters as a sediment source.

Problems of the reliability of data have been discussed at length (e.g. Milliman and Meade, 1983; Walling and Webb, 1983; Jansson, 1988) (table 7.8). The one factor which is frequently overlooked, however, relates to the reliability of the sediment discharge rating curve. The sediment contained in rivers may be derived from a number of sources, including the bed and banks, landslide debris and slopewash. Amount or yield is often correlated with discharge but, in detail, the relationship is qualified by the availability of sediments within a catchment, the effectiveness of water in reaching them and the energy available for entrainment (Grenney and Heyse, 1985; Rodda, 1985). It is routine practice for hydrologists to assume that the latter is of overriding importance, allowing the derivation of a functional relationship between sediment discharge and water discharge. However, in reality there is a complex relationship between sediment concentration and discharge as an example from Papua

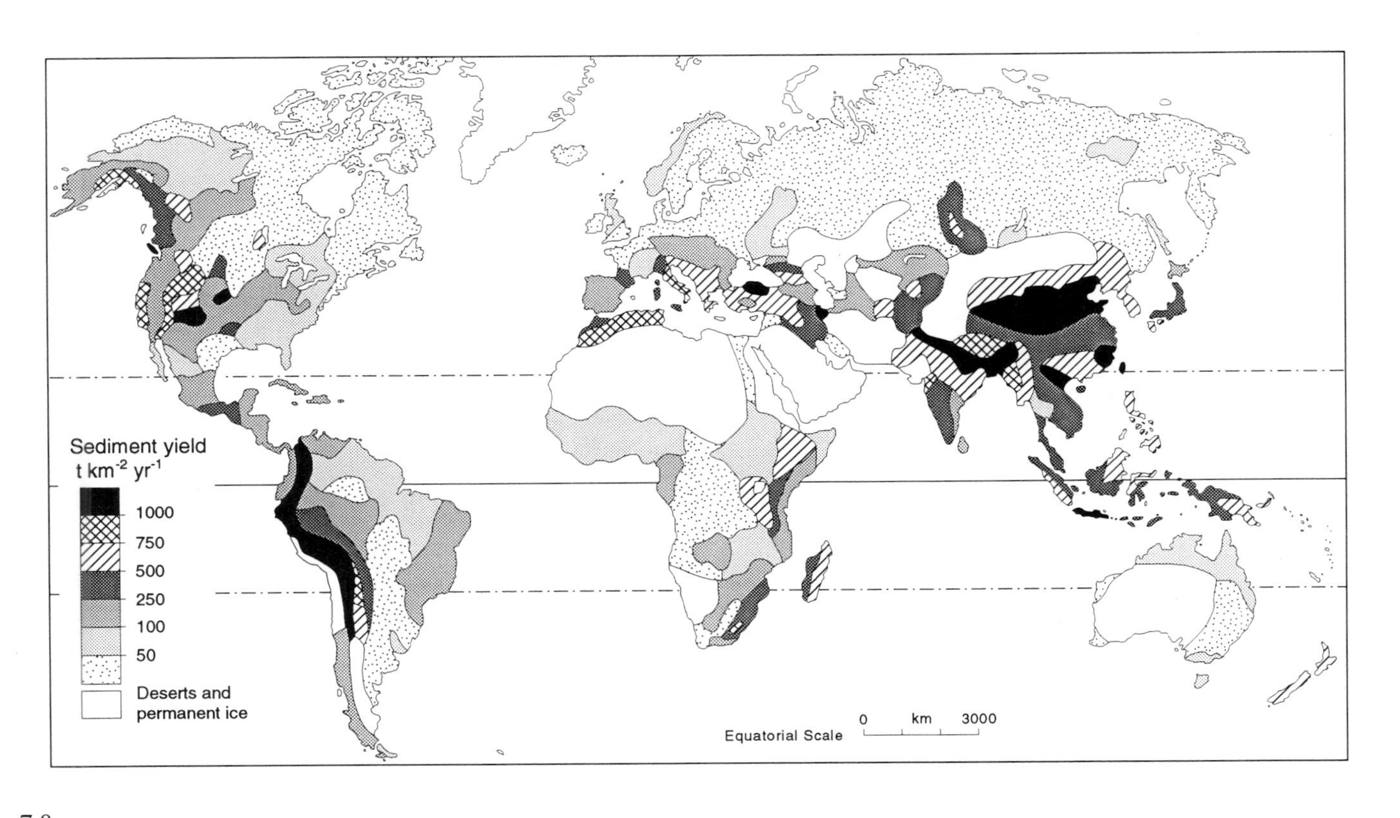

Figure 7.8
Global sediment yields (from Walling and Webb, 1983).

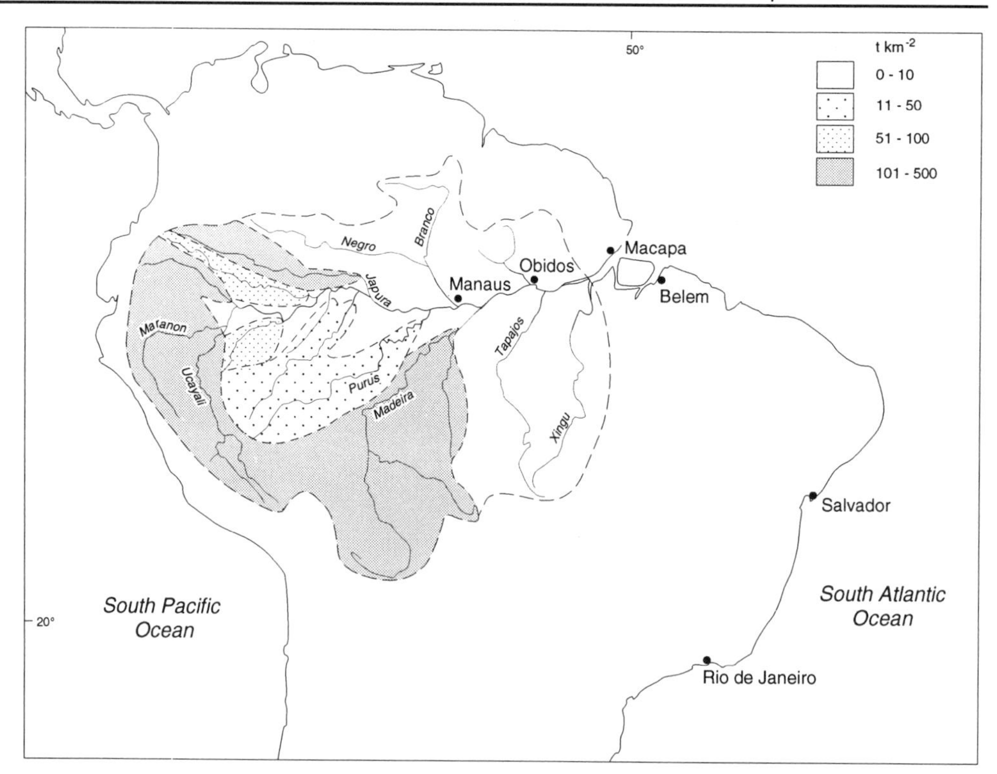

Figure 7.9
Sediment yield in the Amazon basin (based on Gibbs, 1967).

New Guinea demonstrates (Spencer and Douglas, 1985). Here sediment loads at baseflow and at medium discharges are low since the large bed material is too coarse to be transported and stream depth is insufficient to initiate bank erosion. However, sediment concentrations rise rapidly as soon as flow velocities become sufficient to begin bed erosion and expose the finer material trapped below. There is also an added complication, namely large and intermittent pulses of sediment associated with landslides, triggered by prolonged and intense rainstorms or by the undermining of slopes at peak flow. Walling (1977, 1978) has shown that in southwest England rating curves probably underestimate sediment load by up to 80 per cent compared with results from continuous monitoring techniques (e.g. turbidity meters). Reiger and Olive (1988) note a persistent underestimation of

Table 7.8 Problems associated with the derivation of sediment yield

1	Deficiencies in sampling equipment and technique for estimating discharge
2	Deficiencies in laboratory analysis of collected sediment
3	Inconsistencies over whether mineral only or mineral and organic matter should be included
4	Problem of observation and range of flows over which measurement occurs
5	Theoretical and empirical problems associated with the establishment of a sediment rating curve

sediment values using a rating curve for Nigerian rivers. They also question the theoretical basis of the log–log regression relationship between sediment and discharge.

Sediment concentrations at a given discharge are also likely to be different, during an individual storm (figure 7.10) or on a seasonal basis (figure 7.11), according to whether stage is rising or falling. The direction of the hysteresis will vary according to the source of sediments and the presence of surface flow within the catchment. Temple and Sundborg (1972) illustrate the complexity of sediment discharge relationships in Tanzania's largest river, the Rufiji River, during a sixteen year period. While their results demonstrate a reasonable relationship between daily flow and weekly suspended sediment data, they note that sediment concentrations are highest on the rising limb of sudden flash floods, particularly when baseflow is low. On a seasonal time scale, baseflow sediment concentrations are at a maximum up to two months earlier than the maximum baseflow discharge. However, they fall more quickly

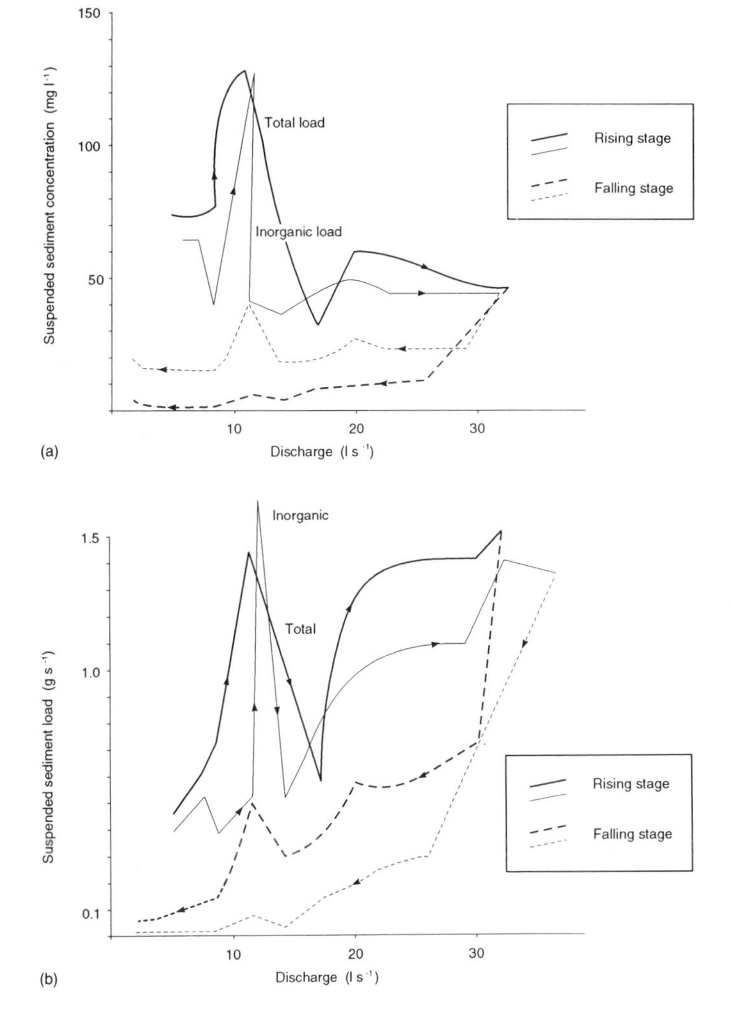

Figure 7.10
Relationship between suspended sediment and discharge for a single flood hydrograph at Pont Cassé, Dominica: (a) suspended sediment concentration; (b) suspended sediment load.

such that, from the end of April to June, sediment yield is relatively low while river discharge remains high (figure 7.11).

For the much smaller (19 km^2) Morogoro river catchment in Tanzania, Rapp et al. (1972) analysed the particle size distribution of suspended sediment during storm hydrographs in an effort to discover its source. They report that the initially low baseflow sediment concentrations rose rapidly with discharge with the sediment and discharge peaks almost coinciding. The median grain size of the material was taken to be indicative of a kinematic wave of channel water with eroded and incorporated material. During the main and falling stages, concentrations were lower and particle size smaller; this was assumed to represent material freshly washed from the slopes. Oyebande (1981) has also used particle size to indicate sediment source. He relates the large size of the bulk of the sediment from the Challawa River in Nigeria at peak flow to the relative importance of gully erosion in the granitic headwaters. Finer material is thought to be derived from sheetwash.

Figure 7.11
Relationship between suspended sediment and discharge on an annual basis: the Rufiji River at Stiegler's Gorge, Tanzania, 1959–60 (from Temple and Sundborg, 1972).

7.4 Solute Transport in the Humid Tropics

Despite the fact that, on a world scale, solute concentrations are generally very low, dissolved minerals form an important component of the humid tropical denudation system. Notwithstanding the problems involved with measuring solute load and comparing

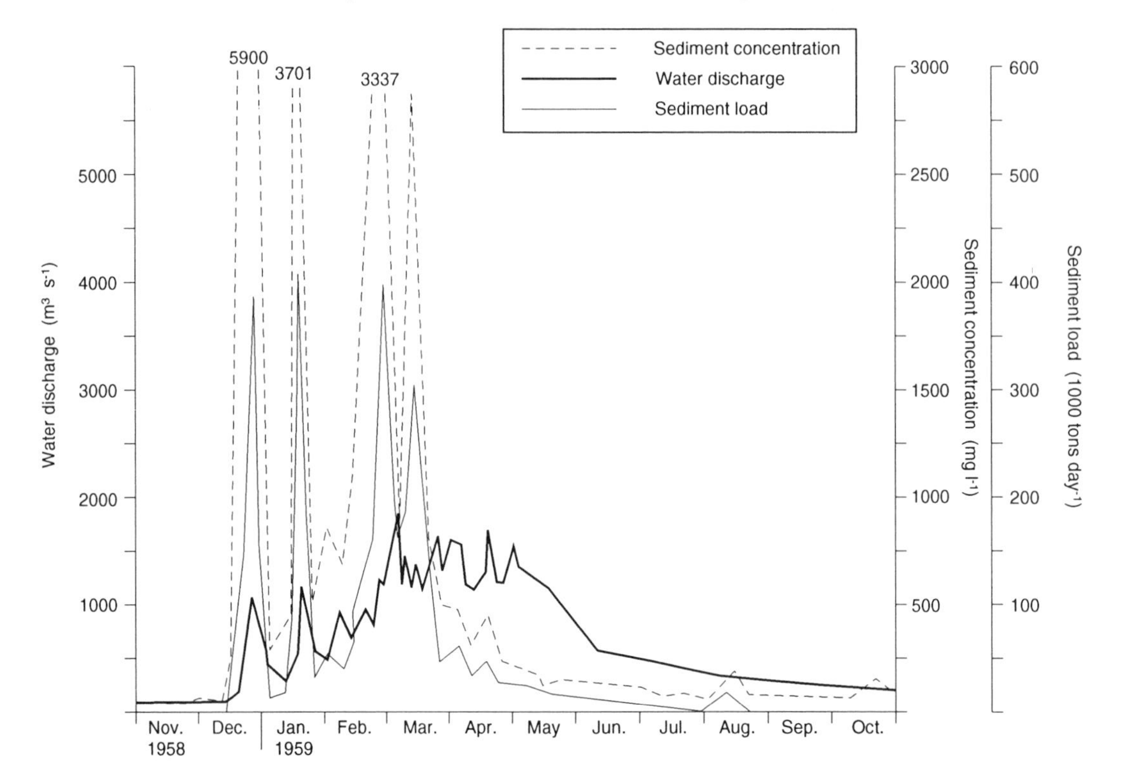

results (table 7.9), data for several large rivers indicate that solutes contribute up to 85 per cent of the total sediment on an annual basis. Also, an even greater contribution is likely during low flow conditions when suspended sediment concentrations are often negligible (table 7.10). Chemical denudation may therefore make a significant contribution to overall denudation. However, like bedload, solute load is routinely dismissed as being insignificant and is often excluded from denudation estimates.

The importance of solutes derives from the fact that, although solute concentrations are often low, this is compensated by the high discharge of humid tropical rivers. The overriding importance of runoff to rates of chemical denudation is noted in several studies (e.g. Turvey, 1975; Leigh, 1982; Biksham and Subramanian,

Table 7.9 Problems associated with empirical solute data

1	Deficiencies in techniques for estimating discharge
2	Deficiencies in laboratory analysis of collected water samples
3	Incommensurate nature of instantaneous and seasonal data
4	Failure to include contribution from silica when using conductivity to estimate total dissolved solutes
5	Reliability of constant *k* in conductivity versus total dissolved solute relationship
6	Inconsistencies in the use of gross and net solute figures (the latter are adjusted to take account of the atmospheric input of solutes)

Table 7.10 Sediment loads of some tropical rivers

River	Area (10^3 km^2)	Runoff (mm)	Solute load ($m^3\ km^2\ yr^{-1}$)	Solute load (%)	Suspended load ($m^3\ km^2\ yr^{-1}$)	Suspended load (%)
Meybeck (1976, 1979)						
Amazon	6300	840	17.5	37.0	29.8	63.0
Brahmaputra	580	990	49.1	8.7	517.0	91.3
Congo (Zaire)	4000	294	4.4	46.8	5.0	53.2
Ganges	975	360	29.4	12.7	202.6	87.3
Magdalena	240	930	44.1	10.5	377.4	89.5
Mekong	795	690	28.3	14.7	164.1	85.3
Niger	1125	165	3.4	13.1	22.6	6.9
Orinoco	950	948	19.6	36.4	34.3	63.6
Parana	2800	192	7.5	33.2	15.1	66.8
Zambesi	1340	159	4.3	13.2	28.3	86.8
Reading (1986)						
Dominica						
Blenheim			5.7	56.4	4.4	43.6
Geneva			14.2	28.1	36.3	71.9
O'hara			12.0	31.2	26.5	68.8
Batalie			11.2	29.9	26.5	70.1
Macoucherie			8.5	26.9	23.1	73.1
Padu			51.6	36.2	90.0	63.8
Pont Casse			2.3	34.3	4.4	65.7
Biksham and Subramanian (1988)						
India						
Godavari			21690[a]	8.0	257900[a]	92.0

[a] 10^3 t yr^{-1}.

1988). In northeast Queensland, Douglas (1973) describes a situation where streams with the lowest mean solute concentration but the highest mean annual runoff have the highest mean dissolved load.

For solution to take place, water must pass solute sources at a speed which allows incorporation to take place. Thus throughflow, rather than the more rapid surface flow, tends to be more concentrated with solutes. The overall significance of throughflow to total and chemical denudation has been investigated by Leigh (1982) in peninsular Malaysia (section 7.2.2). Solutes here contribute 5 per cent of surface wash sediment but throughflow solutes make up 15–23 per cent of sediment in total (i.e. surface wash and throughflow). Leigh (1982) again comments upon the importance of runoff, noting that variations in rates of solute transport were influenced more by the volumes of throughflow than by variations in the solute concentration.

The origin of solutes, their rate of acquisition and their transport through the catchment to the river will vary from catchment to catchment, within catchments and with time (i.e. discharge), according to the interplay of a large number of dependent variables. Models of these relationships provide a convenient framework within which they can be investigated (figure 7.12). However, as the following examples illustrate, relationships are often complex and site or time specific.

During the 1960s, it was noted that the silica content of tropical rivers was often much greater than that of streams in temperate latitudes (e.g. Corbel, 1957, 1964; Rougerie, 1960; Livingstone, 1963; Strakov, 1967; Douglas, 1969) (table 7.11). This fact is often cited as evidence of the exaggerated importance of chemical weathering in the humid tropics. Explanations centre on the relative solubility of silica in humid tropical environments. Laboratory studies suggest that the solubility of silica increases with temperature and, following a review of empirical data from temperate and tropical areas, Hem (1970) concluded that temperature has a strong influence on the silica content of the world's streams. In contrast Davis (1964), Carbonnel (1965) and Douglas (1969) suggested that air temperature has little effect on silica mobility. For example, Douglas (1969, p. 12) concluded that 'the temperature factors affecting weathering rates and thus silica concentrations in drainage waters are less important in total geological work done in removing silica than contrasts in precipitation and other factors affecting runoff'.

Several other variables are now known to be important in the control of silica mobility (Thomas, 1974). The effect of pH is well established, at least under laboratory conditions (figure 7.13). However, Davis (1964) considers that neither salinity nor pH has any measurable influence on natural silica concentrations. Huang and Kellar (1972) have suggested that humic acids may play a part in bringing silica into solution and Turvey (1975) has discussed the importance of rainforest litter as a source of silica, since the high

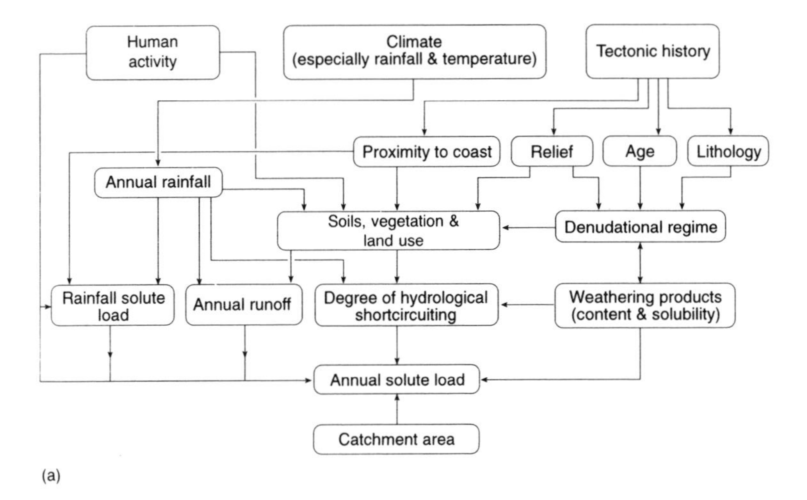

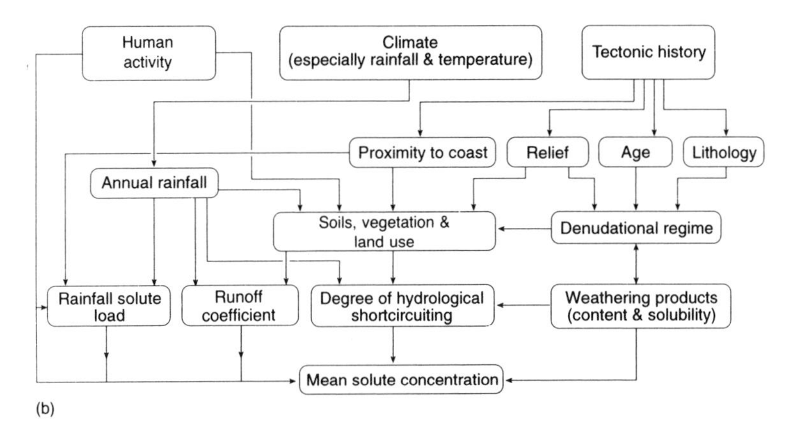

Figure 7.12
Factors controlling the solute content of rivers: (a) solute load; (b) solute concentration (adapted from Walsh, 1980).

concentrations of silica in litter may be in a potentially more soluble form than in the soil. Lithology also plays a part and silica concentrations of streams draining sandstone and shales (Douglas, 1978; Walsh, 1982) are generally lower than those of streams draining granitic and basaltic rock (Douglas, 1968, 1969; Weninger, 1968; Turvey, 1975; Reading, 1986). This possibly reflects the presence of silica in the relatively insoluble form quartz in the former and its occurrence within the more easily weathered silicate minerals such as feldspar in the latter. What is clear from these results is that the silica cycle is highly complex and that geomorphologists still lack an understanding of its detailed functioning.

Within any catchment, there are likely to be a number of sediment sources and sinks. For example, in the Amazon Gibbs (1967) shows that some 86 per cent of solutes (plus around 82 per cent of suspended material) are derived from approximately 12 per cent of the catchment, i.e. the Andes and its foothills. Biksham and Subramanian (1988) demonstrate the importance of lithology

Table 7.11 Dissolved solids in selected tropical rivers

Catchment	*TDS*	*Silica*	*% silica*	*Source*
A. Small catchments				
Malaysia	75	15	20	Douglas (1969)
Gombak	48	25	48	Douglas (1967)
Sungei Pasir	35	8	23	Singapore Water Department
Tebrau	45	10	22	(in Douglas, 1969)
Scudai				
Papua New Guinea				
Ei Creek	75	31	41	Turvey (1974)
Ivory Coast				
Me	85	15	18	Turvey (1974)
Agneby	92	14	15	Turvey (1974)
Queensland				
Babinda	33	10	30	Douglas (1967)
Davies	46	14	30	Douglas (1967)
Barron	65	12	18	Douglas (1969)
Behana	34	12	35	Douglas (1969)
Freshwater	52	14	27	Douglas (1969)
Mary	43	7	16	Douglas (1969)
Millstream	45	10	22	Douglas (1969)
Wild	66	14	21	Douglas (1969)
Nitchaga	48	18	37	Douglas (1969)
Guyana				
Essequibo	34	16	46	Livingstone (1963)
Demara	82	41	50	Livingstone (1963)
Dominica				
Various	35–261	17–66	11.4–42.5	Reading (1986)
B. Large catchments				
India				
Godavari	95–547	3–31	3–30	Biksham and Subramanian
Tributaries	42–455	1–31	0.6–12	(1988)
Africa				
Niger	49	10	20	Livingstone (1963)
Benue	83	15	18	Rougerie (1960)
Senegal	34	9	26	Rougerie (1960)
South America				
Orinoco	54	8	15	Livingstone (1963)
Amazon (at Obidos)	43	11	25	Livingstone (1963)
South East Asia				
Mekong	198	15	8	Livingstone (1963)
C. Extra-tropical rivers				
St Lawrence	166	6	3	In Douglas (1969)
Hudson	173	5	3	In Douglas (1969)
Mississippi	221	6	3	In Douglas (1969)
Rio Grande	881	30	3	In Douglas (1969)
Columbia	191	13	7	In Douglas (1969)
Yukon	268	13	5	In Douglas (1969)
Volga	458	12	3	In Douglas (1969)
Plate	103	19	19	In Douglas (1969)
Thames (at Walton)	368	8	2	In Douglas (1969)

TDS, total dissolved solids.

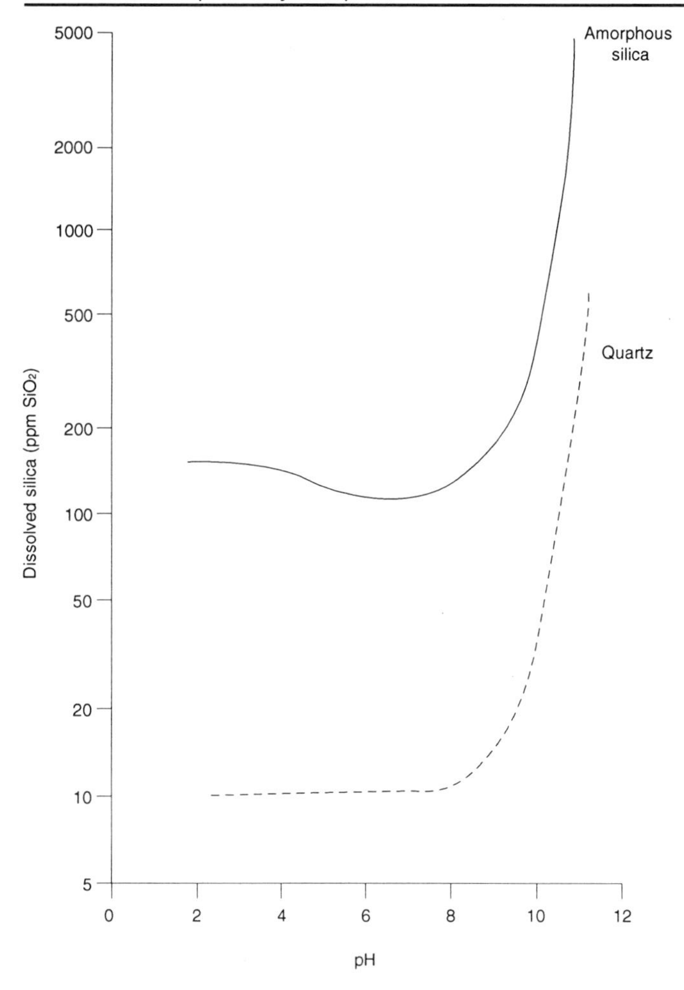

Figure 7.13
Solubility of silica at 25 °C.

in the Godavari basin in India. Here more than 85 per cent of the dissolved load is derived from the 40 per cent of the catchment underlain by the Deccan Traps. The remaining 15 per cent comes from the granites and other hard rocks. An area of sedimentary rocks covering approximately 7 per cent of the catchment close to its outlet in the Bay of Bengal is shown to be a solute sink, depleting around 5 per cent of the calcium and 54 per cent of the sodium. This is possibly due to the selective ionic exchange from the river water to the sediments. In contrast, this same area contributes around a third of the river's suspended sediment at its outlet. The release of solutes along floodplains has also been implicated as important in the formation of clay minerals and even the formation of duricrusts (e.g. Lamotte and Rougerie, 1962).

In common with suspended sediment, the various chemical constituents of solute load often react differently to changes in discharge on a seasonal or individual storm basis. This reflects the fact that different amounts and types of soluble minerals are derived from different locations within the catchment and enter solution at different rates. Hysteresis can be used as a tracer to investigate hillslope processes. Clockwise loops tend to suggest that storm water flushes out solute-rich water while anticlockwise loops (figure 7.14) can be explained by the expansion of the drainage net into solute-rich waters.

In general, however, the tendency is for solute concentrations to fall with increasing discharge. Silica concentrations are the least predictable (e.g. Davis, 1964; Edwards, 1973; Edwards and Liss, 1973) possibly because of the maintenance of silica concentrations by biological buffering mechanisms and sorption reactions. In the Barron river, north Queensland, Douglas (1973) found that, following Hurricane Flora in 1974, silica concentrations initially followed the fall in total solute concentrations with increasing discharge but then rose sharply. Douglas suggests the rise may be due to the transport of silica derived from particulate matter to the river.

Figure 7.14
Variations in solutes with discharge during a flood peak at Pont Cassé, Dominica: (a) silica (SiO_2) concentration; (b) chlorides concentration; (c) gross total dissolved solids concentration; (d) solute load; —, rising stage; - - -, falling stage.

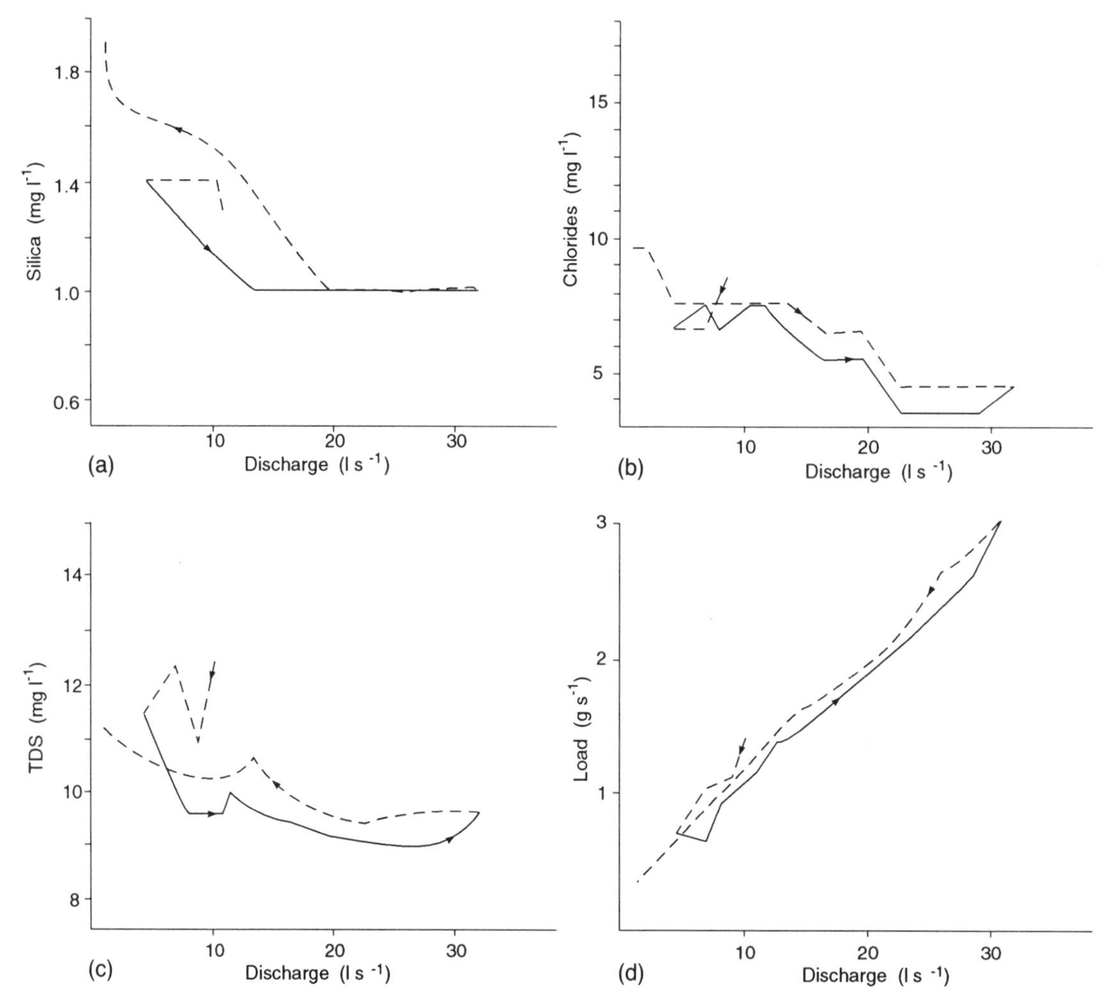

7.5 High Magnitude Events and Large-scale Changes in Humid Tropical Catchments

In the humid tropics considerable importance has been attached to the role played by high magnitude episodic events (section 8.3). Case studies indicate that the denudation caused by a single storm in the wet tropics may approach the mean annual erosion rate (Wolman and Gerson, 1978). The role of episodic mass movements has frequently been emphasized in areas of high relief (Simonett, 1967; So, 1971; Reading, 1986), while surface wash is considered more important in seasonal areas (Temple and Rapp, 1972). The significance of the humid tropical hydrological system is generally seen as indirect, through its control of runoff processes, in particular the surface wash–throughflow ratio and the development of positive pore water pressures on hillslopes. The traditional belief that streams in the humid tropics are unimportant as denudational agents has been qualified in terms of geological age and relief. The existence of stream systems where flood forms persist over time suggests that there should also be a temporal qualification. Gupta (1984) bemoans the lack of empirical data but postulates that high magnitude floods are extremely important as channel and valley forming events in certain areas. Catchments most likely to have river networks with at least semi-permanent flood morphologies are those where high magnitude precipitation occurs with a low recurrence interval. These are in monsoonal areas or areas which lie in the path of tropical cyclones, particularly if the land constitutes a significant orographic barrier. In such environments high magnitude events are considered essential for the initiation of episodes of mass movement and large-scale changes in channel morphometry. In an investigation of cyclone effects in Dominica, West Indies, Reading (1986) also comments upon the rapid rate of recovery of baseflow and sediment discharge to pre-cyclone levels.

According to Gupta (1984), river systems in which high magnitude events are important formative factors include one or more of the following physiographic characteristics:

1 valley flat storage areas containing abundant amounts of sand-size or coarser alluvium;
2 an unusually wide flood channel cross-section (geological and topographical conditions permitting) recognizable by the absence of vegetation or vegetation of a specific type;
3 vegetation in lower parts of the valley flats indicating periodic inundation;
4 a distinct floodplain terrace, low enough to be at least partly submerged during large flood events;
5 bars, flood channels and chute bars in the channel or on the floodplain or terrace whose material is coarser than the rest of the alluvium;
6 upper parts of the valley displaying high velocity flow sedimentary structures;

7 landforms related to the high frequency lower magnitude flows confined to lower parts of the channel, smaller in size and made of finer material.

An extreme hydrological event has a dramatic and disruptive effect on the physical landscape in the short term. However, notwithstanding the effects of climatic change, the long-term development of the physical environment (including the soils and vegetation) will have taken account of the frequency and intensity of such events. Their effects on the human landscape are almost always viewed in the short term and are undesirable. For example, floods are frequently viewed as at best an inconvenience and at worse a major hazard to life and well-being. In an ever-increasing number of catchments, the hydrological regime is (to a greater or lesser extent) managed for the purposes of flood control and water supply and frequently for hydroelectric power generation. In most cases, the objective of human intervention is to even out variations in discharge and provide a steady but reliable flow. However, natural river channels are shaped by the superposition of the influences of various discharges such that no single discharge will produce the same form.

The presence of a dam and reservoir across a natural channel will affect all aspects of a catchment's hydrology. Downstream of the dam, river stage, discharge and flow durations will be altered and thus also the sediment-carrying characteristics. These, in turn, will be reflected by changes in channel morphology. Alterations of the flow regime, and in particular stage, will also affect fluvial processes on the valley-side slopes since the level of flow in the main channel constitutes the local base level to which processes work. Thus decreases in stage downstream of a dam may be expected to activate or increase the rate of soil loss on adjacent slopes. Rates of erosion may also increase upstream of the dam as a result of increased human activity around the reservoir. The deposition of sediment in the reservoir is inevitable. However, excessive sedimentation is probably the single most common failure of large dams in the tropics. The sediment deposited in the reservoir is often unavailable to the streams below the dam outlet and river bed scour below the dam frequently results in erosion of agricultural land and buildings.

Case studies of the effects of dams on the physical and human environment in humid tropical regions are numerous. In addition to the published reports of, for example, Oyebande (1981), Chettri and Bowonder (1983), Olofin (1984) and Brabben (1987), many others are unpublished in government departments or in consultants' offices. Their common theme is that the hydrological impact of a dam is greater than that anticipated at the design stage, which frequently reduces the expected life of water management projects (table 7.12). Dam construction is also associated with a host of other health, socio-cultural and economic problems and the success of any large-scale water management scheme in the humid tropics, as elsewhere, requires the appreciation and effective

Table 7.12 Reservoir sedimentation

Country	Reservoir	Completion date	Test date	Loss in total capacity (%)	Siltation rate (ha m/100 km)[a]	
					Assumed	Actual
India[b]	Bhakra	1959	1974	5.6	4.3	6.1
	Maithon	1956	1974	10.7	1.6	13.1
	Panchet	1956	1974	9.2	2.5	10.0
	Mayurakshi	1955	1974	9.2	3.6	16.4
	Tungabhadra	1953	1974	9.4	4.3	6.4
	Nizamsagar	1931	1974	62.0	0.3	6.8
	Lower Bhawani	1953	1974	4.2	–	4.2
	Matatilla	1958	1974	0.8	–	15.2
Nigeria[c]	Challawa	To be completed		20 in 50 years		
	Dutsi-ma	To be completed		20 in 50 years		
	Yola (Benue)	To be completed		21 in 25 years		

[a] Hectare metres per 100 square kilometres catchment area.
[b] Chettri and Bowonder, 1983.
[c] Oyebande, 1981.

management of a host of interdependent physical and human variables. Hydrological change must be expected but the aim is to make a compromise between the physical requirements of a drainage system and the needs of humans. This must be underpinned by an understanding of the workings of the physical system.

Gaps in our knowledge are clearly evident and there is an urgent requirement for more information, derived from long-term and repeatable catchment-based investigations. It is only from a position of such understanding that we can expect accurately to predict and plan for the changes which are likely to ensue from interference with the natural system, from human activity or the result of climatic change (section 9.9).

EIGHT

Environmental Resources and Hazards in the Humid Tropics

8.1 Resources and Resource Utilization

The term 'humid tropics' embraces a huge variety of landscapes and ecosystems containing an enormous array of resources and hazards. They occur in the form of materials (e.g. ores, minerals, timber and water), landforms (e.g. hills, lakes and areas of beauty) and processes (e.g. landslides and cyclones) and are important to local, national and international economies. A fine line separates resources from hazards; often it is the timing, magnitude or intensity which determines the classification.

A number of natural resources and hazards develop or acquire their distinctive value/threat characteristics only under humid tropical conditions (e.g. tropical hardwoods, bauxite, tropical cyclones). Due to the complications of tectonics and climatic change, however, not all tropical resources lie within regions currently classified as humid tropical (section 1.1).

The exploitation of finite and renewable resources within the humid tropics is a hugely important topic, with global as well as regional implications. Similarly, the establishment of hazard reduction strategies and disaster preparedness principles is extremely complex. Both are largely beyond the scope of this book. This chapter describes a number of humid tropical resources and hazards and introduces some of the environmental consequences of their occurrence and exploitation. Some of the environmental and management issues associated with resource exploitation are discussed further in chapter 9.

8.2 Mineral Resources

Many rocks, soils and minerals can be utilized as raw materials for construction or manufacture or have a supplementary financial value as a commodity. Those of particular interest here are ones which have accumulated or concentrated under humid tropical weathering regimes and hydrological activity (plate 8.1).

Plate 8.1
Aerial view of a bauxite mine in the polygonal karst region of Jamaica, West Indies.

8.2.1 Ores as products of humid tropical weathering processes

Under a humid tropical weathering regime and well-drained conditions, alkalis and alkaline earth minerals and then silica are almost completely leached. Remaining parent minerals weather into kaolinite and amorphous or crystallized ferruginous oxyhydroxides (e.g. aluminium hydroxide and iron oxides) (section 6.2). The residual material constituting this deeply weathered profile is known as saprolite. The relative importance of residual free iron and aluminium minerals within the profile is distinguished by the terms lateritic (iron rich) and bauxitic (alumina rich) although the term laterite is also frequently used in a generic sense to describe residual, metal-enriched deposits.

The term bauxite is often treated as an economic term to describe deposits which have an economic value as aluminium ore. Bauxite is a collective term for aluminium-rich mineral mixtures formed from the decomposition of aluminium silicate parent material. Residual 'lateritic' bauxites, developed largely on volcanic or intrusive rocks, and karstic bauxites, which occur on carboniferous rocks, make up the two most important types. Under current levels of technology and exploration bauxite reserves are thought to amount to around 38,000 million tons, enough for at least another 500 years at current levels of consumption. Approximately 90 per cent of the reserves lie within the tropics and a substantial

Table 8.1 Major world reserves of bauxite (1981 estimate)

	Bauxite (million tonnes)	*Percentage of World total*
Guinea	8200	28.6
Brazil	5000	17.4
Australia	4400	15.5
Jamaica	2032	7.1
India	1600	5.6
Guyana	1016	3.5
Cameroon	1016	3.5
Greece	900	3.1
Ghana	780	2.7
Indonesia	710	2.5
Surinam	498	1.7
Yugoslavia	400	1.4
Total	26592	92.6

Source: Edwards and Atkinson, 1986

proportion within countries which are politically unstable (table 8.1). The consumer countries of the industrialized world have little (e.g. USA, France) or no (Germany, Japan) reserves. The availability of bauxite therefore includes important strategic and political elements. Tropical bauxites are also currently favoured by the aluminium industry since they contain the aluminium oxides gibbsite and bohemite which are more easily processed than the aluminium oxide diaspore found, for example, in southeast European bauxites. Furthermore tropical deposits are also often mineable by relatively inexpensive open-cut methods (e.g. plate 8.1).

8.2.2 Hypogene deposits and supergene enrichment

The weathering from the primary rock minerals to the lateritic mineral assemblage follows a well-defined sequence (sections 4.2 and 6.2) and the number of stages involved is closely related to the aggressiveness of leaching. Sluggish leaching results in a greater number of stages. In brief, primary minerals are replaced by 2:1 clays and these by 1:1 clays (kaolins) with oxides and hydroxides of iron and aluminium developing late in the sequence. Passing upwards in the profile leaching becomes progressively more extensive and eventually reaches a point where the fabric of the saprolite collapses thus allowing the concentration of metal-rich segregations (e.g. pisoliths). In some cases enrichment can occur as a result of continuous short-lived mobility as minerals are repeatedly mobilized by solution, move down the profile and are then precipitated lower in the profile. This repeated solution and deposition mechanism is often the main cumulative mechanism in the case of nickeliferous and manganiferous laterites (Nahon et al., 1985).

Ollier (1984) uses the terms hypogene and supergene to differentiate ores formed by these two processes (figure 8.1).

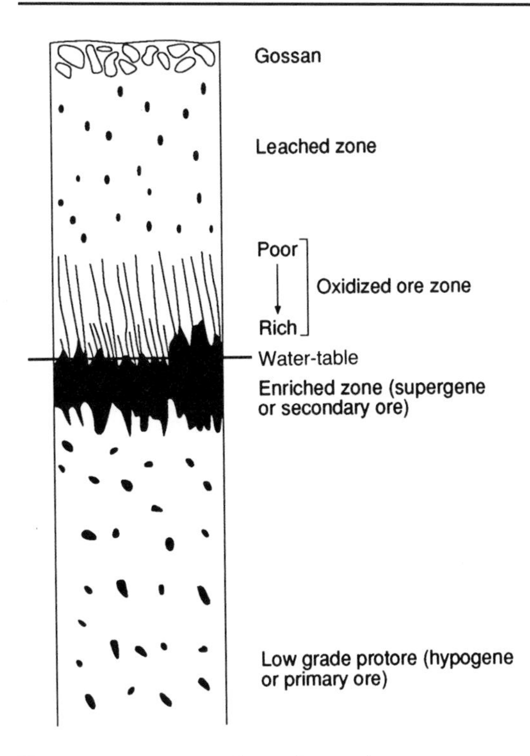

Figure 8.1
Ore zones within a weathered profile (from Ollier, 1984, p. 131).

Hypogene ores are those formed as water ascends up the profile and supergene ores are those which have been enriched by mobilization and precipitation. Supergene enrichment is especially important in sulphide ores (e.g. chalcocite). Mineralized but unconcentrated (low-grade) deposits are known as protores. The removal of valueless material by landscape denudation may result in the formation of a gossan, a mixture of limonite and quartz, which is often taken as an indicator to prospectors of an enriched supergene ore further down the profile.

8.2.3 Relative and absolute enrichment by weathering and denudation

There is fair agreement in the scientific literature that laterites form under climatic conditions at present found within a belt lying between 30 °N and 30 °S of the equator (McFarlane, 1991). Laterites currently found outside this zone are regarded as 'fossil', formed under earlier humid tropical conditions or prior to a shift in the positions of the continents in relation to the climatic belts.

The accumulation in deeply weathered profiles of iron, aluminium and also copper and nickeliferous minerals is now generally considered to occur from overhead (McFarlane, 1976). The development of residua in most major interfluve situations is thought to derive from the consumption of formerly overlying rock during the process of landscape denudation (figure 8.2). This is a case of vertical or *relative enrichment* (d'Hoore, 1954). Horizontal

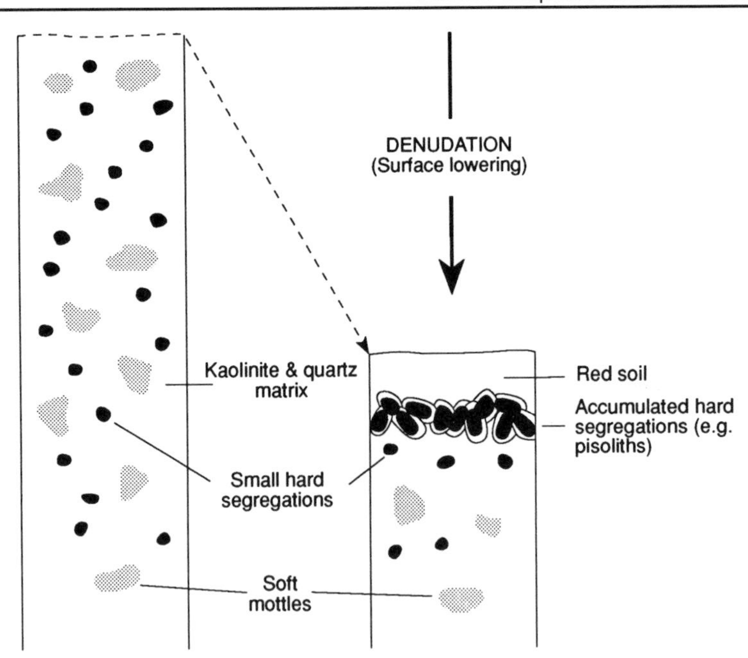

Figure 8.2
Removal of quartz and kaolinite matrix results in surface lowering and residual accumulation (e.g. of pisoliths) as a stone layer (from McFarlane, 1991).

or *absolute enrichment* refers to the accumulation of mechanically transported material and solutes from a topographically upslope source to slope-bottoms. Detrital laterites are genetically unrelated to the profiles they overlie with surface materials having been mechanically transported to the site. The discontinuity between the disturbed colluvium and the underlying *in situ* material may be indicative of an interruption in profile development by erosion and deposition (Eswaran et al., 1981).

8.2.4 Placers as transported mineral deposits

Placer deposits are formed by the mechanical concentration of minerals released by weathering from source rocks in which their concentration is normally sub-economic (Edwards and Atkinson, 1986). They provide the world's major supplies of tin, diamonds, niobium, tantalum, zirconium and titanium and important supplies of gold and platinum. The fundamental requirement for placer development is an appropriate source rock. However, the global distribution of placer deposits (figure 8.3) is largely a product of the variation in geomorphological processes currently, or in the recent geological past, acting at the earth's surface (Sutherland, 1985).

First the heavy mineral must be released from its source or host rock by weathering. The removal of clays and lighter quartz material through denudation may leave the heavier minerals at the surface to form an *eluvial* placer similar, in many ways, to other residual ores such as bauxite, lateritic iron and nickel. Eluvial gold and diamonds are found in lateritic crusts. However, these deposits are often difficult to process owing to problems of crushing the

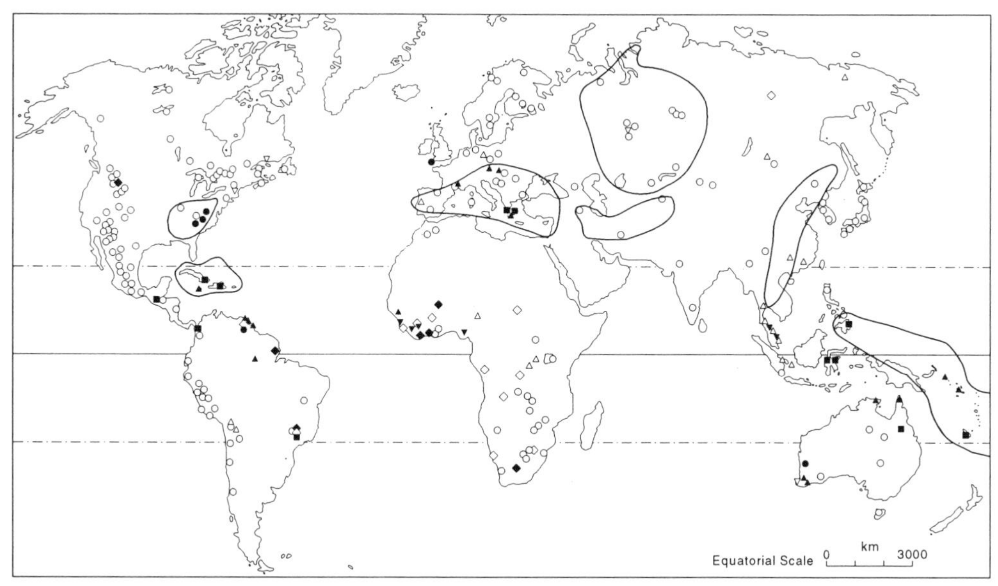

Figure 8.3
Major mineral resources including weathering deposits and placers formed under humid tropical conditions.

duricrust and separating the precious minerals in the presence of iron.

Eluvial materials are commonly transferred downslope and become concentrated as *colluvial* or *alluvial* placers. The former includes material found in stone lines, debris flows and mudslides. By far the most important, economically, however, are the alluvial deposits. For example, important gold and diamond reserves are found in alluvial fan deposits which are periodically reworked to concentrate the heavy grains. Uplift and/or subsidence can result in thick sedimentary accumulations containing several placer layers, as in the Witwatersrand Basin in South Africa. In contrast, the reworking of higher terraces and channel gravels may result in accumulation within a single layer, such as in the channel placer deposits found in Sierra Leone and Ghana (Thomas, 1988). Although gravel reworking causes the transport of heavy grains downstream, or from one terrace to another, the placer is progressively enriched by reworking.

In the humid tropics, there is evidence of large-scale transfer of sediment from upper to lower slopes on the Jos Plateau in northern Nigeria, in southern and central Africa, in parts of the Zambian plateau and in Sierra Leone (Thomas, 1988). Some of the most important diamond placers found in West Africa occur as coarse basal gravels buried in deep bedrock channels several metres below the current level of river flow. It is thought that these are 'pluvial' deposits of the Late Pleistocene containing some of the colluvial slope material which was transferred downslope by mudflows.

In equatorial southeast Asia, e.g. peninsular Malaya, Sumatra, Kalamantan and the intervening islands, rich heavy mineral placers, especially tin and gold, are found as Pleistocene alluvium deposits. Many of these deposits appear to have formed as braided river plains or alluvial fans during major depositional episodes during the Quaternary. It is thought that channel cutting and rapid sedimentation occurred as a result of abrupt changes in climate associated with transitions from cool stadials to warmer and wetter interstadials. In Kalamantan, neotectonics may also have played an important part in accelerating the transfer of sediment downslope (Thomas, 1988).

8.2.5 Environmental consequences of material extraction

The commercial extraction of material resources spawns a multitude of complex economic and social repercussions which are beyond the scope of this text. The exploitation of materials also invariably results in changes to surrounding environmental resources, such as water and air, flora and fauna, which are traditionally regarded as 'free goods' with little or no market value. Economically motivated decisions on production and consumption therefore frequently pay little attention to the well-being of the environment. This is particularly true in the humid tropics, especially in developing countries, where indifference and neglect of the environment have been tolerated by governments whose environmental policies are prejudiced by the financial rewards from mining.

Extraction invariably involves disruption of plants and animals, soils, vegetation, hydrology, air quality and landscape value. This has a profound effect on the culture and life-styles of indigenous populations. The environmental costs are often difficult to calculate in financial terms and are rarely, if ever, added to the sale price of materials.

In India, pollution and environmental degradation have reached alarming proportions from a combination of poverty, deforestation and industrial development without adequate environmental safeguards (Govind, 1989). The extraction industry is only partly to blame but unregulated extraction of coal, limestone and other minerals has degraded large areas of soil and forest and increased sediment discharge and the siltation of water pipes, e.g. at Makum (Assam) and Banganapalle (Andhra Pradesh).

Southeast Asia is the major tin-producing region of the world (plate 8.1). Exploitation takes place in alluvial deposits along the coast and inshore by gravel pumping or dredging respectively. Gravel pumping involves removing the topsoil and liquefying unconsolidated deposits, allowing the tin ore and other heavy solids to settle out before washing away the remainder into streams and rivers. This sediment settles downstream and at the coast. Abandoned mines are difficult to rehabilitate because removal of the topsoil frequently results in accelerated erosion during the monsoon period.

Dredging entails removing the tin-rich sediment with a bucket or suction pipe, gravitational separation and the release of tailings back into the sea. Sediment plumes in the Andaman Sea off Thailand are large enough to be seen by satellite imagery. Studies carried out between 1981 and 1986 show that tailings have smothered large areas of coral off the coast of Phunket Island (Thailand's major tourist resort). Water turbulence during the monsoon regularly removes some sediment but the reefs have not, in general, recovered (Chansang, 1988). Mangroves and macro-benthic fauna on the seabed are also known to be affected but, like the corals, they may be able to recover if conditions for their growth are restored quickly.

Alluvial sediments in areas such as the Amazon Basin are dredged for gold. The gold particles are amalgamated with mercury and as part of the process 5–30 per cent of the mercury is lost or discharged directly into the river. Studies by Malm et al. (1990) in the Maderia catchment, southwest Amazon, indicate heavy mercury contamination in the area (figure 8.4). Up to 1000 times natural background levels of mercury were found in the water along tributaries of the Maderia while fish were found to contain up to five times the Brazilian safety limit of mercury. Levels of mercury in the hair of local Indians and gold miners were sufficiently high to lead to severe health effects due to mercury poisoning.

Our record of environmental protection in the humid tropics is poor. The OECD countries have concluded that, in the long term, environmental protection and economic development are not only compatible but are interdependent and mutually reinforcing. However, for the poor countries of the humid tropics there is some reluctance to invest in fixed-cost pollution control assets. Many believe the initiative must be taken by the richer countries further north (see sections 9.17).

8.3 Environmental Hazards

Physical events (e.g. storms, floods, earthquakes and volcanic eruptions) become environmental hazards when they pose a threat to humans and what they value: namely life, social and economic well-being, material goods and landscape (Perry, 1981). Traditionally, these events have also been seen as 'acts of God' suggesting that humans have no part in creating the hazards or mitigating their effects. However, as human influence is ubiquitous over the globe, there are intricate relationships between 'acts of God' and 'acts of Mankind'. Smith (1992) describes environmental hazards as lying at the interface between the 'natural events system' and the 'human use system' (figure 8.5). Smith (1992) stresses that human sensitivity to environmental hazards is complex but can be summarized as representing a combination of the following:

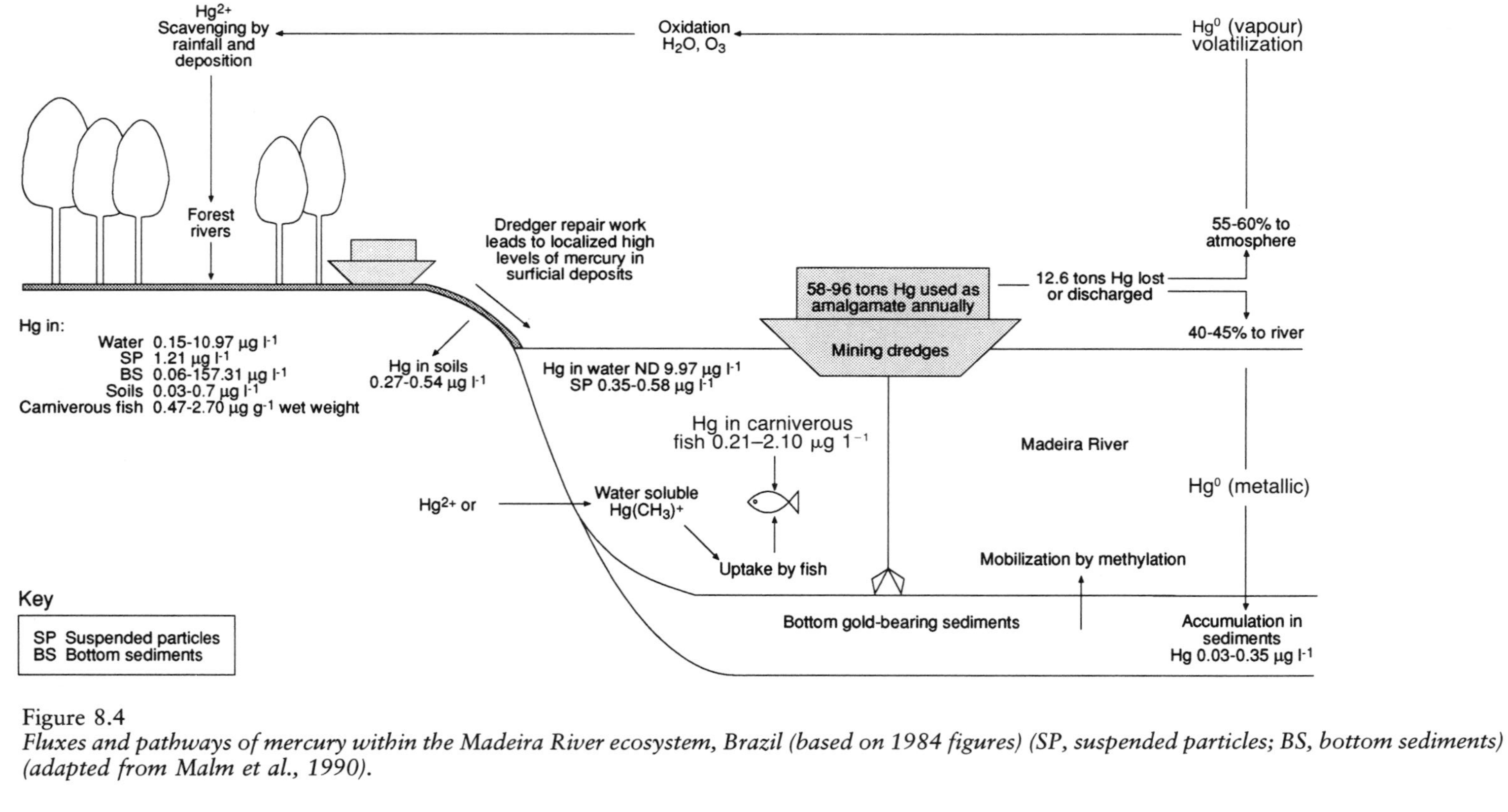

Figure 8.4
Fluxes and pathways of mercury within the Madeira River ecosystem, Brazil (based on 1984 figures) (SP, suspended particles; BS, bottom sediments) (adapted from Malm et al., 1990).

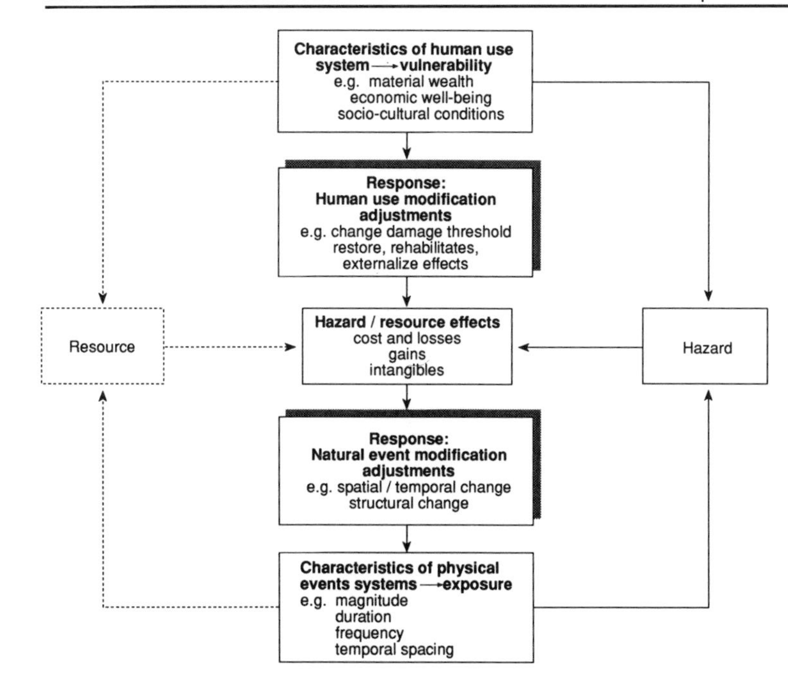

Figure 8.5
Environmental hazards as an interaction between a human use system and a natural events system (adapted from Burton et al., 1978).

1 physical exposure, reflecting the magnitude and timing of a physical event;
2 human vulnerability, reflecting the breadth of social and economic tolerance available at the same site.

Munchener Ruck's (1978) world map of exposure to a variety of extreme natural events (figure 8.6) reveals that many tropical countries have a potentially high hazard risk. In the developing countries, characteristics of the human-use system mean that the realization of a hazard all too often results in acute human suffering and loss of life. Highly urbanized industrial societies suffer, in material terms, the greatest damage to buildings, industry and the physical infrastructure.

8.4 Tectonic Events as Environmental Hazards

8.4.1 Earthquakes

In global terms earthquakes directly kill more people per year, on average, than any other hazard (Smith, 1992). Figure 8.7 illustrates the distribution of earthquake epicentres in 1970. It shows marked linear concentrations, the most pronounced as belts around the Pacific Ocean, with a conspicuous off-shoot extending into the eastern Indian Ocean. Seismicity from the southwestern Pacific to central America relates to the boundaries of the Indo-Australian, Eurasian and Philippine tectonic plates, whereas Caribbean/central

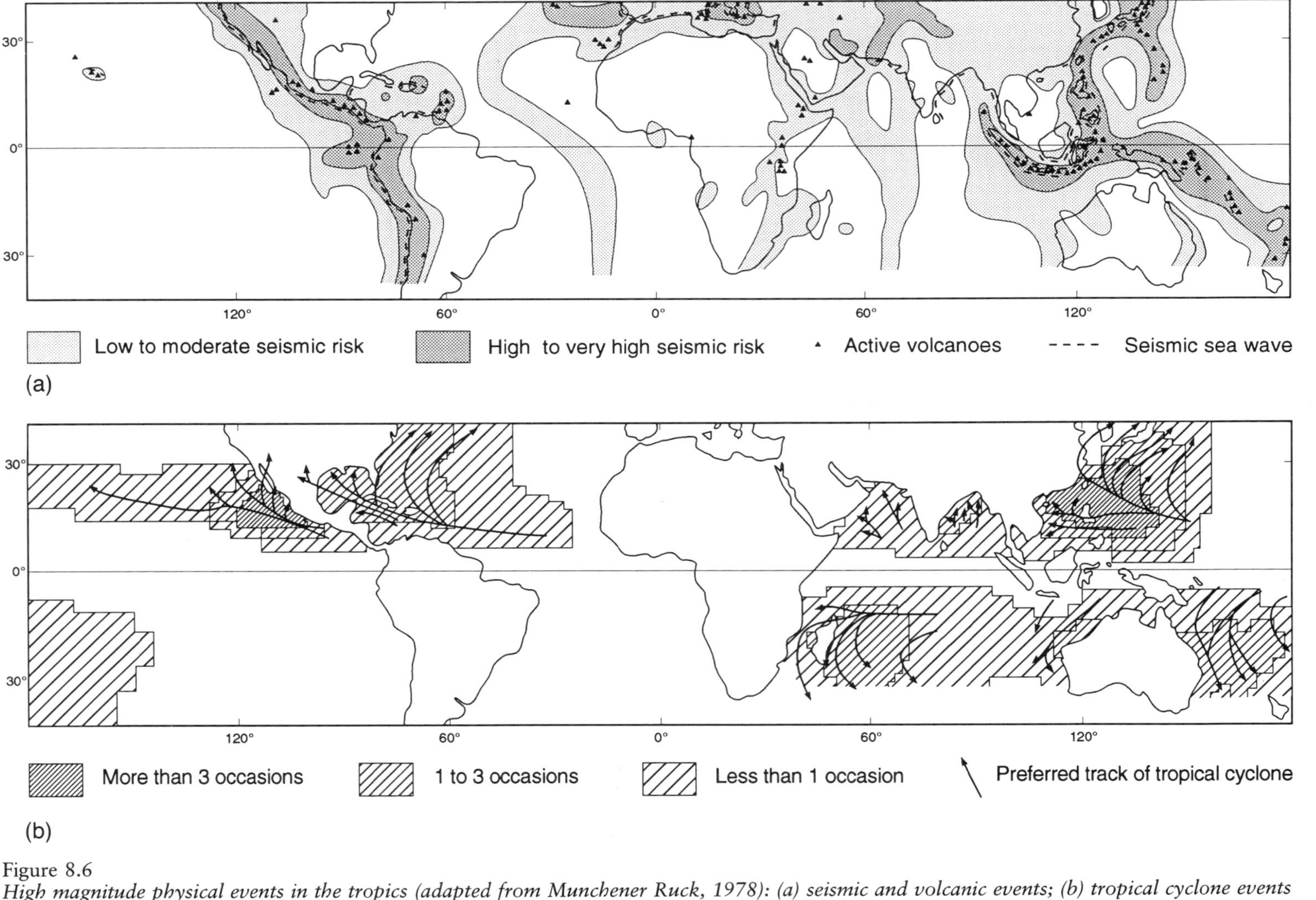

Figure 8.6
High magnitude physical events in the tropics (adapted from Munchener Ruck, 1978): (a) seismic and volcanic events; (b) tropical cyclone events (average number of occasions each year on which tropical cyclone winds at sea reach or exceed Beaufort force 8).

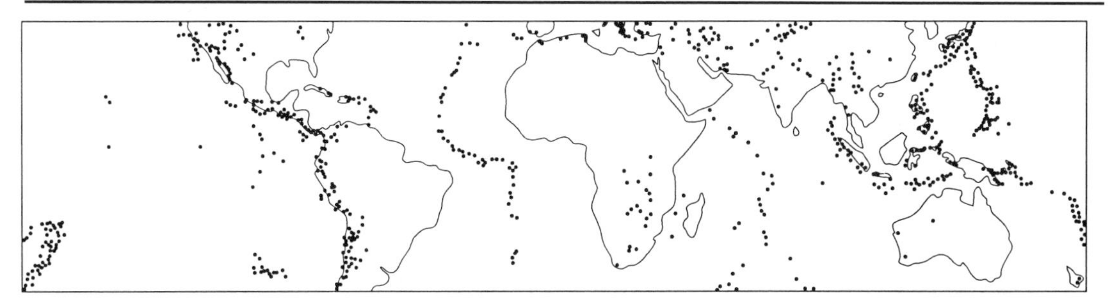

Figure 8.7
The location of tropical earthquakes.

American earthquakes are associated with the American, Nazca, Coco and Caribbean plates (figure 8.8).

The strongest earthquakes usually occur at the margins of colliding plates, where oceanic crust is subducted below continental crust (figure 8.9). They are therefore virtually absent from almost all of Africa and the Amazon but are relatively commonplace in the humid tropical Pacific belt and West Indies (figure 8.7). The largest magnitude earthquake recorded in the humid tropics reached 8.9 on the Richter scale, at Sumba, Indonesia, in August 1977. However, one of the most catastrophic earthquakes ever recorded devastated Guatemala in February 1976 with a magnitude of 'only' 7.5 on the Richter scale. The Guatemalan earthquake killed at least 22,000 people, injured nearly 75,000 and made over 1 million inhabitants homeless (Hewitt, 1983).

Energy release (as measured by the Richter magnitude) and hazard impact are only loosely linked. The main environmental hazard associated with earthquakes, which correlates directly with damage to structures, is ground shaking, a term used to describe the

Figure 8.8
Major crustal plates and movements (adapted from Goudie, 1989).

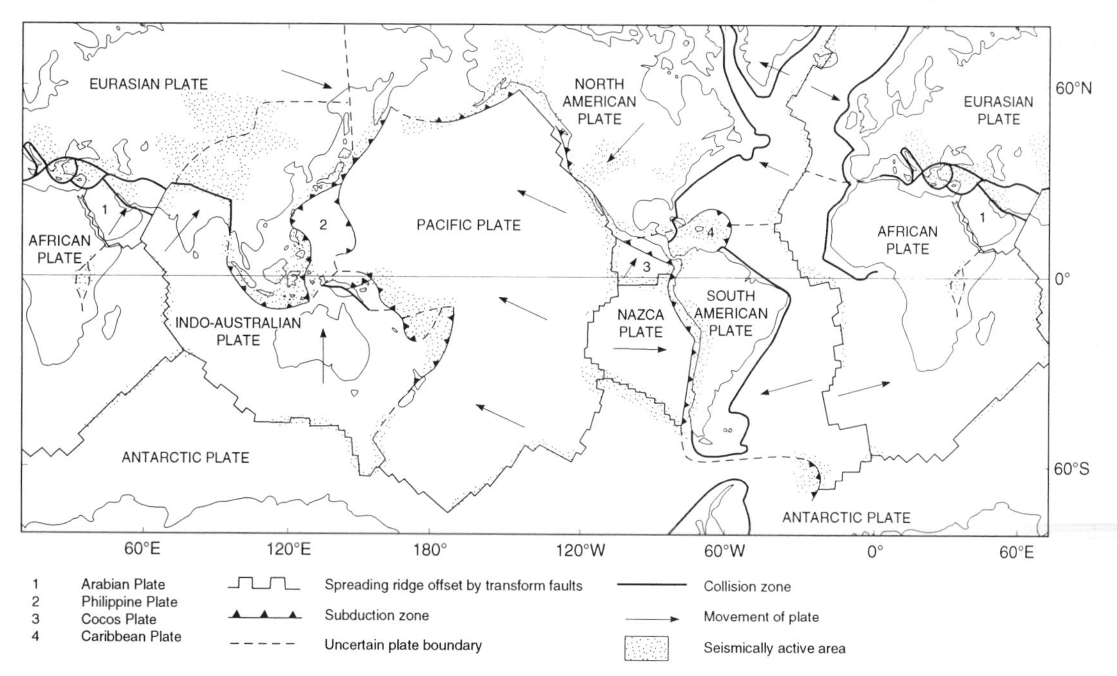

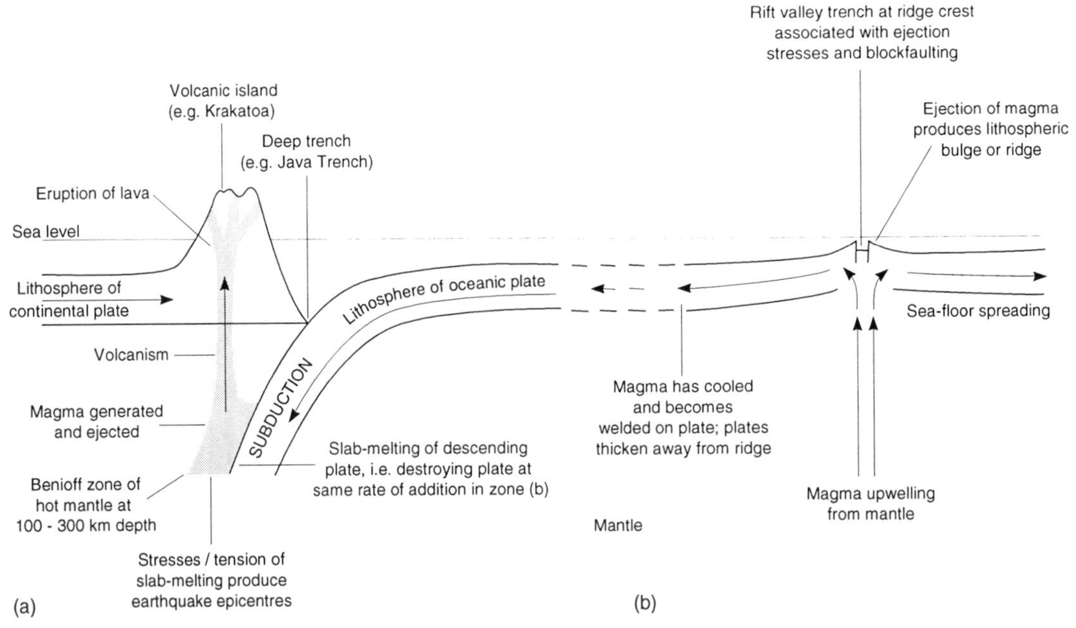

Figure 8.9
Diagrammatic section of ocean floor topography in (a) convergent zones and (b) divergent zones (not drawn to scale).

vibration of ground during an earthquake. Ground shaking is a function of the earthquake magnitude, epicentre depth and the properties of the surficial materials. It was the nature of the surface materials which amplified the tremors and exacerbated the destruction caused by the 1976 Guatemalan earthquake. In San Salvador in 1986, 1500 people died, 10,000 were injured and 250,000 were made homeless as a modest (5.4 magnitude) earthquake produced a three second tremor which was magnified five times as it passed upwards to the ground surface through approximately 25 m of unconsolidated volcanic ash (Smith, 1992). Certain types of clays and waterlogged sediments catastrophically lose strength and structure, in a process known as liquefaction, under earthquake vibrations. These materials, described by engineers as 'quick' or 'sensitive', often exacerbate an earthquake's effect (plate 8.2). For example, liquefaction of alluvial and colluvial material of the Liganea plain during Jamaica's 1692 earthquake resulted in the submergence of Port Royal, Jamaica's former capital, into Kingston harbour. The effects of the 1985 earthquake in Mexico City were also intensified by liquefaction, but also by the fact that many of the poorest inhabitants had built homes on this sensitive material. Factors such as population density and building construction are often at least as important as the earthquake's magnitude in governing the extent of its economic damage and the death toll (plate 8.3).

The modified Mercalli intensity scale (table 8.2) provides a qualitative measure of the ground shaking which, as well as being of more direct relevance to the earthquake hazard, also allows evaluation of historical earthquakes. The scale includes observed

Plate 8.2
Ground subsidence caused by the liquefaction of alluvial material during an earthquake, Port Royal, Jamaica. Much of Port Royal, the island's former capital, now lies submerged in Kingston harbour as a result of earthquake subsidence.

Plate 8.3
Wooden buildings are able to flex and absorb some of the energy of an earthquake or hurricane. As a result they may suffer less damage than rigid concrete buildings.

(and subjective) human perceptions of the tremor and the extent of physical damage to buildings.

8.4.2 Volcanoes

Volcanoes are associated with both diverging and converging (subducting) plate margins (figure 8.8) but the most destructive eruptions in the region have been recorded in the latter zones (e.g. Tambora, Indonesia, in 1815 and Mt Pelé, Martinique, in 1902). Here, melting induced in the mantle above the Benioff zone produces magma that rises buoyantly to the surface (figure 8.9).

Volcanic eruptions are the source of multiple potentially hazardous phenomena. Pyroclastic falls and flows, lava flows and air-borne debris are the primary hazards, whereas secondary

Table 8.2 Abridged modified Mercalli intensity scale

Average peak velocity ($cm\,s^{-1}$)		*Intensity value and description*	*Average peak acceleration*
	I	Not felt except by a very few under exceptionally favourable circumstances	
	II	Felt only by a few persons at rest, especially on upper floors of buildings. Delicately suspended objects may swing	
	III	Felt quite noticeably indoors, especially on upper floors of buildings, but many people do not recognize it as an earthquake. Standing automobiles may rock slightly. Vibration like passing truck. Duration estimated	
1–2	IV	During day felt indoors by many, outdoors by few. At night some awakened. Dishes, windows, doors disturbed; walls make creaking sound. Sensation like heavy truck striking building. Standing automobiles rock noticeably	0.015*g*–0.02*g*
2–5	V	Felt by nearly everyone, many awakened. Some dishes, windows and so on broken; objects overturned. Disturbance of trees, poles and other tall objects sometimes noticed. Pendulum clocks may stop	0.03*g*–0.04*g*
5–8	VI	Felt by all, many frightened and run outdoors. Some heavy furniture moves; a few instances of fallen plaster and damaged chimneys. Damage slight	0.06*g*–0.07*g*
8–12	VII	Everybody runs outdoors. Damage negligible in buildings of good design and construction; slight to moderate in well-built ordinary structures; considerable in poorly built or badly designed structures; some chimneys broken. Noticed by persons driving cars	0.10*g*–0.15*g*
20–30	VIII	Damage slight in specially designed structures; considerable in ordinary substantial buildings with partial collapse; great in poorly built structures. Panel walls thrown out of frame structures. Fall of chimneys, factory stacks, columns, walls, monuments. Heavy furniture overturned. Sand and mud ejected in small amounts. Changes in well water. Persons driving cars disturbed	0.25*g*–0.30*g*
45–55	IX	Damage considerable in specially designed structures; well-designed frame structures thrown out of plumb; great in substantial buildings, with partial collapse. Buildings shifted off foundations. Ground cracked conspicuously. Underground pipes broken	0.50*g*–0.55*g*
>60	X	Some well-built wooden structures destroyed; most masonry and frame structures destroyed with foundations; ground badly cracked. Rails bent. Landslides considerable from river banks and steep slopes. Shifted sand and mud. Water splashed, slopped over banks	>60*g*
	XI	Few, if any (masonry), structures remain standing. Bridges destroyed. Broad fissures in ground. Underground pipelines completely out of service. Earth slumps and land slips in soft ground. Rails bend greatly	
	XII	Damage total. Waves seen on ground surface. Lines of sight and level distorted. Objects thrown into the air	

g, acceleration due to gravity ($9.8\,m\,s^{-2}$).

hazards triggered by volcanic activity include mudflows or lahars (section 8.4.3), landslides (section 8.5) and tsunamis (section 8.4.4). The volcanic hazard occurs because areas subject to volcanic activity are settled by people or used for agriculture. These people risk the hazard in order to take advantage of the volcano's value as a resource or out of ignorance of the hazard. For example, under humid tropical conditions, volcanic material can rapidly weather to produce fertile soils (chapters 5 and 6) which are often well watered by rainfall orographically increased by the volcano's relief. Furthermore and especially when volcanicity has not been evident in historical times, these fertile soils have attracted agricultural activity and intensive settlement. Settlement close to volcanoes known to be active is also commonplace in parts of the humid tropics, the consequence of severe population pressures which dictate that all available land must be utilized. In Indonesia, for example, a large proportion of the population is forced to tolerate the volcanic hazard since there is only a limited amount of 'safe' land. Not surprisingly, almost two-thirds of all volcanically related deaths occur in Indonesia (Smith, 1992).

Volcanoes at convergent plate margins produce the most explosive eruptions (Blong, 1984). Many volcano-related deaths are associated with explosive eruptions which eject a turbulent mixture of hot gases and pyroclastic material (fragments of rock) into the atmosphere. Pyroclastic flows, at temperatures of up to 1000 °C, flow downhill at speeds of up to $100\,km\,h^{-1}$ and may travel 30 or 40 km from their source.

A measure termed the volcanic explosivity index (VEI), suggested by Newell and Self (1982), indicates that the most explosive eruption in historic times occurred in the Java trench at Tambora, Indonesia, in 1815. It has been assigned a VEI of 7 and its total energy has been estimated at 8.4×10^{19} joules. Ten thousand deaths were directly attributed to this eruption, although a further 82,000 people died from associated disease and starvation (Blong, 1984). The actual VEI does not, of course, directly relate to the hazard level of the eruption since the hazard also relates to the sensitivity (and proximity) of the population likely to be affected. For example, the survival rate at Tambora has been estimated at 71 per cent. In comparison a mere 0.01 per cent survived when Mt Pelé erupted with a much lower explosivity in Martinique in 1902. Approximately 28,000 people were killed in Martinique, mainly in the town of St Pierre, some 6 km away at the foot of the volcano (plate 8.4). They were literally incinerated by a great flash of searing heat as a nuée ardenté rolled down the side of the volcano towards the town, at speeds estimated at up to $33\,m\,s^{-1}$ and at temperatures over 700 °C. The air blast which preceded the cloud was probably of sufficient strength to demolish buildings. Ejected ash and rock debris showered the surrounding area and there are accounts of ships being sunk several kilometres out at sea. The Mt Pelé eruption completely devastated an area of $58\,km^2$.

As a global, long-term average, around 78 per cent of deaths due to volcanicity are due to pyroclastic flows (i.e. the nuée ardenté), 13.7 per cent to lahars (section 8.4.3) and 6.4 per cent to tephra (ash) ballistic projectiles. However, if we remove the Mt Pelé eruption from these figures, the respective distribution becomes 43 per cent, 36 per cent and 17 per cent (Blong, 1984). Falls of ash tend to cause a great deal more disruption, injury and damage than death. They cause choking dust, bury soils and buildings and may contain toxic chemicals which pollute land and water. Between

Plate 8.4
Mt Pelé, on the island of St Vincent, West Indies. Houses forming part of the rebuilt town of St Pierre, destroyed by a nuee ardente during the volcano's last eruption in 1902, are visible on the far shoreline.

1963 and 1965, for example, ashfalls from the Puerto Rican Irazu volcano resulted in over $150 million of damage to the coffee crop and agricultural land. Light falls of ash, on the other hand, may supply soil nutrients and help counteract the effects of leaching.

Lava flows are generally only life-threatening when the lava is highly fluid and mobile. For example, the eruption of the Nyirangongo volcano in Zaire in 1977 released a lake of highly fluid lava creating a wave of lava which killed seventy-two people and destroyed over 400 homes (Smith, 1992). Although lava flows *per se* make a very small contribution (0.5 per cent) to deaths by volcanic hazards, they have been responsible for sterilizing large tracts of agricultural land and creating food shortages and famine.

Indonesia and the Caribbean represent the most hazardous volcanic environments, experiencing 67 per cent and 13 per cent of eruption deaths around the world between 1600 and 1982. However, the Caribbean fatalities occurred mostly on two days in 1902 following the Mt Pelé devastation. Other humid tropical areas experiencing volcanic hazards are Central America (2.3 per cent of global volcanic deaths as above), the Philippines (1.6 per cent) and Papua New Guinea (1.5 per cent) (Blong, 1984).

8.4.3 Lahars

Lahars are mudflows comprising volcanic material triggered by massive amounts of water from rainstorms, the collapse of a crater lake or the melting of snow and ice. Due to their speed and unpredictability, they are the second most life-threatening volcanic hazard after pyroclastic flows (section 8.4.2). *Lahar* is a word of Indonesian origin which indicates the importance of these phenomena in the humid tropics. However, it is in the South

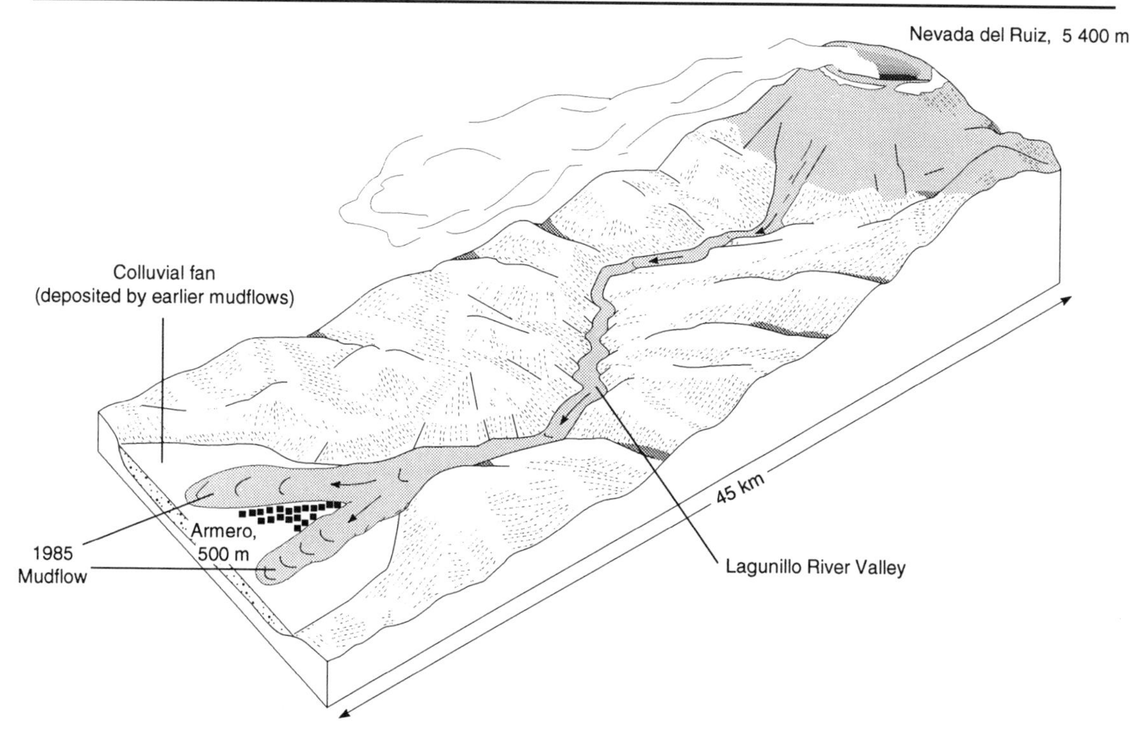

Figure 8.10
The 1985 lahar from Mt Nevada del Ruiz (adapted from Rapp and Nyberg, 1991).

American continent that they have caused some of the worst disasters. For example, the northern part of the Andean mountain chain from central Colombia to southern Ecuador has a history of large and spectacular lahars. They are triggered by volcanic activity from some twenty active volcanoes many of which are structurally weak and permanently snow capped. The worst volcanic disaster in the world (excluding Mt Pelé) occurred as a result of lahars following the eruption of the Nevado del Ruiz volcano in Colombia in 1985. The volcano is known to have triggered several large lahars in historical times, notably in 1595 and 1845. However, in the last century settlement in the area has increased dramatically. In 1985, an eruption resulted in widespread and rapid melting of glacial ice; a large lahar swept down the Lagunillas valley and 25,000 people died as 3–8 m of muddy slurry engulfed the town of Armero located on a colluvial fan at the mouth of the valley 45 km downstream (figure 8.10).

8.4.4 Tsunami

Tectonic displacement of the seabed, due to shallow focus earthquakes or volcanic eruptions, generates large sea waves or tsunami which cause a significant number of deaths in low-lying areas of the humid tropics (plate 8.5). In the past 100 years, around 400 tsunami have resulted in the death of some 50,000 coastal inhabitants in twenty-two countries in the circum-Pacific region. Almost a quarter of all Pacific tsunami are thought to originate

Plate 8.5
A tsunami hazard warning notice, Japan (Photo: J. B. Reading).

along the Japan–Taiwan island arc. In eastern Honshu, a 10 m tsunami has a return period of ten years. In 1933, a 1 in 70 years event produced a 24 m wave which resulted in over 3000 deaths, the injury of over 1000 people and the loss of over 7000 homes along the Sanriku coast (Horikawa and Shuto, 1983).

Studies of Pacific tsunami have demonstrated that seismic sea-wave generation is due more to the type of earthquake than the Richter magnitude *per se*. For example, major wave turbulence is associated more with subduction or thrust faulting, with characteristic downward movement into the crust. Here, the ocean floor is moved vertically and displaces the water into violent perturbations up to 12 m in height (Whittow, 1980). However, table 8.3 indicates that measurements of tsunami magnitude can be related to the earthquake magnitude. For example, sea waves exceeding 5 m in their coastal run-up are only associated with earthquakes exceeding Richter scale 8.

Such tsunami were recorded with the Philippines' earthquake of 1976 which accounted for the drowning of 3000 people. However, the most devastating and also the best studied volcanogenic tsunami of all time occurred in 1883, following the eruption of Krakatoa in the Sunda Strait, Indonesia. These tsunami were caused by

Table 8.3 The relationship between earthquake magnitude and tsunami magnitude and maximum run-up

Earthquake magnitude	*Tsunami magnitude*	*Maximum run-up (m)*
6	Slight	0.5–1.75
7	0	1–1.5
7.5	1	2–3
8	2	4–6
8.25	3	8–12

Source: Whittow, 1980

pyroclastic flows impacting on the sea, by rockfalls, by atmospheric shockwaves and as the result of caldera subsidence (Blong, 1984). The resultant sea waves are thought to have reached 30 m height which swamped the coasts of Java and Sumatra (more than 100 km from the eruption) and drowned over 30,000 people (Whittow, 1980).

It is interesting to note that the Seismic Sea Wave Warning System was inaugurated in 1946 after an Aleutian earthquake caused many tsunami drownings and resulted in many million dollars of damage in California and Hawaii. It has claimed reasonable success in terms of warnings and coastal evacuations although, since tsunami travel so rapidly, arrival times along coasts close to the eruption are impossible to forecast. Indeed, the best success has been attributed to warnings for Hawaii associated with tsunami originating around the Pacific margin.

8.5 Landslide Hazards

The movement downslope of surface material under the influence of gravity can be triggered by seismic events, atmospheric events or progressive weathering (chapter 4) and there are few countries where landslides do not occur. However, some areas are particularly prone to landslides owing to the combination of soils, rock type, topography, intense and prolonged rainfall, rapid land-use change, high population density and economic constraints (plate 8.6). Poor, overpopulated humid tropical regions are therefore in a high risk category and the landslide hazard in many areas is currently increasing as land hunger forces settlement on inherently unstable slopes.

The greatest landslide hazard results from rapid and usually highly fluid mass movements which are difficult to predict and even

Plate 8.6
Landslide hazard, Trinidad, West Indies (Photo: A. Brass).

more difficult to escape. Seismically triggered landslides are often on a gigantic scale, involving the mobilization of several square kilometres of material. One of the most catastrophic earthquake-triggered landslides in more recent times was experienced in the humid tropics in Peru in 1970. Following one of the strongest earthquake shocks on record, the summit ice cap on Mt Huascarán collapsed, creating a gigantic debris flow down the Rio Shacsha valley. The speed of flow was over 320 km h^{-1}, and within four minutes the flow had obliterated settlements in the valley and the flourishing town of Yungay was buried under millions of tonnes of mud and rock. In total, the Huascarán disaster claimed 25,000 victims and the destruction was almost unbelievable, possibly surpassing in magnitude such catastrophic events as at St Pierre in 1902 and Pompeii in AD 79 (Whittow, 1980).

Earthquakes and tropical cyclones often trigger numerous smaller landslides which are capable of destroying individual buildings, burying agricultural land and blocking transport routes. In total, these smaller landslides add up to a significant amount of material moved and damage caused and exacerbate damage relief operations.

8.6 Atmospheric Extremes as Hazards

8.6.1 Droughts

Humans adapt their activities to the mean moisture conditions and irregular deficiencies in this expected rainfall amount result in drought. Therefore, although droughts are normally associated with regions of low rainfall, they can also occur in areas which receive much higher annual totals.

Most meteorological definitions of drought relate to a continuous period without rainfall or, more meaningfully, they associate drought with long periods of anomalously low rainfall. Although meteorological definitions pay no attention to water availability and give no indication of the hazard impact, they should not be overlooked since droughts begin with, or are compounded by, rainfall deficiencies which can be explained by reference to anomalies in the global circulation. For example, in Köppen's Am and Aw climatic zones (figure 1.2) the dry season may from time to time be extended by unusual atmospheric developments which can lead to drought. For example, in the Am zone, the failure or late arrival of the monsoon rains occurs due to a breakdown of essential upper–lower tropospheric coupling (section 2.5, figure 2.14(d)). In the Aw zone, an abnormal weakening of the ITCZ in the high-sun season can extend the dry season until water deficiency becomes a serious problem.

Apart from these seasonal extensions of dry conditions in the Am and Aw zones, there is increasing evidence that anomalies in expected rainfall within the humid tropics can result from oscillations in the large-scale interaction of the atmosphere and ocean. For example, the reversed Walker circulation (figure 2.17(b)) associated with El Niño is responsible for pronounced anticyclogenesis and intense dryness over the normally humid southwest Pacific and northeast Australia. Even failure of the Indian monsoon can be related to a weakening of the land–sea circulation systems. For example, the 1987 drought in Fiji (the worst one in over a century) was attributed to a strong ENSO episode in that year (Reddy, 1989).

The strong 1982–3 El Niño coincided with dryness and drought conditions as far apart as Africa, India, Australia and northern South America, and there is increasing evidence that the situation in the tropical Pacific (i.e. with El Niño or La Niña) may be teleconnectively linked to climatic conditions elsewhere in the tropics and even beyond. For example, Trenberth et al. (1988) have linked the drought of 1988 across North America to unusually low sea surface temperatures in the Pacific (i.e. La Niña) causing a northward displacement of the ITCZ southeast of Hawaii and the development of anticyclonic conditions over mid-west America. Links have also been suggested between Pacific and Atlantic sea surface temperatures and reduced rainfall across the Sahel of Africa. However, the linkages are complex and contain numerous feedbacks in operation which are not yet fully understood (Walsh and Reading, 1991).

8.6.2 Floods

Floods, like droughts, are natural and irregular events. However, on a global scale, floods are also the most common of all environmental hazards, because flood-prone areas are attractive for human settlement and agriculture. This attractiveness often relates to the agricultural productivity of these areas and floods are often essential to maintain the fertility of the soils for agriculture. At once the conflict between environmental resource and hazard becomes apparent. River flooding is a characteristic and widespread problem in parts of the humid tropics, a consequence of the intense and prolonged nature of rainfall in the region (section 3.6.3) and the nature of human settlement and land use (section 8.10 below). Some of the worst floods occur in Asia, where on average each year about 4 million ha of land are damaged and 17 million people are affected (Smith, 1989).

The most serious flooding tends to be associated with extremely prolonged and intense rainfall events following the development of synoptic-scale weather disturbances, especially tropical cyclones (section 2.4.2) and monsoon depressions (section 2.5). Some flood events may also be associated with large-scale atmospheric processes. For example, Pearce (1988) suggests that the 1988 floods in Bangladesh and Sudan can be linked to La Niña in the tropical Pacific (section 2.7).

Terrain factors (such as slope gradient and length), vegetation status and land use affect the magnitude of these floods and human-induced changes in the catchment are often largely responsible for increasing the frequency or magnitude of localized flooding. Urbanization and deforestation are often cited as the causes of increased flooding. However, the population's vulnerability to the flood hazard is equally increased by poverty and land hunger. For example, torrential monsoon rains produce regular floods in India and Bangladesh. The flood hazard here occurs because people choose or are forced to live on flood plains and in low-lying areas. The flood hazard is exacerbated by poor land use practices upstream (especially deforestation), which increase the speed and magnitude of the flood wave.

One of the most devastating monsoon floods occurred in the Ganges delta and (what is now) Bangladesh in October 1955. Over 1700 people drowned and 5.5 million crop acres were lost at a cost of £63 million. Twenty to thirty per cent of the country is regularly inundated by floodwater; however, in 1988 the figure rose to 46 per cent and 1500 people lost their lives. Typically the poorest in society suffer most since they are forced to live on the most marginal land and are unable to afford flood protection. Remedial measures to deal with such disasters are extremely basic in most developing countries. These poor economies can do little more than bear the loss since they lack the necessary levels of investment and technology to adopt suitable corrective or preventative measures. Improved meteorological forecasting and communications may help mitigate flood disasters but sustainable economic development is the key to reducing the flood hazard.

8.6.3 Tropical cyclones

Approximately 15 per cent of the world's population live in cyclone-prone regions and the eighty or so tropical cyclones recorded each year are on average responsible for around $1500 million damage and 15,000 deaths (Smith, 1989). Tropical cyclones bring torrential rain, extreme winds and storm surges (section 2.4.2) to coastal parts of the tropics and the combination of these atmospheric elements often increases the overall hazard. The greatest threat from tropical cyclones occurs along densely populated low-lying coasts (e.g. Bangladesh, Gulf States of America) and in small island states (e.g. the West Indies, Fiji, the Philippines and Japan).

Within the West Indies, the impact of cyclones on poorly developed agricultural regions (e.g. Hispaniola, Windward and Leeward Islands) is often overshadowed by the massive financial losses sustained by urban areas along the Gulf coast, Florida, and in the US Virgin Islands. For example, the most intense Atlantic storm on record, Hurricane Gilbert, tracked across the Caribbean, through the Gulf of Mexico and into northern Texas in September 1989. With a minimum central pressure of 884 mb, recorded soon after the storm passed over Jamaica, Hurricane Gilbert was also to become the region's costliest hurricane. Jamaica suffered billions of dollars of damage to its buildings and physical infrastructure and over 6000 people were left homeless (plate 8.7). In Mexico the hurricane caused over $2 billion of damage. Further north, in the USA, which was away from the most severe weather conditions, property damage was estimated at around $56 million, mainly as a result of more localized tornadoes. However, due to the success of hurricane warning and disaster preparedness procedures, relatively few people lost their lives. A total of 318 deaths has been directly attributed to the hurricane. The worse disaster occurred in Monterez, Mexico, when 140 people were killed as buses carrying evacuees from the storm were overturned by floodwater.

By comparison, in the densely populated and relatively undeveloped Indian subcontinent, mortality is unfortunately often much greater. The greatest disasters have occurred in the Ganges Delta which, due to its overpopulation, low elevation and mobile deltaic substrate, is extremely vulnerable to any extreme physical event. In addition, the delta is located at the head of the Bay of Bengal which has a high frequency of cyclone activity (i.e. more than 20 per cent frequency, as indicated in figure 2.10).

The two worst cyclone disasters on record occurred in 1970 and 1977 and were responsible for an estimated 1 million and 100,000 deaths respectively in India and Bangladesh. The 1977 cyclone was one of the most severe storms recorded in the Bay of Bengal when 160 km h^{-1} winds and 6 m surge waves struck the low-lying coast line of Andhra Pradesh, southeastern India. Entire villages were obliterated in the Kistna and Godavari river deltas and many coastal towns were severely damaged, with most casualties caused by collapsing buildings. In some rural areas, more than 90 per cent of the population are thought to have perished. Ironically, a hurricane warning was issued some 48 hours before the storm

Plate 8.7
The destruction of buildings at Hector's River, Jamaica, by Hurrican Gilbert in September 1988. The piles of rubbish in the foreground were houses before the hurricane struck. (Photo: Jack Tyndale Biscoe.)

struck Andhra Pradesh. However, it was either ignored by the local population or they were unable to take the necessary evasive action since road networks are poorly developed and transport facilities are inadequate and inefficient.

It is promising to note, however, that advanced warnings of cyclones, better preparedness and more adequate evacuations are now operational in Bangladesh. At the end of April 1991, the area was again affected by an intense cyclone. Winds in excess of 200 km h^{-1}, sea surges of 6 m and widespread flooding in the southeastern regions (including the port of Chittagong) again caused considerable death and destruction, particularly in poor rural communities. However, losses were apparently reduced by improved satellite-based cyclone forecasts and by the evacuation of some threatened residents to shelters built above the height of the tidal surges.

Meteorological forecasting of cyclone tracks forms an essential element in disaster preparedness and is especially useful for prompting evasive action and reducing deaths. For example in the West Indies region the US Weather Bureau's National Hurricane Centres constantly monitor all atmospheric disturbances. They provide up to three-day warnings of approaching storms based on ground-based and satellite data using complex computer-based forecasts. Furthermore, many coastal urban areas act upon these regularly updated forecasts within sophisticated emergency action plans. Local inhabitants are made aware of the actions they should take and are given adequate opportunity and time to carry them out.

With realistic investment (from overseas aid particularly), less sophisticated warning systems can also reduce the enormous losses suffered in the developing world. In Fiji, for example, the Nadi Hurricane Warning Centre was established in 1980 following the disasters of Hurricane Meli in 1979. Criticisms at that time were levelled against the government for the low level of preparedness and (especially) for the inefficiency of emergency services following the horrific event (Thompson, 1981). However, the success of the Nadi centre has been evident in the 1980s because despite an increased frequency of cyclones (figure 2.9) the number of deaths per storm has fallen dramatically (Thompson, 1986).

One final point should be made about high-magnitude physical events, and especially atmospheric extremes. An improvement in our understanding of the physical processes controlling them will allow us to make better predictions about their timing and magnitude and to determine more accurately their potential as hazards. This knowledge can then be integrated into strategies which also address human aspects of sensitivity to reduce these hazards effectively. More research in the physical sciences, and particularly within the tropics, is essential and geographers are focally placed to utilize this information within holistic hazard reduction strategies.

8.7 Soils and Vegetation Resources and Sustainable Management

Sustainable management of the soil and vegetation resources in the humid tropics is an important issue which requires an understanding of the responses of both soil and vegetation to land use changes. It was shown in chapter 4 that many humid tropical soils have inherent management problems, the solutions to which depend on an understanding of the pedological processes and the functioning of the soil–water–plant system. The management of humid tropical vegetation, whether for sustainable forestry, nature conservation or agriculture, requires a similar understanding of its dynamics and relationships to soil, climate and hydrology. Much of the fundamental research into humid tropical soils and vegetation has been stimulated by management concerns, and because so much is currently being written about vegetation destruction and environmental degradation in the humid tropics it is appropriate to examine both soil and vegetation responses to environmental change.

8.8 Soil Erosion and Sediment Yield

Chapters 6 and 7 include reference to the action of geomorphological and hydrological processes in the humid tropics under conditions of undisturbed vegetation (e.g. sections 6.4 and 7.3). However, the situation differs when the vegetation is disturbed as both soil erosion on slopes and sediment yields from catchments usually increase dramatically in response to changes in the partitioning of the forest hydrological cycle and the soil properties (Bruijnzeel, 1990). Observations of soil losses after forest disturbance were made as early as 1836 in Malaya by James Low (1836/1972) and in 1908 in Sierra Leone by Lane-Poole (Millington, 1988). The main geomorphological responses to vegetation disturbance are accelerated slope erosion (splash erosion, sheetwash and rill erosion), gullying, river channel erosion and mass movements.

8.8.1 Slope erosion and gullying

Forest ecosystems, and those related to them (e.g. fallow regrowth) or that mimic them (e.g. shifting cultivation, certain types of plantations and agroforestry systems), exhibit low erosion rates compared with other types of land use (Wiersum, 1984; Bruijnzeel, 1990) (table 8.4).

The main factors leading to increased slope erosion in the agroecosystems reviewed in table 8.4 are the destruction of the tree canopy and the repeated disturbance of the ground vegetation and litter. Under relatively undisturbed conditions erosion rates are low (see for example the first four land use types in table 8.4) and they only increase slightly when the understorey is removed and the

Table 8.4 Rates of slope erosion (t ha^{-1} yr^{-1}) under different land uses in the humid tropics

Land use	*Minimum*	*Median*	*Maximum*
Natural forests	0.03	0.3	6.2
Fallow regrowth phase in shifting cultivation system	0.05	0.2	7.4
Plantations	0.02	0.6	6.2
Multistorey tree gardens	0.01	0.1	0.15
Tree crops with a cover crop or mulch	0.1	0.8	5.6
Cropping phase of shifting cultivation systems	0.4	2.8	70.0
Agricultural intercropping in young forest plantations	0.6	5.2	17.4
Tree crops, clean weeded	1.2	48.0	183.0
Forest plantations with litter removed or burned	5.9	53.0	105.0

Source: data from Wiersum, 1984

canopy is left more or less intact. However, when the litter layer is destroyed erosion rates increase. This occurs slowly at first because of the residual effect of decaying soil organic matter on aggregate stability (Wiersum, 1985) but then, as this pool of organic material is used up and there is little replacement (e.g. under tree crops and forests with clean weeding, litter removal and burning), the rates rise dramatically (Bailly et al., 1984; Wiersum, 1984; Bruijnzeel and Bremmer, 1989). The role of the tree canopy in reducing erosion includes protection of the soil surface from direct rainfall, although splash erosion from intercepted water dripping from trees can be a locally important agent of splash erosion (Mosley, 1982; Wiersum, 1985; Vis, 1986; Brandt, 1988), and the provision of organic matter by leaf fall (Wiersum, 1985).

Gullies often form in the humid tropics where overland flow or rillwash from cleared agricultural or grazing land increases markedly in depth (Bergsma, 1977; Brunsden et al., 1981; Haigh, 1984b; Rijsdijk and Bruijnzeel, 1990). However, under a forest cover gullies often result from soil exposure after treefalls (Ruxton, 1967; Turvey, 1974), extreme rainfall (Herwitz, 1986) or the collapse of subsurface pipes (Morgan, 1986).

8.8.2 Mass movements

Attributing specific mass movements to vegetation disturbance in the humid tropics is difficult. As discussed above (sections 8.6 and 6.4.2) most mass movements appear to be related to intense rainfall and seismic activity (Starkel, 1972; Prasad, 1985; Ramsay, 1986, 1987; Whitehouse, 1987). The occurrence of shallow, transitional landslides, earthflows and mudflows has been linked to vegetation disturbance (Manandhar and Khanal, 1988; Brass, 1992) because at shallow depths the effect of root binding on soil stability is important (Starkel, 1972; Ziemer, 1981; O'Loughlin, 1984). However, causal linkages between deep, rotational landslides and vegetation disturbance have not been satisfactorily proven. Mass movements have also been related to terrace agriculture and various forms of construction. Poor control of terrace drainage can lead to mass movements (Marston, 1989; Rijsdijk and Bruijnzeel, 1990) and slumping often occurs on terrace risers (Euphrat, 1987). Mass movements caused by construction activities related to irrigation canals (Ramsay, 1986), around reservoirs when water levels are lowered (Rudra, 1979; Carson, 1985; Galay, 1987; Mahmood, 1987) and road building (Haigh, 1984a; Henderson and Wittha-wawatchutikul, 1984; Aitken and Leigh, 1992) have been recorded in the humid tropics.

8.8.3 Sediment yield

The effect of land use conversion on soil erosion is not always apparent in measurements of catchment sediment yield. This is because sediment production sites in a catchment are often far fewer than areas of temporary sediment storage both on slopes (e.g. upslope of tree trunks, in litter, in depressions and at basal slope concavities) and in river channels (e.g. meander point bars, floodplain and swamp deposition) (Deitrich and Dunne, 1978; Millington, 1981). Therefore, whilst the effects of erosion on a slope are often noticeable very quickly (Shrestra, 1988) those downstream may take many years to become apparent (Goswani, 1985; Pearce, 1986). This time lag is greater in large catchments where the sediment delivery ratios are low (Walling, 1983) because of the increased buffering effect provided by temporary storage. However, attributing changes in sediment yield solely to deforestation is problematic because it ignores the effects of extreme rainfall events and long-term trends related to climatic variability.

Despite these fluctuations in sediment yield, the undersampling of bedload, and variations in river sampling networks, researchers have been able to link increases in sediment yields at a number of river gauging stations to periods of deforestation in the humid tropics (figure 8.11). For example, AUSTEC (1974, quoted in Aitken and Leigh, 1992) demonstrated that sediment yield in the forested Sungai Pahang catchment in Malaysia was 2.7 million tonnes before 1900, but during the peak of disturbance in the catchment in 1975 it was estimated to be 8.5 million tonnes. Further problems in examining sediment yield trends concern whether the river section upstream of a sampling station is aggrading or degrading and the effects of extreme seismic events which can cause large numbers of mass movements in a short time span (Goswani, 1985). Consequently, variations in sediment yield over time need to be examined with caution.

Comprehensive studies of erosion rates, sediment transport and sedimentation have been carried out in catchments in southeast Asia (Rijsdijk and Bruijnzeel, 1990) and Sierra Leone (Millington, 1981). The latter study compared three small drainage basins with

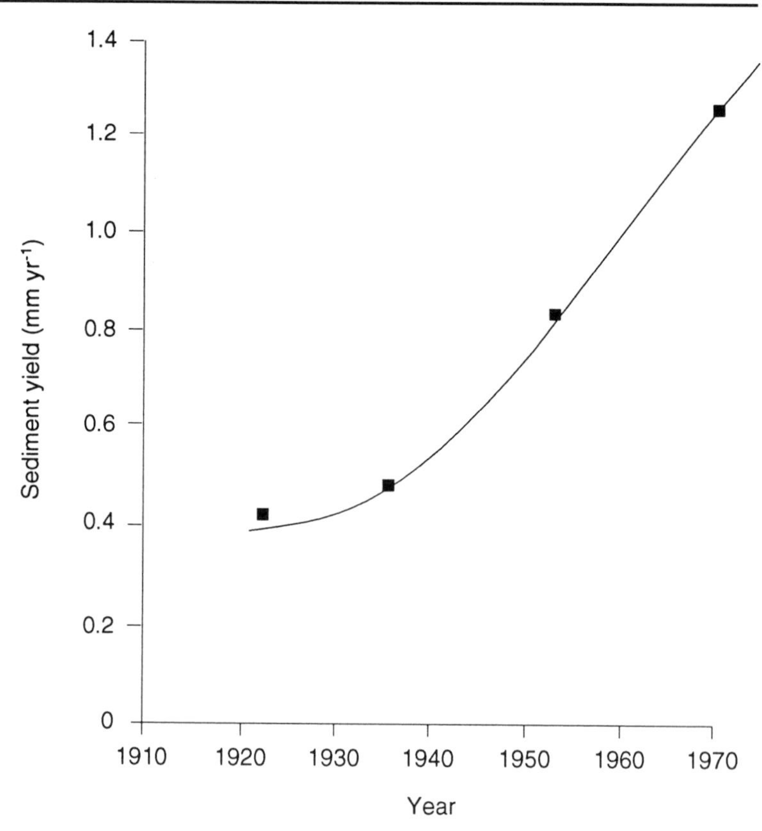

Figure 8.11
Acceleration of erosion inferred from sediment yield at Meilktila Lake, Burma (based on work by Gear).

different land uses and was able to ascribe differences in reservoir siltation rates to measured erosion rates under different land uses (table 8.5). The main consequences of increased sediment yield are increased downstream siltation in reservoirs and irrigation works (e.g. Millington, 1981; Gupta, 1983; Woolridge, 1986; Mahmood, 1987), on floodplains and in estuaries. Changes in water quality also occur and the combined effect has been shown to have both ecological and human impacts (Aitken and Moss, 1975; Colchester, 1989).

The influence of such anthropogenic activity on sediment yields is most clearly seen in geologically stable areas (van Dijk and Vogelzang, 1948; Douglas, 1967; Dunne, 1979; Lam, 1979; Bailly et al., 1984; Fritsch and Sarrailh, 1986). In tectonically active, geologically young areas it is far more difficult to analyse because of the naturally higher and more variable sediment yields (Pain and Bowler, 1973; Carson, 1985; Whitehouse, 1987).

8.8.4 Soil disturbance, erosion and sediment yield due to commercial logging activities

Increased soil erosion due to logging is well known (Gilmour, 1971; Megahan and Kidd, 1972; Liew, 1974; Rothwell, 1978) and erosion rates of up to 266 m^2 ha^{-1} have been recorded in Sabah

Table 8.5 Erosion rates and sediment yields in two small basins in Sierra Leone

Land use	Mean erosion rate ($t\ km^2 yr^{-1}$)	Bambara Spring		Mountain Torrent	
		Area (km^2)	Soil loss (t)	Area (km^2)	Soil loss (t)
Bare rock	0	0.019	0	0	0
Bush fallow	0.05	0	0	0.542	0.027
Cultivation[a]	21.24	0	0	0.060	1.274
Grassland	7.45	0.015	0.112	0.239	1.781
Forest	0.05	0.725	0.036	0.155	0.008
Settlement[b]	18.75	0	0	0.209	3.919
Settlement[c]	7.45	0.043	0.32	0	0
Total		0.802	0.468	1.205	7.009
Sediment yield (predicted)			0.584		5.817
Sediment yield (measured)			0.07		2.00

[a] From field survey data, upland cultivation areas accounted for 10 per cent of the area of bush fallow; mean erosion rates are an average of those for rice and cassava cultivation.
[b] This settlement class comprises earth roads and gardens.
[c] This settlement class comprises mainly grassland of the university campus.

Source: Millington, 1981

(Liew, 1974). Recent work in Borneo has explored the actual mechanisms at work in the humid tropics in more detail (Douglas et al., 1990, 1992). Douglas et al. found that sediment yields from logged catchments exceeded those of nearby undisturbed catchments and were related to three sequential periods of logging activity – road construction, logging adjacent to roads, and high lead and tractor-hauled logging (figure 8.11). The main post-logging sediment sources were gullies which formed along the snig tracks of winched logs and the log assembly and loading zones. Immediately after logging, sediment accumulated on the slopes and in the streams, and during this period large amounts of sediment were transported by runoff events of moderate magnitude. However, after a time much higher magnitude discharges were needed to move the remaining sediment which had been trapped in the channel system. Reductions in sediment yields in logged areas can be achieved by proper road construction and alignment, and cross drains on the roads (Baharuddin, 1988).

8.9 Soil and Vegetation Interactions after Disturbance

The on-site effects of vegetation disturbance are not simply confined to soil loss through slope erosion. Soils may undergo physical, chemical or biological degradation and all of these processes are, at least in part, related to vegetation succession after disturbance.

In chapter 5 the relatively closed nutrient cycle typical of forests on low fertility soils was noted, as well as the tendency toward more open cycles on higher fertility soils. Vegetation disturbance interrupts the nutrient cycle in three ways: (i) by the loss of nutrients tied up in the vegetation that is lost either by export or burning; (ii) by increased nutrient leaching; and (iii) by nutrient losses through soil erosion.

8.9.1 Soil nutrient responses to logging

The increase in organic litter during and after logging leads to a delayed release of nutrients to the soil. The time period over which the nutrients are released depends on the type of litter and the soil microclimate. Gillman et al. (1985) noted no overall loss in nitrogen and exchangeable cations in soils in a logged forest in Queensland, although different microenvironments caused by logging did show changes. Organic carbon levels, however, decreased by 15 per cent. Zulkifli Yusop (1989) noted significant changes in the nutrient content of streamflow from catchments with different types of logged dipterocarp forest in Malaysia. In the first year after logging streamflow nitrate levels increased by 300 per cent in a forest logged using a winch set-up, and by 180 per cent in an area of supervised logging. Nitrate levels returned to pre-logging levels in the second year, with a quicker return to these levels being noted in the supervised logged forest. Increases in potassium were also noted, although the levels still exceeded pre-logging levels in the second year. In Surinam Poels (1987) compared nutrient levels in streams from an undisturbed forest with those in streams from a forest that had been logged two years previously and then refined (i.e. non-commercial tree species had been poisoned and the lianas cut down). Calcium, magnesium and potassium levels were all higher in the logged forest than in the undisturbed forest, and increased again when the logged forest was refined. In both examples the increased nutrient levels can be related to the large amounts of decaying litter produced by logging and subsequent forestry operations.

8.9.2 Nutrient responses to forest fires

Forest fires have been recently recognized as a major disturbance factor in tropical forests but little work has been carried out on how they affect soil nutrient losses. In Sabah, Grip (1986, quoted in Bruijnzeel, 1990) noted that nutrient levels in streams draining areas burnt two years previously were greater than those from unburnt forests. Uhl et al. (1982) and Uhl and Jordan (1984) examined soil nutrient levels in burnt and unburnt forest plots in Venezuela. In forests developed on ferralsols, magnesium, potassium and nitrate levels increased dramatically after burning, but recovered to normal levels after about two years. Similarities exist between the behaviour of the major cations on the poorer *Caatinga* forests developed on podzols and forests on ferralsols, although the patterns were less marked.

8.9.3 Nutrient responses during conversion of forest to pasture

In Central America and Amazonia much forest has been converted to pasture for rearing beef cattle (Lanly, 1982; Fearnside, 1987; Uhl et al., 1988). These areas are often quoted as examples of how luxuriant forest can be rapidly turned into 'desert'. This appears to occur because the sustainability of the pasture is severely reduced by soil infertility and the invasion of vigorous weeds (Alvin, 1978; Buschbacher, 1986). The economic life of such pastures ranges from four to eight years. However, such findings are equivocal; Buschbacher (1987) has shown that lightly grazed pastures may cause less environmental damage than shifting cultivation in some circumstances.

The nutrient dynamics of pasture conversion and abandonment have been studied by various researchers (e.g. Buschbacher, 1984; Uhl et al., 1988). In southern Venezuela Buschbacher (1984, 1987) compared soil nutrient levels in pastures derived from mature and 15-year-old secondary forest. The total ecosystem nutrient stocks of calcium and potassium changed little between mature forest and pasture, although magnesium decreased and nitrogen increased. However, nutrient stocks in the vegetation component of the ecosystem decreased at the expense of the soil where, initially, calcium, magnesium and phosphorus levels in the upper 15 cm increased as cut-and-burnt debris was incorporated. Nevertheless, a proportion of the nutrients remained trapped in the undecomposed debris, were volatilized during burning or were leached. The levels of most nutrients declined markedly during the first year of pasture. The decline became slower in the second year, and the level of some nutrients even increased in the second year due to wood decomposition. Over three years the levels of all of the nutrients except nitrogen declined. Interestingly the net primary productivity of pasture was almost equal to that of the mature forest owing to efficient nutrient capture by the dense fibrous root networks of grass.

Once the vegetative debris has been decomposed the nutrient levels decline rapidly and, unless the pastures are fertilized, they are often abandoned. Whilst the overall decline in nutrient levels may promote reduced pasture productivity and their subsequent abandonment, Serrão et al. (1979) suggest that the lack of available phosphorus may be the most important influence on the declining productivity of pastures.

Soils and vegetation can recover after pasture abandonment if the intensity of the initial forest clearance has not been too severe (e.g. clearance by machinery before grass seeding) (Uhl et al., 1988). However, if it was particularly destructive and involved soil compaction, the loss of woody debris and seed bank removal, regrowth may be hindered. The time taken for soil nutrient levels to attain the levels found under undisturbed forest depends on site characteristics and the rate at which the logged debris decomposes.

After 2–8 years of regrowth topsoil nutrient levels generally bear little relationship to the previous intensity of grazing or the regrowth biomass. The different trends in soil nutrient level behaviour are shown in figure 8.12. Germane to the topic of succession after abandonment is nutrient cycling during grazing. Nutrient recycling by cattle urea and faeces is different from that by litterfall in forests. Specifically, the nutrients are more mobile, recycling is faster, and the distribution is patchy (McNaughton, 1976). In southern Venezuela Buschbacher (1987) found that after 3.5 years pastures from secondary forest had nutrient stocks that were similar to those of nearby 3 year old shifting cultivation plots. The indication was that recovery to secondary forest was feasible, although it may proceed at a slower rate owing to slower seed dispersal mechanisms in pastures than forests.

8.9.4 Nutrient status under plantation cropping

Measurements of the nutrient status of soils under plantation agriculture are rare, and those available are fraught with a variety of errors (Bruijnzeel, 1990). The approach taken by most researchers has been to calculate the nutrient balance over the period of the first rotation of the tree (i.e. the period from planting to harvesting); an example of such a 'balance-sheet' is shown in table 8.6. From such balance-sheets it has been found that

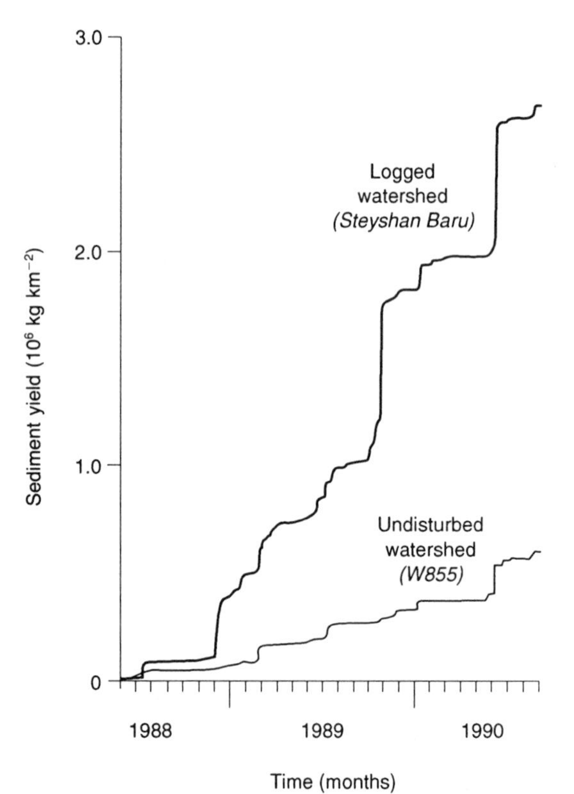

Figure 8.12
Cumulative curves for sediment yield in logged and undisturbed catchments in Sabah (after Douglas et al., 1992, figure 2).

Table 8.6 Nutrient 'balance-sheet' ($kg\ ha^{-1}$) for an ideally stocked *Agathis dammara* plantation under a variety of management conditions and a 40-year rotation period on andosols in Java

	P	*K*	*C*	*Mg*
Inputs:				
Rainfall	35	385	395	160
Weathering	200	1575	3060	1565
Total	235	1960	3455	1725
Outputs				
Leaching	30	880	1160	1220
Leaching[a]	2	25	15	15
Soil erosion[b]	2	15–25	20–140	10–40
Crop harvesting[c]	20	40	20	7.5
Stemwood harvest	105	435	660	155
Stemwood and bark harvest	145	560	1295	250
Stemwood, bark and branch harvest	205	855	1680	330
Total tree[d]	350	1655	3315	705
Balances				
Bole harvest	0.85	0.78	0.74	0.88
Bole and branch	1.10	0.93	0.86	0.92
Total tree	1.72	1.34	1.33	1.14
Soil nutrients 0–100 cm[e]	90–110	540–1340	750–4130	340–1220
Net loss as a percentage of reserves[f]	155–189	59–122	25–151	20–72

[a] This assumed an increase in water yield equal to rainfall interception in a mature plantation, a doubling of potassium concentrations and no changes in calcium, magnesium and phosphorus concentrations in streamflow.
[b] The first three years of forest growth are as an agroforestry system known as 'taungya'; the erosion rate is assumed to be $1\ cm\ yr^{-1}$.
[c] Based on upland rice and maize crops, with straw burnt on-site.
[d] Does not include the undergrowth.
[e] All in the readily available state.
[f] In the case of the total free harvest.

Source: Bruijnzeel and Wiersum, 1985

particular nutrient deficiencies limit biomass productivity in the second and subsequent rotations. Hase and Folster (1983) found that calcium was a limiting factor to teak growth on loamy soils in southern Venezuela. Bruijnzeel and Wiersum (1985) found that phosphorus was a limiting factor on *Agathis dammara* plantations on andosols in Java. Russell (1983) found that most of the major cations would be depleted in two to four pine and *Gmelina* rotations on sandy acrisols in Brazil.

8.10 The Effects of Shifting Cultivation on Soils and Vegetation

The responses of soil and vegetation to shifting cultivation and related farming systems in the humid tropics has been well documented (e.g. Jordan, 1987a). Although all such farming systems involve a period of cultivation followed by a longer fallow regrowth period (Watters, 1971; Ruthenberg, 1980; Savage et al., 1982) (plate 8.8), the effects on soil and vegetation differ owing to

(a)

(b)

(c)

(d)

Plate 8.8
The main phases of shifting cultivation: (a) cut wood is left to dry; (b) the slash is burnt; (c) the slash is reduced to partially burnt wood and ashes (note the patchy distribution of ash in this 'poor' burn); (d) a cultivated field after one year. All plates are from locations in Sierra Leone.

variations in site productivity and the intensity of leaching (Andriesse and Schelhaas, 1987).

Crop yields during a cultivation phase generally decline. For example, early research by Grist (1953) noted that rice yields in Malaya declined from 1.8–2.24 t ha^{-1} in the first year of cultivation to 0.9–1.34 t ha^{-1} in the third year. Declining crop yields have since been recorded in many shifting cultivation systems, but there is debate about what brings about this decline. The main arguments appear to be

1 declining soil fertility, particularly the limiting effects of phosphorus which is fixed in the soil and not available to crop plants (Sanchez, 1973; Arnason et al., 1982; Scott, 1987) (Jordan (1987b) goes further, arguing that lack of available phosphorus is the only important cause and that there is no decline in soil fertility);
2 deterioration of soil structure and texture (Arnason et al., 1982);
3 changes in the soil fauna;
4 increased soil erosion (Roose, 1982);
5 increased leaching (Roose, 1982);
6 the influx of insects, diseases and animal pests;
7 the influx of weeds which compete with crops for nutrients (Lambert and Arnason, 1980; Moody, 1982);
8 the change from broad-leaf weeds in the first year to grasses in subsequent years (Velasco et al., 1961; Moody, 1975; Kunstanter and Chapman, 1978).

Some authors consider the last two factors to be the most important ones in influencing farmers to abandon plots (e.g. Whitmore, 1975), whilst those working in South America favour declining soil fertility and/or lack of available phosphorus. The overall crop yield and its decline is also sensitive to the length of the fallow regrowth. During the fallow period the regenerating vegetation extracts and immobilizes nutrients from the soil at the expense of soil nutrient stocks (Scott, 1987). At the same time soil structure is ameliorated by the incorporation of decaying organic debris and the activity of soil fauna such as earthworms. A decline in the fallow period almost always leads to declining crop yields and the invasion of grass and herbaceous vegetation (Scott, 1987).

8.10.1 Changes in nutrient levels

During the different phases of a shifting cultivator's farming cycle (figure 8.13) soil nutrient levels are subject to a series of changes, although the extent to which these occur may vary with the soil management practices (Roose, 1982). After vegetation has been cut, there is a slight increase in available nutrients as they are released by the decaying green vegetation (Ewel et al., 1981; Jordan, 1987b). A

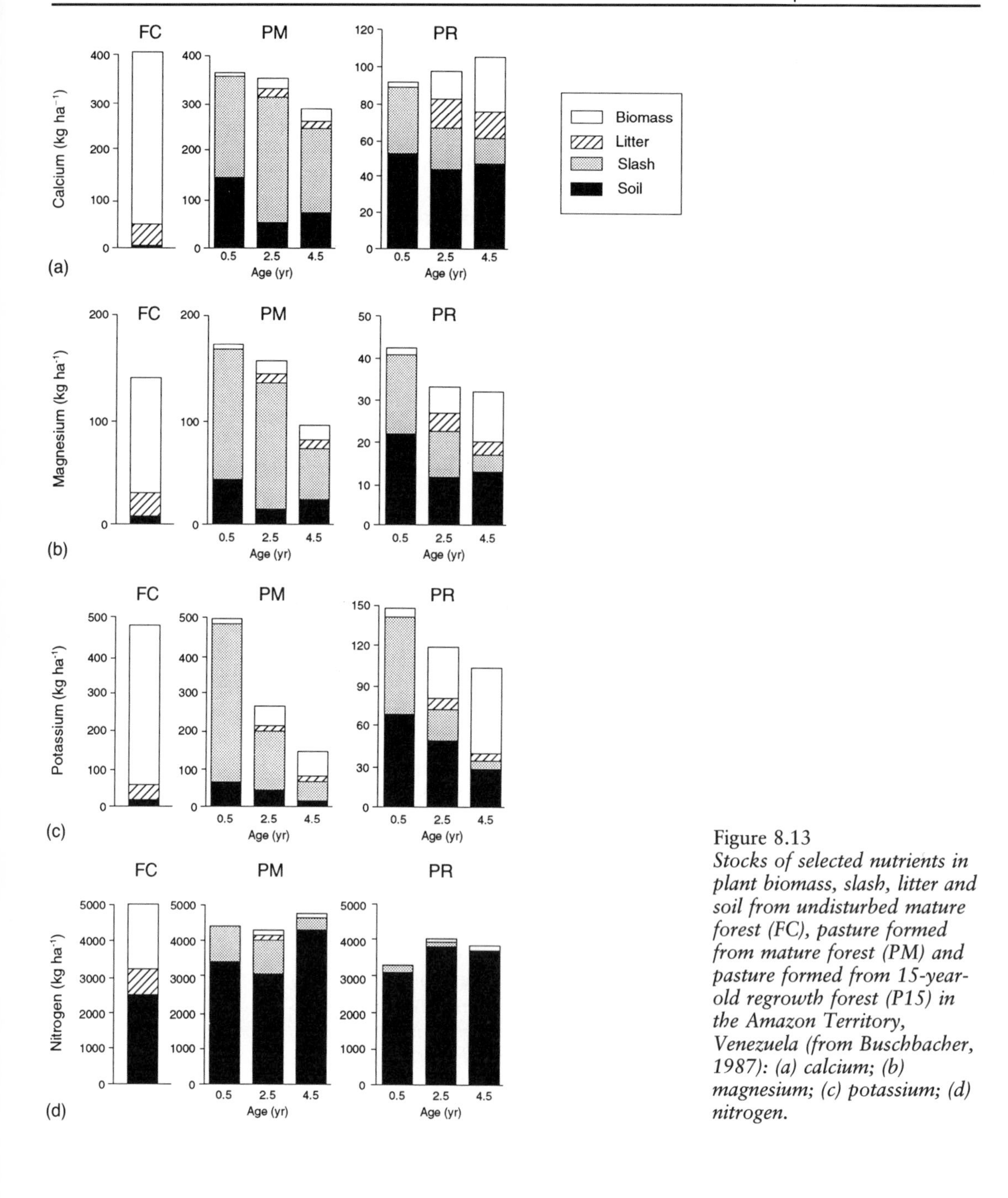

Figure 8.13
Stocks of selected nutrients in plant biomass, slash, litter and soil from undisturbed mature forest (FC), pasture formed from mature forest (PM) and pasture formed from 15-year-old regrowth forest (P15) in the Amazon Territory, Venezuela (from Buschbacher, 1987): (a) calcium; (b) magnesium; (c) potassium; (d) nitrogen.

large nutrient flux occurs when the dried vegetation is burnt (figure 8.14), although the actual amount varies with factors such as the amount of combustible biomass, vegetation type and the temperature of the burn (Bartholomew et al., 1953; Nye and Greenland,

	Month	Jan.	Feb.	March	April	May	June	July	Aug.	Sept.	Oct.	Nov.	Dec.
	Weather	Heavy rain		Rain slackening	Rains cease		Cold			Hot		Early rains	Mid rains
	Food supply	Hunger months	Cucurbits ripe	Subsidiary crops ripe	Early millet ripe			Ample food supply			Less food	Getting scarce	Hunger month
Men's work	Millet farms etc.		Fencing farms					Tree cutting			Firing farms		Sowing main crop
	Subsidiary crops	Mound digging and sowing							Dry-weather sowing			Mound digging and sowing	
	Other activities	Setting fishing traps and nets	Building houses (new village)			Setting fish weirs				Fish poisoning Net hunting		Fish spearing and trapping on flats	
Women's work	Farms			Harvesting: Maize	Early millet	Main crop millet			Piling branches		Firing farms		Sowing main crop
	Gardens	Mound digging and sowing					Havesting groundnuts		Dry-weather sowing			Mound digging and sowing	
	Other activities	Collecting mushrooms and caterpillars		Collecting caterpillars					Collecting wild spinaches Fish poisoning			Fruits	Collecting mushrooms

Figure 8.14
A typical shifting cultivation farming calendar for Bemba agriculture in Zambia (from Grove and Klein, 1979, p. 55, figure 46).

1960; Seubert et al., 1977; Ewel et al., 1981; Folster et al., 1983; Stromgaard, 1984; Jaffre, 1985; Andriesse, 1987). There is little loss of organic matter during a burn because of the accumulation of charcoal and the leaching of small organic matter particles from the ash (Stromgaard, 1991), but most importantly burning does not destroy soil organic matter (Nye and Greenland, 1960; Kyuma et al., 1985; Andriesse and Schelhaas, 1987) until very high temperatures are reached (Raison, 1979). However, during the burn there is a partial loss of volatile elements such as nitrogen, the proportion lost increasing with the burn temperature. Further losses can occur after burning in windy conditions as dry ashes are blown away. However, once it rains the ashes are washed into the soil, the nutrients are quickly released, and soil nutrient levels and pH increase (figure 8.15). The increase in pH is particularly sensitive to soil and rainfall characteristics, being greatest and most persistent on fertile soils and under moderate rainfall conditions (Popenoe, 1960; Nye and Greenland, 1960; Zinke et al., 1970; Brinkmann and Nascimento, 1973; Sanchez, 1976; Seubert et al., 1977; Lal and Cummings, 1979; Ewel et al., 1981). The nutrients in the soil are very prone to leaching during cultivation owing to the lack of vegetation cover and the generally higher rainfall totals during cultivation. However, Jordan (1987b) found that the level of nutrients in the soil remains more or less constant, suggesting that leaching losses are offset by the nutrient flow from decaying vegetation. The behaviour of individual cations varies with soil type, but the general trends are as follows.

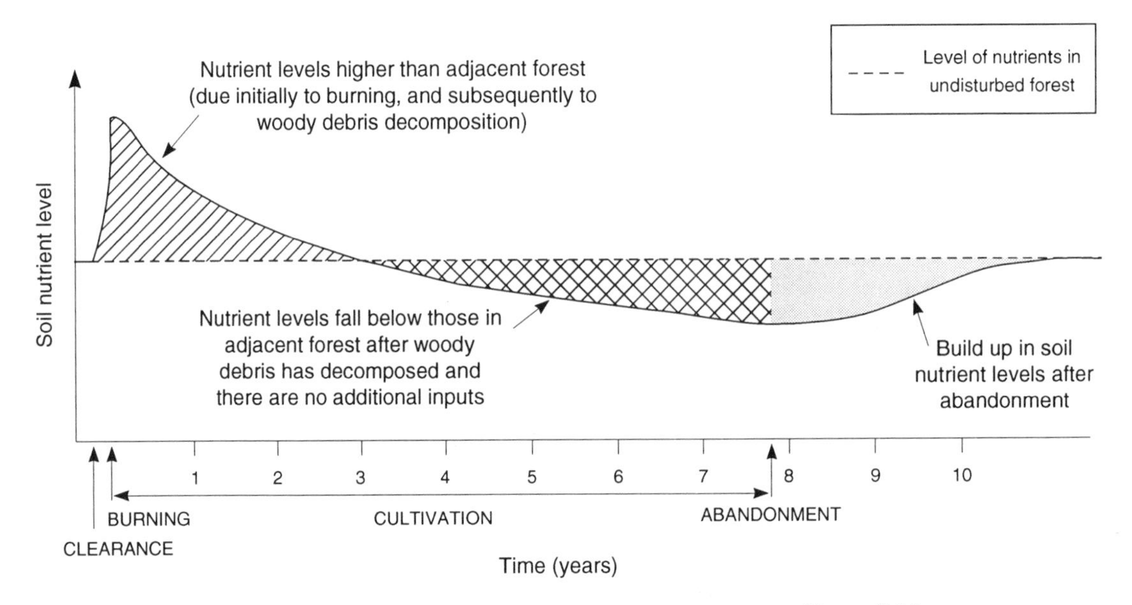

Figure 8.15
Hypothetical topsoil nutrient levels in humid tropical shifting cultivation systems.

1 There is a post-burn increase in the main cation levels which rapidly decline (Zinke et al., 1970; Aweto, 1981; Ramakrishnan and Toky, 1981; Andriesse and Schelhaas, 1987; Stromgaard, 1991).
2 The decline in potassium and sodium levels is usually more rapid than that of calcium and magnesium (Sanchez, 1976; Jordan, 1987b).
3 Nutrients that are leached from the topsoil are often fixed by clay minerals in the subsoil and are then extracted by plant roots (Sanchez, 1976; Andriesse and Schelhaas, 1987).
4 Most of the phosphorus complexes with iron and aluminium after the burn and becomes immobilized and, although phosphorus levels in the topsoil remain high, little is available to plants (Jordan, 1987b; Saldarriaga, 1987). However, Adedeji (1984) and Andriesse and Koopmans (1984) have recorded large increases in the surface layers of the topsoil (down to depths of 5 cm) which they attribute to the accumulation of burnt biomass and the mineralization of organic phosphorus at high temperatures. There remains a possibility that phosphorus fixation is less pronounced in sandy soils (Sanchez, 1976; Sanchez et al., 1983).
5 Nitrogen levels decline over the cultivation cycle due to volatilization loss, nitrification after the burn and crop uptake; carbon and sulphur are also volatilized and decline during cultivation (Ayanaba et al., 1976; Ewel et al., 1981; Ramakrishnan and Toky, 1981; Jordan, 1987b).

In a comprehensive study of the impacts of slash and burn on a Costa Rican humid tropical forest Ewel et al. (1981) examined the behaviour of nitrogen, phosphorus, potassium, calcium, magnesium, sulphur and carbon, as well as burn temperatures, seed banks and mycorrhizae levels in both the vegetation and soil. Of particular interest are the proportions of the various elements in the different parts of the ecosystems and their pathways during the study. Phosphorus, calcium and magnesium were conserved during the burn, although small amounts of calcium and phosphorus had been lost during decomposition of the cut material. Wind blow, erosion and leaching losses were high (40–51 per cent of the three cations) leaving only 20–45 per cent of the amount of these elements in the pre-burn vegetation and soil at the start of cultivation (figure 8.16). Different behaviour was recorded for the volatile elements (carbon, nitrogen and sulphur) (figure 8.16). Large amounts of the above-ground stocks were lost in the burn (39–47 per cent), with relatively small amounts being lost due to decomposition (0–10 per cent) and wind blow, erosion and leaching (14–16 per cent); the amounts remaining ranged from 11 to 13 per cent (figure 8.16(b)). The behaviour of these elements was similar in the top 30 mm of the soil (figure 8.16(c)) but the effects of the burn on these nutrients were hardly felt below 30 mm. High potassium concentrations in the ash

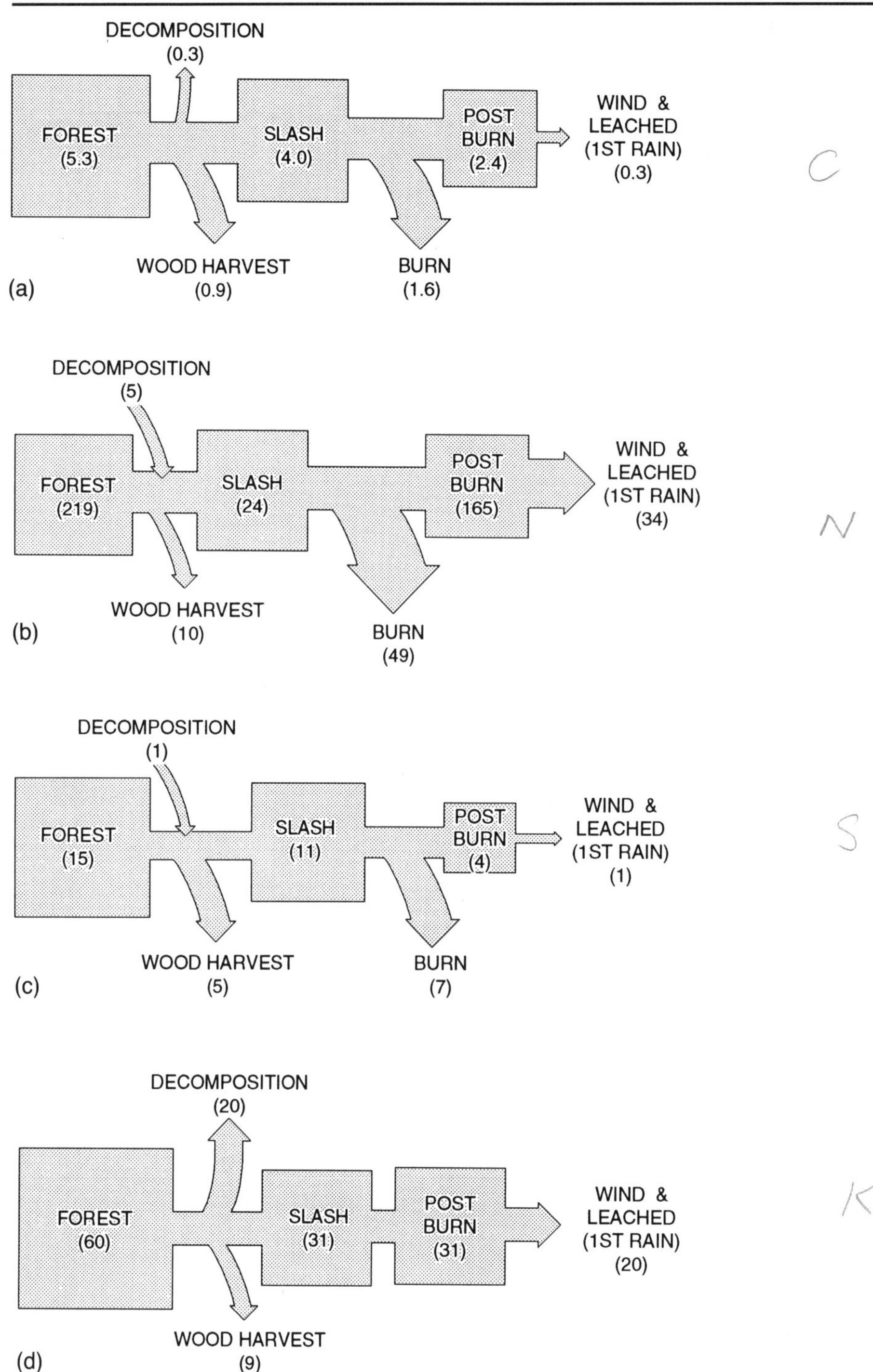

Figure 8.16
Summary of storages and losses of the four most mobile elements during wood harvest, decomposition, burning and post-burn erosion (from Ewel et al., 1981): (a) carbon (kg m^{-2}); (b) nitrogen (g m^{-2}); (c) sulphur (g m^{-2}); (d) potassium (g m^{-2}).

were noted (figure 8.16) and, unlike carbon, nitrogen and sulphur which are concentrated in the soil at all times, the maximum potassium is concentrated in the wood, fuel or ash. Seed banks were reduced to 76 per cent of their former level during the period between vegetation cutting and burning, and a further 39 per cent of the seed bank was lost during the burn; in total 63 per cent of seeds were lost. This type of analysis can be extended further and some workers (e.g. Nakano, 1978; Saldarriaga, 1987; Scott, 1987) have examined the behaviour of the major elements from the final harvest of a cultivation cycle through the fallow regrowth to the start of the next cultivation cycle (figure 8.17).

The amounts of soil nutrients available are not the only factors affecting their uptake by vegetation (Jordan, 1987b; Uhl, 1987). Many of the early regrowth trees and shrubs have extremely efficient nutrient uptake mechanisms (Vitousek, 1984) and after a few years a high proportion of the total nutrient pool has been immobilized in the plants and the soil nutrient levels are similar to those in undisturbed forest (Uhl and Jordan, 1984). In particular they appear to be able to mobilize phosphorus which is not available to crops (Saldarriaga, 1987).

In many tropical soils nutrients are highly concentrated in the topsoil. Consequently, soil erosion during the cultivation and early fallow phases can have particularly severe effects on the overall soil nutrient status and land productivity. This has been vividly illustrated in Tanzanian ferralsols by Moberg (1972) (figure 8.18), and the effects of simulated erosion of lixisols on maize and cowpea yields and root development in Nigeria was illustrated by Lal (1976) (figure 8.19). Cultivation periods longer than one year often show marked increases in surface runoff and soil erosion in subsequent years (Ramakrishnan and Toky, 1981; Hurni, 1982; Mishra and Ramakrishnan, 1983; Sato et al., 1984; Das and Maharjan, 1988). However, soil productivity must be seen in a more comprehensive manner than simply nutrient status; in particular the water retention capacity and root penetration ability of soils is important. For example, Arnason et al. (1982) noted that the erosion of finer silt and clay during cultivation phases in *milpa* cultivation in Belize reduced the water retention capacity in sandy-loams to the detriment of maize production.

8.10.2 Changes in the soil fauna

The decline in soil fauna (e.g. decomposer organisms which maintain soil fertility, and earthworms which ameliorate soil structure) is very important (plate 8.9) but is under-researched in the context of shifting cultivation (Aina, 1984; Barois et al., 1987; Lal, 1988; Lavelle, 1988; Lavelle and Pashanasi, 1989; Mulongoy and Bedoret, 1989; Lavelle and Fragoso, 1992). Nevertheless, the impacts of shifting cultivation on soil fauna can be great. Soils under cleared vegetation in Nigeria exhibited a 50–60 per cent

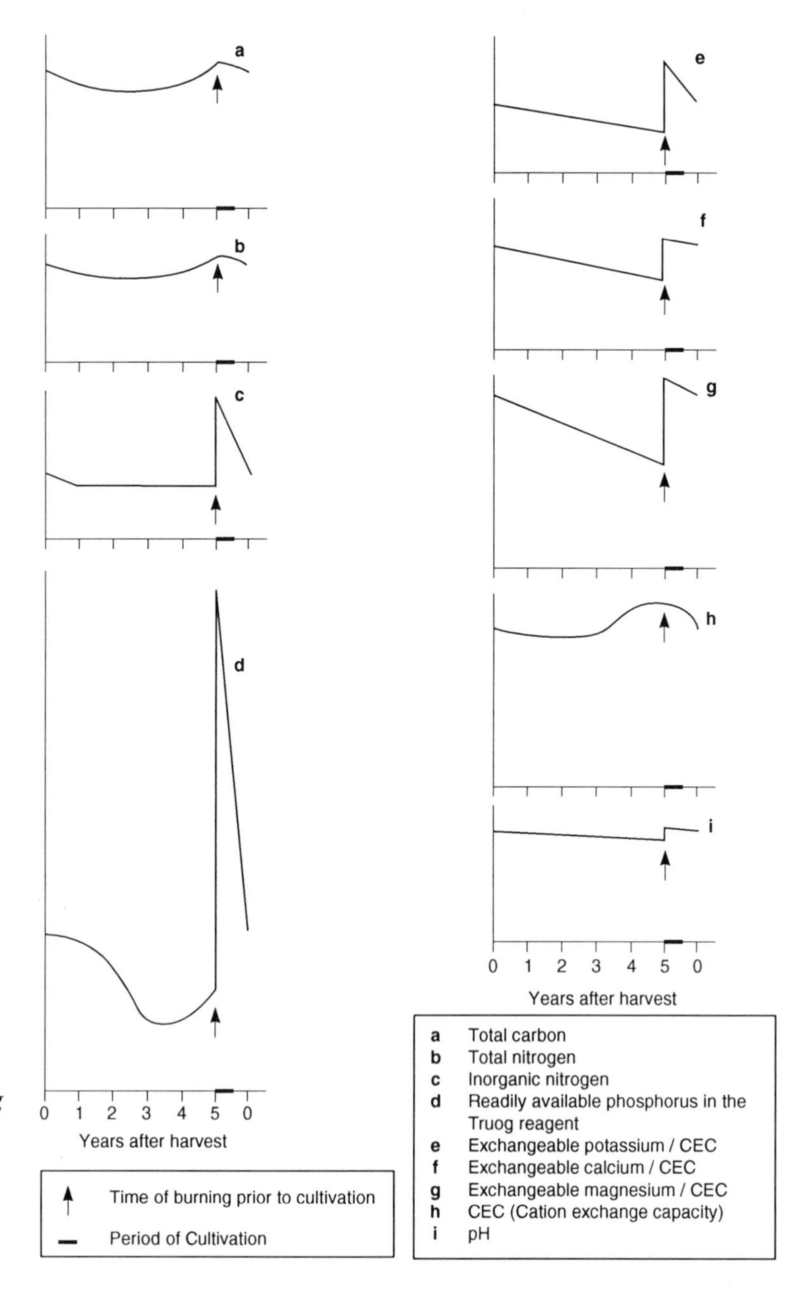

Figure 8.17
Quantitative trends in topsoil (0–5 cm) chemistry in a shifting cultivation system (adapted from Nakano, 1978, p. 434, figure 2).

decline in the main surface and subterranean faunal elements in the first growing season. These changes were attributable to changes in soil microclimate and the loss of niches at the soil surface. After the first year the plots were invaded by ants and spiders and a change in the earthworm species was noted (Critchley et al., 1979).

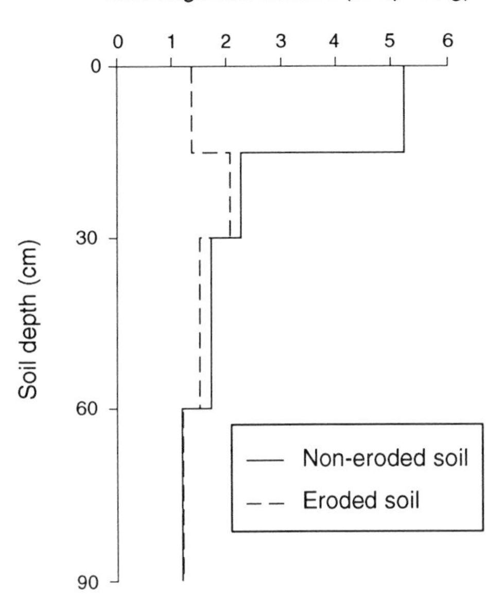

Figure 8.18
Variation in the exchangeable calcium level at different depths in eroded and non-eroded ferralsols from Tanzania (data from Moberg, 1972; Millington, 1992, p. 235, figure 14.2c).

8.10.3 Vegetation succession in shifting cultivation systems

Under 'normal length' fallow systems the cultivated fields are initially colonized by herbs. These are followed by ephemeral grasses which are replaced by fast-growing woody species, growing from either suckers or seeds of pioneer species (plate 8.8, figures 8.20 and 8.21). These species vary considerably; for example in central Luzon (Philippines) *Lithocarpus jordanae* and *Symplocos*

Plate 8.9
Evidence of the role of soil fauna in turning over humid tropical soils – a termite mound in Gambia.

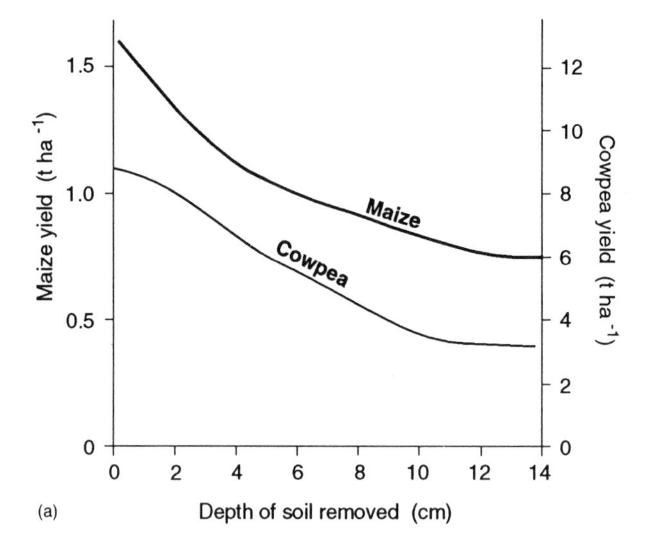

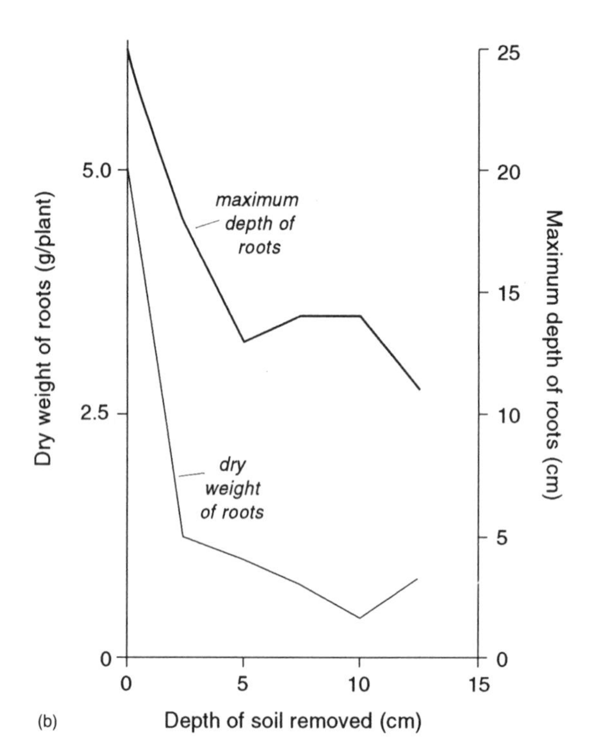

Figure 8.19
(a) Effects of soil removal on maize and cowpea yields on Nigerian lixisols; (b) influence of soil removal (simulated erosion) on root development in maize (adapted from Lal, 1976).

luzonensis sprout freely from felled stumps (Kowal and Kassam, 1978), whereas species of Macaranga are common in lowland successions in Papua New Guinea (Whitmore, 1975); severe burns, however, can reduce the sprouting ability of stumps (Jordan, 1987b). Pure stands of pioneer trees are not uncommon. If the fallow periods become too short, tree regrowth is replaced by shrubs and, if these cannot be sustained, grass fallows become

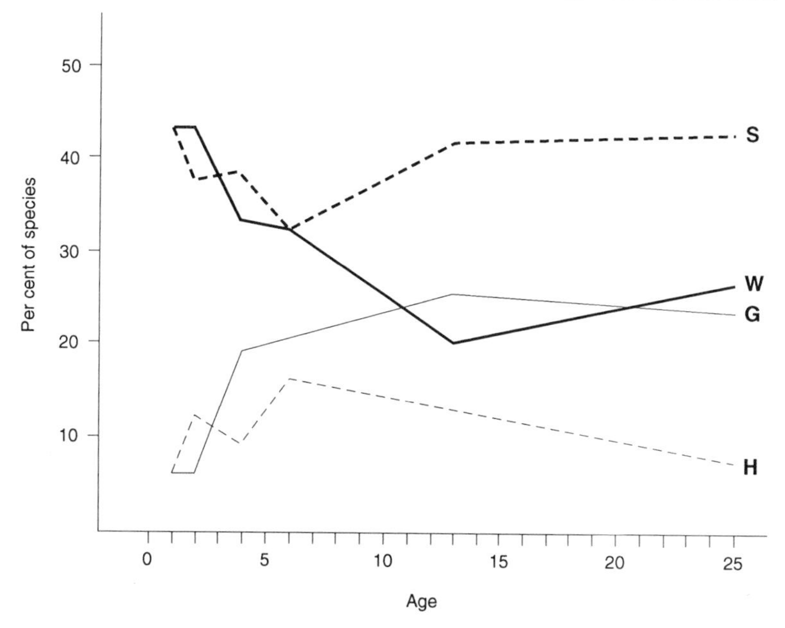

Figure 8.20
Percentage of species represented by arborescent woody plants (W), shrubs (S), herbs (H) and grasses (G) in fallows of different ages in Zambia (after Stromgaard, 1984, figure 3).

Figure 8.21
The transition from primary forest to cultivation to secondary forest, with biomass stock determinations from Peru. The biomass studies for primary forest are approximate and based on allometric relationships derived from trees smaller than those in the primary forest plot that was surveyed (after Scott, 1987, p. 41, figure 4.3).

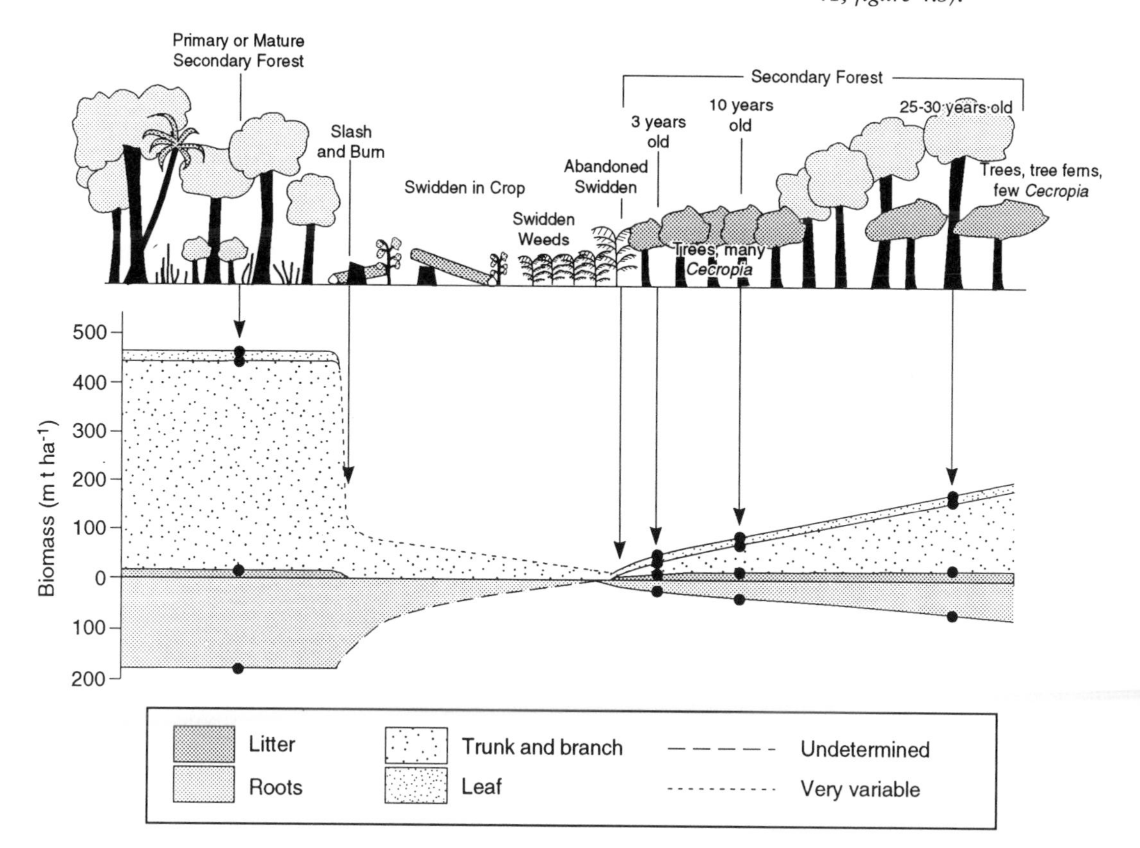

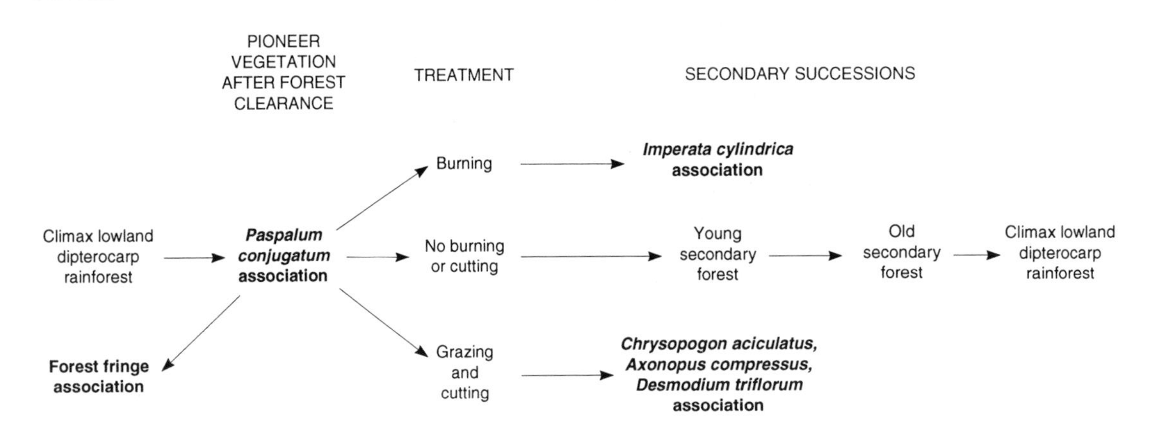

Figure 8.22
Summary of the four main grassland associations (shown in bold) derived from lowland dipterocarp rainforest by different management systems in Pahang, Malaya (after Whitmore, 1975).

dominant. Grass fallows are common in the monsoonal tropics where vegetation dries out and burning is an important management tool, but they also evolve through forest clearance followed by grazing (figure 8.22). In the Gran Pajonal of Peru Scott (1987) has recorded fallows that take fifteen years to form a reasonable secondary forest; however, in some areas where the fallows are constantly burnt, fire-resistant species take over and an *Andropogon*-dominated grass fallow evolves (figure 8.23). In some parts of southeast Asia it has been noted that a single year's farming on steep slopes can cause so much erosion that only scrub vegetation can recolonize the fields (Virgo and Ysselmuiden, 1979). Purata (1986) examined vegetation succession in forty abandoned fields in Mexico and found four distinct successional communities with different floristic and physiognomic characteristics. The particular successional pathways appeared to be related to the surrounding forest type, length of cultivation, regrowth age and the proportion

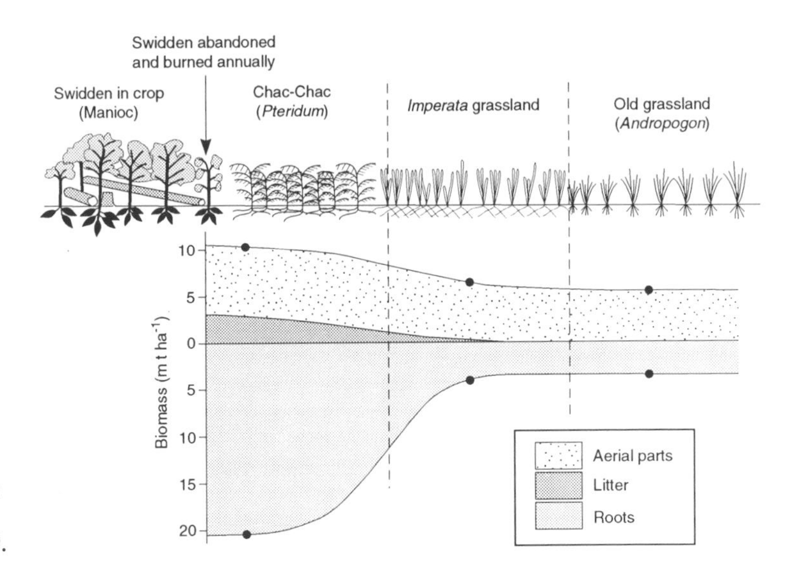

Figure 8.23
The transition from shifting cultivation to grassland following frequent burning in the Gran Pajonal, Peru (after Scott, 1987, p. 41, figure 4.3).

of the field perimeter forested. Uhl et al. (1981) noted that plant density in fallow regrowth increased markedly four months after burning because early colonizing plants began to set seed and continued seeding until the tenth month. Between the tenth and twenty-second month vegetation density was relatively stable but the overall stand height increased and a 5 m high canopy of *Cecropia* spp. formed. The earliest colonizers were grass and forb seeds blown onto the field, whereas the first woody species came from the seed bank that survived the burn. This regeneration of vegetation appeared to be dependent on the transfer of nutrients from the soil to vegetation (Scott, 1987).

NINE

Environmental Issues Facing the Humid Tropics

The environmental issues currently facing the humid tropics are wide ranging in their nature and extent. Many are acute and geographically specific; increasingly, however, our attention is being drawn to problems of a global scale. Most are the result of a combination of interlinked and interdependent factors and have no single or simple solution. This chapter focuses upon two of the most important global-scale environmental issues currently affecting the humid tropics – deforestation and climatic change.

9.1 Deforestation – Trends and Prospects

Reports in the popular press and on television and scientific studies all indicate that deforestation, and in particular the destruction of humid tropical forests, is an extensive problem with local, regional and global consequences. Some of these issues have been reported in other chapters. We do not wish to expand on these here, but we shall attempt to analyse the global nature of tropical deforestation in three ways: (i) to analyse the spatial patterns and temporal trends of tropical deforestation; (ii) to examine the arguments for tropical forest conservation; and (iii) to examine tropical forests as part of the global climate system.

9.2 Forest Areas and Deforestation Rates

Much has been written about the spatial variability and temporal fluctuations in tropical deforestation rates (e.g. Persson, 1974; Sommer, 1976; Myers, 1980; Lugo and Brown, 1982; Melillo et al., 1985). Useful analyses are provided by a number of authors (e.g. Mather, 1990; Whitmore, 1990). Most of the quantitative data

Table 9.1 Deforestation rates for closed tropical forest[a] in selected countries

	Deforestation rates (10^3 ha yr^{-1} (%))		
Country	*1981–5*	*Recent*	*Reference*
Tropical America			
Brazil[b]	1480 (0.4)	8000 (2.2)	Setzer et al. (1988)
Costa Rica	65 (4.0)	124 (7.6)	Sader and Joyce (1988)
Tropical Africa			
Cameroon[c]	80 (0.4)	100 (0.6)	JIM (1988)
Tropical Asia			
India[c]	147 (0.3)	1500 (4.1)	Vohra (1987)
Indonesia	600 (0.5)	900 (0.8)	World Bank (1988)
Mynamar	105 (0.3)	677 (2.1)	Kyaw (1987)
Philippines	92 (1.0)	143 (1.5)	PFMB (1988)
Thailand[d]	379 (2.4)	397 (2.5)	RTFD (1986)
Vietnam	65 (0.7)	173 (2.0)	Quy (1988)

[a] Closed tropical forests refer to forests with a high crown density and little grass at the forest floor.
[b] This calculation only refers to the 'Legal Amazon'.
[c] Calculated (by World Resources Institute, 1990) from total stock numbers for the appropriate references.
[d] Includes both open and closed forests.

Source: partly adapted from World Resources Institute, 1990

used by these scientists was based on publications from the 1970s and, until recently, the most authoritative estimates were those produced by FAO (Lanly, 1982). More recent studies suggest that deforestation rates are greater than previously thought (table 9.1). For example, the World Resources Institute (1990) calculated that approximately 20.4 million hectares of tropical forest is being lost each year, 79 per cent more than the 1982 FAO estimate. The annual forest loss equates to an area equivalent to the size of Panama: other such emotive and equally meaningless areal comparisons abound! This is partially because the rates are spatially very variable (table 9.2).

A fundamental difficulty in establishing the areas of forest and rates of forest loss relates to the *accuracy* of the data. Information sources range from outdated government statistics through to up-to-date maps derived from satellite imagery. Six factors in particular make estimates problematic and should be borne in mind when using these statistics:

Table 9.2 Mean annual rates of deforestation, 1981–5

Region	*Undisturbed productive closed forest*	*Logged productive closed forest*	*All closed forests*
Tropical America	0.29	2.80	0.64
Tropical Africa	0.19	2.41	0.61
Tropical Asia	0.39	2.14	0.60
All tropics	0.28	1.98	0.62

Values are percentages of the forest area in 1980.

Source: adapted from Lanly, 1982

1 variations in the definition of forest and 'forest land' between different countries;
2 variations in the definition of forest destruction, ranging from complete clearance at one end of the spectrum to selective felling at the other;
3 variations in the techniques used to undertake forest inventories, because they are often specific to the purpose of the inventory;
4 most inventories concentrate on commercial timber in forests and neglect many non-commercial trees, as well as other plants;
5 many estimates are for individual countries and may refer to all types of forest vegetation, not just humid tropical forests;
6 data are often withheld by governments and companies for strategic reasons.

Notwithstanding these difficulties, sections 9.2.1–9.2.3 provide a 'best available' indication of deforestation in the humid tropics in the early 1990s.

9.2.1 Forest cover and deforestation rates in Latin America

Humid tropical forest cover in Latin America extends from Mexico to northern Argentina, although the bulk of this is located in the Amazon Basin. The active fronts of deforestation are mainly around the fringes of the Amazon Basin; in forest outliers in Brazil, Paraguay and Venezuela; and in many parts of Central America, northwest South America and the West Indies (figure 9.1).

The political and technical problems surrounding estimates of forest cover are well illustrated by the Brazilian experience. In Brazil recent estimates of annual deforestation rates range from 1.7 to 8 million hectares. Setzer et al. (1988) used satellite imagery from 1987 and arrived at an estimate of 8 million hectares per year. Their study has been criticized however, since the rate of deforestation in 1987 was probably anomalously high because (i) it was the final year of tax credits for land clearance; (ii) the government was actively discussing land reforms aimed at unimproved land; and (iii) the deforested areas were identified by smoke plumes from forest fires. This is probably more extensive than the actual area cleared, but also does not identify the areas cleared without burning.

Subsequent estimates made by the same group after government measures to reduce deforestation had been introduced were lower, e.g. 4.8 million hectares in 1988 (Neto, 1989) and a further decline of 40–50 per cent on the 1988 figure in 1989 (World Resources Institute, 1990). Satellite imagery of this area obtained subsequently by European Union researchers suggests that the Brazilian legislation has been very effective and that the focus of deforestation has shifted to the Bolivian Amazon.

Figure 9.1
Percentage of land area of different nations occupied by forest; all forest types are included (from Soussan and Millington, 1992, p. 81, figure 5.1).

In terms of the area deforested, the spectrum of estimates is again wide. At the top end Fearnside (1990) projected that 35 million hectares of the Legal Amazon would be deforested by 1989 whilst at the other end the Brazilian Space Research Institute (INPE) estimated that the deforested area would only be 17 million hectares (Pereira da Cunha, 1989).

9.2.2 Forest cover and deforestation rates in Africa

The extent of humid tropical forest in continental Africa has recently been mapped by Millington et al. (1994). The most extensive areas of humid tropical forests are restricted to the Zaire Basin and parts of West and Central Africa (particularly Liberia, southern Ivory Coast, south and west Ghana, Cameroon and Equatorial Guinea) (figure 5.3). Little remains elsewhere in west Africa or on the Indian Ocean islands. The extent of humid tropical forest as well as wood growing stocks and sustainable yields for different parts of Africa are given in table 9.3.

Lanly (1982) and Westoby (1989) estimate that the mean annual deforestation rates for closed forests in tropical Africa between 1980 and 1985 were just under half a per cent of the forest area in 1980, a lower estimate than either tropical America or Asia. Deforestation rates in central Africa are generally assumed to be lower than those in Latin America or southeast Asia (e.g. Myers, 1980). However, the apparently lower rate may be partly due to the fact that there have been fewer studies of deforestation in central Africa compared with other humid tropical areas. Detailed studies of deforestation in west Africa (e.g. Millington, 1988) and Madagascar (Battistini and Verin, 1972; Bourdière, 1972) have shown that deforestation is as severe as in parts of Latin America and southeast Asia.

In this context, the *main* problem in Africa at the present time is the loss of savanna woodlands in east, west and southern Africa rather than humid tropical deforestation since some 11.5 million hectares were lost between 1980 and 1985. Nevertheless, the reduction in humid tropical forests areas in west Africa (Cameroon, Ghana, Guinea, Ivory Coast, Liberia and Nigeria) as well as Madagascar is a cause for concern (figure 9.1).

Table 9.3 Humid tropical forest areas, growing stocks and yields in Africa, excluding islands

Region[a]	*Area (km^2)*	*Growing stock (million tonnes)*	*Sustainable yield (10^3 t yr^{-1})*
West African Coast	64,611	1584	17,219
Central Africa	1,550,844	17,526	2,226,337
Eastern Africa	202,455	1557	153,265

[a] West African Coast: Guinea Bissau, Guinea, Sierra Leone, Liberia, Ivory Coast, Ghana, Togo, Benin, Nigeria.
Central Africa: Cameroon, Equatorial Guinea, Gabon, Congo, Zaire.
Eastern Africa: Uganda, Burundi, Rwanda, Kenya, Tanzania.

Source: after Millington et al., 1994

9.2.3 Forest cover and deforestation rates in Asia, Australasia and Oceania

Current estimates of forest cover in Asia (e.g. India, Indonesia, Mynamar and the Philippines) are based on interpretations of satellite images and are higher than those based on older (and less reliable) methods. The reasons for the disparities have provoked heated debates in many countries, particularly India. However, it is likely that human activities and natural factors such as forest fires (e.g. those in Borneo in 1982–3; see section 5.4.3) have contributed, at least in part, to the shortfall. It is interesting to note that estimates of deforestation in Thailand derived from 1985 satellite imagery showed only a slight increase over those derived from similar imagery in 1978 (table 9.1).

Two points can be made about deforestation in all three areas. First, deforestation rates are increasing in each area, and second, the most active areas of forest loss are patches in a mosaic of cleared and forested areas.

9.2.4 Mapping tropical forests and monitoring deforestation from space

The ability to map and monitor forest areas accurately and to provide estimates of deforestation from remotely sensed data is an important tool for foresters, environmentalists and planners (Myers, 1988). Nevertheless, complications do arise with the use of such data. For example, there is a trade-off between cost and spatial accuracy. Weather satellite data are most useful for mapping entire continents and large regions because of its coarse spatial resolution and daily acquisition (Millington et al., 1989, 1992; Cross, 1990), while earth-resource satellite data cover smaller areas less frequently but at a much finer spatial resolution. Other problems exist, e.g. obtaining cloud-free and smoke-free imagery in the humid tropics has hindered accurate forest mapping (Cross, 1991; Millington et al., 1992). However, the advent of operational satellites with microwave (radar) sensors (e.g. on the European ERS-1 and the Japanese Fuyo-1 satellites) provide cloud penetration capabilities and initial studies (Leysen et al., 1994) suggest that these data complement those from optical and near-infra red sensors (e.g. Landsat TM and NOAA AVHRR). Clearly whilst satellite imagery has overcome some of the problems of accurate forest area estimation it is far from a faultless tool.

9.3 Hydrological Consequences of Deforestation

The processes of logging and timber removal or the conversion of forest to other land uses such as pasture, arable or urban involves significant disruptions to nutrient cycles and water balances. The precise nature of these changes has been the focus of much scientific study and debate. Unfortunately, studies conducted at varying scales, using different methods of approach and directed at different research questions make spatial comparisons difficult (Anderson and Spencer, 1991) and, on occasions, rhetoric and semantic fuzziness has led to the misunderstanding and misinterpretation of important principles (Hamilton, 1985).

Runoff processes are invariably affected by vegetation removal and ground surface disturbance. For example, under a forest canopy soil infiltration rates are usually high and overland flow is rare; soil losses are therefore usually low (section 7.2.1). When surface cover is removed raindrops are allowed to impact directly on the soil, loosening fine soil particles and blocking soil pores. This results in reductions in infiltration rates and increases in overland flow, soil erodibility and soil loss. Wiersum (1985) describes increases in the soil loss following the experimental removal of trees and undergrowth in a Java rainforest. Sediment yields rose from 0.03 kg m^{-2} under undisturbed forest to 0.08 kg m^{-2} with intact litter cover and 1.59 kg m^{-2} with the removal of all surface cover. The greatest sediment loads occurred with intact canopy but no litter or undergrowth, because of the large drop size and splash impact of drops falling from the canopy. For Maraca island, Nortcliff et al. (1990) describe a runoff coefficient of 6 per cent in forested and partly cleared plots (with remaining litter) compared with 16 per cent in completely cleared plots.

The methods used to clear forest can have a dramatic effect on the level of soil loss. Compaction of the soil by vehicles and associated reductions in infiltration rates can extend deep into the soil profile (van der Weert, 1974) and persist for years after logging (Malmer, 1990). Damage by machines compounds the problems introduced by the removal of loss of litter cover and on sloping ground can lead to massive soil erosion during heavy rainfalls. Mechanical clearance methods which cut the trees above ground levels result in less erosion than methods which bulldozer the vegetation, exposing roots and bare soil. Soil losses are also often high along access routes to logged areas. Skid tracks and logging roads often become eroded into deep gullies which increase the length of ephemeral channels leading into the drainage network. Gilmore (1971) has reported a sixfold to twelvefold increase in sediment loss, mainly from skid tracks, in Australia. Anderson and Spencer (1991) describe erosion levels of 10.8–12.9 t ha^{-1} yr^{-1} and runoff of 149–189 m^3 ha^{-1} $month^{-1}$ for skid roads in Indonesia. These figures compare with little runoff or erosion under undisturbed forests.

Information on rates of nutrient loss in association with sediment removal is lacking but they are likely to be significant (Anderson and Spencer, 1991). The effects of forest clearance on water yield is also unclear. Oyebande (1988) provides a summary of the results of studies which have taken place and reports typical increases in maximum and mean annual stream flow of between 450 and

400 mm yr^{-1} on clearance. This relates to an increase in water yield of up to 6 mm yr^{-1} per percentage reduction in forest cover (above a 15 per cent threshold). The catchments included in Oyebande's (1988) review are generally between 3 and 35 ha, and Anderson and Spencer (1991) caution the scaling up of results to large catchments because of the significant effect that a small-scale event, such as a single large tree fall, can have on a small catchment.

With reductions in infiltration following soil degradation there may be less opportunity for water-table recharge leading to diminished dry season river flows (Bruijnzeel, 1986). Some studies, however, report increases in groundwater levels (Boughton, 1970). Logged catchments usually show greater stormflow volumes and higher and earlier peak flows than comparable undisturbed catchments (Douglas and Swank, 1975). Flooding may increase locally but the effect of changes in discharge often become insignificant as the water moves down the catchment and as scale increases.

Much of the information on the hydrological effects of forest clearance relates to small catchments within which monitoring, and in some circumstances forest disturbance, can easily be managed. Furthermore most studies occur over short time scales and therefore represent only a 'snapshot' of conditions. Most logged areas are converted to another use, such as grazing or agriculture, under which discharge extremes and sediment loss are likely to moderate.

9.4 Climatological Consequences of Deforestation

It has been recognized for some time (SMIC, 1971) that deforestation is responsible for important changes in heat balances, water budgets and ecological balances on a variety of scales. Deforestation affects a significant range of interrelated climatic elements at the earth's surface, namely aerodynamic roughness, albedo, net radiation/temperature and the energy balance transfer (i.e. the partitioning of net radiation into sensible and latent heat fluxes). Furthermore, these changes can modify atmospheric circulation systems either through a direct influence of *in situ* free convection or, in the case of latent heat energy conversions, at an appreciably different altitude, longitude and latitude from the source of evaporation.

The replacement of tropical forests by savannah grasslands has only a small effect on albedo, with decreases estimated from 0.24 to 0.18 in the wet season and from 0.18 to 0.14 in the dry season (Budyko, 1971). However, if desertification follows the deforestation, then the associated increase in surface albedo (to 0.28) can seriously disrupt the regional radiation balance. Surface temperatures decrease, following the increased solar radiation reflection, which cools the atmosphere with increased anticyclonic subsidence and reduced free convection, leading to reduced precipitation. The associated decreases in vegetation cover, as the rains fail, further increases the surface albedo and, through positive feedback loops, accentuates the anticyclogenesis and drought conditions (Charney, 1975; Kemp, 1990).

Apart from distinctive radiation balance changes, the individual energy balance fluxes change dramatically with desertification following the replacement of large latent heat sources (LE) by even larger sensible heat transfers (H). For example, tropical forest LE fluxes are more than three times the H fluxes (Bowen ratio 0.33) compared with semi-desert H fluxes which are about five times the LE fluxes (Bowen ratio 5.6).

The above energy balance changes will influence the development of sub-synoptic scale disturbances over tropical land masses. The sensible heat dominance will intensify surface heating and the potential for free convection. However, the diminished latent heat fluxes will lessen the water vapour content of the air and cause a reduction in the potential for the latent heat of condensation developing in free convective systems. The weakening of this heat conversion source (through potential and kinetic energy transfers) will affect the intensity of the ITCZ and the breakdown of the trade wind inversion. The repercussions for synoptic-scale disturbances (section 2.4) are less clear but SMIC (1971) suggested that the replacement of latent heat sources might have a significant effect on the generation and dissipation of easterly waves (section 2.4.1). Newell (1971) thought that the replacement might even affect the dynamics of the general circulation through a series of non-linear interactions.

The connection between tropical deforestation and enhanced anticyclogenesis/desertification is also emphasized by the associated dust-loading changes (SMIC, 1971). Deforestation leads to increased runoff, decreased soil moisture storage and increased soil desiccation. Periodic dust storms concentrate particulate matter in the troposphere which alters the regional radiation balance. The increasing dust-veil effect reduces incoming solar radiation and accentuates surface cooling. Contact-chilling in the lower troposphere increases anticyclonic subsidence and aridity. Desertification is accentuated: for example, Bryson (1971) argued that the Rajputana Desert in western India has developed in this way following deforestation several thousand years ago (and the elimination of a freshwater lake). 'Bryson's studies remind us that numerical simulations of man-made changes in the tropics must include the effects on the tropospheric radiation balance induced by windblown dust' (SMIC, 1971). Some twenty years later, we still await this inclusion and the quantitative confirmation of this important relationship.

9.5 Causes of Deforestation

Lowland tropical forests represent one of the final terrestrial frontiers open to exploitation by humankind, and on many parts of the globe rainforests are rapidly being converted to other land uses or depleted of some of their most valuable assets. Rolling back the tropical forest frontier may have analogies with the loss of Mediterranean forests in the Greek and Roman eras, the loss of temperate European woodlands over a long period and the rapid exploitation of North American forests in the nineteenth century. Appealing though these analogies may be, commentators such as Mather (1990) caution, quite rightly, against straightforward comparisons. In this section and those following we shall examine the global aspects of tropical deforestation and then look at three case studies to see if a general model of tropical deforestation exists.

The countries with the most rapid rates of humid tropical deforestation are difficult to categorize since they vary considerably in demographic and economic factors: there is no 'typical' country profile. This is because there are many reasons for the conversion of humid tropical forest; however, six direct causes are clearly identifiable (though their grouping and ranking varies according to different authorities): (i) conversion to permanent agricultural land (arable cropland, tree crop plantations, forest plantations and pastureland); (ii) timber extraction; (iii) shifting cultivation, (iv) infrastructure development, e.g. roads, towns, reservoirs and mines; (v) charcoal production for iron ore smelting; (vi) areas where local resource use (e.g. for fodder, fuelwood, fruits and other forest products) begins to exceed the supply.

In almost every case it is impossible to ascribe one cause of deforestation because, as will be shown in the following case studies, the primary causes interact and the dominant causes vary with time. Nevertheless, there are some general trends which are currently in force. In Latin America, much of the current deforestation can be attributed to land clearance for pastoral and arable agriculture, charcoal production (in Brazil) and clearance for roads, mines and reservoirs. The main causes in Africa appear to be logging (although this is decreasing in importance) and the demand for land for shifting and plantation agriculture. In southeast Asia closed forest destruction has mainly been caused by land clearance for plantations, agricultural land in planned migration and resettlement projects, and logging. A trend common in both parts of Latin America (especially western and southern Amazonia) and southeast Asia (especially Kalimantan and the Philippines) is the, often illegal, invasion of forest along access roads provided for logging and oil exploration. Such broad generalizations hide a complex pattern which can only be unravelled by referring to individual countries in most instances.

9.5.1 Sierra Leone

In the 1500s most of Sierra Leone was covered by large areas of monsoonal humid tropical forest. Evidence provided by archaeologists and anthropologists suggests that population densities at this time were very low, and that most people practised slash-and-burn agriculture with (presumably) only a localized impact on the forest. Agricultural change was initiated after the first wave of European colonization in the mid-sixteenth century. Impacts on the forest initially remained low as scattered trading posts and mission posts were founded on the coast and along some of the major rivers (figure 9.2).

Deforestation began in earnest in the nineteenth century with the demand for ship timbers from the British Navy which was based on the Freetown Peninsula to control the slave trade. Timber extraction began in 1816, with loggers working along lines of penetration afforded by the major rivers. By 1860, however, a lack of adequate wood supplies meant that the timber trade had

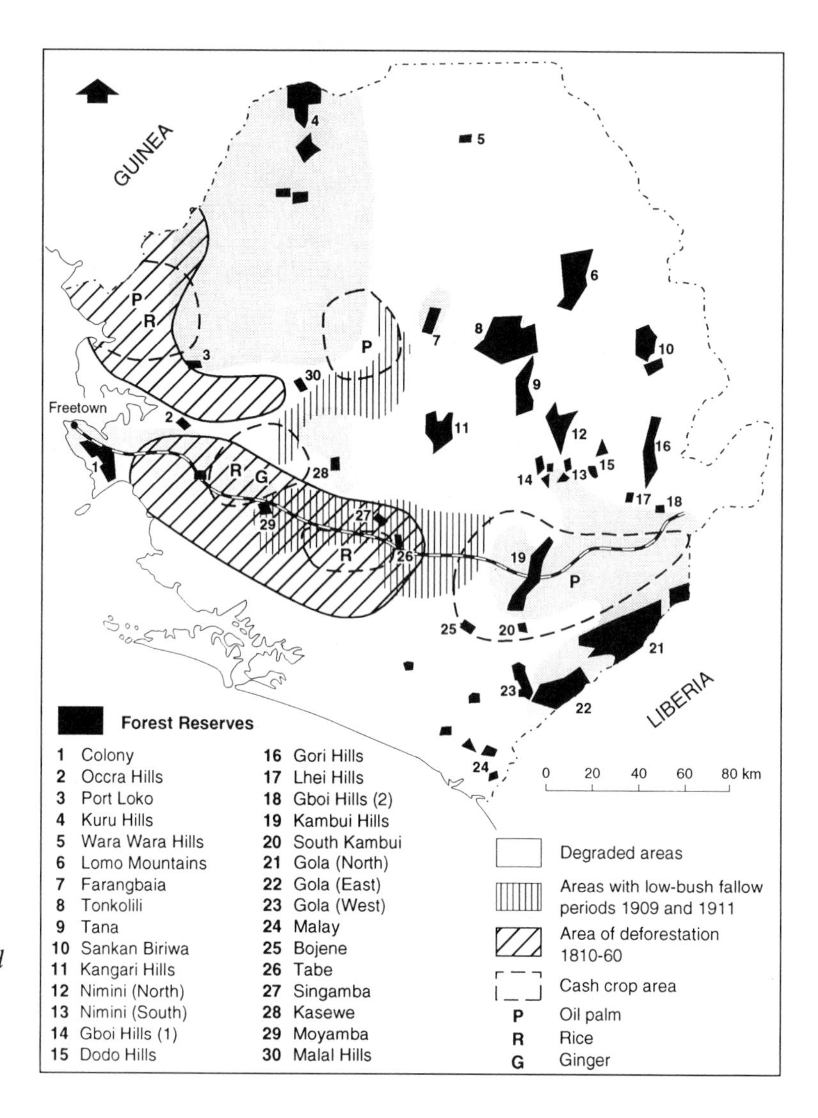

Figure 9.2
Sierra Leone: deforestation and cash cropping, 1910–11, and forest reserves created since 1911 (from Millington, 1988, p. 230, figure 11.1).

dwindled in volume (Cole, 1968) (figure 9.2). Simultaneously in Europe there was a rising demand for the kernel oil from the oil palm (*Elaeis guineensis*), a fire-resistant tree native to west African forests. Consequently, after logging, the cleared land was colonized by an influx of subsistence farmers, who found that the oil palm thrived in both their upland rice fields and the fallow regrowth (plate 9.1), and provided an easily marketable cash crop at the many European trading posts.

The potential of Sierra Leone to produce other cash crops, particularly cocoa, coffee and wild rubber, was recognized by the colonial administrators in the late nineteenth century. A railway

Plate 9.1
*Isolated fire-tolerant oil palm trees (*Elaeis guineensis*) are commonly found in shifting cultivation systms in West Africa.*

Plate 9.2
The Sierra Leone Railway penetrated into the interior of the country at the turn of this centry. Trading houses of European companies such as Paterson Zochonis and CFAO were established adjacent to the stations along the line to buy and sell imported goods. Two former trading houses can be seen behind the now abandoned platforms at Segbwema in eastern Sierra Leone.

line was built into the interior at the turn of the century to tap this potential. Forests were converted to agricultural land along the railway line (figure 9.2, plate 9.2), and land conversion was so rapid in the first two decades of the twentieth century that reservation of the remaining forests became an important issue (Millington, 1988). The first forest reserves were created in 1913 (figure 9.2). Nonetheless, forest exploitation has continued throughout this century, mainly in response to the demand for agricultural land and, to a lesser extent, timber extraction and diamond mining. The agricultural policies which have reduced the forest area in the last seventy years have included the clearance of freshwater and mangrove swamp forests for rice cultivation and the settlement of nomadic Fulani pastoralists in the north of the country. The conversion of forest to agricultural land, along with some (un)controlled logging in eastern Sierra Leone, now constitutes the main national causes of deforestation. Fuelwood extraction, whilst found throughout the country, is only a localized problem around the large towns, particularly Freetown (Inglis, 1988).

In Sierra Leone deforestation is not a recent phenomenon; it has a long history spanning at least 250 years (Cole, 1968; Millington, 1988). Moreover, during this time deforestation rates, though not known precisely, clearly have fluctuated according to the demands of both local and international trade. Two further points can be made concerning the causes of deforestation. First, land originally logged was subsequently colonized by farmers, a phenomenon that is happening currently in parts of Malaysia and Indonesia (see section 9.3.3). Second, forest destruction during the last seventy years has been stimulated by a variety of causes, e.g. agricultural policies, timber extraction, fuelwood exploitation.

9.5.2 The Brazilian Amazon

Forest clearance in the Brazilian Amazon is a comparatively recent phenomenon compared with that experienced in Sierra Leone. Up to the end of the nineteenth century, the Amazon was exploited, almost exclusively, at low levels of intensity by Amerindians. Their patterns of resource exploitation were mainly controlled by resource availability; river valleys and wetlands were used most intensively, whilst the *terra firme* was cultivated by modest slash-and-burn agriculture (Eden, 1989). Early European colonization followed a similar pattern of resource exploitation, with little attempt at commercial agriculture. For example, between 1839 and 1912 the main commercial activity, rubber tapping, used only 'wild' forest trees and did little, if anything, to promote forest clearance.

Serious forest destruction followed the establishment of twenty agricultural colonies between Belém and Bragança (the *Zona Bragantina*) in the early part of the twentieth century (Peneteado, 1967; Sioli, 1973) (figures 9.3 and 9.4). Settlers from France and Spain and *nordestinos* (migrants from drought-prone, northeast Brazil) were given plots of land in this area from the late nineteenth century onwards to grow food for the burgeoning population of Belém. The farming system was based on forest clearance and burning, but crop yields fell rapidly due to declining soil fertility and weed invasion (see section 8.10.1). The settlers, unable to produce enough food on the plots allocated to them, cleared more forest, but the result was the same: the productivity of the newly cleared land declined and the settlers moved on once more. This led to an unanticipated expansion in the area of land cleared, which was exacerbated by the timber and fuel needs of the railway built to serve the region. By the late 1920s, the entire area had been reduced to a mosaic of cultivation and forest regrowth.

From the late 1920s to the mid-1950s, forest destruction was again muted (figure 9.4). Most activity focused on the establishment of rubber plantations at Fordlandia and Belterra by the Ford Motor Company in 1927 and 1934 respectively, and by Pirelli and Goodyear in the 1950s. The success of these plantations, however, was limited by soil erosion and weed invasion.

Currently parts of the Brazilian Amazon (most notably Mato Grosso, Rondônia, Acre and southern Pará) are reeling under some of the most rapid deforestation rates in the world (table 9.1). The limited extent of deforestation until the 1950s and the high deforestation rates currently prevailing in the region point to the fact that much of the forest destruction in the Brazilian Amazon is of recent origin (figure 9.4). There are five main causes of the current wave of forest destruction in the region (figure 9.3).

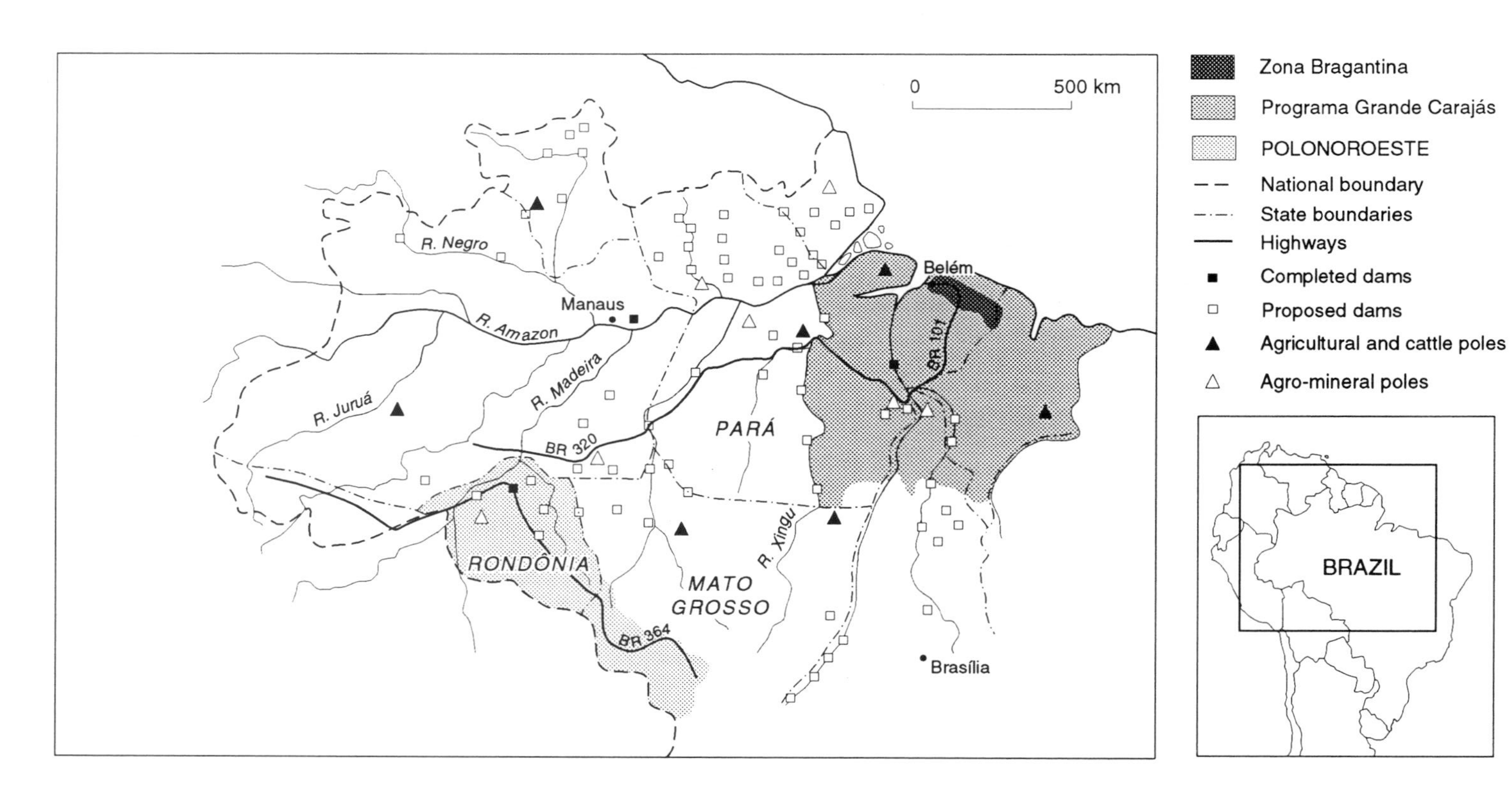

Figure 9.3
The Brazilian Amazon: economic development and deforestation (adapted after Soussan and Millington, 1992, p. 88, figure 5.2).

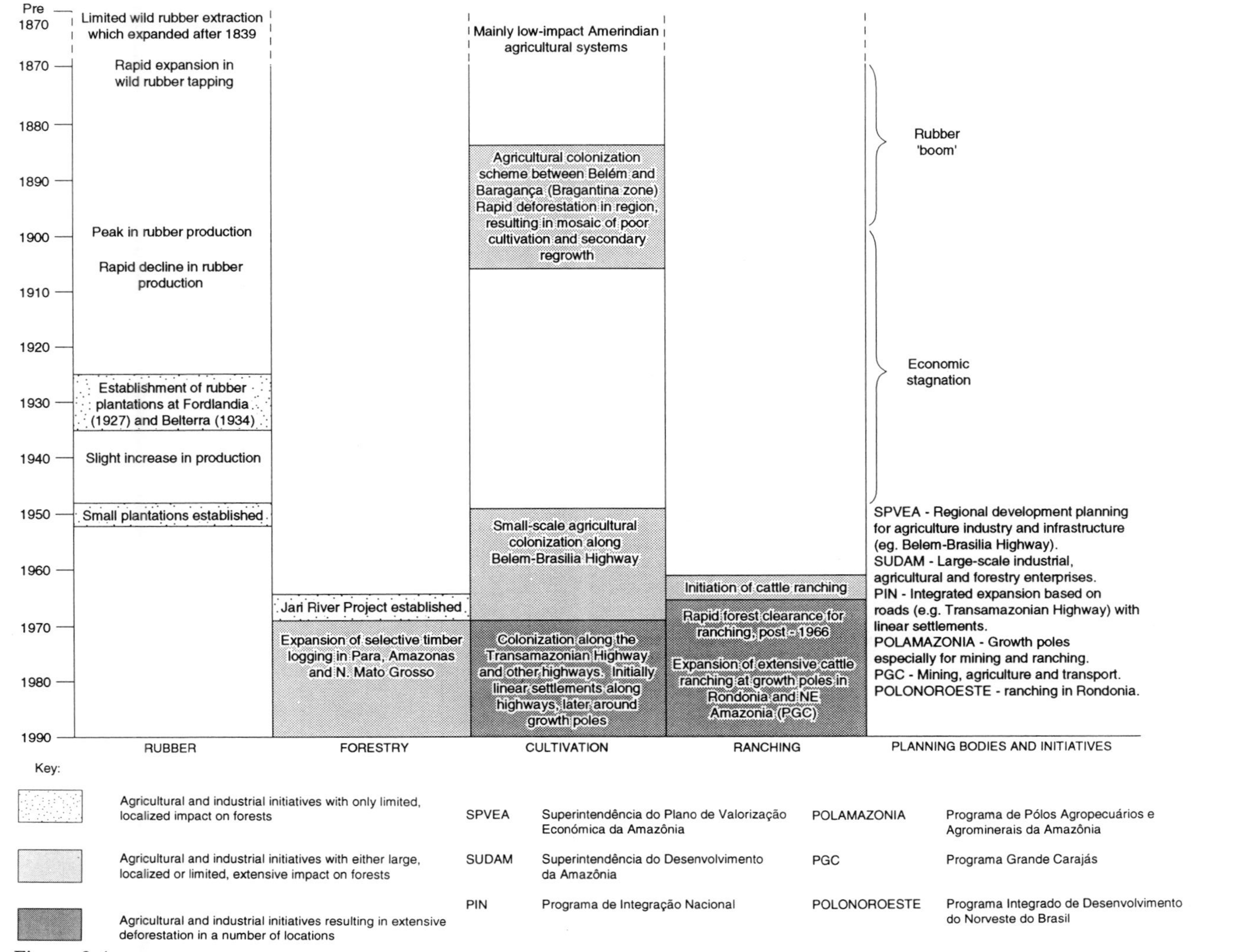

Figure 9.4
Chronology of economic activity and deforestation in the Brazilian Amazon, 1870–1990.

1 Agricultural colonization along highways and agricultural growth poles. This is fuelled partially by the influx of landless migrants and partly by land speculation. A worrying trend is that when the original colonists sell land deforestation rates increase markedly (Fearnside, 1990). Most of the land cleared is used for commercial crop production.
2 Conversion of forest to pastures for cattle ranching, again along highways and around cattle growth poles in eastern and southeastern Pará and northern Mato Grosso Provinces. Much of this clearance is fuelled by land speculation which may not be sustainable in the medium to long term (Fearnside, 1990).
3 Mining around planned agro-mineral poles, the most pertinent example being the Greater Carâjas Project in southeastern Amazonia, which includes the exploitation of enormous iron deposits, a 900 km long railway to the Atlantic seaboard and extensive deforestation (to be followed later by plantations) to produce charcoal to smelt the ore. A major concern is freelance gold mining by *garimpeiros*, which is causing localized deforestation throughout the Amazon and leads to the added environmental problem of mercury contamination downstream of the mined areas.
4 Large hydroelectric power schemes, the most notorious being the Tucuruí on the Rio Tocantins.
5 Forestry which is mainly restricted to Pará, Amazonas and northern Mato Grosso Provinces.

The exploitation of the Brazilian Amazon has partly been planned by the Brazilian government through a series of projects and initiatives aimed at economic development (figure 9.4). The aim of these schemes has been to encourage migration from the overcrowded cities and drought-ridden northeast and settlement in the Amazon: as Westoby (1989) puts it: 'the absurd Amazon dream of marrying the "men with no land" to the "land with no men". Studies of deforestation in Rondônia clearly show that extensive forested areas can rapidly be converted into other land uses, a process which Fearnside (1984) called explosive deforestation. In the early 1970s only scattered areas in the state were disturbed. However, road development along the World Bank-funded Cuiba-Porto Velho (BR364) Highway, and branches off it, followed by surges of migrants has led to an exponential acceleration in the area deforested (figure 9.5) (Fearnside, 1986; Malingreau and Tucker, 1987). A similar picture is emerging in Acre where development is occurring along the extension of BR364 and other newly constructed roads, although at the present time there is a greater proportion of forest remaining than in Rondônia. In Mato Grosso deforestation is mainly found along roads in areas where ranching is most concentrated (figure 9.3).

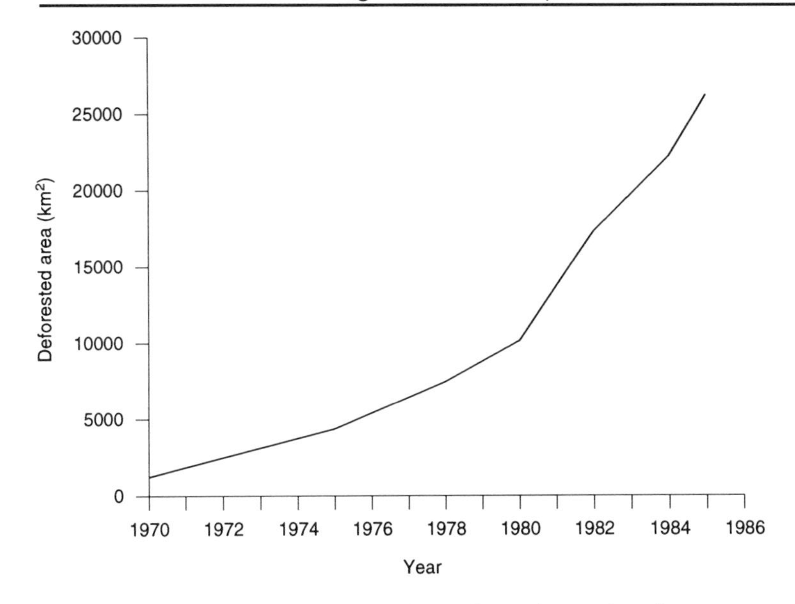

Figure 9.5
Deforestation in Rondonia (after Malingreau and Tucker, 1987; adapted from Soussan and Millington, p. 89, figure 5.4).

Fearnside (1990) has shown that there has also been near-exponential acceleration in deforestation in Amazonas, Mato Grosso, Pará and Roriama since the mid-1970s, and that the destruction of forest in these areas may well occur within a generation as has already happened with the drier forests in south–central states of Brazil.

In light of the rapacious exploitation of the Amazonian forests, international pressure on the Brazilian government became intense in the second half of 1988. This pressure was exacerbated by the shooting of hundreds of rubber tappers in Acre who were in conflict with cattle ranchers and, ultimately, their union leader Chico Mendes. Initial Brazilian government reaction drew condemnation of outside interference (*The Economist*, 11 March 1989). But later, more encouraging signs became evident. For example, the Amazonian nations have signed the Amazon Pact which recognizes the genetic and biological importance of the region and is aimed at sustainable forest use. A special commission on the Amazon environment has been formed. Furthermore, Brazil announced a conservation master plan for the Amazon and ended tax credits for land clearance; and the International Symposium on the Amazon in 1989 recommended international funding and research efforts to protect the Amazon. Nevertheless plans are still in hand for a highway from the Amazon to the Pacific to exploit new timber areas for Pacific Rim markets, and the Grand Carájas mining project (see above) threatens large areas of forest.

In the Brazilian Amazon, the main themes that emerge are as follows.

1 Deforestation is mainly a recent phenomenon, although there have been periods of rapid, localized deforestation in the past.

2 It has been promoted by government policies, not the least of which is the creation of new states within the Amazon Basin. Government policies aimed at controlling deforestation have generally been ineffective (Fearnside, 1990).
3 A number of causes at any one time are responsible for deforestation.
4 The deforestation process can be divided into two components: (i) the establishment of new foci of deforestation, and (ii) the expansion of areas in already existing foci.
5 Land speculation, which is tied in with the fact that land titles are given to those that clear forest, is a major driving force in Amazonian deforestation. The land speculation opportunities are many, and studies conducted by Maher (1979) and Hecht et al. (1988) have illustrated that profits from pure land speculation exceed those of agricultural production.

9.5.3 Peninsular Malaya and Borneo (Malaysia, Indonesia and Brunei)

Rapid forest exploitation in this region began about two centuries ago, initially stimulated by the colonial powers and subsequently by independent governments; the colonial impact on southeast Asian forests is summarized concisely by Whitmore (1990). Prior to this various forms of hunter-gathering and shifting cultivation existed in the region (and still does in a dwindling number of areas) (Whitmore, 1975). As in the example from Sierra Leone (section 9.5.1), the early colonists had little impact on the forests as they only traded in spices and other forest products (Aitken and Leigh, 1992).

However, the introduction of firearms undoubtedly caused an increase in hunting and the extermination of some animals and the decimation of other animal populations. Another colonial import, plantation agriculture, which is practised for a wide variety of crops, has replaced much forest. Moreover, weeds, often brought in from outside southeast Asia, have managed to get a foothold in the region through plantation agriculture (Whitmore, 1975).

Although the most rapid economic development in the low latitude tropics is occurring in this region, many countries are still, to a large extent, dependent on land resources for food crop production, and tree crops and timber for export. From the 1960s onwards the main causes of deforestation have been mining, forestry and shifting and permanent cultivation (Whitmore, 1975; Furtado, 1979; Brookfield and Byron, 1990; Aitken and Leigh, 1992). Strip mining, involving vegetation removal, washing soils to obtain ore and the accumulation of sterile tailings, has been a problem in Malaysia (Furtado, 1979). Secondary forest takes up to sixty years to establish itself on tin tailings (Palaniappan, 1974) and pollutants and silt washed down rivers are a problem throughout the region.

In Java, northern Sumatra and the west coast of peninsular Malaysia, significant amounts of deforestation has taken place prior to 1941. Contemporaneous deforestation is most rampant in the heartland of the Malay Peninsula and Borneo (Brookfield and Byron, 1990), and there appear to be three principal causes:

1 shifting cultivation, which is particularly important in Borneo;
2 the extension of permanent cultivation, often promoted by government settlement schemes;
3 logging, which has expanded enormously since the early 1960s.

Shifting cultivators probably clear more land each year in this region than loggers, but at low cultivation densities it hardly poses a problem. However, the extension of shifting cultivation along roads built to harvest timber has expanded the scope for cultivation. It is in these new areas that much damage is being done by immigrant farmers with little local environmental knowledge raising cash crops (Kartawinata and Vadya, 1984). In peninsular Malaysia, the clearance of forest for agricultural colonization has seriously depleted the remaining dry and swamp forests (Brookfield, 1991) (figure 9.6). The situation is better in Borneo where 75 per cent of Kalimantan is still classed as forest. Here there has been in-migration of approximately half a million people between 1971 and 1985 and the resulting settlement patterns are very uneven.

The proportion of the world's traded timber originating in this region has risen rapidly since the end of the Second World War. The region's timber resources first began to play an important role in world trade in the 1950s and, with the exception of hardwoods from swamp forests, their rise in importance has mirrored the demand from Japan, Taiwan and South Korea. Pringle (1975) calculated that whilst timber removals from all tropical forests increased by 80 per cent between 1953 and 1967, the increase from southeast Asia was 126 per cent. This dramatic rise is partially due to their proximity to the market, but the higher volume of merchantable timber per hectare in this area than is generally found in Africa and Latin America has been of more importance (Kumar, 1986). This is reflected in timber production statistics (figure 9.7) which show that the combined Indonesian, Malaysian and Brunei share of the world exports of non-coniferous tropical timber has risen from 17 per cent in 1965 to between 27 and 33 per cent since 1973 (figure 9.8); over the same time period the total world production doubled (Brookfield and Byron, 1990). Two further rather alarming facts are worth noting. First, since the mid-1970s over half of the world's hardwood timber exports have come from Borneo, and second, there was a very marked rise in timber exports in 1986 that appears to have been sustained in subsequent years.

Since the mid-1960s, there have been significant shifts in the timber products exported. World sawlog production has declined

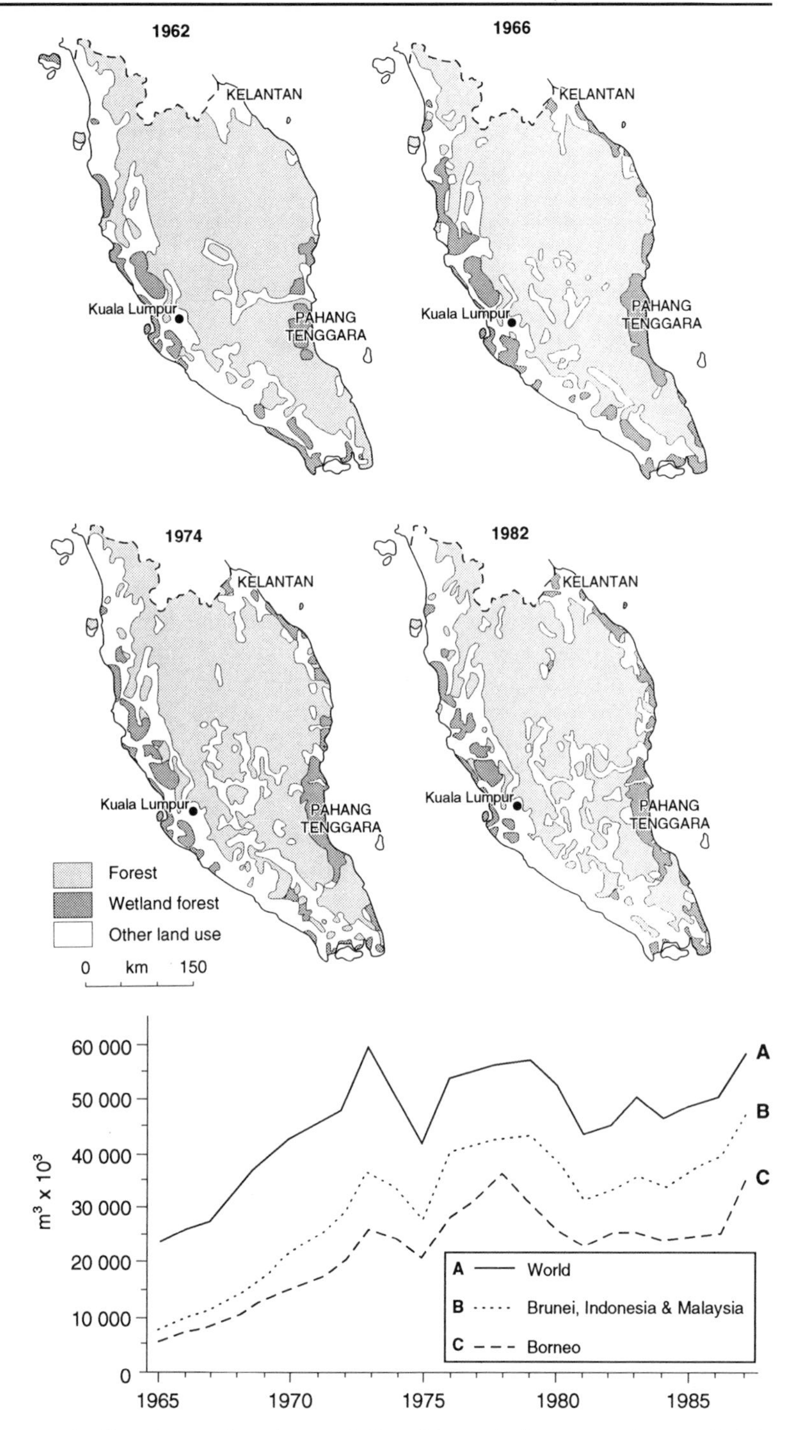

Figure 9.6
Forested areas on the Malay Peninsula at various dates between 1962 and 1982 (after Brookfield and Byron, 1990).

Figure 9.7
Export of all non-coniferous tropical hardwoods, using conversions to roundwood equivalents, in thousand cubic metres (from Brookfield and Byron, 1990).

markedly since the late 1970s. Nevertheless, southeast Asia's proportion of the world total has remained relatively constant as the main areas of sawlog exports have shifted from one part of the region to another. In the late 1980s most sawlog exports

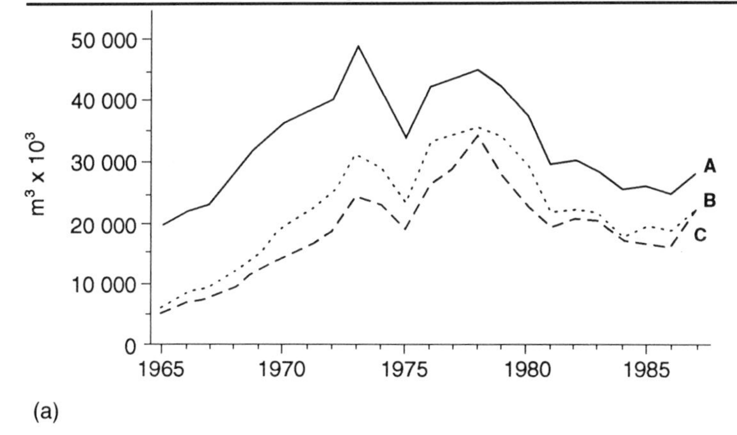

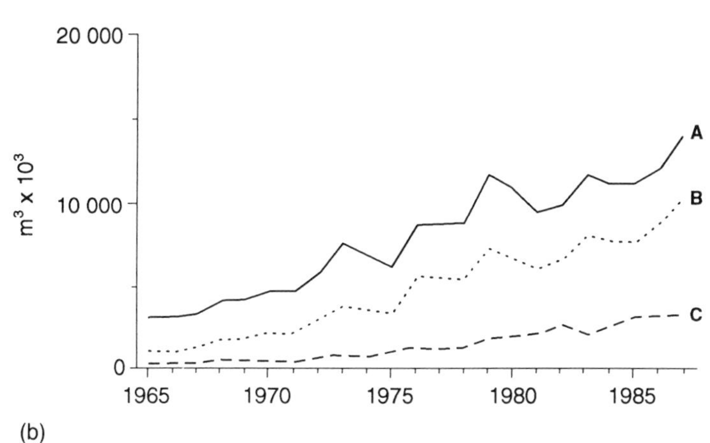

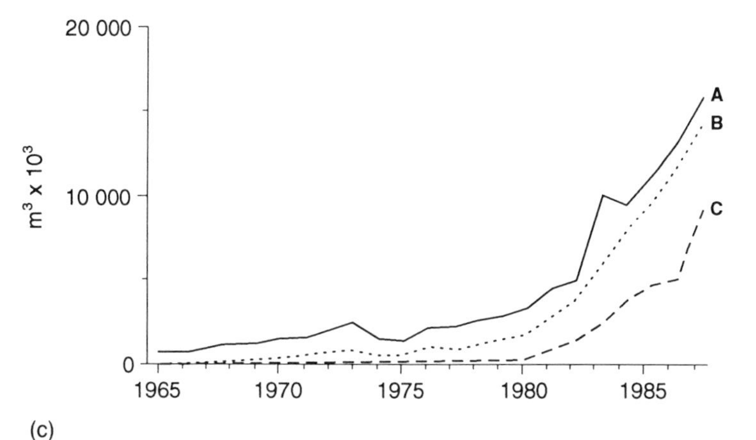

Figure 9.8
Southeast Asia: exports of tropical hardwoods between 1965 and 1987 (curve A, world; curve B, Brunei, Indonesia and Malaysia; curve C, Borneo (after Brookfield and Byron, 1990): (a) exports of non-coniferous tropical hardwood sawlogs; (b) exports of non-coniferous tropical hardwood sawn timber; (c) exports of non-coniferous tropical hardwood as plywood.

emanated from Sabah and Sarawak, but companies are already beginning to exploit the islands further east. The decline in sawlog exports has been brought about by government policies aimed at increasing domestic timber processing. This has resulted in increases in the sawn timber and plywood exports from the region (figure 9.8).

Whitmore (1990) suggests two reasons why timber production is much more important in southeast Asia than in either Africa or Latin America. First, despite their high diversity, the dipterocarps that dominate the region can be grouped into a few classes for sale. Such a grading system failed in West Africa. Second, unlike southeast Asian timber, that from Amazonia is dark, heavy and siliceous and does not meet the most important modern market requirements, i.e. for plywood, veneers and light construction.

It appears that, despite forest policies and research into afforestation, deforestation in southeast Asia is out of control. In Indonesian Borneo, cutting policies and concession areas are being violated, forest inspections are ineffective and there is illegal logging. The expansion of logging tracks into the forest has created a similar situation of explosive deforestation to that found in Rondônia. Timber demand in peninsular Malaysia probably is not being met locally. Furthermore, the timber mills established to process the region's timber and reduce sawlog exports have expanded beyond their sustainable limits and increased the rates of deforestation.

The study of deforestation in southeast Asia makes an interesting comparison to the case studies of Sierra Leone and the Brazilian Amazon. Although deforestation has a long history in some areas (e.g. peninsular Malaysia) in other areas it is a relatively new phenomenon. Although there appears to be a major cause of deforestation – timber production – there are also a host of other, often locally important, reasons. Moreover there is a synergistic relationship between timber extraction and the expansion of agricultural land.

9.5.4 Destruction of mangroves

Compared with the main types of humid tropical forests, for which much data on areal extent and deforestation rates exist, very little is known about the extent of mangrove forests. Despite the fact that mangrove communities are valuable ecosystems providing breeding grounds for many fish and crustacea and generally have a high resource potential, they appear to be being rapidly destroyed in many areas (figure 9.9; table 9.4). In addition to the causes of destruction listed in table 9.4 oil spillage and thermal water pollution also have detrimental effects on mangroves. Changes in global sea level due to global warming may represent an additional threat in the future (section 9.9).

9.6 Deforestation – A Summary

The three regional case studies of deforestation (Sierra, Leone, the Brazilian Amazon, and Borneo and the Malay Peninsula) can be usefully compared to analyse the past and present trends in deforestation.

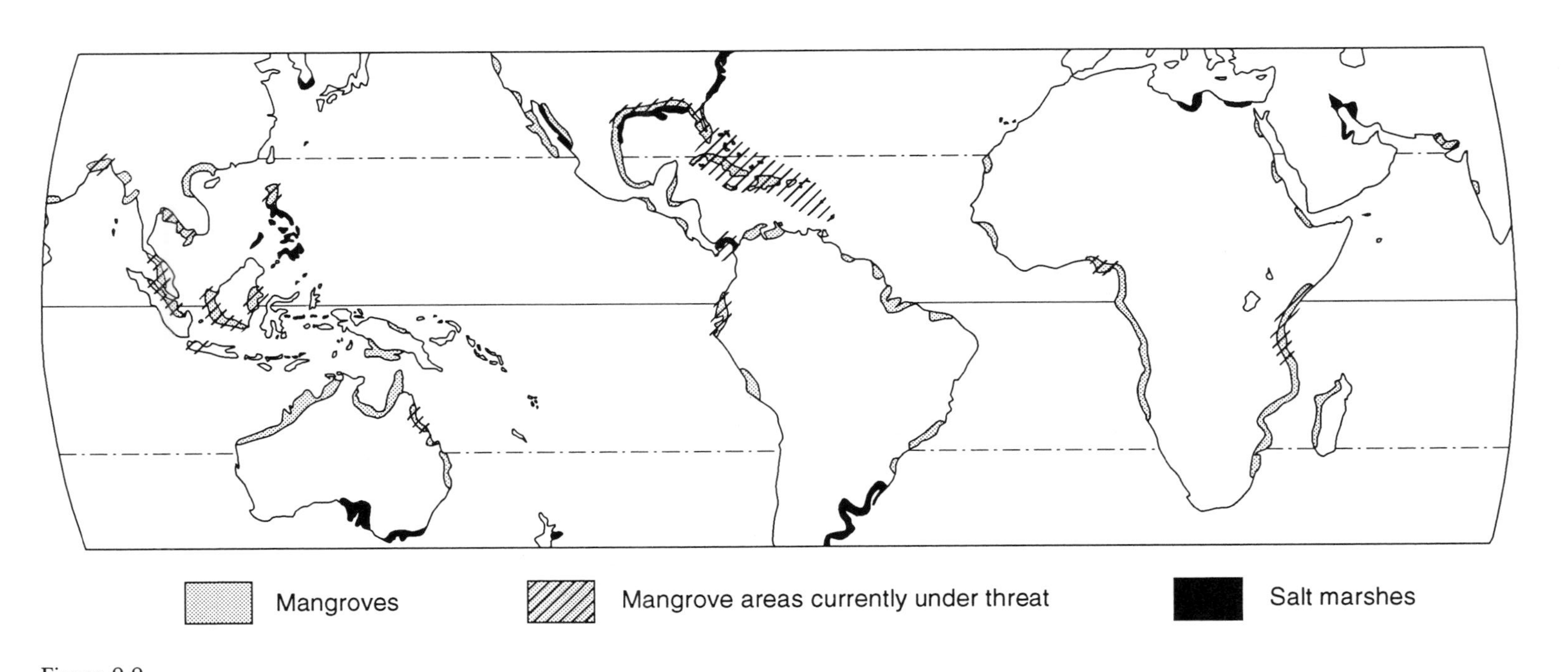

Figure 9.9
Areas of mangrove disturbance in the humid tropics.

Table 9.4 Main areas of mangrove destruction in the humid tropics

Area	*Causes*
Niger, Delta, Nigeria	Exploitation of timber, fuel and fodder Urban expansion
Kenya and Tanzania	Exploitation of timber and fuel Tourism development
Sundarbans, India and Bangladesh	Exploitation of timber, fuel and fodder Expansion of fishponds
Malaysia and Gulf of Thailand	Expansion of fishponds and shellfish ponds Exploitation for woodchips Agricultural land expansion, especially for coconuts in Thailand Alluvial mining and mine tailings
Philippines	Expansion of timber, fishponds, fuel and tannin Mine tailings
Indonesia	Exploitation of timber, woodchips (especially Sabah) and fuel Mining Expansion of fishponds and coconut plantations
Singapore	Expansion of prawn farming and mud crab fattening ponds Urban expansion
Queensland, Australia	Urban expansion Tourism development
Caribbean	Tourism development
Panama	Expansion of fishponds and shellfish ponds
Ecuador	Expansion of fishponds and shellfish ponds

Source: Lean et al., 1990; Fortes, 1988

In all three areas there is a history of deforestation dating back to the early parts of this century, and far further back in the case of Sierra Leone and southeast Asia. Deforestation then is not simply a recent phenomenon, and the history of deforestation represents an important avenue of study if we are to understand the causes and consequences of humid tropical deforestation. Notwithstanding this, deforestation rates are currently higher in many areas than they have been in the historical past, the exception in the three case studies being Sierra Leone. Furthermore, in a few areas in the Amazon and Indonesia deforestation is a relatively new phenomenon.

Sierra Leone is typical of areas that have experienced severe deforestation in the past and that are now areas of mixed secondary forest regrowth and agriculture. Similar areas are found in other parts of West Africa, East Africa, Madagascar, India and southeast Asia, i.e. areas with a long history of colonization. However, parts of Central America, Nigeria, Borneo, New Guinea, the Solomon Islands and Cambodia provide interesting comparisons, being areas where intensively farmed agricultural land has reverted to mature forest when civilizations have collapsed or the people have moved elsewhere (Whitmore, 1990).

Parts of the Brazilian Amazon and Borneo represent the active front of tropical forest destruction – areas where deforestation is occurring on a large scale and at high rates for the first time. Similar areas are found in other parts of Amazonia, Central America, central Africa and many of the Pacific Islands.

It is also apparent that deforestation rates show significant fluctuations over time, fluctuations which are a function of both local and international patterns of trade. Even in the Brazilian Amazon, where deforestation is mainly a recent phenomenon, there have been periods with high deforestation rates in the past.

It is clear from these three studies that there are many reasons for deforestation, and that these reasons vary both over time and from place to place, even within a region. For example, in southeast Asia, timber extraction is the primary cause of deforestation now, but in the past this has been less important in some areas than land clearance for agriculture and mining. There is a interesting synergism between the primary cause of forest destruction and agricultural expansion which is apparent in all three case studies. In Sierra Leone deforestation to provide the navy with timber was followed by agricultural land colonization in the 1800s. A similar pattern of deforestation and agricultural expansion is currently being experienced in parts of Kalimantan and Rondônia. History is clearly repeating itself, but the lessons of history do not appear to have been learnt.

9.7 Conservation of Plant Resources

Conservation of the plant resource base is one of the main weapons used in the battle against tropical deforestation. There are two strands to the argument:

1. plants which are only used locally but could have a wider market (the field of the ethnobotanist) (see for example the review by Prance, 1990);
2. the conservation of germplasm. This is often termed the conservation of biodiversity (Wilson, 1988) and, as such, its importance was highlighted at the United Nations Conference on Environment and Development held in Rio in 1992.

The conservation of germplasm is important because the products of bioengineering are being increasingly used to meet the demands of an expanding population for food from crops and animals, industrial crops, chemicals and drugs (Mannion, 1992). Vital to this is the conservation of germplasm – the genetic potential of living organisms – and genetic resources in a diversified form. At one time genetic diversity was generally assumed to be greatest in centres of diversity such as the Vavilov Centres of Diversity (figure 9.10). Taking this line of argument, in the Vavilov Centres all of the

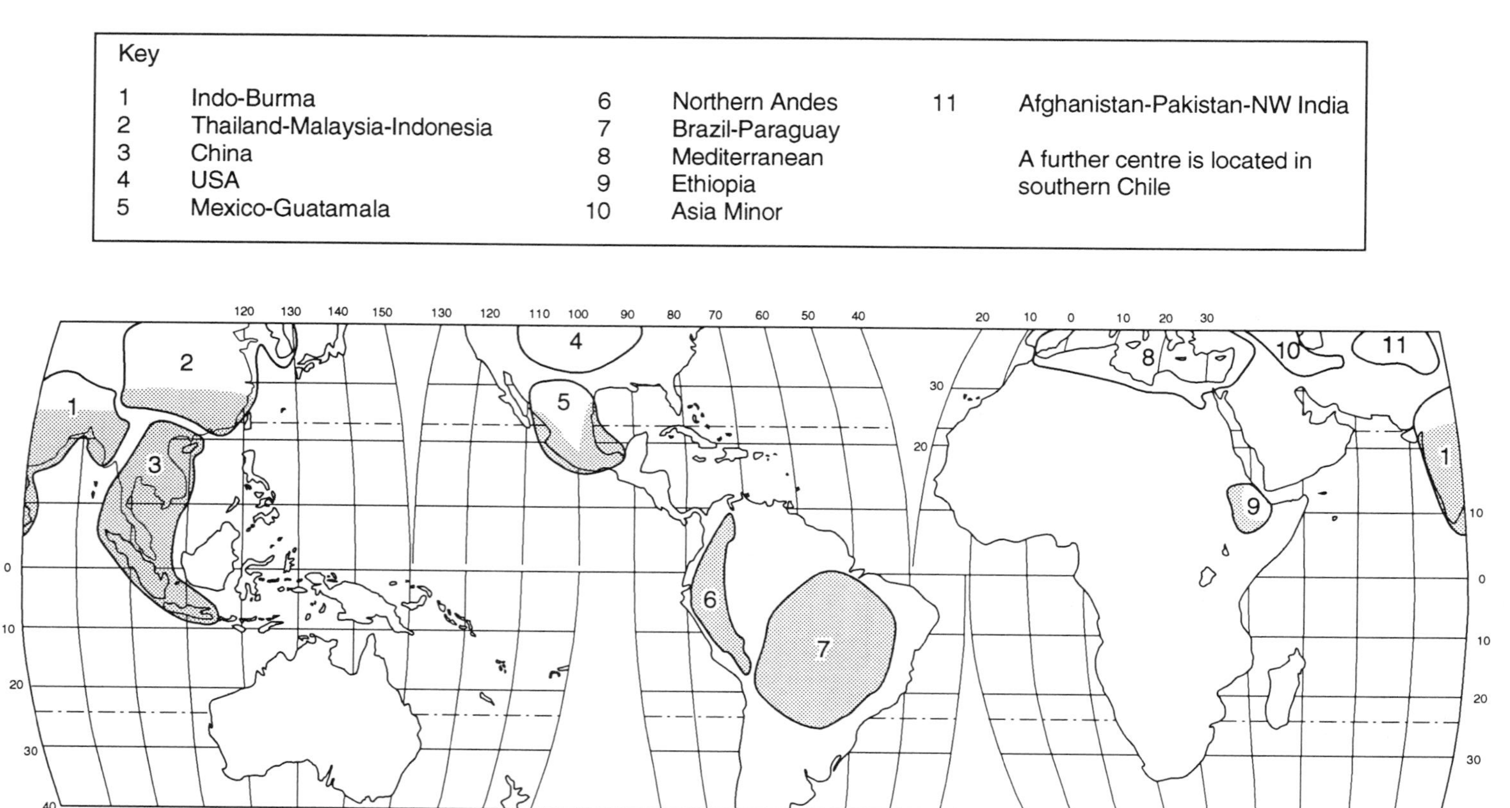

Figure 9.10
Vavilov Centres (only humid tropical centres are shaded): 1, Indo-Burma; 2, Thailand, Malaysia, Indonesia; 3, China; 4, USA; 5, Mexico, Guatamala; 6, northern Andes; 7, Brazil, Paraguay; 8, Mediterranean; 9, Ethiopia; 10, Asia Minor; 11, Afghanistan, Pakistan, northwest India; a further centre is located in southern Chile (based on various sources).

world's crops and animals have been domesticated and have diversified (table 9.5), and consequently these areas hold large genepools of the main crops. It should be pointed out, however, that, although other schemes for centres of origin and diversification exist, the basic argument with respect to germplasm conservation is the same. It is clear from recent ethnobotanical research that other areas are also important, e.g. Madagascar (Rasoanaivo, 1990).

A number of these centres are in the humid tropics and contain a variety of habitats threatened by processes such as the destruction of natural vegatation, forest fragmentation, agricultural expansion and changes in agricultural technology; once destroyed the germplasm in these is lost forever. The crux of the problem lies in the fact that crops evolve and diversify in response to environmental stimuli, e.g. pests, diseases and climatic changes, by genetic change. For this to happen requires a large genepool spread over both the domesticated and wild relatives of domesticated plants, particularly the ancient crop varieties which are still grown in some areas. But as the wild plants are destroyed and the ancient crops are replaced by new varieties the potential for genetic evolution declines.

Research carried out in the Peruvian Amazon by researchers from Missouri Botanical Gardens has shown that even within these areas there are large variations in biodiversity. Phillips (1993) has shown that edible fruit productivity is greater on alluvial soil forests than adjacent *terra firma* sandy-soil forests. Phillips also shows that the overall levels of fruit production can be much lower than reported elsewhere (e.g. Peters et al., 1989; Peters and Hammond, 1990) owing to species composition. The dominance of fruit-bearing palms which do not drop fruit, thereby making collection difficult,

Table 9.5 Major crops originating in the humid tropics and their associated Vavilov Centres of diversity

1	*Brazil and Paraguay* Brazil nut, cacao, cashew, cassava, groundnut, mate, para rubber and pineapple
2	*Northern Andes* (Bolivia, Ecuador and Peru) Cacao, edible roots and tubers (e.g. arracacha, oca and ullucu), papaya, quinine
3	*Central America* (Guatemala and Mexico) Cacao, cashew, guava, papaya, sapoldilla, sweet potato
4	*Ethiopia* Banana, coffee, okra
5	*India and Burma* Betel nut, betel pepper, cowpea, eggplant, hemp, jute, citrus, black pepper, rice, sugar cane, taro, yam
6	*Southeast Asia* (Indonesia, Malaysia and Thailand) Banana, betel palm, breadfruit, citrus, coconut, ginger, sugar cane, tung, yam
7	*China* Cowpea, orange, paper mulberry, soybean, sugar cane, tea

reduces the overall value of the forest if it is to be used solely for fruit collection. More worrying is the fact that forest fruit collection, and here we could expand the argument to the collection of other minor forest products (see for example the review by Whitmore, 1990), has a lower productivity than most other forms of agricultural production in the humid tropics in the short term. The way forward is to diversify and expand the use of these products (Prance, 1990) and to ensure that extractors receive high prices for their produce.

The genetic resource base is found in four situations: (i) domesticated animals and crops; (ii) wild relatives of domesticated animals and crops; (iii) wild species which are used by people; and (iv) wild species that are currently not used.

The second group, the wild relatives, have traditionally been seen as holding the greatest potential. For instance, genetic material from wild varieties of groundnuts from Amazonia were used to breed groundnuts which are resistant to leafspot; similar advances have been made for coconuts, oil palm, rubber and tomatoes (Frankel and Hawkes, 1981). However, plant breeders are now turning their attention to the wild species and the genetic resources of humid tropical forest are being promoted as a reason to save their further destruction (Wetterberg et al., 1976; De Vos, 1977). In this context it is important to note that almost all researchers agree that the best way to conserve the genepool is to reserve large areas of forest in which there is a minimum number of breeding individuals (plants or animals). Other methods such as conservation outside the forest (e.g. in botanical gardens, zoos or as parts of organisms in seed banks) are far less satisfactory (Whitmore, 1990). The research agenda in this area is clearly laid out and a detailed discussion is not possible here. The initiative of the International Board for Plant Genetics Research in grading crops into priorities for germplasm collection is one such step being taken (IBPGR, 1981) (table 9.6).

9.8 Deforestation – Ways Forward

The widespread global concerns over tropical deforestation have resulted in public pressure on governments and policy makers to tackle the issues that lead to this unacceptable (at least in the eyes of many people in developed countries) rape of the forest. Such pressure has resulted in a number of recent initiatives to halt deforestation. We can do little more than draw the reader's attention to these because a detailed discussion falls outside the remit of this book and many are, as yet, unproven.

Following the categories established by Whitmore (1990), the response can be broken down into four: (i) aid agencies, (ii) non-governmental organizations (NGOs), (iii) national agencies, and (iv) the International Timber Trade Organisation (ITTO).

Table 9.6 Germplasm collection priorities for humid tropical food crops

Priority	*Americas*	*Africa*	*Asia*	*Pacific Basin*
First	Cassava Maize *Phaseolus* beans Sweet potato Tree fruits and nuts Vegetables	Cassava Coffee Finger millet Rice (*O. glaberrina*) Sweet potato Vegetables Starchy banana and plantain	Coconut Maize Finger millet Rice Sugarcane Sweet potato Vegetables	Coconut Breadfruit Sugarcane Starchy banana and plantains Taro Yam
Second	Cocoa Groundnut Oil palm Cowpea	Banana Bambara groundnut Groundnut Soybean	Banana Cassava Chickpea	*Vinca* spp. Winged bean

Third and fourth priority crops are not listed here

Source: data from IBPGR, 1981

In 1985 a number of aid agencies grouped together to formulate the Tropical Forestry Action Plan (TFAP) as a way of co-ordinating activities; priority areas are listed in table 9.7. Though the international aid in this sector doubled between 1984 and 1987 in the light of the TFAP, it has been criticized for focusing too strongly on commercial forestry considerations (Caufield, 1987) and not enough on the requirements of conservation or the indigenous forest dweller (McDermott, 1988; Westoby, 1989). Shiva (1987) viewing it as a device to exacerbate forest destruction suggested it be renamed 'The Action Plan for Tropical Forest Destruction'. Most aid agencies now require environmental impact analyses to be carried out before development, although the strength of negative results from environmental impact assessments may be weakened in the face of ready finance (Elliott, 1988). An encouraging trend is the increasing recognition by aid agencies of

Table 9.7 Tropical Forestry Action Plans: priority areas

FAO Committee on Forest Development in the Tropics
Forestry in land use
Forest-based industrial development
Fuelwood and energy
Conservation of tropical forestry ecosystems
Institutions (administration, research, training, extension)
WRI Task Force
Rehabilitation of upland watersheds and semi-arid lowlands
Forest management for industrial use
Fuelwood and agroforestry
Conservation of forest ecosystems

Source: after Mather, 1990

indigenous resource management systems in the humid tropics as the building blocks for sustainable forestry and agroforestry, thereby providing avenues for development.

The World Conservation Strategy (IUCN, 1980) is in many respects the NGOs' agenda for rain forest conservation. The main activities in the conservation sphere have been the following:

1 the production of IUCN Red Data Books on endangered species;
2 Debt-for-Nature swaps where NGOs purchase part of national debts at highly discounted rates in return for forest areas being conserved (the first swap was in Bolivia when Conservation International acquired US$65,000 of the country's debt for US$100,000 and the government established the Reserva de la Biosfera de la Beni (Palca, 1987; Walsh, 1987); similar swaps have subsequently been made in Costa Rica and Ecuador (Dunne, 1989));
3 embargoes in consumer nations on timber forests that are not exploited in a sustainable manner (Secrett, 1987; McDermott, 1988);
4 local protests concerning developments at a national level, most notably in Brazil (Whitmore, 1990) and Malaysia (Aitken and Leigh, 1984).

In the last two decades national agencies have created over 3000 reserves and parks in the humid tropics. However, many of these parks and reserves fail to function because of understaffing and undercapitalization. Tree felling and poaching clearly cannot be controlled if the commitment to a national park is solely on paper. Biosphere Reserves are areas with a pristine core (in the case of the humid tropics a forest type) surrounded by a buffer zone that is managed for sustainable production and often includes agriculture and forestry. Such reserves are registered with UNESCO, which has designated thirteen areas of tropical forest as World Heritage Areas under the World Heritage Convention (Whitmore, 1990). A further trend, particularly evident in Brazil, is to zone forest utilization according to function, e.g. production forestry, extractive reserves where minor forest products are harvested, nature reserves and reserves for indigenous peoples.

The ITTO was established in 1985 at the behest of the UNCTED and includes thirty-six wood-producing nations, twenty-four consumer nations and various NGOs (Oldfield, 1989). The aim of the International Timber Trade Agreement (ITTA) is to promote sustainable forest use and to conserve tropical forests along with their genetic resources and their environments, whilst at the same time promoting the expansion and diversification of the international tropical timber trade. The ITTO's priorities are (i) an improvement in forest management and wood utilization, (ii) improved timber market intelligence, (iii) to increase the proportion of wood processing in producer nations, (iv) improved marketing and distribution and (v) the encouragement of industrial timber afforestation through plantation forestry. What effectiveness the ITTA and ITTO will have in halting or accelerating humid tropical deforestation remains to be seen.

9.9 The Changing Atmosphere

During the last decade, concern has grown over possible increases in global temperatures, due, at least in part, to increased amounts of carbon dioxide and other 'greenhouse gases' being released into the atmosphere as a by-product of human activity. There is similar concern over the depletion of stratospheric ozone associated with the use of chlorofluorocarbons (CFCs) and, more recently, related to the release of bromines.

From the geological and palaeoecological record there is convincing evidence that the world's climate has been very different in the geological past and that these changes have affected the humid tropics (section 5.8). What is so alarming about possible contemporary climatic change is, first, the speed of change and, secondly, the fact that for the first time human activity may be directly responsible for human well-being. Undoubtedly, the impact of any atmospheric change is potentially one of the most serious environmental issues to affect the humid tropics over the next few decades.

Evidence that the climate is already changing is, at present, equivocal but there is at least a partial understanding of why and how it may change and the likely impacts this will have upon various regions.

9.10 Evidence of Contemporary Atmospheric Change

There is irrefutable evidence that, since the industrial revolution, human activities have resulted in the release of new or increased amounts of 'greenhouse gases' and substances which can destroy ozone. Furthermore, there is evidence that during the past century the earth has warmed by approximately 0.5 °C on average and that, in the past few decades, concentrations of springtime polar stratospheric ozone have been reduced. However, at the present time there is little evidence directly linking global warming with an anthropogenetically enhanced greenhouse effect. Indeed, sceptics argue that the assumed overall 0.5 °C rise in global temperatures recorded during the past century may essentially relate to the positioning of most recording stations in areas of expanding urbanization and associated heat island effects. Similarly, there is as yet no firm evidence that the earth is suffering increased levels of ultraviolet radiation due to the depletion of stratospheric ozone (Thompson, 1989).

Future scenarios of global climatic change (section 9.12) are based on subjective climate modelling which is characterized by gross simplification and limited tectonic, oceanic and cloud representation. For example, measurements of sea-level changes in the South Pacific reveal a rise of 1.6 mm per year between 1900 and 1975 and an accelerated rate of rise since 1944. For the period 1946–85, mean sea level has risen at a rate of 2.59 mm per year in the Auckland area of New Zealand. Glacial melt due to global warming could theoretically account for about half of this rise and thermal expansion of the oceans for the remainder. However, the recognition of sea-level changes is complicated by land uplift or subsidence due to isostatic and eustatic processes in tectonically active regions. Nevertheless, for whatever reason, the sea is rising and there are numerous low-lying areas within the humid tropics which are likely to experience the effects of increased flooding, rising water tables and salinization of soils.

9.11 Causes of Climatic Change

On a geological time scale, climates change in response to changes in the terrestrial geography (e.g. due to mountain building), continental drift and cyclical variations in the earth's orbit (e.g. the Milankovitch cycles). Evidence of these long-term changes are easily found in the geological record and can often be seen as relic landforms and palaeosols. The global climate is also thought to be affected by astronomical cycles of shorter periodicity. For example, conspicuous variations in solar energy output (i.e. sunspot cycles) occur at regular intervals, particularly every 11 and 100 years. Solar temperatures during dark cool spots (sunspot minima) and bright hot flares (sunspot maxima) can vary up to 1100 °C and could theoretically influence the intensity of insolation received, particularly within the tropics, and hence modify the energy balance and ITCZ activity (section 2.2). Correlations between sunspot cycles and tropical temperatures, however, are uncertain. For example, tropical temperatures showed a positive correlation with sunspot flares between 1930 and 1950 but an inverse relationship from 1875 to 1920. Current American research by the George C. Marshall Institute reveals that the twentieth century has been dominated by sunspot maxima and high temperatures. This could account for the 0.5 °C measured increase in mean global temperatures this century (section 9.10). Forecasts for the next century suggest enhanced sunspot minima and the return of cooler conditions. This cooling could balance the predicted global warming from an accelerated greenhouse effect and reduce the impact of the projected environmental change.

The inconclusive sunspot–temperature relationships indicate that short-term atmospheric changes involve relatively subtle mechanisms associated with changes in solar radiation intensity *and* changes in atmospheric composition. The latter includes changes in trace gases and particulate matter which control the radiation balance and heat transfers at the earth's surface, and hence also dictate the intensity of synoptic-scale weather disturbances. There is a growing awareness, however, of the difficulty in separating natural change from that produced or accelerated by human activities such as deforestation (section 9.4) and gaseous release. Also, the complexity of both negative and positive feedback must be considered since these control the environmental impact of any change.

9.11.1 The greenhouse effect and global warming

The 'greenhouse' gases carbon dioxide, methane and nitrous oxides occur naturally in the atmosphere. However, over the past 150 years their concentrations have been greatly increased by a range of human activities (Thompson, 1989), especially the combustion of fossil fuels. Figure 9.11(b) illustrates the increase

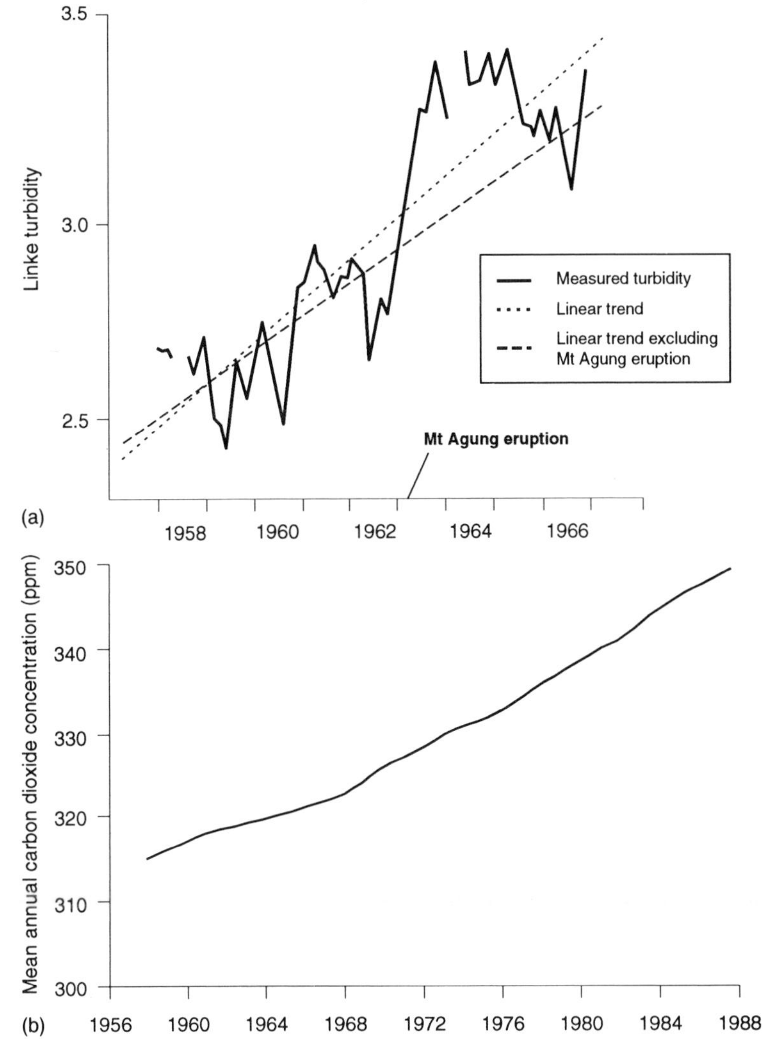

Figure 9.11
Atmospheric composition changes at Mauna Loa, Hawaii: (a) turbidity levels (Bryson and Kutzbach, 1968); (b) mean annual carbon dioxide concentrations (Gribbin, 1986).

in carbon dioxide at Hawaii, from 316 ppm in 1956 to 350 ppm in 1988. At present, carbon dioxide is increasing by 1.5 per cent per annum and at this rate the gas concentration will increase by 150 per cent over the next century compared with a mere 14 per cent over the last century. Furthermore, predictive modelling based upon world energy needs (Rowntree, 1990) suggests that carbon dioxide levels will reach 540 ppm by the year 2050, i.e. double the pre-industrial level of 1850. When we include anthropogenic increases in methane and nitrous oxides and the increasing presence of CFCs (a range of 'greenhouse' gases which do not occur naturally in the atmosphere), then it seems likely that an enhanced greenhouse effect could indeed result in global warming.

Counteracting this trend are particulate matter such as dust and sulphate aerosols released from natural sources (e.g. volcanoes) and human activities (e.g. fossil-fuel combustion). Such aerosols act as a shield in the troposphere and the associated reduction in atmospheric transparency results in a greater scattering of solar radiation back to space. This occurs at the expense of insolation; hence surface temperatures are lowered. Bryson (1968) estimated that a global turbidity rise of only 3–4 per cent could lower surface temperatures by 0.4 °C.

The humid tropics experience regular volcanic eruptions and the associated build-up of dust shields (section 8.4.2) and indeed the increased volcanicity in southeast Asia in the 1960s (e.g. Mt Agung, Indonesia, in 1963 and Mt Awu in the Philippines in 1966) could account for the recognized global cooling at that time. It must be noted that the volcanic eruption/climate association is exceedingly complex. For example, during the 1980s the global warming trend continued despite major volcanic eruptions, including El Chichón in Mexico in April 1982. Indeed, the dust-loading associated with the El Chichón eruption was evidently short lived since 1983 proved to be one of the warmest years on record worldwide. However, during the eruption of Mount Pinatubo in the Philippines in May 1991 a large ash cloud was produced and large quantities of dust and sulphate aerosols were injected into the stratosphere; this is thought to have temporarily halted the current warming trend.

The production of particulate matter by the combustion of fossil fuels exacerbates natural dust concentrations. Even though the humid tropics are remote from major combustion source regions, figure 9.11(a) shows that Hawaii experienced increased dust-loading during the late 1950s and early 1960s. Excluding the dust emission from Agung in 1963 (the broken line in figure 9.11(a)), it is apparent that the background turbidity increase over the period was about 30 per cent. Applying the above Bryson (1968) factor to this increase worldwide, the global temperatures would fall by 3.4 °C. This has not happened in the last two decades, perhaps because the dust-loading effect is being overwhelmed by the enhanced greenhouse effect.

9.12 Future Global Warming in the Humid Tropics: Predictive Approaches

Empirical evidence suggests that, during the past century, there has been an increase in concentrations of tropospheric greenhouse gases which could have been responsible for an overall 0.5 °C temperature rise. Various future trace-gas levels, and their possible impact on future global warming, have been simulated by numerous global climate models (GCMs). Many simulations have been based on a projected doubling of carbon dioxide from pre-industrial levels. However, very different global and regional scenarios are predicted by the various models and by using different projected gas increases. Despite the complexity of modern GCMs they remain gross simplifications of the global atmosphere. Their limitations are discussed by Rowntree (1990) and are mostly associated with simple ocean heating representation, simplistic ocean–atmosphere and land–atmosphere coupling and basic cloud characteristics. Problems of predicting small-scale (country-wide) changes from GCMs have also been highlighted. Furthermore, Hansen et al. (1988) warn that the changes in climate predicted by GCMs may be askew by as much as 50 per cent. Even the same model may predict very different scenarios with a small change in its base data. Hence, 'best estimate' figures produced by the latest models are usually qualified by 'possible ranges'. Despite the subjectivity and simplicity of GCMs they remain one of our most valued predictive tools. Furthermore, the models themselves and their predictions are continuously being revised along with our knowledge of global systems. The latter part of this chapter examines some of the atmospheric changes predicted by GCMs over the next century and the associated environmental consequences for the humid tropics.

9.13 The Predictions of Global Climate Models

All GCMs agree that the effects of increased carbon dioxide in the atmosphere will not be uniform around the globe. However, they differ in their prediction of the pattern, magnitude and even the direction of change (e.g. figure 9.12 and table 9.8). Models based upon a doubling of carbon dioxide levels indicate an average annual global warming of between 3.5 and 5.3 °C and increased global precipitation of between 7 and 15 per cent (Rowntree, 1990). The temperature increases in the humid tropics would be mostly less than 4.5 °C (figure 9.12) and indeed would be a relatively modest 1.5 °C at 10 °N (table 9.8). Rainfall changes are more difficult to predict but it appears from table 9.8 that there would be an increase in subtropical regions, especially between 10 °N and 20 °N, as a result of an intensification of the ITCZ over the Caribbean, the Philippines, southern India and the Sahel.

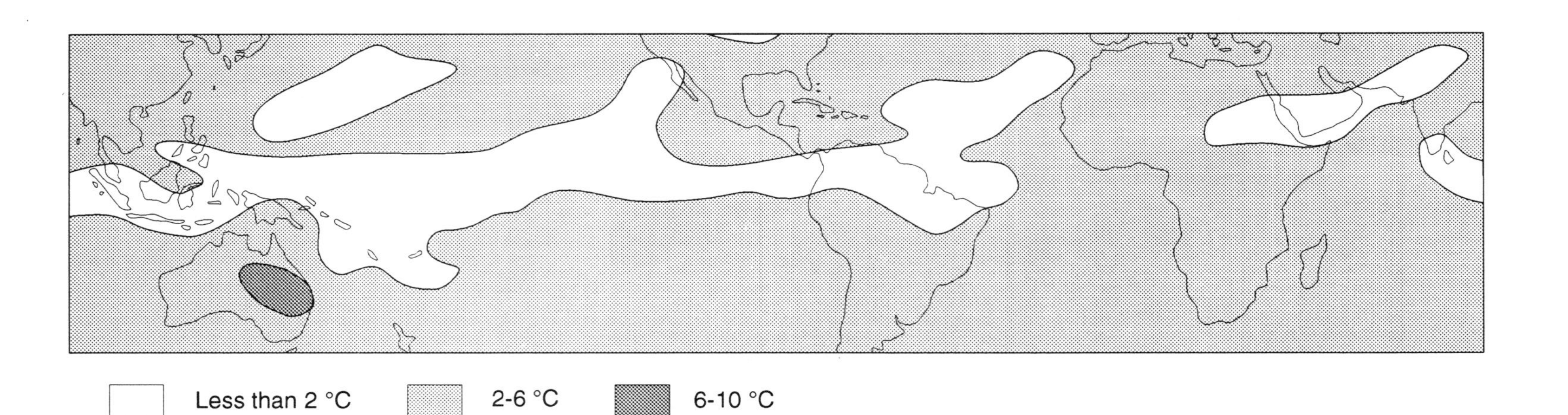

Figure 9.12
Computer model prediction of greenhouse warming in the tropics (from Gribbin, 1987).

Table 9.8 Probable effects of increased carbon dioxide content in the tropical atmosphere

Latitude	*Average annual change in surface temperature (°C)*	*Change in precipitation (percentage)*
30 °N	+4.5	0
20 °N	+2.5	+20
10 °N	+1.5	+20
Equator	+3.0	0
10 °S	+4.0	−20
20 °S	+4.5	− 5
30 °S	+4.0	+ 5

Source: Perry and Perry, 1986

Conversely, humid tropical countries around 10 °S would experience a 20 per cent decrease in rainfall. Thus, Brazil, Peru and Zaire, for example, would suffer more regular and severe droughts. A doubling of carbon dioxide levels may also be associated with increased and more intense tropical cyclones. For example Emanuel (1987) suggests that the latter may increase by 40–50 per cent due to increases in available kinetic energy. Eyre and Gray (1990), however, found no evidence of an increase in cyclone frequency or severity as a result of the observed global warming and increases in sea surface temperature during the last century or at the present time for the three cyclone regions studied (Caribbean, Eastern Pacific and Australia).

The climatic changes predicted by the GCMs are well beyond the range of climates experienced in the past 2–3 million years. For example, a global mean temperature rise of 3 °C (which the 'warming' consensus agrees is likely to happen over the next century) would represent even warmer conditions than those experienced in the last interglacial 120,000 years ago. However, sceptics argue that the impact of feedback mechanisms (especially increasing cloud cover), sunspot minima and dust shields must not be ignored and collectively they could overwhelm the projected enhanced greenhouse effect and reintroduce global cooling.

9.14 Sea-level Rise due to Predicted Global Warming

The response of the environment to these dramatic atmospheric changes would be wide ranging. Perhaps the most serious consequences would be associated with the rise of sea level following sea water expansion and ice sheet ablation with a global warming trend. Such a rise (section 9.10) has been recorded since 1900, at a rate equal to 10–15 cm per century. Once again, predictions for the future vary. With a doubling of carbon dioxide, Gribbin (1988) predicts an acceleration of sea level rise to 8 mm per year and a total rise of approximately 1 m by the end of the next century. The United Nations Environment Programme (UNEP, 1989) suggests that sea level will rise by 130–350 mm before the year 2030 and by up to 2 m within a century. The IPCC (1990) propose a 'best estimate' of 580 mm (range 220–1060 mm) for a rise in sea level by the year 2090. Within the humid tropics these forecast rises would have devastating consequences, causing seawater inundation along the low-lying coastlines of Thailand, throughout Indonesia and in a host of low-lying coral islands including the Maldives, Kiribati, Tuvalu and Tokelau. Bangladesh, already impoverished and extremely vulnerable to a range of natural hazards, would suffer particularly severely. In addition to the loss of fertile agricultural land and human suffering, this would result in the destruction of important marine ecosystems (especially mangrove swamps and coral reefs) (Stoddart, 1990; Spencer, 1994) and the seawater contamination of groundwater reserves.

9.14.1 The effect on mangroves

The predicted rise in sea level is yet another threat to the mangrove forests (section 9.5.4) which occupy the tidal zone and protect the coastal rim of many humid tropical regions (section 6.6.3). Although mangroves survive periodic immersion by seawater, they require sea-level stability in order to maintain their unique ecosystem. Evidence from the Pacific and Florida (cited by Ince, 1990) reveals that, in the past, large mangrove swamps did not exist in the most recent geological period of sea level rise (some 3000 years BP). Increased storminess associated with global warming and the more effective coastal erosion of offshore reefs would result in the net removal of sediment from the mangrove swamps and the ultimate failure of the mangrove ecosystem. Ince (1990) cited that mangrove swamps accumulate at a rate of up to 120 mm per century, a figure well below the predicted sea-level rises linked to global warming discussed above (Gribbin, 1988; IPCC, 1990).

9.14.2 The effect on coral reefs

In common with mangroves, coral reefs also require exacting physical conditions in order to grow (section 6.6.1). Reefs are unable to cope with rising sea levels since, even though tropical corals can survive in water as deep as 30 m, new coral accumulates at a finite rate. Large flat reefs appear to be able to grow at a rate of 7 mm per year (compared with the forecasted 8–11 mm sea-level rises) although so-called 'catch-up' reefs in Mexico have annual growth rates of up to 12 mm (Ince, 1990). Rapidly increasing water depths (at a rate exceeding natural coral accumulation) reduce the amount of sunlight reaching the coral; more precisely, they reduce the amount of light available for photosynthesis by the polyp's symbiotic algae and growth is curtailed. Furthermore, the optimum temperature for coral growth is a uniform 29 °C throughout the surface waters. Temperature increases of just 1–2 °C above this optimum destroy the algae and lead to so-called 'coral bleaching'. Such temperature increases are tolerated for only short periods of time (i.e. the high-sun season) but the long-term global warming proposed (table 9.8) would be fatal. Exacerbating the effects of deteriorating conditions, the predicted rise in tropical storm activity would increase the physical damage done to reefs at a time when essential growth rates are declining.

9.14.3 Sea-level rise in island states: a case study of the Maldives Republic

Coral reefs, like mangroves, act as natural sea defences and are vital for the continued existence of low-lying island states such as the Maldives (some 3 m above present-day sea level at their highest point). The Republic of the Maldives is naturally obsessed with the fear of rising sea level, especially since the majority of its 180,000 population, and virtually the whole of its infrastructure and economy, lie only between 80 cm and 2 m above sea level. It is hoped that coral growth would continue to keep pace with modest sea-level rises although associated 'coral bleaching' and increased sea erosion could well remove and destroy coral at a rate faster than it is capable of growing (Ince, 1990). In the Maldives, the breaching of the coral sea defences would eliminate staple crops, like salt-sensitive mango and taro, and forests would be destroyed. With sea-level rises in excess of 1.2 m, the airport on Male would be submerged, which would restrict tourism, a vital part of the Maldivian economy (Ince, 1990). Many homes and businesses would be flooded and indeed, according to UNEP (1989), a 2 m sea-level rise would inundate about 50 per cent of this island republic. This would lead to a mass exodus of islanders fleeing their disappearing homeland and would create a new class of environmental refugee. In the Maldives, as in other threatened countries, mitigation of the effects of sea-level rise will require clear and careful coastal management. If predictions of the rate of sea-level rise are correct, then the construction of adequate and sustainable sea defences will become a priority in the next two decades or so. These defences should aim to supplement and reinforce the natural protection afforded by coral reefs and mangroves in harmony with their ecosystem constraints. Unfortunately, the financial cost of any engineering response will greatly exceed the financial resources of many developing countries, unless massive overseas aid is forthcoming. As an alternative, many humid tropical countries may be obliged to sacrifice rather than save vulnerable areas.

9.14.4 Sea-level rise in continental lowlands: a case study of Bangladesh

The threat of the loss of valuable farmland and property under rising sea levels is not just confined to island states like the Maldives. All coastal lowlands, and especially river deltas, on the continental margins are also at risk. However, deltaic areas in southeast Asia (like that of the River Ganges in Bangladesh) would bear the most serious consequences, since they are very heavily populated. Figure 9.13 illustrates the threat to Bangladesh of proposed sea-level rises. A 1.5 m rise would inundate about 15 per cent of all the land area (and about 20 per cent of the farmland, with 20 per cent less agricultural production) and would displace about 20 per cent of the nation's 100 million population (Parry, 1990). Even the 'best estimate' of the IPCC (1990) of a sea-level rise about 50 cm would still inundate large areas of Bangladesh farmland (figure 9.13). This would accentuate the existing massive migration of rural peoples to urban areas well above sea level, especially Dhaka. However, it must be noted that the above inundations will be complicated or offset by delta aggradation/ extension following the predicted increases in rainfall and sediment transfers from the headwaters of the Ganges.

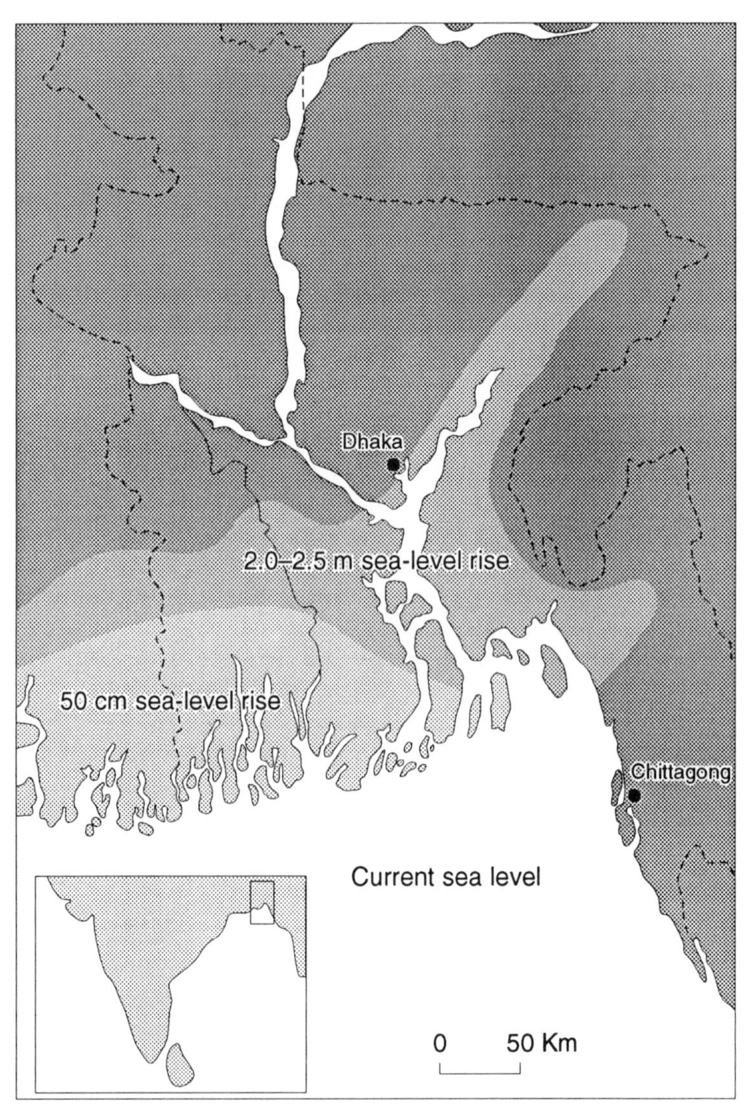

Figure 9.13
Proposed sea-level changes and the inundation of Bangladesh (from UNEP, 1989).

9.14.5 Sea-level rise and water resources

In addition to the direct loss of farmland and property from inundation, agriculture and the population would also suffer from the associated increased saltwater intrusion into surface water and particularly groundwater reserves. Saltwater contamination by rising sea levels will probably be felt most acutely on small islands, where the supply of underground freshwater is often critical for sustainable habitation. Freshwater supplies are frequently concentrated in a lens-shaped zone under such islands, fed by rainwater percolation (figure 9.14). The lens is maintained by the density contrasts between seawater (1.025 g ml^{-1}) and fresh water (1.0 g ml^{-1}), which allows the fresh water to float on the seawater (Woodroffe, 1989). It appears that these freshwater lenses can only accumulate under islands more than 400 m across, assuming that the geology is suitable (i.e. permeable rocks exist) and that surface rainfall is adequate (Ince, 1990).

It is difficult to predict how rising sea levels will affect the vital freshwater aquifers of small islands; however, measured variations due to periodic tidal changes indicate that they will be seriously and adversely affected (Woodroffe, 1989). The obvious effect will be associated with the initial reduction of the size of small islands which will contract the area of the lens they support. In some extreme cases, the island will shrink to below the critical size (400 m across) required to support it (Ince, 1990).

Two of the major ways in which groundwater may be affected are illustrated in figure 9.14. First, with a reduction in available reserves abstraction rates may become unsustainable, causing upconing of brackish water from the transition zone and increasing

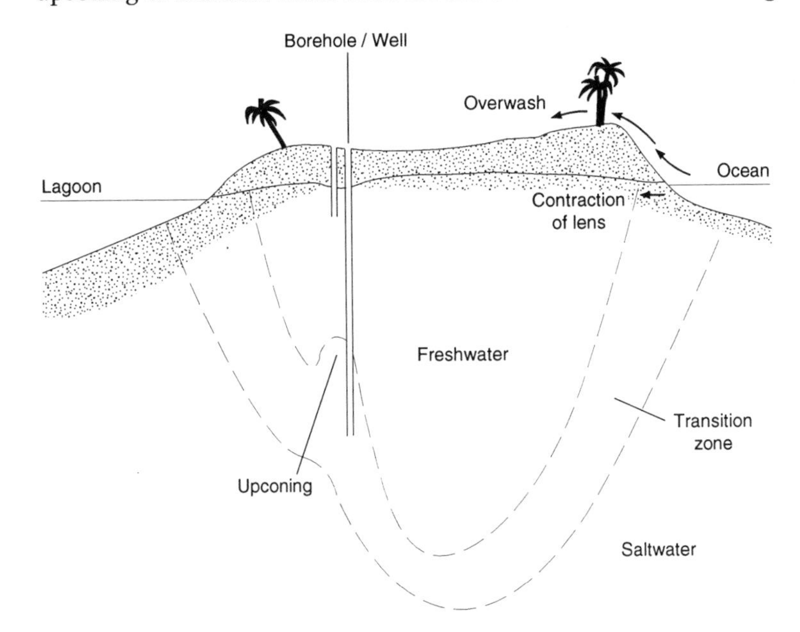

Figure 9.14
Salt-water intrusion into a freshwater lens (after Ince, 1990)

salinization of deep wells and boreholes. For example, it has been estimated that a lowering of the overall water table by 25 mm pulls up the base of the lens (and the height of salinization) forty times as far (i.e. 1 m). Second, contamination of groundwater may occur because of overwash and subsequent percolation of seawater during severe storms (whose intensity and frequency may increase).

9.15 Global Warming and Agricultural Potential

Unfortunately, no comprehensive study has been made of the impact of global warming on the agricultural potential of the humid tropics. However, a significant amount of information is available for the semi-arid tropics and mid-latitude regions. The effects of global warming at low latitudes are much more difficult to estimate because agricultural production potential is largely a function of the amount and distribution of precipitation. In fact, there is little agreement (or confidence) in recognizing the ways in which rainfall may be affected by global warming (despite table 9.8 predictions). Consequently, very few estimates are currently available of how crop yields might respond to a range of climate-change scenarios in low latitude regions (Parry, 1990).

Tropical ecosystems and agriculture would almost certainly be modified by global warming, since the speed of the predicted temperature rise is estimated at around ten times that of biological evolution. Details of the changes, however, remain speculative. Norton (1985), for example, has estimated that every 1 °C temperature increase would extend the range of plants by some 200 m elevation. A 1 °C temperature rise would also possibly extend tropical ecosystems polewards by 100–150 km. The shift of tropical agricultural zones is unclear although at latitude 50 ° it is assumed that a 1 °C warming would induce a 200–400 km poleward shift of cropping zones.

Shifts in agricultural zones, however, would be compounded by the effects of other atmospheric uncertainties, in particular those of rainfall and carbon dioxide, and indirectly by changes in pest and disease infestations. For example, increased carbon dioxide *per se* could influence ecosystems and agriculture in a number of ways. It has been estimated that every 10 ppm increase in carbon dioxide levels would enhance plant growth by between 0.5 and 2 per cent (Thompson, 1989). Such increases would also raise the rate of photosynthesis (the so-called 'fertilization effect'), especially in crops like rice and soya bean. However, the increased carbon dioxide levels would also have negative effects for agriculture such as more vigorous weed growth. Tropical parasites and diseases may also proliferate and spread into middle latitudes as they become less temperate.

The complexity of the situation is illustrated by Panturat and Eddy's (1989) attempts to model changes in agricultural productivity in the humid tropics on the basis of (albeit inconsistent) GCM predictions of rainfall patterns, the variable that most determines crop yield in the area (Parry, 1990). They suggest that intensification of the southeast Asian monsoon (following global warming and ITCZ strengthening) would lead to increased summer rainfall and, possibly, reduced winter rainfall. In their study, rice production in Thailand was correlated with rainfall changes of +5 per cent (summer) and −11 per cent (winter) and, despite compensating irrigation changes, resulting potential rice yields were estimated to decrease by up to 7 per cent. Conversely, the beneficial direct effect of enhanced CO_2 was estimated to increase yields by 4–13 per cent. However, yield trends are further confused by temperature increases, resulting in a more rapid growth but lower overall yields and greater pest infestations.

The picture is certainly complex and may take years to resolve. In the meanwhile, it may be prudent to assume that crop-water availability could decrease in *some* tropical regions and that, under these circumstances, there could be sustained regional dislocation of access to food supplies (Parry, 1990).

9.16 International Concern and Action to Alleviate the Effects of Climate Change

The atmospheric and environmental consequences of carbon dioxide doubling and significant global warming would be serious and wide ranging in the humid tropics. Even though the evidence for such changes is as yet inconclusive, the climatic model predictions have forced the developed world into action. For example, fundamental to more accurate and detailed climatic-change predictions are the improvements in our network of base-line measurements of atmospheric elements (both surface and upper air) and environmental factors discussed in this section. If the enhanced greenhouse effect is a reality, we should soon have evidence of greatly increased infrared counter-radiation and cooling in the lower stratosphere. This evidence requires the implementation of a rigorous monitoring project and will only be possible with considerable funding following a high level of intergovernmental and interdisciplinary co-operation. It is promising to report that some progress is being made following the recommendations adopted by the Second World Climate Conference in 1990. For example, a number of important studies are being addressed by international programmes such as the World Climate Programme (WCP) and the International Geosphere Biosphere Programme (IGBP).

The first major UK-funded programme, the Terrestrial Initiative in Global Environmental Research (TIGER), is designed to improve our understanding of processes associated with the main greenhouse gases and to assist climate modellers to reduce predictive uncertainties. Within the humid tropics TIGER will concentrate on carbon fluxes, water–energy exchange and surface radiation balances within the rainforests (initially of Amazonia and the

Cameroons). This project will also contribute to WCP and IGBP experiments aimed at improving the representation of land-surface processes within GCMs.

Efforts to unravel the technical aspects of the climatic change debate have been centred, understandably, upon the developed world's scientific community. In the light of their findings, developed-nation's governments have initiated campaigns for international political co-operation to reduce possible deleterious effects. In this respect, there have been a number of important political milestones. These include 'The 1987 Montreal Protocol on Substances Which Deplete the Ozone Layer' and its revision in 1990 and recommendations made by the Intergovernmental Panel on Climate Change (IPCC) meeting in Geneva in 1990 (and revised in 1992). The Montreal protocol requires signatories to cease production of CFCs and reduce other ozone-damaging chemicals within a strict time scale. The Geneva meeting agreed to stabilize emissions of carbon dioxide at 1990 levels by the year 2000, an effective reduction at this time of 30 per cent.

Many countries in the humid tropics have been involved in establishing these important global protocols, mainly in an attempt to avoid a future environmental catastrophe in their own regions. However, in many developing nations the consequences of signing the protocols have seemed at least as costly (politically and in the short and medium terms) as the consequences of climatic change. Many still fear the effect of emission controls on their future social and economic development. For example, a well-publicized aim of the Indian government is to have a refrigerator in every Indian home by the year 2000. To these ends, CFCs represent a cheap and effective refrigerant and their replacement with expensive ozone-friendly CFC alternatives would increase the price of appliances beyond that generally affordable. Developing nations have stood firm in reiterating their development dilemmas to some effect. The total ban on CFC production by the year 2000 was achieved at the 1990 Montreal protocol review only when India and other developing nations were assured of financial aid regarding the use of costly refrigerant alternatives.

The continuing development of many tropical nations will be an ongoing process over future decades as they deservedly strive to improve their economies and living standards and acquire some of the benefits enjoyed by the developed nations. Energy consumption will have to accelerate in order to sustain this development and the dependence will be on available, basic fossil fuels, which will aggravate the carbon dioxide emission. Developing countries argue (and justify) their case against castigation. They point out that the economic fortunes of many developed nations have arisen as a result of centuries of squandering resources and contaminating every facet of the earth–atmosphere system and that developed nations remain the major polluters today. Furthermore, many of the pollution sources within the developing countries are multinational companies with headquarters based in the developed world.

9.17 The Future in the Humid Tropics

The above discussion has attempted to outline the major facets of the respective global debates; however, it has also demonstrated the complexity of the situation. A lack of quantitative evidence is a fundamental shortcoming; so too is our incomplete knowledge of the workings of global or even regional-scale environmental systems. The resources to improve these situations are almost entirely held by industrial nations based in the mid-latitudes. However, even with the much needed research, it is likely that our knowledge base will remain incomplete for some decades, by which time it may be too late to recover from the deleterious consequences. The political will to address the potential problems within an international framework appears to be increasing. For example, the United Nations conference on Environment and Development (the Earth Summit) held in Rio de Janeiro in 1992 included government delegations from almost all nations plus around 10,000 representatives of non-government organizations, indigenous associations and environmentalists. The agenda for Rio included a world convention on climate, which would legally commit countries to massive energy-saving programmes in the future to avoid global warming. It also included a declaration on forests and biodiversity which would help safeguard the humid tropical biosphere. Unfortunately, opposition to targets and dates by countries with a vested interest in continuing their exploitation of the environment emasculated these and most of the other conference aims. Nevertheless, the Earth Summit did expound the fact that change is essential for sustained development. However, the ultimate success or failure of political negotiations will depend upon considerations of, and agreement on, issues such as sovereignty, responsibility for off-site effects, universal adoption of rules, equity and trade regulations and the funding of scientific research and technology transfer. The role of the scientific community is to monitor and advise to the best of their capability.

Further Reading

Chapter 1

Barrow, C. 1987: *Water Resources and Agricultural Development in the Tropics*. Harlow: Longman.
Douglas, I. and Spencer, T. 1985: *Environmental Change and Tropical Geomorphology*. London: Allen & Unwin.
Holdridge, L. S. 1967: *Life Zone Ecology*. Tropical Science Centre, San José, Costa Rica, 206 pp.
Jackson, I. J. 1989: *Climate, Water and Agriculture in the Tropics*. Harlow: Longman.

Chapter 2

For a textbook which provides a succinct summary of many key points see

Barry, R. G. and Chorley, R. J. 1992: *Atmosphere, Weather and Climate*, 6th edn. London: Routledge.

Specific articles for greater detail on some of the points raised include

Bigg, G. R. 1990: El Ninõ and the Southern Oscillation. *Weather*, 45 (1), 2–8.
Gray, W. M. 1979: Hurricanes: their formation, structure and likely role in the tropical circulation. In D. B. Shaw (ed.), *Meteorology over the Tropical Oceans*, Bracknell, Royal Meteorological Society, 155–218.
Reynolds, R. 1985: Tropical meteorology. *Progress in Physical Geography*, 9, 157–86.

Chapter 3

Ayoade, J. 1983: *Introduction to the Climatology of the Tropics*. New York: Wiley.

Fein, J. S. and Stephens, P. L. 1987: *Monsoons*. New York: Wiley.
Hayward, D. and Oguntoyinbo, J. 1987: *The Climatology of West Africa*. London: Hutchinson.
Nieuwolt, S. 1977: *Tropical Climatology*. New York: Wiley.
Riehl, H. 1979: *Climate and Weather in the Tropcs*. New York: Academic Press.

Chapter 4

There are textbooks on general tropical soil science or specific soil types which cover the material in chapter 4 in more detail. These include the following.

General textbooks

Mohr, E. D., Van Baren, F. A. and Von Schuylerlorgh, J. 1972: *Tropical Soils: A Comprehensive Study of Their Genesis*. The Hague: Mouton-Ichitar Baru-Van Hoeve.
Sanchez, P. A. 1976: *Properties and Management of Soils in the Tropics*. New York: Wiley.
Young, A. 1976: *Tropical Soils and Soil Survey*. Cambridge, Cambridge University Press.

Monographs on specific soils

Andriesse, J. P. 1988: The nature and management of tropical peat soils. *FAO Soils Bulletin*, 59.
Andriesse, W. and Scholten, J. J. 1982: *Acri-Orthic Ferralsol (Haplic Acrorthox) Jamaica*, Soil Monograph Paper 6. Wageningen: International Soil Museum.
Dent, D. 1986: *Acid Sulphate soils: a Baseline for Research and Development*. Wageningen: International Institute of Land Reclamation and Improvement (ILRI), Publication 35.
Ponnamperuma, F. N. 1972: The chemistry of submerged soils. *Advances in Agronomy*, 24, 29–96.
Pons, L. J. and Zonneveld, I. S. 1965: *Soil Ripening and Soil Classification*. Wageningen: International Institute for Land Reclamation and Improvement, Publication 14.
Scholten, J. J. and Andriesse, W. 1982: *Humic Acrisols (Orthoxic Palehumult) Jamaica*, Soil Monograph Paper 5. Wageningen: International Soil Museum.
Wada, K. 1985: Volcanic soils. *Advances in Soil Science*, 2, 174–229.

Chapter 5

Chapman, V. J. (ed.) 1977: *Wet Coastal Ecosystems*. Amsterdam: Elsevier.

Golley, F. B. (ed.) 1983: *Tropical Rain Forest Ecosystems*. Amsterdam: Elsevier.
Jacobs, M. 1987: *The Tropical Rain Forest*. Berlin: Springer.
Longman, K. A. and Jeník, J. 1987: *Tropical Forest and its Environment*, 2nd edn. London: Longman.
Whitmore, T. C. 1990: *An Introduction to Tropical Rain Forests*. Oxford: Clarendon.

Chapter 6

Bird, E. C. F. 1969: *Coasts*. Boston, MA: MIT Press.
Derbyshire, E. 1979: *Geomorphology and Climate*. London: Wiley.
Faniran, A. and Jeje, L. E. 1983: *Humid Tropical Geomorphology*. London: Longman.
Ford, D. C. and Williams, P. W. 1989: *Karst Geomorphology and Hydrology*. London: Unwin.
Guilcher, A. 1990: *Coral Reef Geomorphology*. Chichester: Wiley.
Ollier, C. D.. 1984: *Weathering*. London: Longman.
Thomas, M. F. 1974: *Tropical Geomorphology*. London: Macmillan.

Chapter 7

Balek, J. 1983: *Hydrology and Water Resources in the Tropical Regions. Amsterdam: Elsevier.*
Douglas, I. and Spencer, T. 1985: *Environmental Change and Tropical Geomorphology*. London: Allen & Unwin.
Lal, R. and Russell, E. W. (eds) *Tropical Agricultural Hydrology*. London: Wiley.

Chapter 8

Aitken, S. R. and Leigh, C. H. 1992: *Vanishing Rain Forests: The Ecological Transition in Malaysia*. Oxford: Oxford University Press.
Bruijnzeel, L. A. 1990: *Hydrology of Moist Tropical Forests and Effects of Conversion: A State of the Knowledge Review*. Paris: UNESCO.
Jordan, C. F. (ed.) 1987: *Amazonian Rain Forests. Ecosystem Disturbance and Recovery*. Berlin: Springer.
Savage, J. M., Goldman, D. P., Janos, D. P., Lugo, A. E., Raven, P. H., Sanchez, P. A. and Wilkes, H. G. 1982: *Ecological Aspects of Development in the Humid Tropics*. Washington, DC: National Academy Press.
Smith, K. 1992: *Environmental Hazards*. London: Routledge.

Chapter 9

Aitken, S. R. and Leigh, C. H. 1992: *Vanishing Rain Forests. The Ecological Transition in Malaysia*. Oxford: Clarendon.
Fearnside, P. M. 1990: Deforestation in the Brazilian Amazonia. In G. M. Woodwell (ed.), *The Earth in Transition: Patterns and Process of Impoverishment*, Cambridge: Cambridge University Press, 211–35.
Houghton, J. T., Jenkins, G. J. and Ephraums (eds) 1990: *Climate Change*, IPCC Scientific Assessment. Cambridge: Cambridge University Press.
Parry, M. 1990: *Climate Change and World Agriculture*. London: Earthscan Publications.
Stoddart, D. R. 1990: Coral reefs and islands and predicted sea-level rise. *Progress in Physical Geography*, 14(1), 521–36.
Woodroffe, C. D. 1990: The impact of sea-level rise on mangrove shorelines. *Progress in Physical Geography*, 14(1), 483–520.

Bibliography

Preface

Faniran, A. and Jeje, L. E. 1983: *Humid Tropical Geomorphology*. London: Longman.
Jackson, I. J. 1977: *Climate, Water and Agriculture in the Tropics*, 1st edn. Harlow: Longman.
Nieuwolt, S. 1977: *Tropical Climatology*. London: Wiley.
Richards, P. W. 1952: *The Tropical Rain Forest*, 1st edn. Cambridge: Cambridge University Press.
Sanchez, P. A. 1976: *Properties and Management of Soils in the Tropics*. New York: Wiley.
Thomas, M. F. 1974: *Tropical Geomorphology*. London: Macmillan.
Thomas, M. F. 1994: *Geomorphology in the Tropics*. Chichester: Wiley.
Tricart, J. 1972: *Landforms of the Humid Tropics and Savannas*. London: Longman.
Young, A. 1976: *Tropical Soils and Soil Survey*. Cambridge, Cambridge University Press.

Chapter 1

Barrow, C. 1987: *Water Resources and Agricultural Development in the Tropics*. Harlow: Longman.
Birot, P. 1959: *Géographie Physique Générale de la Zone Intertropicale*. Paris: C.D.U.
Brown, S., Lugo, A. E. and Liegel, B. 1980: *The Role of Tropical Forests in the World Carbon Cycle*. US Department of Energy, CONF-800350, UG11, 156 pp.
Chang, J. and Lau, L. S. 1982: *Definition of the Humid Tropics*. Unpublished memorandum, University of Hawaii.
Douglas, I. and Spencer, T. 1985: *Environmental Change and Tropical Geomorphology*. London: Allen & Unwin.
Environmental Development Action 1981: *Environment and Development in Africa*. Oxford: Pergamon.
Fosberg, F. R., Garnier, B. J. and Kuchler, A. W. 1961: Delineation of the humid tropics. *Geographical Review*, 51 (3), 333–47.

Garnier, B. J. 1958: Some comments on defining the humid tropics, Research Note 11, Department of Geography, University of Ibadan.
Gourou, P. 1966: *The Tropical World*, 4th edn, Translated from French by S. A. Beaver and E. D. Laborde. London: Longman.
Gupta, A. 1984: Urban hydrology and sedimentation in the tropics. In J. E. Costa and P. J. Fleisher (eds), *Developments and Applications in Geomorphology*, Berlin: Springer.
Holdridge, L. S. 1967: *Life Zone Ecology*. Tropical Science Centre: San José, Costa Rica, 206 pp.
Huntington, E, 1915: *Civilization and Climate*. New York: Wiley.
Huntington, E. 1945: *Mainsprings of Civilization*. New Haven, CT: Yale University Press.
Jackson, I. J. 1989: *Climate, Water and Agriculture in the Tropics*. Harlow: Longman.
Köppen, W. 1936: Das geographische System der Klimate. In *Handbuch der Klimatologie*, vol. 1, Berlin: Borntager.
Lewis, L. A. and Berry, L. 1988: *African Environments and Resources*. London: Unwin Hyman.
de Martonne, E. 1946: Géographie zonale: la zone tropicale. *Annales de Geographie*, 55, 1–18.
Mink, J. F. 1983: Groundwater hydrology in agriculture in the humid tropics. In R. Keller (ed.), *Hydrology of Humid Tropical Regions with Particular Reference to the Hydrological Effects of Agriculture and Forestry Practice*, IAHS Publ. 140, London: Wiley.
Nieuwolt, S. 1977: *Tropical Climatology*. London: Wiley.
Ooi Jin Bee 1983: *Natural Resources in Tropical Countries*. Singapore: Singapore University Press.
Persson, R. 1974: *World Forest Resources. Review of World's Forest Resources in the Early 1970s*. Stockholm: Department of Forest Survey, Royal College of Forestry.
Pielke, R. A. 1990: *The Hurricane*. London: Routledge.
Riehl, H. 1979: *Climate and Weather in the Tropics*. London: Academic Press.
Savage, J. M., Goldman, D. P., Janos, D. P., Lugo, A. E., Raven, P. H., Sanchez, P. A. and Wilkes, H. G. 1982: *Ecological Aspects of Development in the Humid Tropics*. Washington, DC: National Academy Press.
Thomson, J. 1882: Notes on the basin of the River Rovuma, East Africa. *Proceedings of the Royal Geographical Society, NS*, 4, 65–79.
Tricart, J. 1972: *Landforms of the Humid Tropics, Forests and Savannas*, trans. C. J. K. de Jonge. London: Longman.

Chapter 2

Asnani, G. C. 1968: The equatorial cell in the general circulation. *Journal of Atmospheric Science*, 25, 133–4.
Barry, R. G. and Chorley, R. J. 1987: *Atmosphere, Weather and Climate*, 5th edn. London: Methuen.
Barry, R. G. and Chorley, R. J. 1992: *Atmosphere, Weather and Climate*, 6th edn. London: Methuen.
Bigg, G. R. 1990: El Niño and the southern oscillation. *Weather*, 45 (1), 2–8.
Case, B. and Mayfield, M., 1990: Atlantic hurricane season of 1989. *Monthly Weather Review*, 118, 1165–77.

Chang, J. H. 1972: *Atmosphere Circulation Systems and Climates*. Honolulu: Oriental Publishing.
Chapas, L. C. and Rees, H. R. 1964: Evaporation and evapotranspiration in southern Nigeria. *Quarterly Journal of the Royal Meteorological Society*, 90, 313–19.
Dobson, T. 1853: *Australian Cyclonology*. Publisher not known.
Fletcher, R. D. 1945: The general circulation of the tropical and equatorial atmosphere. *Journal of Meteorology*, 2, 167–74.
Gray, W. M. 1979: Hurricanes: their formation, structure and likely role in the tropical circulation. In D. B. Shaw (ed.), *Meteorology over the Tropical Oceans*. Bracknell: Royal Meteorological Society, 155–218.
Hamilton, M. G. 1987: Monsoons – an introduction. *Weather*, 42 (6), 186–93.
Hayward, D. and Oguntoyinbo, J. 1987: *The Climatology of West Africa*. London: Hutchinson.
Henderson-Sellers, A. and Robinson, P. J. 1987: *Contemporary Climatology*. Harlow: Longman.
Henry, W. K. 1974: The tropical rainstorm. *Monthly Weather Review*, 102, 717–25.
Jackson, I. J. 1989: *Climate, Water and Agriculture in the Tropics*, 2nd edn. London: Longman.
Kerr, I. S. 1976: Tropical storms and hurricanes in the southwest Pacific, November 1939 to April 1969. New Zealand Meteorological Service Misc. Publ. 148, Wellington, New Zealand.
Keunen, P. H. 1955: *Realms of Water*. London: Cleaver-Hulme.
Pearce, F. 1988: El Ninõ makes it a warm, warm world. *New Scientist*, 120 (1638), 29.
Pielke, R. A. 1990: *The Hurricane*. London: Routledge.
Reading, A. J. 1990: Caribbean tropical storm activity over the past four centuries. *International Journal of Climatology*, 10, 365–76.
Reddy, S. 1989: The 1989 drought in Fiji. *Meteorological Event Report 89/2*, Fiji Meteorological Service.
Reynolds, R. 1985: Tropical meteorology. *Progress in Physical Geography*, 9, 157–86.
Riehl, H. 1954: *Tropical Meteorology*. New York: McGraw-Hill.
Thompson, R. D. 1981: Incidence and extent of tropical cyclones in the Fiji Islands. *Marine Observer*, L1, 189–95.
Thompson, R. D. 1986: Hurricanes in the Fiji area: causes and consequences. *New Zealand Journal of Geography*, 81, 7–12.
Thompson, R. D. 1989: Short-term climatic change: evidence, causes, environmental consequences and strategies for action. *Progress in Physical Geography*, 13, 315–47.
Thompson, R. D., Mannion, A. M., Mitchell, C. W., Parry, M. and Townshend, J. R. G. 1986: *Processes in Physical Geography*, Harlow: Longman.
Zangvil, A. 1975: Temporal and spatial behavior of large-scale disturbances in tropical cloudiness deduced from satellite brightness data. *Monthly Weather Review*, 103, 904–20.

Chapter 3

Ayoade, J. 1976: Evaporation and evapotranspiration in Nigeria. *Journal of Tropical Geography*, 43, 9–19.

Ayoade, J. 1983: *Introduction to the Climatology of the Tropics*. New York: Wiley.
Balek, J. 1983: *Hydrology and Water Resources in the Tropical Regions*. Amsterdam: Elsevier.
Barry, R. G. and Chorley, R. J. 1992: *Atmosphere, Weather and Climate*, 6th edn. London: Methuen.
Bruijnzeel, L. A. 1983: Evaluation of runoff sources in a forested basin in a wet monsoonal environment: a combined hydrological and hydrochemical approach. In R. Keller (ed.), *Hydrology of Humid Tropical Regions with Particular Reference to the Hydrological Effects of Agriculture and Forestry Practice*, IAHS Publ. 140, London: Wiley, 165–74.
Brutsaert, W. H. 1965: Evaluation of some practical methods of estimating evapotranspiration in arid climates at low latitudes. *Water Resources Research*, 1, 187–91.
De Bruin, H. A. R. 1983: Evaporation in humid tropical regions. In R. Keller, (ed.), *Hydrology of Humid Tropical Regions with Particular Reference to the Hydrological Effects of Agriculture and Forestry Practice*, IAHS Publ. 140, London: Wiley, 299–311.
Doorenbonbos, J. and Pruitt, W. O. 1977: Crop water requirements. FAO Irrig. Drain. Paper 24, Rome, 144pp.
Drochon, A. 1976: Données climatologiques au Sol et un altitude pour la station d'Abidjan. *Notes Tranductives et Informatives Selectionnés de la DEM 55*, Dakar, ASECNA.
Edwards, K. A., Blackie, J. R., Cooper, S. M. 1981: Results of East African catchment experiments 1958–1974. In R. Lal and E. W. Russell (eds), *Tropical Agricultural Hydrology*, Chichester: Wiley, 163–88.
Garnier, B. J. 1961: Mapping the humid tropics: climatic criteria. *Geographical Review*, 51, 339–46.
Geiger, R. 1965: *The Climate Near The Ground*. Harvard, MA: Harvard University Press.
Griffiths, J. F. 1972: *Climates of Africa*. Amsterdam: Elsevier.
Gunston, H. and Batchelor, C. H. 1983: A comparison of the Priestly–Taylor and Penman methods for estimating reference crop evapotranspiration in tropical countries. *Agricultural Water Management*, 6, 65–77.
Hayward, D. and Oguntoyinbo, J. 1987: *The Climatology of West Africa*. London: Hutchinson.
Henry, W. K. 1974: The tropical rainstorm. *Monthly Weather Review*, 102, 717–25.
Hudson, N. W. 1971: *Soil Conservation*. London: Batsford.
Jackson, I. J. 1989: *Climate, Water and Agriculture*. Singapore: Longman.
Kayane, I. 1971: Hydrological regions in monsoon Asia: a climatological approach. In B. M. Yoshimo (ed.), *Balance of Monsoon Asia*. Honolulu: University of Hawaii Press, 287–300.
Lockwood, J. G. 1974: *The Physical Geography of the Tropics*. London: Oxford University Press.
Miller, A. A. 1971: *Climatology*. London: Methuen.
Mohr, E. C. J. and Van Baren, F. A. 1959: *Tropical Soils*. Amsterdam: Interscience.
Monteith, J. L. 1965: Evaporation and environment. In *State and Movement of Water in Living Organisms, Symp., Society for Experimental Biology*, 19, 205–34.
Monteith, J. L. 1981: Evaporation and surface temperature. *Quarterly Journal of the Royal Meteorological Society*, 107, 1–27.

Nieuwolt, S. 1977: *Tropical Climatology*, New York: Wiley.
Nullet, D. and Giambelluca, T. W. 1990: Winter evaporation on a mountain slope, Hawaii. *Journal of Hydrology*, 112, 257–65.
Obasi, G. O. P. 1972: Water balance in Nigeria. *Quarterly Meteorological Magazine*, 2 (2), 91–127.
Ojo, O. 1969: Potential evapotranspiration and the water balance in West Africa: an alternative method of Penman. *Archiv für Meteorologie, Geophysik und Bioklimatologie*, B17, 239–60.
Ojo, O. 1970: The seasonal march of the spatial patterns of global and net radiation in West Africa. *Journal of Tropical Geography*, 30, 48–62.
Oliver, J. E. and Hidore, J. J. 1984: *Climatology: an Introduction*. Columbus, OH: Charles Merrill.
Penman, H. L. 1948: Natural evaporation from open water, bare soil and grass. *Proceedings of the Royal Society*, A193, 120–45.
Pople, W. and Mensah, M. A. 1971: Evaporation as the upwelling mechanism in Ghanaian coastal waters. *Nature*, 233, 18–20.
Priestly, C. H. B. and Taylor, R. J. 1972: On the assessment of surface heat flux and evaporation using large-scale parameters. *Monthly Weather Review*, 100, 81–92.
Rakhecha, P. R., Mandal, B. N. and Ramana Murthy, K. V. 1985: Analysis of hourly rainfall distribution of Karanja catchment. *Transactions of the Institute of Indian Geographers*, 7 (2), 95–103.
Riehl, H. 1954: *Tropical Meteorology*. New York: McGraw-Hill.
Sellers, W. D. 1965: *Physical Climatology*. Chicago, IL: Chicago University Press.
Sengele, N. 1981: Estimating potential evapotranspiration from a watershed in the Loweo Region of Zaire. In R. Lal and E. W. Russell (eds), *Tropical Agricultural Hydrology*, London: Wiley.
State Hydrological Institute, 1974: *Atlas of World Water Balance*. Paris: UNESCO.
Swami, K. 1970: Importance of daily and synoptic climatic analyses in ecological studies: an example from Nigeria. *Climate Bulletin*, 8, 40–57.
Thom, A. S. and Oliver, H. R. 1977: On Penman's equation for estimating regional evaporation. *Quarterly Journal of the Royal Meteorological Society*, 103, 345–57.
Thornthwaite, C. W. 1931: The climates of North America according to a new classification. *Geographical Review*, 21, 633–55.
Thornthwaite, C. W. 1948: An approach toward a rational classification of climate. *Geographical Review*, 38, 35–94.
Thornthwaite, C. W. and Mather, J. R. 1955: The water balance. *Publications in Climatology*, 8 (1).
Trewartha, G. T. 1968: *Introduction to Climatology*, New York: McGraw-Hill.
Van Bavel, C. H. M. 1966: Potential evaporation: the combination concept and its experimental verification. *Water Resources Research*, 2, 455–67.
Walker, H. O. 1962: Weather and climate. In J. B. Wills (ed.), *Agriculture and Land Use in Ghana*, London: Oxford University Press.

Chapter 4

Agbu, P. A., Ojanuga, A. G. and Olson, K. A. 1989: Soil–landscape relationships in the Sokoto-Rima Basin, Nigeria. *Soil Science*, 148 (2), 132–9.
Allan, W. 1965: *The African Husbandsman*. Edinburgh: Oliver & Boyd.

Allbrook, R. F. and Ratcliffe, D. J. 1988: Some physical properties of andepts from the Southern Highlands, Papua New Guinea. *Geoderma*, 41, 107–22.

Allison, L. E. 1947: Effects of microorganisms on the permeability of soils under prolonged submergence. *Soil Science*, 63, 439–50.

Anamosa, P. R., Nkedi-Kizza, P., Blue, W. G. and Sartain, P. R. 1990: Water movement through an aggregated, gravelly oxisol from Cameroon. *Geoderma*, 46, 263–81.

Andriesse, J. P. 1969–70: The development of the podzol morphology in the tropical lowlands of Sarawak (Malaysia). *Geoderma*, 3, 261–79.

Andriesse, J. P. 1974: *Tropical Lowland Peats of South-East Asia*, Department of Agricultural Research Communication 63. Amsterdam: Royal Tropical Institute.

Andriesse, J. P. 1988: *The Nature and Management of Tropical Peat Soils*, FAO Soils Bulletin 59. Rome: FAO.

Andriesse, W. and Scholten, J. J. 1982: *Acri-Orthic Ferralsol (Haplic Acrorthox) Jamaica*, Soil Monograph Paper 6. Wageningen: International Soil Museum.

Andriesse, J. R., van Breemen, N. and Blokhuis, W. A. 1973: The influence of mud lobsters (*Thalassina anomala*) on the development of acid sulphate soils in mangrove swamps in Sarawak. In H. Dost (ed.), *Acid Sulphate Soils, Proceedings of the 1st International Symposium*, Publication 18, Wageningen: Institute of Land Reclamation and Improvement, vol. 2, 11–39.

Anthony, E. J. and Marius, C. 1984: Géomorphologie, sediments et sols de la Baie de Sherbro (Sierra Leone meridionale). *Cahiers ORSTOM, Serie Pedologie*, 21 (1), 97–108.

Babalola, O. and Lal, R. 1977: Subsoil gravel horizons and maize root growth. I: Gravel concentration and bulk density effects. *Plant and Soil*, 46, 337–46.

Bachik, A. T., Raveendran N., Wong, S. P. and Pushparajah, E. 1985: Manganese toxicity symptoms in *Hevea*. In A. T. Bachik and E. Pushparajah (eds), *Proceedings, International Conference on Soils and Nutrition of Perennial Crops*, Kuala Lumpur: Malaysian Society of Soil Science, 67–74.

Beadle, L. C. 1960: The swamps of Uganda. *Shell Public Health and Agricultural News*, 3.

Beinroth, F. H., Uehara, G. and Ikawa, H. 1974: Geomorphic relations of oxisol and ultisols on Kauai, Hawaii. *Proceedings of the Soil Science Society of America*, 38, 128–31.

Bleeker, P. 1983: *Soils of Papua New Guinea*. Canberra: CSIRO.

Bleeker, P. and Parfitt, R. L. 1974: Volcanic ash and clay mineralogy at Cape Hoskins, New Britain, Papua New Guinea. *Geoderma*, 11, 123–35.

Boast, R. 1990: Dambos: a review. *Progress in Physical Geography*, 14, 153–77.

Botelho da Costa, J. V. and Cardoso Franco, E. P. 1965: Note on the concepts of ferrallitic soils and oxisols. *Pedologie*, Special No. 3, 181–4.

Bouckaert, W., De Dapper, M. and Alies, R. 1984: Soils and landscape in inland peninsular Malaysia. In *Proceedings 5th ASEAN Soil Conference, Bangkok, (10–23 June, 1984)*, Bangkok: Department of Land Development, C8.1–C8.25.

Boughey, A. S. 1957: Ecological studies of tropical coastlines. I: The Gold Coast, West Africa. *Journal of Ecology*, 45, 665–87.

Brammer, H. 1971: Coatings on seasonally-flooded soils. *Geoderma*, 5, 5–16.

Bravard, S. and Righi, D. 1988: Characteristics of clays in an oxisol–spodosol toposequence in Amazonia (Brazil). *Clay Mineralogy*, 23, 279–89.

Bravard, S. and Righi, D. 1989: Geochemical differences in an oxisol–spodosol toposequence of Amazonia, Brazil. *Geoderma*, 44, 29–42.

Brinkmann, R. 1970: Ferrolysis, a hydromorphic soil forming process. *Geoderma*, 3, 199–206.

Brinkmann, R. and Pons, L. J. 1968: A pedo-geomorphological classification and map of the Holocene sediments in the coastal plain of the three Guianas. *Soil Survey Paper 4*, Netherlands Soil Survey Institute.

Briones, A. A. and Veracion, J. G. 1965: Aggregate stability of some red soils of Luzon. *Philippines Agriculture*, 49, 153–67.

le Brusq, J. Y., Loyer, J. Y., Mougenot, B. and Carn, M. 1987: Nouvelles paragenèses à sulfates d'aluminium, de fer, et de magnesium, et leur distribution dans les sols sulfates acides du Senegal. *Science du Sol*, 25 (3), 173–84.

Buchanan, F. 1807: *A Journey from Malabar through the Countries of Mysore, Canara and Malabar*, vol. 2. London: East India Company.

Buol, S. W. and Sanchez, P. A. 1986: Red soils in the Americas: morphology, classification and management. In *Proceedings, International Symposium on Red Soils*, Beijing: Science Press and Amsterdam: Elsevier, 14–43.

Buol, S. W., Hole, F. D. and McCracken, R. J. 1980: *Soil Genesis and Classification*. Ames, IA: Iowa State University Press.

Cagauan, B. and Uehara, G. 1965: Soil anisotropy and its relationship to aggregate stability. *Proceedings of the Soil Science Society of America*, 29, 198–200.

Calvert, C. S., Buol, S. W. and Weed, S. B. 1980: Mineralogical characteristics and transformations of a vertical rock–saprolite–soil sequence in the North Carolina Piedmont. I: Profile morphology, chemical composition and mineralogy. *Journal of the Soil Science Society of America*, 44, 1096–1103.

Chenery, E. M. 1954: Acid sulphate soils in Central Africa. *Transactions, 5th International Congress of Soil Science, Leopoldville*, vol. 4, 195–8.

Cochrane, T. T. 1986: The distribution, properties and management of acid mineral soils in tropical South America. In *Proceedings, International Symposium on Red Soils*, Beijing: Science Press, and Amsterdam: Elsevier, 77–89.

Collinet, J. 1969: Contribution à l'étude des 'stone-lines' dans la région de Moyen-Ogooué (Gabon). *Cahiers ORSTOM, Serie Pedologie*, 7, 3–42.

Colmet-Daage, F. and Gautheyrou, J. 1974: Soil associations on volcanic material in tropical America with special reference to Martinique and Guadelope. *Tropical Agriculture (Trinidad)*, 51 (2), 121–8.

Coulter, J. K. 1957: Development of peat soils in Malaysia. *Malaysian Agricultural Journal*, 40, 188–99.

Daniels, R. B., Perkins, H. F., Maajek, B. F. and Gamble, E. E. 1978: Morphology of discontinuous phase plinthite and criteria for its field identification in the Southern United States. *Journal of the Soil Science Society of America*, 42, 944–9.

Delvaux, B., Herbillon, A. J. and Vielvoye, L. 1989: Characterization of a weathering sequence of soils derived from volcanic ash in Cameroon: taxonomic, mineralogical and agronomic implications. *Geoderma*, 45, 375–88.

Dent, D. 1986: *Acid Sulphate Soils: a Baseline for Research and Development*, Publication 35. Wageningen: International Institute of Land Reclamation and Improvement (ILRI).

Dent, F. J. 1980: Major production systems and soil-related constraints in South-East Asia. In *Soil-related Constraints to Food Production in the Tropics*, Los Banos (Philippines): International Rice Research Institute, 79–100.

Dijkermann, J. C. and Miedema, R. 1988: An usult–aqult–tropept catena in Sierra Leone, West Africa, 1: Characteristics, genesis and classification. *Geoderma*, 42, 1–27.

Dost, H. (ed.) 1973: *Acid Sulphate Soils*, Publication 18, 2 vols, Wageningen: International Institute of Land Reclamation and Improvement (ILRI).

Dost, H. and van Breemen, N. (eds) 1982: *Proceedings of the Bangkok Symposium on Acid Sulphate Soils*, Publication 31. Wageningen: International Institute for Land Reclamation and Improvement (ILRI).

Dudal, R. and Soepraptohardjo, M. 1960: Some considerations of the genetic relationship between Latosols and Andosols in Java, Indonesia. *Transactions, 7th International Congress of Soil Science (Madison)*, vol. 4, 229–37.

Edelmann, C. H. and van der Voorde, P. K. J. 1963: Important characteristics of alluvial soils in the tropics. *Soil Science*, 95, 258–63.

El-Swaify, S. A. 1980: Physical and mechanical properties of oxisol. In B. K. G. Theng (ed.), *Soils with Variable Charge*, New Zealand: DSIR Soil Bureau, 303–24.

Embrechts, J. and De Dapper, M. 1987: Morphology and genesis of hill-slope pediments in the Febe area (South Cameroon). *Catena*, 14, 31–43.

Escobar, G., Juardo, R. and Guerrero, V. 1972: Proiedades fisicas de algunos suelos derivados de ceniza volcanica del Altiplano de Pasto, Narino, Colombia. *Turriabla*, 22, 338–46.

Eswaran, H. 1970: Micromorphological indicators of pedogenesis in some tropical soils derived from basalts from Nicaragua. *Geoderma*, 7, 15–31.

Eswaran, H. and Ragumohan, G. 1973: The microfabric petroplinthite. *Proceedings of the Soil Science Society of America*, 37, 79–82.

Eswaran, H. and Stoops, G. 1979: Surface textures of quartz in tropical soils. *Journal of the Soil Science Society of America*, 43, 420–4.

Eswaran, H. and Sys, C. 1976a: Physiographic and chemical characteristics of the Quoin Hill toposequence. *Pedologie*, 26 (2), 152–67.

Eswaran, H. and Sys, C. 1976b: Micromorphological and mineralological properties of the Quoin Hill toposequence. *Pedologie*, 26 (3), 280–91.

Eswaran, H. and Tavernier, R. 1980: Classification and genesis of oxisols. In B. K. G. Theng (ed.), *Soils with Variable Charge*, New Zealand: DSIR Soil Bureau, 427–42.

Eswaran, H., Stoops, G. and Sys, C. 1977: The micromorphology of gibbsite forms in soils. *Journal of Soil Science*, 28, 136–43.

Eswaran, H., Ikwara, H. and Kimble, J. M. 1986: Oxisols of the world. In *Proceedings, International Symposium on Red Soils*, Beijing: Science Press, and Amsterdam: Elsevier, 90–123.

FAO/UNESCO 1977a: *Soil Map of the World*, vol. III, *Mexico and Central America*, Rome: FAO, and Paris: UNESCO.

FAO/UNESCO 1977b: *Soil Map of the World*, vol. IV, *South America*. Rome: FAO, and Paris: UNESCO.

FAO/UNESCO 1977c: *Soil Map of the World*, vol. VI, *Africa*. Rome: FAO and Paris: UNESCO.

FAO/UNESCO 1977d: *Soil Map of the World*, vol. VII, *South Asia*. Rome: FAO, and Paris: UNESCO.

FAO/UNESCO 1977e: *Soil Map of the World*, vol. IX, *Southeast Asia*. Rome: FAO, and Paris: UNESCO.

FAO/UNESCO 1977f: *Soil Map of the World*, vol. X, *Australasia*. Rome: FAO, and Paris: UNESCO.

FAO/UNESCO 1988: *Soil Map of the World*, Revised Legend. Rome: FAO, and Paris: UNESCO.

Fauck, R. 1963: The sub-group of leached ferruginous tropical soils with concretions. *African Soils*, 8, 407–30.

Floor, J. and Muyesu, C. C. 1986: Wetland soils in Kisii District, and their suitability for agriculture and other land uses. *Proceedings, 7th Annual Meeting of the Soil Science Society of East Africa, Arusha, 1985.*

Fosberg, F. R. 1954: Soils of the northern Marshall Atolls; with special reference to the Jemo Series. *Soil Science*, 78, 99–107.

Fox, J. E. D. and Hing, T. T. 1971: Soils and forest on an ultrabasic hill north-east of Ranau, Sabah. *Journal of Tropical Geography*, 32, 38–48.

Geiger, L. and Nettleton, W. D. 1979: Properties and geomorphic relationships of some soils of Liberia. *Journal of the Soil Science Society of America*, 43, 1192–8.

Godefroy, J. and Dormoy, J. 1983: Dynamique des éléments minéraux fertilisants dans les sols des bananeraies martiniquais. Deuxieme partie. *Fruits* 38 (6), 451–9.

Gomez-Rivas, C., Carillo-Pachon, I. F. and Estrada, G. 1982: Adsorption de potasio en andosoles de la zona cafetera 1. *Cenicafe (Colombia)*, 33 (4), 104–28.

Guedez, J. E. and Langohr, R. 1978: Some characteristics of pseudo-silts in a soil-toposequence of the Llanos Orientals (Venezuela). *Pedologie*, 28 (1), 118–31.

Haantjens, H. A. 1967: Part V, Pedology of the Safia-Pongani area. In *Lands of the Safia-Pongani Area, Papua New Guinea*, CSIRO land Resources Series 17, 98–141.

Hammond, R. F. 1971: *Survey of Peat Deposits on Makandudu, Milandu and Forkaidu Islands in Milandummandulu Atoll.* Report to the Government of the Maldives, FAO TA3013. Rome: FAO.

Harrassowitz, H. 1926: Laterit. *Fortschritte der Geologie und Paleontologie*, 4, 253–566.

Hassan, M. and Tavernier, R. 1986: Identification legend for soil series developed in the Upper Pliocene deposits of Bangladesh. *Pedologie*, 36 (1), 45–57.

Hesse, P. R. 1961: Some differences between the soils of *Rhizophora* and *Avicennia* mangrove swamps in Sierra Leone. *Plant and Soil*, 14 (4), 335–46.

Hossner, L. R., Freerouf, J. A. and Folsom, B. L. 1973: Solution phosphorus concentration and growth of rice in flooded soils. *Proceedings of the Soil Science Society of America*, 37, 548–55.

Humbel, F. X. 1988: Relations entre la porosité et le pédoclimat dans des sols du domaine ferrallitique au Cameroun. In M. Latham (ed.), *Management of Acid Soils*, 39–52.

Islam, M. A. 1966: Soils of East Pakistan. In *Scientific Problems of the Humid Tropical Zone: Deltas and their Implications*, Paris: UNESCO, 83–7.

Jordan, H. D. 1964: The relation of vegetation and soil to development of mangrove swamps for rice-growing in Sierra Leone. *Journal of Applied Ecology*, 1, 209–12.

Kellogg, C. E. 1949: Preliminary suggestions for the classification and nomenclature of Great Soil Groups in tropical and equatorial regions. *Commonwealth Bureau of Soil Science, Technical Communication*, 46, 76–85.

Kita, K. and Kawaguchi, K. 1960: The effects of the reduction of the soil under waterlogged conditions and the dehydration of the reduced soil upon soil structure. *Journal of Science Soil Manure (Japan)*, 31, 355–79, 495–8.

Koenigs, F. F. R. 1963: The puddling of clay soils. *Netherlands Journal of Agricultural Science*, 11, 145–56.

Latham, M. 1983: Les oxydisols dans quelques milieux insularies du Pacifique – Etapes de leur formation. *Science du Sol*, 21 (3–4), 163–71.

Lepsch, I. F., Buol, S. W. and Daniels, R. B. 1977a: Soils–landscape relationships in the Occidental Plateau of Sao Paulo State, Brazil. I: Geomorphic surfaces and soil mapping units. *Journal of the Soil Science Society of America*, 44, 104–9.

Lepsch, I. F., Buol, S. W. and Daniels, R. B. 1977b: Soils–landscape relationships in the Occidental Plateau of Sao Paulo State, Brazil. II: Soil morphology, genesis and classification. *Journal of the Soil Science Society of America*, 44, 109–15.

Lévêque, A. 1969: Le problème des soils à nappes de gravels: observations et réflexions préliminaires pour le socle granitogneissique au Togo. *Cahiers ORSTOM, Serie Pedologie*, 7, 43–69.

Lugo-López, M. A. and Juárez, J. 1959: Evaluation of the effects of soil moisture and other soil characteristics upon the aggregate stability of some tropical soils. *Journal of the Agricultural University of Puerto Rico*, 43 (4), 268–72.

Martini, J. A. 1969: Geographic distribution and characteristics of volcanic ash soils in Central America. In *UNDP/FAO/IICA Panel sobre Suelos Derivados Volcanicas de America Latina*, Turriabla, Costa Rica, 1–19.

Martini, J. A. and Luzuriaga, C. 1989: Classification and productivity of six Costa Rican andepts. *Soil Science*, 147 (5), 326–38.

Millington, A. C. 1985: Soil erosion and agricultural land use in Sierra Leone. Unpublished D. Phil. thesis, University of Sussex.

Millington, A. C., Helmisch, F. and Rhebergen, G. J. 1985: Inland valley swamps and bolis in Sierra Leone: hydrological and pedological considerations for agricultural development. *Zeitschrift für Geomorphologie, Suppl.-Bd.*, 52, 201–22.

Milne, G. 1936: Normal erosion as a factor in soil profile development. *Nature*, 138, 548–9.

Mizota, C. and Capelle, J. 1988: Characterization of some andept and andic soils in Rwanda, Central Africa. *Geoderma*, 41, 193–209.

Mizota, C. and van Reeuwijk, L. P. 1987: *Chemical and Clay Mineralogy of 'Andisols' and Related Soils from Diverse Climatic Regions*, Monograph 2. Wageningen: International Soil Reference and Information Centre.

Moberg, J. P. 1972: Some fertility problems in the West Lake region of Tanzania, including the effects of different forms of cultivation on the fertility of some ferralsols. *East African Agricultural and Forestry Journal*, 37, 35–46.

Mohr, E. D., Van Baren, F. A. and Von Schuylerlorgh, J. 1972: *Tropical Soils: A Comprehensive Study of Their Genesis*. The Hague: Mouton-Ichitar Baru-Van Hoeve.

Moniz, A. C. and Buol, S. W. 1982: Formation of an oxisol–ultisol transition in Sao Paulo, Brazil. I: Double-water flow model of soil development. *Journal of the Soil Science Society of America*, 46, 1228–33.

Moore, P. D. and Bellamy, D. J. 1974: *Peatlands*. London: Elek Science.

Morelli, M., Igue, K. and Fuentes, R. 1971: Effect of liming on the exchange complex and on the movement of calcium and magnesium. *Turriabla*, 21, 317–22.

Muller, J. P. and Bocquier, G. 1986: Dissolution of kaolinites and accumulation of iron oxides in lateritic-ferruginous nodules: mineralogical and microstructural transformations. *Geoderma*, 37, 113–63.

Nossin, J. J. 1962: Coastal sedimentation in northeastern Johore (Malaya). *Zeitschrift für Geomorphologie*, 6, 296–316.

Ojanuga, A. G. 1978: Genesis of soils in the metamorphic forest region of southwestern Nigeria. *Pedologie*, 28 (1), 105–17.

Paramanathan, S. and Eswaran, H. 1980: Morphological properties of oxisol. In B. K. G. Theng (ed.), *Soils with Variable Charge*, New Zealand: DSIR Soil Bureau, 35–44.

Paramanathan, S. and Eswaran, H. 1984: Problem soils of Malaysia. Their characteristics and management. In *Ecology and Management of Problem Soils in Asia*, Taiwan: Food and Fertilizer Technology Centre for the Asian Pacific Region, 1–13.

Parfitt, R. L., Childs, C. W. and Eden, D. N. 1988: Ferrihydrite and allophane in four andepts from Hawaii and implications for their classification. *Geoderma*, 41, 223–41.

Patrick, W. H. Jr and Khalid, R. A. 1974: Phosphate release and sorption by soils and sediments: effect of aerobic and anaerobic conditions. *Science*, 186, 53–5.

Patrick, W. H. Jr and Mahapatra, I. C. 1968: Transformation and availability to rice of nitrogen and phosphorus in waterlogged soils. *Advances in Agronomy*, 20, 323–59.

Patrick, W. H. Jr and Mikkelsen, D. S. 1971: Plant nutrient behaviour in flooded soil. In R. A. Olsen (ed.), *Fertilizer Technology and Use*, 2nd edn, Madison, WI: Soil Science Society of America, 187–215.

Piggott, C. J. 1968: *A Soil Survey of the Seychelles*, Technical Bulletin 2. Ministry of Overseas Development (Land Resources Division).

Ponnamperuma, F. N. 1965: Dynamic aspects of flooded soils and the nutrition of the rice plant. In International Rice Research Institute (ed.), *Dynamics of Flooded Soils*, Baltimore, MD: Johns Hopkins University Press, 295–328.

Ponnamperuma, F. N. 1972: The chemistry of submerged soils. *Advances in Agronomy*, 24, 29–96.

Ponnamperuma, F. N. 1977a: Physicochemical properties of submerged soils in relation to fertility. International Rice Research Institute, Research Papers Series 5.

Ponnamperuma, F. N. 1977b: Behaviour of minor elements in paddy soils. International Rice Research Institute, Research Papers Series 8.

Pons, L. J. and Driessen, P. M. 1975: Reclamation and development of waste land on oligotrophic peat and acid sulphate soils. In *Proceedings of Symposium on Development of Problem Soils in Indonesia, Jakarta.*

Pons, L. J. and Zonneveld, I. S. 1965: *Soil ripening and soil classification.* Publication 14, Wageningen: International Institute for Land Reclamation and Improvement.

Pons, L. J., van Breemen, N. and Driessen, P. 1982: Coastal sedimentary environments influencing the development of potential soil acidity. In *Acid Sulfate Weathering*, Special Publication 18, Madison, WI: Soil Science Society of America, 1–18.

Pushparajah, E. and Bachik, A. T. 1987: Management of acid tropical soils in southeast Asia. In ISBRAM (International Board for Soil Research and Management), *Management of Acid Tropical Soils for Sustainable Agriculture: Proceedings of an ISBRAM Workshop, Bangkok*, 13–41.

Quantain, P. 1974: Hypothèses sur la genèse des andosols en climat tropical: évolution de la pedogènese 'initiale' en milieu bien drainé sur roches volcaniques. *Cahiers ORSTOM, Serie Pedologie*, 12 (1), 3–12.

Reading, A. J. 1991: The stability of tropical residual soils from Dominica, West Indies. *Engineering Geology*, 31, 27–44.

Rhebergen, G. J. 1980: Swamp classification. In *Proceedings of a Seminar on the Role of the Land and Water Development Division of the MAF in Agricultural Development in Sierra Leone*, Paper 17, Freetown, Sierra Leone: Ministry of Agriculture.

Righi, D., Bravard, S., Chauvel, A., Ranger, J. and Robert, M. 1990: *In situ* study of soil processes in an oxisol–spodosol sequence of Amazonia (Brazil). *Soil Science*, 150 (1), 438–45.

Salati, E. and Vose, P. B. 1984: Amazon Basin: a system in equilibrium. *Science*, 225, 179–86.

Sanchez, P. A. 1972: Nitrogen fertilization and management of tropical rice. North Carolina Agricultural Experimental Station, Technical Bulletin 213.

Sanchez, P. A. 1976: *Properties and Management of Soils in the Tropics.* New York: Wiley.

Sanchez, P. A. 1987: Management of acid soils in the humid tropics of Latin America. In ISBRAM (International Board for Soil Research and Management), *Management of Acid Tropical Soils for Sustainable Agriculture: Proceedings of an ISBRAM Workshop, Bangkok*, 63–108.

Sanchez, P. A. and Cochrane, T. T. 1980: Soil constraints in relation to major farming systems in tropical America. In *Priorities for Alleviating Soil-related Constraints to Food Production*, Los Banos (Philippines): International Rice Research Institute, 107–39.

Sanchez, P. A., Gichuru, M. P. and Katz, L. B. 1982: Organic matter in major soils of the temperate and tropical regions. In *Proceedings 12th International Congress of Soil Science*, 1, 99–114.

Schellman, W. 1964: Zur lateritischen Verwitterung von Serpentinit. *Geologisches Jahrbuch*, 81, 645–78.

Scholten, J. J. and Andriesse, W. 1982: *Humic Acrisols (Orthoxic Palehumult) Jamaica*, Soil Monograph Paper 5. Wageningen: International Soil Museum.

Schwartz, D. 1988: Some podzols on Bateke Sands and their origins, People's Republic of Congo. *Geoderma*, 43, 229–38.

Schwartz, D., Guillet, B., Villemin, G. and Toutain, F. 1986: Les alios humiques des podzols tropicaux du Congo: constituants, micro- et ultrastructure. *Pedologie*, 36 (2), 179–98.

Shamshuddin, J., Mokthar, Nik and Tessens, R. 1984: Surface charge properties of weathered soils in peninsular Malaysia. In A. T. Bachik and E. Pushparajah (eds), *Proceedings of the International Conference on Soils and Nutrition of Perennial Crops*. Kuala Lumpur: Malaysian Society of Soil Science.

Siefferman, G. 1973: Les sols de quelques régions volcaniques du Cameroun: variations pédologiques et minéralogiques du milieu équatorial au milieu tropical. *Memoirs ORSTOM*, 66.

Smyth, A. J. and Montgomery, R. J. 1962: *Soils and Land Use in Central Western Nigeria*. Ibadan: Government Printer.

Spaargaren, O. C., Creutzeberg, D., van Reeuwijk, L. P. and Van Diepen, C. A. 1981: *Thionic Fluvisol (Sulphic Tropaquent)*, Soil Monograph Paper 1. Wageningen: International Soil Museum.

Stobbs, A. R. 1963: *The Soils and Geography of the Boliland Region of Sierra Leone*. Freetown: Government Printer.

Stocking, M. 1986: Tropical red soils: fertility management and degradation. Discussion Paper 189, University of East Anglia, School of Development Studies.

Stoops, G. 1968: Micromorphology of some characteristic soils of the lower Congo (Kinshasa). *Pedologie*, 28, 110–49.

Stromgaard, P. 1985: Biomass, growth, and burning of woodland in a shifting cultivation area of south central Africa. *Forest Ecology and Management*, 12, 163–78.

Suarez de Castro, F. and Rodriguez, A. 1958: Movimiento del aguna en el suelo. Estudios con lisimetrios monoliticos. *Federacion Nacional de Cafeteros de Colombia Bol. Tec.*, 2 (19), 1–19.

Suszcynski, E. 1984: The peat resources of Brazil. In *Proceedings, 7th International Peat Congress, Dublin*, vol. 1, 468–92.

Tamura, T., Jackson, M. L. and Sherman, G. D. 1953: Mineral content of low humid, humic, and hydrol humic latosols of Hawaii. *Proceedings of the Soil Science Society of America*, 17, 343.

Tavernier, R. and Sys, C. 1986: Red soils of central Africa. In *Proceedings, International Symposium on Red Soils*, Beijing: Science Press; and Amsterdam: Elsevier, 167–84.

Tawonas, D., Panichakul, S., Ratanarat, S. and Masangsul, W. 1984: Problem of laterite soil for field crop production in Thailand. In *Ecology and Management of Problem Soils in Asia*, Taiwan: Food and Fertilizer Technology Centre for the Asian Pacific Region, 50–7.

Tessens, E. and Zuayah, S. 1982: Permanent positive charge in ferralsols. *Journal of the Soil Science Society of America*, 46, 1103–6.

Thomas, M. F. and Thorp, M. B. 1985: Environmental change and episodic etchplanation in the humid tropics of Sierra Leone: the Koidu etchplain. In I. Douglas and T. Spencer (eds), *Environmental Change and Tropical Geomorphology*, London: Allen & Unwin, 239–67.

Thomas, M. F., Thorp, M. B. and Teeuw, R. M. 1985: Paleogeomorphology and the occurrence of diamondiferous deposits in Koidu, Sierra Leone. *Journal of the Geological Society (London)*, 142, 789–802.

Tie, Y. L. and Kueh, H. S. 1979: A review of lowland organic soils of Sarawak. Technical Paper 4, Research Branch, Department of Agriculture, Sarawak, Malaysia.

Trapnell, C. G., Martin, J. D. and Allan, W. 1948–50: *Vegetation–soil map of Northern Rhodesia*, Map 1:1,000,000, with memorandum by C. G. Trapnell. Lusaka: Government Printer.

Turner, F. T. and Gilliam, J. W. 1976: Diffusion as a factor affecting the availability of phosphorus in flooded soils. *Plant and Soil*.

Uehara, G. and Gillman, G. P. 1981: *Mineralogy, Chemistry and Physics of Tropical Soils with Variable Charge Clays*. Boulder, CO: Westview.

Van Der Gaast, S. J., Mizota, C. and Jansen, J. H. F. 1986: Curved smectite in soils from the volcanic ash in Kenya and Tanzania: a low angle X-ray powder diffraction study. *Clays and Clay Mineralogy*, 34, 665–71.
Van Ranst, E. and Doube, M. 1986: Etude comparative de l'évolution minéralogique des sols de meme âge sur trachy-basalte sur tuf trachytique dans les haut-plateaux du Cameroun Occidentale. *Pedologie*, 36 (2), 95–155.
Wada, K. 1985: Volcanic soils. *Advances in Soil Science*, 2, 174–229.
Wall, J. R. D., Hansell, J. R. F., Catt, J. A., Ormerod, E. C., Varley, J. A. and Webb, I. S. 1979: *The Soils of the Solomon Islands*. vol. 1. Resources Development Centre, Technical Bulletin 4.
Wright, A. C. S. 1963: Soils and land use of Western Samoa. *DSIR (New Zealand) Soil Bureau Bulletin*, 22.
Young, A. 1976: *Tropical Soils and Soil Survey*. Cambridge: Cambridge University Press.
Zhao Qi-guo and Shi Hua 1986: On the genesis, classification and characteristics of the soils in tropical and subtropical China. In *Proceedings, International Symposium on Red Soils*, Beijing: Science Press; and Amsterdam: Elsevier, 197–228.
Zijsvelt, M. F. W. and Torlach, D. A. 1975: Soil survey and land use potential of the Ala-Kapiura area, West New Britain, Papua New Guinea. *Research Bulletin*, 17, Department of Agriculture, Stock and Fisheries, Port Moresby.

Chapter 5

Aitken, S. R. and Leigh, C. H. 1992: *Vanishing Rain Forests*. Oxford: Clarendon.
Andaya, B. W. and Andaya, L. Y. 1982: *A History of Malaysia*. London: Macmillan.
Anderson, J. M. and Swift, M. J. 1983: Decomposition in tropical forests. In S. L. Sutton, T. C. Whitmore and A. C. Chadwick (eds), *Tropical Rain Forest: Ecology and Management*, Oxford: Blackwell, 287–309.
Anderson, J. M., Proctor, J. and Vallack, H. W. 1983: Ecological studies in four contrasting lowland rain forests in Gunung Mulu National Park, Sarawak III. Decomposition processes and nutrient losses from leaf litter. *Journal of Ecology*, 71, 503–28.
Anon. 1972: Regenwasseranalysen aus Zentralamazonien, ausgeführt in Manaus, Amazonas, Brasilien, von Dr Harald Ungemach. *Amazoniana (Keil)*, 3, 186–98.
Ashton, P. S. 1964: *Ecological Studies in the Mixed Dipterocarp Forests of Brunei State*. Oxford: Clarendon.
Ashton, P. S. 1971: The plants and vegetation of Bako National Park. *Malayan Nature Journal*, 24, 151–62.
Aubreville, A. 1938: La forêt coloniale: les forêts de l'Afrique occidentale française. *Annals Academie des Sciences Coloniale, Paris*, 9, 1–245.
Ayensu, E. 1974: Plant and bat interactions in west Africa. *Annals of Missouri Botanical Gardens*, 61, 702–24.
Baillie, I. C. and Ashton, P. S. 1983: Some aspects of the nutrient cycle in mixed dipterocarp forests in Sarawak. In S. L. Sutton, T. C. Whitmore and A. C. Chadwick (eds), *Tropical Rain Forest: Ecology and Management*, Oxford: Blackwell, 347–56.

van Balgooy, M. M. J. and Tantra, I. G. M. 1986: The vegetation in two areas in Sulawesi, Indonesia. *Bulletin Penelitian Hutan, Special Issue*, 61pp.

Bazzaz, F. A. and Pickett, S. T. A. 1980: Physiological ecology of tropical succession. *Annual Review of Ecology and Systematics*, 11, 287–310.

Beaman, R. S., Beaman, J. H., Marsh, C. W. and Woods, P. V. 1985: Drought and forest fires in Sabah in 1983. *Sabah Society Journal*, 8, 10–30.

Benson, W. W. 1984: Amazon ant-plants. In G. T. Prance and T. E. Lovejoy (eds), *Amazonia*, Oxford: Pergamon, 239–66.

Bentley, B. L. 1977: Extrafloral nectaries and protection by pugnacious bodyguards. *Annual Review of Ecology and Systematics*, 8, 402–27.

Benzing, D. H. 1983: Vascular epiphytes: a survey with special reference to their interactions with other organisms. In S. L. Sutton, T. C. Whitmore and A. C. Chadwick (eds), *Tropical Rain Forest: Ecology and Management*, Oxford: Blackwell, 11–24.

Bernard, F. 1970: Etude de la litière et desa contribution au cycle des éléments minéraux en forêt ombrophile de Côte d'Ivoire. *Oecologia Plantarum*, 7, 279–300.

Bernhard-Reversat, F. 1975: Recherches sur l'écosystème de la forêt subéquatoriale de basses Côte d'Ivoire. Les cycles des macroéléments. *La Terre et la Vie*, 29, 229–54.

Bibby, C. J., Crosby, M. J., Long, A. J. and Sattersfield, A. J., in press: Applying GIS to identify and map avian areas of endemism. *Journal of Biogeography*.

Blasco, F. 1977: Outlines of ecology, botany and forestry of the mangals of the Indian subcontinent. In V. J. Chapman (ed.), *Wet Coastal Ecosystems*, Amsterdam: Elsevier, 241–60.

Bourliere, F. 1983: Animal species diversity in tropical forests. In F. B. Golley, (ed.), *Tropical Rain Forest Ecosystems*, Amsterdam: Elsevier, 77–91.

Brasell, H. M. and Sinclair, D. F. 1983: Elements returned to forest floor in two rainforest and three plantation pots in tropical Australia. *Journal of Ecology*, 71, 367–78.

Briggs, J. G. 1985: The current Nepenthes situation in Borneo. *Malayan Naturalist*, February, 46–8.

Brinkmann, W. L. F. 1985: Studies on hydrogeochemistry of a tropical lowland forest system. *Geojournal*, 11, 89–101.

Bronchart, R. 1963: Recherches sur la développement de *Geophila renaris* de Wild. et Th. Dur. dans les conditions écologiques d'un sous-bois forestier équatoriale. Influence sur la mise à fleurs d'une perte en eau disponsible du sol. *Memoires de la Société Royale des Sciences de Liège (Series 5)*, 8, 1–181.

Bruijnzeel, L. A. 1983: Hydrological and biogeochemical aspects of man-made forests in south-central Java. Unpublished Ph.D. thesis, Vrije Universiteit Amsterdam.

Brünig, E. F. 1969: The classification of forest types in Sarawak. *Malayan Forester*, 32, 143–79.

Brünig, E. F. 1970: Stand structure, physiognomy and environmental factors in some lowland forests in Sarawak. *Tropical Ecology*, 11, 26–4.

Brünig, E. F. 1971: On the ecological significance of drought in the equatorial wet evergreen (rain) forest of Sarawak (Borneo). In J. R. Flenley (ed.), *The Water Relations of Malesian Forests*, Hull: Department of Geography, University of Hull, Miscellaneous Series 11, 66–88.

Brünig, E. F. 1973: Species richness and stand diversity in relation to site and succession in Sarawak and Brunei (Borneo). *Amazoniana*, 4, 293–320.

Budowski, G. 1965: Distribution of tropical American rain forest species in the light of successional processes. *Turriabla*, 15, 40–2.

Bunning, E. 1947: *In den Waldern Nordsumatras*. Bonn: F. Dummlers.

Bush, M., Jones, P. and Richards, K. 1986: *The Krakatoa Centenary Expedition 1983. Final Report*. Hull: Department of Geography, University of Hull, Miscellaneous Series 3.

Cabrera, A. L. and Willink, A. 1973: Biogéographica de América Latina. *OEA, Série Biologia, Monogr*., 13, 1–120.

Cachan, P. 1963: Sinification écologique des variations microclimatiques verticales dans la fôret sempervirente de Basse Côte d'Ivoire. *Annales de la Faculté des Sciences, Université de Dakar*, 8, 89–155.

Cachan, P. and Duval, J. 1963: Variations microclimatiques verticales et saisonnières dans la fôret sempervirente de Basse Côte d'Ivoire. *Annales de la Faculté des Sciences, Université de Dakar*, 8, 5–87.

Cain, S. A., Oliveria Castro, G. M. de, Pures, J. M. and da Silva, N. T. 1956: Application of some phytosociological techniques to Brazilian rain forest. *American Journal of Botany*, 43 (10), 911–41.

Caldwell, M. M. 1981: Plant response to solar ultraviolet radiation. In O. L. Lange, P. S. Nobel, C. B. Osmund and H. Ziegler (eds), *Physiological Plant Ecology 1*, Berlin: Springer, 169–97.

Carter, G. S. 1934: Reports of the Cambridge expedition to British Guiana, 1933. Illumination of the rain forest at ground level. *Journal of the Linnean Society (Zoology)*, 38, 579–89.

Chapman, V. J. 1975: *Mangrove Vegetation*. Lehre: Cramer.

Chapman, V. J. (ed.) 1977: *Wet Coastal Ecosystems*. Amsterdam: Elsevier.

Chen, Y. 1986: Early Holocene vegetation dynamics of Lake Barrine basin, northeast Queensland, Australia. Unpublished Ph.D. thesis, Australian National University, Canberra.

Chi-Wu Wang, 1965: *The Forest of China with a Survey of Grassland and Desert Vegetation*, Maria Moors Cabot Foundation, Publication 5. Cambridge, MA. Harvard University Press.

Chunkao, K., Tangtham, N. and Ungkulpakdikul, S. 1971: Measurements of rainfall in early wet season under hill- and dry-evergreen, natural teak forests of Thailand. *Kog-ma Watershed Research Bulletin*, 10, Bangkok.

Clements, R. G. and Colon, J. A. 1975: The rainfall interception process and mineral cycling in a montane rainforest in Puerto Rico. In F. G. Howell, J. B. Gentry and M. N. Smith (eds), *Mineral Cycling in Southeastern Ecosystems*. Oak Ridge, TN: US Energy Research and Development Administration, 813–23.

Cole, N. H. A. 1968: *The Vegetation of Sierra Leone*. Njala (Sierra Leone): Njala University Press.

Colinvaux, P. 1987: Amazon diversity in light of the palaeoecological record. *Quaternary Science Reviews*, 6, 93–114.

Collins, N. M. 1983: Termite populations and their role in litter removal in Malaysian rain forests. In S. L. Sutton, T. C. Whitmore and A. C. Chadwick (eds), *Tropical Rain Forests: Ecology and Management*, Oxford: Blackwell, 311–16.

Collins, N. M. and Morris, M. G. 1985: Threatened swallowtail butterflies of the world. In *The IUCN Red Data Book*, Gland: IUCN.

Connell, J. H. 1978: Diversity in tropical rain forests and coral reefs. *Science*, 199, 1302–10.

Corlett, R. T. 1986: The mangrove understorey: some additional observations. *Journal of Tropical Ecology*, 2, 93–4.
Corner, E. J. H. 1964: *The Life of Plants*. London: Weidenfeld & Nicolson.
Corner, E. J. H. 1988: *Wayside Trees of Malaya*, 3rd edn. Kuala Lumpur: Malayan Nature Society.
Cremers, G. 1973: Architecture de quelques lianes d'Afrique tropicale. *Candoella*, 28, 249–80.
Cremers, G. 1974: Architecture de quelques lianes d'Afrique tropicale. *Candoella*, 29, 57–110.
Dantas, M. and Phillipson, J. 1989: Litterfall and litter nutrient content in primary and secondary Amazonian 'terra firme' forest. *Journal of Tropical Ecology*, 5, 27–36.
Davis, D. D. 1962: Mammals of the lowland rain forest of North Borneo. *Bulletin of the National Museum Singapore*, 31, 1–129.
Deitrich, W. E., Windsor, D. M. and Dunne, T. 1982: Geology, climate and hydrology of Barro Colorado Island. In E. G. Leigh, A. S. Rand and D. M. Windsor (eds), *The Ecology of a Tropical Forest. Seasonal Rhythms and Long-term Changes*, Washington, DC: Smithsonian Institution Press, 21–46.
Deshmukh, I. 1986: *Ecology and Tropical Biology*. Oxford: Blackwell.
Doley, D. 1981: Tropical and subtropical forests and woodlands. In T. T. Kozlowski (ed.), *Water Deficits and Plant Growth*, vol. VI, *Woody Plant Communities*, New York: Academic Press, 209–33.
Dolph, G. E. and Dilcher, D. E. 1980: Variation in leaf size with respect to climate in Costa Rica. *Biotropica*, 12, 91–9.
Ducke, A. and Black, G. A. 1945: Notas sobre a fitogeographia de Amazônia Brasileira. *Boletin Tecnico. IAN (Belém)*, 29, 1–62.
Dunn, F. L. 1975: *Rain-forest Collectors and Traders: a Study of Resource Utilization in Ancient and Modern Malaya*, Monographs of the Malaysian Branch of the Royal Asiatic Society. Kuala Lumpur: Council of the Malaysian Branch of the Royal Asiatic Society.
Edwards, P. J. 1977: Studies of mineral cycling in a montane rain forest in New Guinea, II: The production and disappearance of litter. *Journal of Ecology*, 70, 807–28.
Edwards, P. J. 1982: Studies of mineral cycling in a montane rain forest in New Guinea, V: Rates of cycling in throughfall and litterfall. *Journal of Ecology*, 70, 807–28.
Ellenberg, H. and Mueller-Dombois, D. 1967: Tentative physiognomic-ecological classification of plant formations of the Earth. *Berichte des Geobotanischen der Eidgenössischen Technischen Hochschule Stiftung Rübel Zürich*, 37, 21–55.
Erwin, T. L. 1982: Tropical forests: their richness in Coleoptera and other arthropod series. *Coleopterist's Bulletin*, 36, 74–5.
Ewel, J. 1980: Tropical succession: manifold routes to maturity. *Biotropica*, 12, 2–7.
Fassbender, H. W. and Grimm, V. 1981: Ciclos bioquimicos en un ecosistema forestal de los Andes Occidentales de Venezuela, II. Producción y descomposición de los reduos vegetales. *Turriabla*, 31, 39–47.
Flenley, J. R. 1979: *The Equatorial Rain Forest: A Geological History*. London: Butterworth.
Fletcher, N., Oberbauer, S. F. and Strain, B. R. 1985: Vegetation effects on microclimate in lowland tropical forest in Costa Rica. *International Journal of Biometeorology*, 29, 145–55.

Gentry, A. H. 1988: Changes in plant community diversity and floristic composition along environmental and geographical gradients. *Annals of the Missouri Botanical Garden*, 75, 1–34.

Golley, F. B. (ed.) 1983: *Tropical Rain Forest Ecosystems*. Amsterdam: Elsevier.

Golley, F. B., McGinnis, J. T., Clements, R. G., Child, G. I. and Deuver, M. J. 1975: *Mineral Cycling in a Tropical Moist Forest Ecosystem*. Athens, GA: University of Georgia Press.

Gómez-Pompa, A., Whitmore, T. C. and Hadley, M. 1991: *Rain Forest Regeneration and Management*. Paris: UNESCO; and Carnforth: Parthenon.

Gong, W.-K. and Ong, J.-E. 1983: Litter production and decomposition in a coastal hill dipterocarp forest. In S. L. Sutton, T. C. Whitmore and A. C. Chadwick (eds), *Tropical Rain Forest: Ecology and Management*, Oxford: Blackwell, 275–85.

Goulding, M. 1984: Forest fishes of the Amazon. In G. T. Prance and T. E. Lovejoy (eds), *Amazonia*, Oxford: Pergamon, 267–77.

Grimm, V. and Fassbender, W. H. 1981: Ciclos bioquímicos en un ecosistema forestal de los Andes Occidentales de Venezuela, III. Ciclo hidrológico y translocación des elementos químicos con el agua. *Turriabla*, 31, 89–99.

Hall, J. B. and Swaine, M. D. 1976: Classification and ecology of closed-canopy forest in Ghana. *Journal of Ecology*, 64, 913–51.

Hallé, F. and Oldeman, R. A. A. 1970: *Essai sur l'Architecture et la Dynamique de Croissance des Arbres Tropicaux*. Paris: Masson.

Hallé, F., Oldemann, R. A. A. and Tomlinson, P. B. 1978: *Tropical Trees and Forests: an Architectural Analysis*. Berlin: Springer.

Hamilton, A. C. 1976: The significance of patterns of distribution shown by forest plants and animals in tropical Africa for reconstruction of Upper Pleistocene palaeoenvironments. *Palaeoecology of Africa and the Surrounding Islands*, 9, 63–97.

van der Hammen, T. 1974: The Pleistocene changes of vegetation and climate in tropical South America. *Journal of Biogeography*, 1, 3–26.

Heaney, A. and Proctor, J. 1989: Chemical elements in litter in forests on Volcan Brava, Costa Rica. In J. Proctor (ed.), *Mineral nutrients in tropical forest and savanna ecosystems*, Oxford: Blackwell, 255–72.

Herrera, R. 1979: Nutrient distribution and cycling in an Amazon caatinga forest on spodosols in southern Venezuela. Unpublished Ph.D. dissertation, University of Reading.

Holdridge, L. R. 1967: *Life Zone Ecology*. San José, Costa Rica: Tropical Science Center.

Hommel, P. W. F. M. 1990: A phytosociological study of a forest area in the humid tropics (Ujung Kulon, West Java, Indonesia). *Vegetatio*, 89, 39–54.

Hope, G. S. 1976: The vegetational history of Mt. Wilhelm, Papua New Guinea. *Journal of Ecology*, 64, 627–63.

Hubbell, S. P. and Foster, R. B. 1983: Diversity in canopy trees in a neotropical forest and implications for conservation. In S. L. Sutton, T. C. Whitmore and A. C. Chadwick (eds), *Tropical Rain Forest: Ecology and Management*, Oxford: Blackwell, 25–41.

Hyndman, D. C. and Menzies, J. T. 1990: Rain forests of the Ok Tedi headwaters, New Guinea: an ecological analysis. *Journal of Biogeography*, 17, 241–73.

Irmler, U. 1977: Inundation forest types in the vicinity of Manaus. *Biogeographica*, 8, 17–30.

Jacobs, M. 1987: *The Tropical Rain Forest.* Berlin: Springer.

Janzen, D. H. 1976a: The microclimate difference between a deciduous forest and adjacent riparian forest in Guanacaste Province, Costa Rica. *Brenesia*, 8, 29–33.

Janzen, D. H. 1976b: Why are there so many species of insects? In *Proceedings of the International Congress Entomology 1976*, 232–6.

Janzen, D. H. 1985: Mangroves: where's the understorey? *Journal of Tropical Ecology*, 1, 89–92.

Jeník, J. 1978: Roots and root systems in tropical trees: morphologic and ecologic aspects. In P. B. Tomlinson and N. M. Zimmerman (eds), *Tropical Trees as Living Systems*, Cambridge: Cambridge University Press, 323–49.

Johansson, D. 1974: Ecology of vascular epiphytes in west African rain forests. *Acta Phytogeographica Seucia*, 59, 1–29.

John, O. M. 1973: Accumulation and decay of litter and net production of forest in tropical West Africa. *Oikos*, 24, 430–5.

Jones, E. W. 1956: Ecological studies on the rain forest of southern Nigeria, IV: The plateau forest of the Okumu Forest Reserve (Part II). *Journal of Ecology*, 44, 83–117.

Jordan, C. F. 1982: The nutrient balance of an Amazonian rain forest. *Ecology*, 63, 647–54.

Jordan, C. F. 1989: *An Amazonian Rain Forest: The Structure and Function of a Nutrient Stressed Ecosystem and the Impact of Slash-and-Burn Agriculture.* Paris: UNESCO; and Park Ridge, NJ: Parthenon.

Jordan, C. F., Kline, J. R. and Sasscer, D. S. 1972: Relative stability of mineral cycles in forest ecosystems. *American Naturalist*, 106, 237–53.

Jordan, C. F., Golley, F., Hall, J. and Hall, J. 1980: Nutrient scavenging of rainfall by the canopy of an Amazonian rain forest. *Biotropica*, 14, 1–9.

Kellman, M. C., Hudson, J. and Sanmugadas, K. 1982: Temporal variability in atmospheric nutrient influx to a tropical ecosystem. *Biotropica*, 14, 1–9.

Kenworthy, J. B. 1971: Water and nutrient cycling in a tropical rain forest. In J. Flenley (ed.), *The Water Relations of Malesian Forests*, Hull: University of Hull, 49–65.

Kiew, R. 1982: Observations on leaf color, epiphyll cover, and damage on Malayan *Iguanura wallichiana. Principes*, 26, 200–4.

Kiew, R. 1989: Conservation status of palms in Peninsular Malaya. *Malaysian Naturalist*, November, 3–15.

Klinge, H. 1973: Biomasa y materia orgánica del suelo en el ecosistema de la pluviselva centroamazónica. *Acta Ciencia Venezolana*, 24, 174–81.

Kramer, F. 1933: De naturrlijke verjonging in het Goenoneg Gdeh complex. *Tectona*, 26, 156–85.

Kwesiga, F. R. 1985: Aspects of the growth and physiology of tropical tree seedlings in shade. Unpublished Ph.D. thesis, University of Edinburgh.

Lam, P. K. S. and Dudgeon, D. 1985: Seasonal effects on litterfall in a Hong Kong mixed forest. *Journal of Tropical Ecology*, 1, 55–64.

Lawson, G. W., Armstrong-Mensah, K. O. and Hall, J. B. 1970: A catena in moist semi-deciduous forest near Kade, Ghana. *Journal of Ecology*, 58, 371–98.

Lee, D. H. and Lowry, J. B. 1974: Physical basis and ecological significance of iridescence in blue plants. *Nature*, 254, 50–1.

Lee, D. H., Lowry, J. B. and Stone, B. C. 1979: Abaxial anthocyanin layer in leaves of tropical rain forest plants: enhancer of light capture in deep shade. *Biotropica*, 11, 70–7.

Leiberman, D., Leiberman, M., Hartshorn, G. and Peralta, R. 1985: Growth rates and age–size relationships of tropical wet forest trees in Costa Rica. *Journal of Tropical Ecology*, 1, 97–109.

Longman, K. A. and Jeník, J. 1987: *Tropical Forest and its Environment*, 2nd edn. London: Longman.

Lugo, A., Gonzalez-Liboy, J. A., Cintrón, B. and Dugger, K. 1978: Structure, productivity and transpiration of a subtropical dy forest in Puerto Rico. *Biotropica*, 10, 278–91.

Mabberley, D. J. 1983: *Tropical Rain Forest Ecology*. Glasgow: Blackie.

Mackie, C. 1984: The lessons behind East Kalimantan's forest fire losses. *Borneo Research Bulletin*, 16 (7), 63–74.

MacKinnon, J. A. 1972: The behaviour and ecology of the Orang Utan (*Pongo pygmaeus*). Unpublished D.Phil. thesis, University of Oxford.

Madge, D. S. 1965: Leaf fall and litter dispersion in a tropical forest. *Pedobiologica*, 5, 273–88.

Madison, M. 1977: Vascular epiphytes: their systematic occurrence and salient features. *Selkyana*, 2, 1–13.

Malingreau, J.-P., Stephens, G. and Fellows, L. 1985: Remote sensing of forest fires: Kalimantan and North Borneo in 1982–83. *Ambio*, 14, 314–21.

Manokaran, N. 1979: Stemflow, throughfall and rainfall interception in a lowland tropical rain forest in peninsular Malaysia. *Malaysian Forester*, 42, 174–201.

Manokaran, N. 1980: The nutrient contents of precipitation, throughfall and stemflow in a lowland tropical rain forest in peninsular Malaysia. *Malaysian Forester*, 43, 266–89.

Martínez-Yrízar, A. and Sarukhán, J. 1990: Litterfall patterns in a tropical deciduous forest in Mexico over a five-year period. *Journal of Tropical Biology*, 6, 433–44.

Mayr, E. and O'Hara, R. J. 1986: The biogeographic evidence supporting the Pleistocene refuge hypothesis. *Evolution*, 40, 55–67.

McClure, F. A. 1966: *The Bamboos. A Fresh Perspective*. Cambridge, MA: Harvard University Press.

Medway, Lord 1978: *The Wild Animals of Malaya (Peninsular Malaysia) and Singapore*, 2nd edn. Oxford: Oxford University Press.

Meijer, W. 1959: Plantsociological analysis of montane rain forest near Tjibodas, West Java. *Acta Botanica Neerlandica*, 8, 221–91.

Montgomery, G. G. and Sunquist, M. E. 1978: Habitat selection and use by two-toed and three-toed Sloths. In G. G. Montgomery (ed.), *The Ecology of Arboreal Herbivores*, Washington DC: Smithsonian Institution Press, 329–59.

Morgan, D. C. and Smith, H. 1981: Non-photosynthetic responses to light quality. *Encyclopaedia of Plant Physiology (NS)*, 12A, 108–34.

Myers, N. 1984: *The Primary Source, Tropical Forests and Our Future*. New York: Norton.

Nye, P. H. 1961: Organic matter and nutrient cycles under moist tropical forest. *Plant and Soil*, 13, 333–46.

Nye, P. H. and Greenland, D. J. 1960: *The Soil under Shifting Cultivation*. Commonwealth Bureau of Soils, Technical Communication 51.

Odum, H. T. and Pigeon, R. F. (eds) 1970: *A Tropical Rain Forest: a Study of Irradiation and Ecology at El Verde, Puerto Rico*, 3 vols. Washington, DC: US Atomic Energy Commission.

Ogawa, H. 1978: Litterfall and mineral nutrient content of litter in Pasoh Forest. *Malayan Nature Journal*, 30, 375–80.

Ohsawa, M., Nainggolan, P. H. J., Tanaka, N. and Anwar, C. 1985: Altitudinal zonation of forest vegetation on Mount Kerinci, Sumatra: with comparison to zonation in a temperate region of south east Asia. *Journal of Tropical Ecology*, 1, 193–216.

Oldeman, R. A. A. 1978: Architecture and energy exchange of dicotyledonous trees in the forest. In P. B. Tomlinson and M. N. Zimmermann (eds), *Tropical Trees as Living Systems*, Cambridge: Cambridge University Press, 535–60.

Parker, G. G. 1983: Throughfall and stemflow in the forest nutrient cycle. *Advances in Ecological Research*, 13, 58–133.

Pires, J. M. and Prance, G. T. 1984: The vegetation types of the Brazilian Amazon. In G. T. Prance and T. E. Lovejoy (eds), *Amazonia*, Oxford: Pergamon, 109–45.

Poore, M. E. D. 1968: Studies in Malaysian rain forest, 1: The forest on Triassic sediments in Jangka Forest Reserve. *Journal of Ecology*, 56, 143–96.

Prance, G. T. 1979: Notes on the vegetation of Amazonia, III: The terminology of Amazon forest types subject to inundation. *Brittonia*, 31, 26–38.

Proctor, J. 1987: Nutrient cycling in primary and old secondary rainforests. *Applied Geography*, 7, 135–52.

Proctor, J., Anderson, J. M., Chai, P. and Vallack, H. W. 1983a: Ecological studies in four contrasting lowland rain forests in Gunung Mulu National Park, Sarawak, I: Forest environment, structure and floristics. *Journal of Ecology*, 71, 237–60.

Proctor, J., Anderson, J. M. and Vallack, H. W. 1983b: Comparative studies on soils and litterfall in forests at a range of altitudes on Gunung Mulu, Sarawak. *Malaysian Forester*, 46, 60–76.

Purata, S. E. 1986: Floristic and structural changes during old-field succession in the Mexican tropics in relation to site history and species availability. *Journal of Tropical Ecology*, 2, 257–76.

Rai, S. N. and Proctor, J. 1986: Ecological studies on four rain forest sites in Karnataka, India, II: Litterfall. *Journal of Ecology*, 71, 237–60.

Rejmanek, M. 1976: Centres of species diversity and centres of species diversification. *Encyclopaedia of Plant Physiology*, 15, 1077–1105.

Richards, P. W. 1952: *The Tropical Rain Forest*, 1st edn. Cambridge: Cambridge University Press.

Ricklefs, R. E. 1973: *Ecology*. Newton, MA: Chiron.

Rijksen, H. D. 1978: A field study of Sumatran orang utans (*Pongo pygmaeus abelii* Lesson 1827). Ecology, behaviour and conservation. *Mededelingen Landbouwhogeschool Wageningen*, 78 (2), 421 pp.

Riswan, S., Kenworthy, J. B. and Kuswata, K. 1985: The estimation of temporal processes in tropical rain forest: a study of primary mixed dipterocarp forest in Indonesia. *Journal of Tropical Ecology*, 1, 171–82.

Rosen, B. R. 1981: The tropical high diversity enigma – a coral's eye view. In P. L. Forey (ed.), *The Evolving Biosphere*, Cambridge: Cambridge University Press, 103–29.

Rosevar, D. R. 1947: Mangrove swamps. *Farm and Forest*, 8, 23–30.

Salati, E., Dall'olio, A., Matsui, E. and Gat, J. R. 1979: Recycling of water in the Amazon Basin: an isotopic study. *Water Resources Research*, 15, 1250–8.

Salo, J., Kalliola, R., Hakkinen, I., Makinen, Y., Niemala, P., Puhhakka, M. and Coley, P. D. 1986: River dynamics and the diversity of Amazon lowland forest. *Nature*, 322, 254–8.

Sanford, R. L., Jr, Saldarriaga, J., Clark, K. E., Uhl, C. and Herrera, R. 1985: Amazon rain-forest fires. *Science*, 227, 53–5.

Scholander, P. F., van Dam, L. and Scholander, S. I. 1955: Gas exchange in roots of mangroves. *American Journal of Botany*, 42, 92–8.

Schultz, J. P. 1960: *Ecological Studies on Rain Forest in Northern Surinam*. Amsterdam: North-Holland.

Servant, J., Delmas, R., Rancher, J. and Rodriguez, M. 1984: Aspects of the cycle of inorganic nitrogen compounds in the tropical rain forest of the Ivory Coast. *Journal of Atmospheric Chemistry*, 1, 391–401.

Sollins, P. and Drewry, G. 1970: Electrical conductivity and flow rate of water through the forest canopy. In H. T. Odum and R. F. Pigeon (eds), *A Tropical Rain Forest: a Study of Irradiation and Ecology at El Verde, Puerto Rico*. Washington DC: US Atomic Energy Commission, F29–F33.

van Steenis, C. G. G. J. 1958: Basic principles of rain forest sociology. In *Study of Tropical Vegetation*, Paris: UNESCO, 159–65.

van Steenis, C. G. G. J. 1962: The mountain flora of the Malaysian tropics. *Endeavour*, 21 (83–83), 183–93.

Sutton, S. L. 1983: The spatial distribution of flying insects in tropical rain forests. In S. L. Sutton, T. C. Whitmore and A. C. Chadwick (eds), *Tropical Rain Forests: Ecology and Management*, Oxford: Blackwell, 77–91.

Sutton, S. L., Whitmore, T. C. and Chadwick, A. C. (eds) 1983: *Tropical Rain Forests: Ecology and Management*, Oxford: Blackwell.

Swaine, M. D. and Hall, J. B. 1983: Early succession on cleared forest land in Ghana. *Journal of Ecology*, 71, 601–27.

Swaine, M. D. and Hall, J. B. 1988: The mosaic theory of forest regeneration and the determination of forest composition in Ghana. *Journal of Tropical Ecology*, 4, 253–69.

Swaine, M. D., Liebermann, D. and Putz, F. E. 1987: The dynamics of tree populations in tropical forest: a review. *Journal of Tropical Ecology*, 3, 359–66.

Sylvester-Bradley, R., de Oliveira, L. A., de Podesta Filho, J. A. and St John, T. V. 1980: Nodulation of legumes, nitrogenase activity of roots and occurrence of nitrogen-fixing *Azospirillum* spp. in representative soils of Central Amazonia. *Agroecosystems*, 6, 249–66.

Tanner, E. V. J. 1980: Studies on the biomass and productivity in a series of montane forests in Jamaica. *Journal of Ecology*, 65, 883–918.

Tanner, E. V. J. 1981: The decomposition of leaf litter in Jamaican montane rain forests. *Journal of Ecology*, 73, 553–68.

Terborgh, J. 1984: The ecology of Amazon primates. In G. T. Prance and T. E. Lovejoy (eds), *Amazonia*, Oxford: Pergamon, 109–45.

Thom, B. G. 1967: Mangrove ecology and deltaic geomorphology. *Journal of Ecology*, 55, 301–43.

Tixier, P. 1966: *Flore et Vegetation Orophiles de l'Asie Tropicale*. Paris: Société d'édition d'enseignment superier.

Turvey, N. D. 1974: Water in the nutrient cycle of a Papuan rain forest. *Nature*, 251, 414–15.

UNESCO, 1978: *Tropical Rain Forest Ecosystems*. Paris: UNESCO/UNEP/FAO, UNESCO.

Vareschi, V. 1980: *Vegetationsokologie der Tropen*. Stuttgart: Ulmer.

Walker, D. 1982: Speculations about the origin and evolution of Sunda-Sahul rain forests. In G. T. Prance (ed.), *Biological Diversification in the Tropics*, New York: Columbia University Press, 554–75.

Walker, D. and Chen, Y. 1987: Palynological light on tropical rainforest dynamics. *Quaternary Science Reviews*, 6, 77–92.

Walter, H. 1979: *Vegetation of the Earth and Ecological Systems of the Geobiosphere*, 2nd edn, trans. J. Weiser. New York: Springer.

Wang, Chi-Wu 1961: *The Forest of China with a Survey of Grassland and Desert Vegetation*, Maria Moors Cabot Foundation Publication 5. Cambridge, MA: Harvard University Press.

Watson, J. G. 1928: Mangrove forests of the Malay Peninsula. *Malayan Forestry Record*, 6, 1–275.

Watts, I. E. M. 1955: *Equatorial Weather, with Special Reference to Southeast Asia*. London: University of London Press.

West, R. C. 1977: Tidal salt-marsh and mangal formations of middle and South America. In V. J. Chapman (ed.), *Wet Coastal Ecosystems*, Amsterdam: Elsevier, 193–214.

Whitaker, R. H., Bush, M. and Richards, K. 1989: Plant recolonisation and vegetation succession on the Krakatau Islands, Indonesia. *Ecological Monographs*, 59, 59–123.

Whitmore, T. C. 1974: *Change with Time and the Role of Cyclones in Tropical Rain Forest on Kolombangara, Solomon Islands*. Commonwealth Forestry Review Paper 46.

Whitmore, T. C. 1981: *Wallace's Line and Plate Tectonics*. Oxford: Clarendon.

Whitmore, T. C. 1984: *Rain Forests of the Far East*, 2nd edn. Oxford: Clarendon.

Whitmore, T. C. 1990: *An Introduction to Tropical Rain Forests*. Oxford: Clarendon.

Whitmore, T. C., Peralta, R. and Brown, K. 1986: Total species count in a Costa Rican tropical rain forest. *Journal of Tropical Ecology*, 1, 375–8.

Wiebes, J. T. 1976: A short history of fig wasp research. *Gard Bulletin*, 29, 207–32.

Wiegert, R. G. 1970: Effects of ionising radiation in leaffall, decomposition and litter macroarthropods of a montane rain forest. In H. T. Odum and R. F. Pigeon (eds), *A Tropical Rain Forest*, Oak Ridge, TN: US Atomic Energy Commission, H89–H100.

Williamson, G. B. et al. 1983: Drip-tips, drop sizes and leaf drying. *Biotropica*, 15, 232–4.

Woods, P. 1989: Effects of logging, drought, and fire on structure and composition of tropical forests in Sabah, Malaysia. *Biotropica*, 21, 290–8.

Wyatt-Smith, J. 1953: A note on the vegetation of some islands in the Malacca Straits. *Malay Forester*, 16, 1919–205.

Wyatt-Smith, J. 1966: Ecological studies on Malaysian forests I. Malay Forestry Department Research Pamphlet 32

Yoda, K. 1974: Three dimensional distribution of light intensity in a tropical rain forest in W. Malaysia. *Japanese Journal of Ecology*, 24, 247–54.

Yoda, K. 1978: Organic carbon, nitrogen and mineral nutrient stock in the soils of Pasoh Forest. *Malayan Nature Journal*, 30, 229–51.

Chapter 6

Abrahams, A. D. 1972: Drainage densities and sediment yields in eastern Australia. *Australian Geographical Studies*, 10, 19–41.

Arnhert, F. 1970: Functional relationships between denudation, relief and uplift in large mid-latitude drainage basins. *American Journal of Science*, 268, 243–63.

Balazs, D. 1968: Karst regions in Indonesia. *Karszt-Es Barlangkut-Atlas*, 5, 3–61.

Behrmann, W. 1921: Die Oberflachenformen in den feuchtwarmen tropen. *Z. Ges. Erdkd. Berlin*, 56, 44–60.

Bird, E. C. F. 1969: *Coasts*. Boston, MA: MIT Press.

Bishop, W. W. 1966: Stratigraphic geomorphology: a review of some East African landforms. In G. H. Dury (ed.), *Essays in Geomorphology*, London: Heinemann, 139–76.

Bornhardt, W. 1900: Zur Oberflachen-gestaltung und Geologie, Deutsch-Ostafrika. Berlin: Collection Deutsch-Ostafrika.

Brand, E. W. 1985: Landslides and their control in Southeast Asia. In A. S. Balasubramanian, D. T. Bergado and S. Chandra (eds), *Geotechnical Engineering in Southeast Asia*, Southeast Asian Geotechnical Society. Rotterdam: Balkema, 167–88.

Brand, E. W. 1989: Occurrence and significance of landslides in southeast Asia. In E. E. Brabb and Harrod (eds), *Landslides: Extent and Economic Significance*, 303–323.

Bremer, H. 1979: Relief und Böden in den Tropen. *Zeitschrift für Geomorphologie, Supplementband NF*, 9, 249–84.

Bremer, H. 1980: Landform development in the humid tropics, German geomorphological research. *Zeitschrift für Geomorphologie, Supplementband* 36, 162–75.

Brook, G. A. and Ford, D. C. 1978: The origin of labyrinth and tower karst and the climatic conditions necessary for their development. *Nature*, 275, 493–6.

Brook, G. A. and Ford, D. C. 1980: Hydrology of the Nahanni karst, northern Canada and the importance of extreme summer storms. *Journal of Hydrology*, 46, 103–21.

Brunsden, D. and Prior, D. B. 1984: *Slope Instability*. Chichester: Wiley.

Büdel, J. 1948: Das System der Klimatischen Geomorphologie, *Beilrage zur Geomorphologie der Klimazonen und Vorzeit-Klimate V. Verhandlugrn Deutscher Geographentag, Munchen*, 27, 65–100.

Büdel, J. 1957: Die Doppelten Einebnungsflachen in denfeuchten tropen. *Zeitschrift für Geomorphologie, NF*, 1, 201–88.

Büdel, J. 1965: Die relief typen der Flachenspil-zone Sud-Indiens am Ostabfall Dekans gegen Madras. *Coll. Geog. Bonn*, 8.

Büdel, J. 1970: Pedimente, Rumpfflächen und Rückland Steilhänge; deren aktive und passive Rückverlegung in verschiedenen Klimaten. *Zeitschrift für Geomorphologie, NF*, 14, 1–57.

Büdel, J. 1982: *Climatic Geomorphology*, trans. L. Fisher and D. Busche. Princeton, NJ: Princeton University Press.

Carson, M. A. and Kirkby, M. J. 1972: *Hillslope Form and Process*. Cambridge: Cambridge University Press.

Chorley, R. J., Schumm, S. A. and Sugden, D. E. 1984: *Geomorphology*. London: Methuen.

Cooke, R. U. and Doornkamp, J. C. 1990: *Geomorphology in Environmental Management*. Oxford: Clarendon.

Corbel, J. 1957: Les Karsts du nord-ouest de l'Europe. *Mem. Inst. Etudes Rhodaniennes*, 12.

Corbel, J. and Muxart, R. 1970: Karsts des zones tropicales humides. *Zeitschrift für Geomorphologie*, 14, 411–74.

Cotton, C. A. 1942: *Climatic Accidents in Landscape Making*. Wellington: Whitcombe & Tombs.

Crowther, J. 1984: Mesotopography and soil cover in tropical karst terrain, West Malaysia. *Zeitschrift für Geomorphologie, NF*, 28 (2), 219–34.

Crozier, M. J. and Eyles, R. J. 1980: Assessing the probability of rapid mass movement. In *Proceedings of the 3rd Australian–New Zealand Conference on Geomechanics, Wellington*, vol. 2, 47–51.

Cunha, S. B., Machado, M. B. and Mousinho de Meis, M. R. 1975: Drainage basin morphometry on deeply weathered bedrocks. *Zeitschrift für Geomorphologie, NF*, 19, 125–39.

Daly, D. D. 1882: Surveys and explorations in the narive states of the Malayan Peninsular, 1875–82. *Proceedings of the Royal Geographical Society*, 4, 393–412.

Darwin, C. 1842: *The Structure and Distribution of Coral Reefs*. London: Smith, Elder.

Darwin, C. 1890: *On the Structure and Distribution of Coral Reefs: also Geological Observations on the Volcanic Islands and Parts of South America*. London: Ward Lock.

Davis, W. M. 1899: The geographical cycle. *Geographical Journal*, 14, 481–504.

Davison, C. 1889: On the creeping of the soil cap through the action of frost. *Geological Magazine*, 6, 255.

Deere, D. U. and Patton, F. D. 1971: Slope stability in residual soils. In *Proceedings of the 4th Pan-Am Conference on Soil Mechanics*, vol. 1, Puerto Rico, 87–170.

De Martonne, E. 1946: Géographie zonale. Le zone tropicale. *Annales Géographie*, 55, 1–8.

Derbyshire, E. 1976: *Geomorphology and Climate*. London: Wiley.

De Vallego, L. I. G., Jimenez Salas, J. A. and Leguey Jimenez, S. 1981: Engineering geology of the tropical volcanic soils of Lake Laguna, Tenerife. *Engineering Geology*, 17 (1–2), 1–17.

Dixon, H. H. and Robertson, R. H. S. 1971: Some experiences in tropical soils. *Quarterly Journal of Engineering Geology*, 3, 137–50.

Doornkamp, J. C. and King, C. A. M. 1971: *Numerical Analysis in Geomorphology: an Introduction*. London: Edward Arnold.

Douglas, I. 1978: Tropical geomorphology: present problems and future prospects. In C. Embleton, D. Brunsden and D. K. C. Jones (eds), *Geomorphology: Present Problems and Future Prospects*, Oxford: Oxford University Press, 162–84.

Douglas, I. and Spencer, T. (eds) 1985: *Environmental Change and Tropical Geomorphology*. London: Allen & Unwin.

Erhart, H. 1956: *Le Genèse des Sols, en tant que Phénomène Géologique. Esquisse d'une Theorie Geologhimique. Biostasie et Rhexistasie*. Paris: Masson.

Eyles, R. J. 1966: Stream representation on Malayan maps. *Journal of Tropical Geography*, 22, 1–9.

Fairbridge, R. W. 1967: Landslide distribution and earthquakes in the Bewani and Torricelli Mountains, New Guinea. In J. N. Jennings and J. A. Mabbutt (eds), *Landform Studies from Australia and New Guinea*, Cambridge: Cambridge University Press, 64–8.

Faniran, A. and Jeje, L. E. 1971: Landform examples from Northern Nigeria no. 3: differential weathering, 1: Corestones and stonelines. *Nigerian Geographical Journal*, 14 (1), 105–9.

Faniran, A. and Jeje, L. E. 1983: *Humid Tropical Geomorphology*. London: Longman.

Ford, D. C. and Williams, P. W. 1989: *Karst Geomorphology and Hydrology*. London: Unwin.

Fukuoka, M. 1980: Landslides associated with rainfall. *Geotechnical Engineering*, 11, 1–29.

Gerstenhauer, A. 1960: Der tropische Kegelkarst in Tabasco (Mexico). *Zeitschrift für Geomorphologie, NF*, 29 (4), 483–95.

Ghose, B., Pandey, S., Singh, S. and Lal, G. 1967: Quantitative geomorphology of the drainage basins in the central Luni basin in western Rajasthan. *Zeitschrift für Geomorphologie, NF*, 11, 146–60.

Goudie, A. S. 1973: *Duricrusts in Tropical and Subtropical Landscapes*. Oxford: Clarendon.

Goudie, A. S. 1986: *The Human Impact on the Environment*. Oxford: Blackwell.

Goudie, A. S. 1989: *The Nature of the Environment*. Oxford: Blackwell.

Goudie, A. S. and Pye, K. (eds) 1983: *Chemical Sediments and Geomorphology*. London: Academic Press.

Gregory, K. 1976: Drainage networks and climate. In E. Derbyshire (ed.), *Geomorphology and Climate*, London: Wiley, 289–315.

Gunn, J. 1981: Hydrological processes in karst depressions. *Zeitschrift für Geomorphologie*, 25 (3), 313–31.

Hutchingson, J. N. 1968: *The Encyclopedia of Geomorphology*. Encyclopedia of Earth Science Series. vol. 1, ed. R. W. Fairbridge. New York: Reinhold.

Imray, I. 1848: Observations on the characteristics of endemic fever in the island of Dominica. *Edinburgh Medical and Surgical Journal*, 70 (177), 253–87.

Jennings, J. N. 1969: Karst of the seasonally humid tropics in Australia. In O. Stelcl (ed.), *Problems of Karst Denudation*, Brno, 149–58.

Jennings, J. N. 1983: Sandstone pseudokarst or karst. In R. W. Young and G. C. Nanson (eds), *Aspects of Australian Sandstone Landscapes*, Wollongong: Australia and New Zealand Geomorphology Group, 21–30.

Jennings, J. N. 1985: *Karst Geomorphology*. Oxford: Blackwell.

Jennings, J. N. and Sweeting, M. M. 1963: Karst of the seasonally humid tropics in Australia. In O. Stelcl (ed.), *Problems of Karst Denudation*, Brno, 149–58.

Johannes, R. E. 1977: *Coral Reefs in Coastal Ecosystem Management*. New York: Wiley.

Jones, D. E. and Holtz, W. G. 1973: Expansive soils – the hidden disaster. *Civil Engineering, ASCE*, 49–51.

Kessel, R. H. 1973: Inselberg landform elements; definition and synthesis. *Rev. Geom. Dyn.*, 22, 97–108.

Kiersch, G. A. and Treasher, R. C. 1954: Investigations, areal and engineering geology – Folsom Dam Project, central California. *Economic Geology*, 50 (3), 271–310.

King, C. A. M. 1972: *Beaches and Coasts*. London: Arnold.

King, L. C. 1953: Canons of landscape evolution. *Bulletin of the Geological Society of America*, 64, 721–52.

King, L. C. 1957: The uniformitarian nature of hillslopes. *Transactions of the Edinburgh Geological Society*, 17, 81.

Kirkby, M. J. 1967: Movement and theory of soil creep. *Journal of Geology*, 75, 359–78.

Knill, J. L. and Jones, K. S. 1965: The recording and interpretation of geological conditions in the foundations of the Roseires, Kariba and Latiyan dams. *Geotechnique*, 15 (1), 94–120.

Kronbert, B. I., Nesbitt, H. W. and Llamm, W. W. 1986: Upper Pleistocene Amazon deep-sea muds reflect intense chemical weathering of their mountainous source lands. *Chemical Geology*, 54, 283–94.

Kuruppuarachchi, T. A., Senavirathana, Y. B. and Ranatunga, E. R. 1987: An enquiry into the antecedent precipitation conditions related to earth-slip generation in the Belihul Oya catchment, Maturata area. *Proceedings of the 43rd Annual Session of the Sri Lankan Association, Colombo, Sri Lanka.*

Larsen, M. C. and Torres-Sanchez, A. J. M. 1990: Rainfall–soil moisture relationships in landslide prone areas of a tropical rainforest, Puerto Rico. *Proceedings of the International Conference on Tropical Hydrology*, American Water Resources Association, 127–30.

Lehmann, H. 1954: Das Karstphanomen in der Verschiedenen Klimazonen. *Erdkunde*, 8, 114–39.

Lehmann, H. 1970: Kelgelkarst und Tropengrenze. *Tinbinger Geogr. Stud.*, 34, 107–12.

Lewis, A. L. 1974: Slow movement of earth under tropical rainforest conditions. *Geology*, 2, 9–10.

Lewis, L. A. 1976: Soil movements in the tropics – a general model, *Zeitschrift für Geomorphologie, Supplementband, NF*, 24, 132–44.

Linton, D. L. 1955: The problem with tors. *Geographical Journal*, 121, 470–87.

Loughnan, F. C. 1969: *Weathering*, London: Longman.

Mabbutt, J. A. 1961: A stripped landsurface in Western Australia. *Transactions of the Institute of British Geographers*, 29, 101–14.

Machatschek, F. 1955: *Das Relief der Erde: Versucheiner regionalen Morphologia der Erdoberflach*, vol. II, 2nd edn. Berlin: Borntraeger.

Madduma Bandara, C. M. 1974a: Drainage density and effective precipitation. *Journal of Hydrology*, 21, 187–90.

Madduma Bandara, C. M. 1974b: The orientation of straight slope forms on the Hatton Plateau of central Sri Lanka. *Journal of Tropical Geography*, 38, 37–44.

Martini, J. E. 1981: The control of karst development with reference to the formation of caves in poorly soluble rocks in eastern Transvaal. *Proceedings of the 8th International Congress on Speleology*, 4–5.

McDonald, R. C. 1985: Tower karst geomorphology in northern Borneo. *Zeitschrift für Geomorphologie, NF*, 29 (4), 483–95.

McFarlane, M. J. 1983: Laterites. In A. Goudie and K. Pye (eds), *Chemical Sediments and Geomorphology*, London: Academic Press.

Moeyersons, J. 1981: Slumping and planar sliding on hill-slopes in Rwanda. *Earth Surface Processes*, 6, 265–74.

Moeyersons, J. 1988: The complex nature of creep movements on steeply sloping ground in southern Rwanda. *Earth Surface Processes and Landforms*, 13, 511–24.

Morgan, R. P. C. 1976: The role of climate in the denudation system: a case study from West Malaysia. In E. Derbyshire (ed.), *Geomorphology and Climate*, London: Wiley.

Moye, D. G. 1955: Engineering geology for the Snowy Mountain scheme. *Journal of the Institute of Engineers, Australia*, 27, 287–98.

Nunn, P. D. 1994: *Oceanic Islands*. London: Blackwell.

Ollier, C. D. 1960: The inselbergs of Uganda. *Zeitschrift für Geomorphologie, NF*, 4, 43–52.

Ollier, C. D. 1984: *Weathering*. London: Longman.

Otvos, E. G. 1976: 'Pseudokarst' and 'pseudokarst terrain': Problems of terminology, *Geological Society of America Bulletin*, 87, 1021–7.

Panos, V. and Stelcl, O. 1968: Physiographic and geologic control in development of Cuba mogotes. *Zeitschrift für Geomorphologie, NF*, 12, 117–73.

Peltier, L. C. 1950: The geographical cycle in periglacial regions as it is related to climatic geomorphology. *Annals of the Association of American Geographers*, 40, 214–36.

Penck, W. 1924: *Die Morphologische Analyse*. Stuttgart. English translation by H. Czech and K. C. Boswell, *Morphological Analysis of Landforms*, London: Macmillan, 1953.

Petit, M. 1985: A provisional world map of duricrust. In I. Douglas and T. Spencer (eds), *Environmental Change and Tropical Geomorphology*. London: Allen & Unwin, 269–80.

Pitts, J. 1983: The form and causes of slope failures in an area of West Singapore Island. *Singapore Journal of Tropical Geography*, 4 (2), 162–8.

Pouyllau, M. and Seurin, M. 1985: Pseudo-karst dans des roches greso-quartzitiques de la formation Roriama. *Karstologia*, 5, 45–52.

Pugh, J. C. 1966: Landforms in low latitudes. In G. H. Dury (ed.), *Essays in Geomorphology*, London: Heinemann.

Pye, K., Goudie, A. S. and Watson, A. 1986: Petrological influence on differential weathering and inselberg development in the Kora area of central Kenya. *Earth Surface Processes and Landforms*, 11, 41–52.

Reading, A. J. 1986: Landslides, heavy rainfalls and hurricanes in Dominica, West Indies. Ph.D. thesis, University of Wales.

Rougerie, G. 1960: Le façonnement actuel des modèles en Côte d'Ivoire forestière. *Memoires de l'Institut Français d'Afrique Noire*, 58, 542 pp.

Rouse, W. C., Reading, A. J. and Walsh, R. P. D. 1986: Volcanic soil properties in Dominica, West Indies. *Engineering Geology*, 23, 1–28.

Ruxton, B. P. 1958: Weathering and sub-surface erosion in granite at the piedmont angle, Balos, Sudan. *Geological Magazine*, 95, 353–77.

Ruxton, B. P. and Berry, L. 1957: Weathering of granite and associated erosional features in Hong Kong. *Bulletin of the Geological Society of America*, 68, 1263–92.

Sapper, K. 1935: *Geomorphologie der feuchten Tropen*. Berlin: Teubner.

Savigear, R. A. G. 1960: Slopes and hills in West Africa. *Zeitschrift für Geomorphologie Supplementband 1, Morphologie des Versants*, S156–71.

Schumm, S. A. 1956: The role of creep rainwash on the retreat of badland slopes. *American Journal of Science*, 254, 693–706.

Schumm, S. A. 1963: The disparity between present rates of denudation and orogeny. US *Geological Survey Professional Paper* 454-H.

Sczerban, E. and Urbani, F. 1974: Carsos de Venezuela Parte 4, *Boletin de la Sociedad Venezolana Espeleologia*, 5, 27–54.

Sharpe, C. F. S. 1938: *Landslides and Related Phenomena*. New York: Columbia University Press.

Simonett, D. S. 1967: Landslide distribution and earthquakes in the Bewani and Torricelli Mountains, New Guinea. In J. N. Jennings and J. A. Mabbutt (eds), *Landform Studies from Australia and New Guinea*, Cambridge: Cambridge University Press, 64–8.

Skempton, A. W. and Delory, F. A. 1957: Stability of natural slopes in London Clay. In *Proceedings of the 4th International Conference on Soil Mechanics and Foundation Engineering, London*, vol. 11, 378–81.

Smith, D. I. and Atkinson, T. C. 1976: Process, landforms and climate in limestone regions. In E. Derbyshire (ed.), *Geomorphology and Climate.* Chichester: Wiley, 367–409.

Snedacker, S. 1982: A perspective on Asian mangroves. In C. H. Soysa, C. L. Sien and W. L. Collier (eds), *Man, Land and Sea.* Massachusetts.

So, C. L. 1971: Mass movements associated with the rainstorm of July 1966 in Hong Kong. *Transactions of the Institute of British Geographers*, 53, 55–6.

Sowers, G. F. 1954: Soil problems in the southern Piedmont region. *Proceedings of the American Society of Civil Engineers*, 80.

Sowers, G. F. 1963: Engineering properties of residual soils derived from igneous and metamorphic rocks. In *Proceedings of the 2nd Pan-American Congress on Soil Mechanics and Foundation Engineering, Brazil*, vol. I, 39–61.

Sowers, G. F. 1967: Discussion. In *Proceedings of the 3rd Pan-American Congress on Soil Mechanics and Foundation Engineering, Caracas*, vol. III, 135–61.

Spencer, T. and Douglas, I. 1985: The significance of environmental change: diversity, disturbance and tropical ecosystems. In I. Douglas and T. Spencer (eds) *Environmental Change and Tropical Geomorphology*, London: Allen & Unwin, 13–27.

Stoddart, D. R. 1969: Climatic geomorphology: review and reassessment. *Progress in Geography*, 1, 159–222.

Stoddart, D. R. 1971: Coral reefs and islands in catastrophic storms. In J. A. Steers (ed.), *Applied Coastal Geomorphology*, London: Macmillan, 155–97.

Stoddart, D. R. and Cann, J. R. 1965: Nature and origin of beachrock. *Journal of Sedimentary and Pedology*, 35, 243–7.

Stokes, W. L., Judson, S. and Picard, M. D. 1978: Introduction to Geology: Physical and Historical. Oxford: Prentice Hall.

Summerfield, M. A. 1985: Tectonic background to long-term landform development in tropical Africa. In I. Douglas and T. Spencer (eds), *Environmental Change and Tropical Geomorphology*, London: Allen & Unwin, ch. 14.

Sweeting, M. M. 1958: The karstlands of Jamaica. *Geographical Journal*, 124, 184–99.

Sweeting, M. M. 1966: The weathering of limestones, with particular reference to the carboniferous limestones of northern England. In G. H. Dury (ed.), *Essays in Geomorphology*, London: Heinemann, 177–210.

Terzaghi, K. 1958: Design and performance of the Sasumua dam. *Proceedings of the Institute of Civil Engineers*, 9, 369–94.

Thom, B. G. 1967: Mangrove ecology and deltaic geomorphology: Tabasco, Mexico. *Journal of Ecology*, 55, 301–43.

Thom, B. G. 1982: Mangrove ecology: a geomorphological perspective. In B. Clough (ed.), *Mangrove Ecosystems in Australia: Structure, Function and Management*, Canberra: ANU Press, 3–17.

Thom, B. G. 1984: Coastal landforms and geomorphic processes. In S. Snedaker and J. G. Snedaker (eds), *The Mangrove Ecosystem: Research Methods*, Paris: UNESCO, 3–17.

Thom, B. G., Wright, L. D. and Coleman, J. M. 1975: Mangrove ecology and deltaic-estuarine geomorphology: Cambridge Gulf-Ord River, Western Australia. *Journal of Ecology*, 63, 203–32.

Thomas, M. F. 1965: Some aspects of the geomorphology of Domes and Tors in Nigeria. *Zeitschrift für Geomorphologie*, 4, 43–52.

Thomas, M. F. 1974: *Tropical Geomorphology*. London: Macmillan.

Thomas, M. F. 1978: The study of inselbergs. *Zeitschrift für Geomorphologie, NF*, 31, 1–41.

Thomas, M. F. 1994: *Geomorphology in the Tropics*. Chichester: Wiley.

Thomas, M. F. and Thorp, M. B. 1985: Some aspects of the geomorphological interpretations of Quaternary alluvial sediments from Sierra Leone. *Zeitschrift für Geomorphologie, Supplementband NF*, 36, 140–61.

Thomson, J. 1882: Notes on the basin of the River Rovuma, East Africa. *Proceedings of the Royal Geographical Society, NS*, 4, 65–79.

Thorn, C. E. 1988: *An Introduction to Theoretical Geomorphology*. London: Unwin Hyman.

Tinley, K. L. 1985: Coastal dunes of South Africa. *South African National Scientific Programmes Report*, 109.

Tricart, J. 1972: *Landforms of the Humid Tropics and Savannas*. London: Longman.

Twidale, C. R. 1964: A contribution to the general theory of domed inselbergs. *Transactions of the Institute of British Geographers*, 34, 91–113.

Vargas, M. 1963: General discussion. In *Proceedings of the 2nd Pan-American Congress on Soil Mechanics and Foundation Engineering, Brazil*, vol. II, 539–41.

Vargas, M. 1969: Residual soil sampling. In *Proceedings of Special Session 1 on Soil Sampling, 7th International Conference on Soil Mechanics and Foundation Engineering*, Melbourne: International Group on Soil Sampling, 50–3.

Varnes, D. J. 1958: Landslides in engineering practice. *Highway Research Board Special Report*, 29.

Verstappen, H. T. 1960: Some observations on karst development in the Malay Archipelago. *Journal of Tropical Geography*, 14, 1–10.

Viles, H. A. (ed.), 1988: *Biogeomorphology*. Oxford: Blackwell.

Wallace, K. B. 1973: Structural behavior of residual soils of the continually wet highlands of Papua New Guinea. *Geotechnique*, 23 (2), 203–18.

Walsh, R. P. D. 1980: Drainage density and hydrological processes in a humid tropical environment: the Windward Islands. Ph.D. thesis, University of Cambridge.

Walsh, R. P. D. 1985: The influence of climate, lithology and time on drainage density and relief development in the tropical volcanic terrain of the Windward Islands. In I. Douglas and T. Spencer (eds), *Environmental Change and Tropical Geomorphology*, London: Allen & Unwin, 93–122.

Warkentin, B. P. 1974: Physical properties of Caribbean clay soils. *Tropical Agriculture (Trinidad)*, 51 (2), 279–87.

Watson, A. and Pye, K. 1985: Pseudokarstic micro-relief and other weathering features on the Mswati Granite (Swaziland). *Zeitschrift für Geomorphologie, NF*, 29 (3), 285–300.

Wesley, L. D. 1973: Some basic engineering properties of halloysite and allophane clays in Java, Indonesia. *Geotechnique*, 23, 471–94.

Weyland, E. J. 1934: Peneplains and some other erosional platforms. *Bulletin of the Geological Survey of Uganda, Annual Report*, Notes 1, 74, 366.
White, W. B. 1984: Rate processes: chemical kinetics and karst landform development. In R. G. LaFleur (ed.), *Groundwater as a Geomorphological Agent*, London: Allen & Unwin, 227–48.
Williams, P. W. 1972: Morphometric analysis of polygonal karst in New Guinea. *Bulletin of the Geological Association of America*, 83, 761–96.
Williams, P. W. 1985: Subcutaneous hydrology and the development of doline and cockpit karst. *Zeitschrift für Geomorphologie, NF*, 29 (4), 463–82.
Williams, P. W. 1987: Geomorphic inheritance and the development of Tower Karst. *Earth Surface Processes and Landforms*, 12, 453–65.
Williams, P. W. and Downing, R. K. 1979: Solution of marble in the karst of the Pikikiruna Range, northwest Nelson, New Zealand. *Earth Surface Processes*, 4, 15–36.
Wilson, R. C. L. 1983: *Residual Deposits: Surface Related Weathering Processes and Materials*. Oxford: Blackwell.
Wirthman, A. 1977: Erosive Hangentwicklung in verschiedenen Klimaten. *Zeitschrift für Geomorphologie, Supplementband, NF*, 28, 42–61.
Wright, L. W. 1973: Landforms of the Yaruna granite area, Vitu Levu, Fiji: a morphometric study. *Journal of Tropical Geography*, 37, 74–80.
Yamazaki, F. and Takenaka, H. 1965: On the influence of air-drying on Atterberg's limits. *Transactions of the Agricultural Engineering Society of Japan*, 14, 46–8.
Young, A. 1978: A twelve year record of soil movement on a slope. *Zeitschrift für Geomorphologie, Supplementband*, 29, 104–10.
Young, R. W. 1986: Tower karst in sandstone: Bungle Bungle massif, northwestern Australia. *Zeitschrift für Geomorphologie, NF*, 28 (2), 219–34.
Young, R. W. 1988: Quartz etching and sandstone karst. *Zeitschrift für Geomorphologie*, 32, 409–23.

Chapter 7

Baille, I. S. 1975: Piping as an erosional process in the uplands of Sarawak. *Journal of Tropical Geography*, 41, 9–15.
Bakker, J. P. 1957: Zur Enstechung von Pingen, Oricangas und Dellen in der feuchten Tropen, mit besonder Berücksichtingung des Vitzberggebietes (Surinam). *Abb. Georgr. Inst. Fr. Univ. Berlin*, 5, 7–20.
Balek, J. 1983: *Hydrology and Water Resources in the Tropical Regions*. Amsterdam: Elsevier, 271 pp.
Banajee, A. K. 1972: Morphology and genesis of pipe structure in ferralitic soils in Midnapore. *Journal of the Indian Society of Soil Science*, 20, 399–402.
Biksham, G. and Subramanian, V. 1988: Nature of solute transport in the Godavari Basin. *Indian Journal of Hydrology*, 103, 375–92.
Birot, P. 1968: *The Cycle of Erosion in Different Climates*. London: Batsford.
Birot, P. 1970: L'influence du climat sur la sédimentation continentale. Paris: Centre du Documentation Universitaire.
Bonell, M. 1988: Hydrological processes and implications for land management in forests and agricultural areas of the wet tropical coast of north-

east Queensland. In R. F. Warner (ed.), *Fluvial Geomorphology of Australia*, London: Academic Press.

Bonell, M. and Gilmour, D. A. 1978: The development of overland flow in a tropical rainforest catchment. *Journal of Hydrology*, 39, 365–82.

Bonell, M., Gilmour, D. A. and Sinclair, D. F. 1981: Soil hydraulic properties and their effect on surface and subsurface water transfer in a tropical rainforest catchment. *Hydrological Sciences – Bulletin des Sciences Hydrologiques*, 26 (1, 3), 1–18.

Bonel, M., Cassells, D. S. and Gilmour, D. A. 1982: Vertical and lateral soil water movement in a tropical rainforest catchment. In E. M. O'Loughlin and L. Bren (eds), *National Symposium on Forest Hydrology (Proceedings of the Melbourne Symposium, May 1982*, Canberra: Institute of Engineers, Australia/Forest Hydrology Working Group, Australian Forestry Council, 30–8.

Bonell, M., Gilmour, D. A. and Cassells, D. S. 1983a: Runoff generation in tropical rainforests of northeast Queensland, Australia, and the implications for land use management. In R. Lal and E. W. Russell (eds.), *Tropical Agricultural Hydrology*, London: Wiley, 287–97.

Bonell, M., Gilmour, D. A. and Cassells, D. S. 1983b: A preliminary survey of the hydraulic properties of rainforest soils in tropical north-east Queensland and their implications for the runoff processes. In J. de Ploey (ed.), *Rainfall Simulation, Runoff and Soil Erosion, Catena Special Supplement*, 4, 57–77.

Bonell, M., Cassells, D. S. and Gilmour, D. A. 1983c: Vertical soil water movement in a tropical rainforest catchment in northeast Queensland. *Earth Surface Processes*, 8 (3), 253–72.

Brabben, T. E. 1987: Reservoir developments and their impact on the Brantas river, Indonesia. Hydraulics Research Report, Engineering Management Group, Appropriate Development Panel.

Bruijnzeel, L. A. 1983: Evaluation of runoff sources in a forested basin in a wet monsoonal environment: a combined hydrological and hydrochemical approach. In R. Keller (ed.), *Hydrology of Humid Tropical Regions with Particular Reference to the Hydrological Effects of Agriculture and Forestry Practice*, IAHS Publ. 140, London: Wiley, 165–74.

Brunig, E. F. 1975: Tropical ecosystems: state and targets of research into the ecology of humid tropical systems. *Plant Research and Development*, 1, 22–38.

Brutsaert, W. H. 1965: Evaluation of some practical methods of estimating evapotranspiration in arid climates at low latitudes. *Water Resources Research*, 1, 187–91.

Büdel, J. 1957: Die Flachenbildung in den feuchten Tropen und die Rolle fossiler solcher Flachen in anderen Klimatzonen. *Verhan dt. Geogr. Tags*, 31, 89–121.

Burton, C. K. 1964: The older alluvium of Johore and Singapore. *Journal of Tropical Geography*, 18, 30–42.

Cailleux, A. 1959: Etudes sur l'érosion et la sédimentation en Guyane. *Mem. Serv. Carte Géol. Fr.*, 49–73.

Carbonnel, J. P. 1965: Sur les cycles de mise en solution du fer et de la silice en milieu tropical. *Comptes Rendus de l'Academie des Sciences, Paris*, 260, 4043–8.

Chettri, R. and Bowonder, B. 1983: Siltation in Nizamsagar reservoir: environmental management issues. *Applied Geography*, 3, 193–204.

Corbel, J. 1957: L'érosion chimique des granites et silicates sous climates chauds. *Revue de Géomorphologie Dynamique*, 8, 4–8.

Corbel, J. 1964: L'érosion terrestre, étude quantitative. *Annales de Géographie*, 73, 385–412.

Coster, C. 1938: Bovengrondsche afstrooming en erosie op Java. *Tectona*, 31, 613–719.

Dabin, B. 1957: Note sur le fonctionnement des parcelles expérimentales pour l'étude de l'érosion à l'érosion d'Adiopodioume (Côte d'Ivoire). Mimeo, Decrét. Permanent Bureau Sols AOF, Dakar, 16 pp.

Davis, S. N. 1964: Silica in streams and ground water. *American Journal of Science*, 262, 870–91.

De Bruin, H. A. R. 1983: Evapotranspiration in humid tropical regions. In R. Keller (ed.), *Hydrology of Humid Tropical Regions with Particular Reference to the Hydrological Effects of Agriculture and Forestry Practice*, IAHS Publ. 140, London: Wiley, 229–311.

Dedkov, A. P. and Mozzherin, V. I. 1964: *Eroziya i Stok Nanosov na Zemle*. Izdatelstvo Kazanskogo Universiteta.

Doorenbos, T. and Pruitt, W. O. 1977: Crop water requirements. *Irrigation and Drainage Paper 24*, Rome: FAO.

Douglas, I. 1967: Man, vegetation and the sediment yield of rivers. *Nature*, 215, 925–8.

Douglas, I. 1968: Erosion in the Sungei Gombak catchment, Selangor, Malaysia. *Journal of Tropical Geography*, 26, 1–16.

Douglas, I. 1969: The efficiency of humid tropical denudation systems. *Transactions of the Institute of British Geographers*, 46, 1–16.

Douglas, I. 1973: Rates of sedimentation in selected small catchments in Eastern Australia. University of Hull Occasional Papers in Geography, 21.

Douglas, I. 1978: The impact of urbanisation on fluvial geomorphology in the humid tropics. *Geo-Eco-Trop*, 2, 229–42.

Douglas, I. and Spencer, T. 1985: Present-day processes as a key to the effects of environmental change. In I. Douglas and T. Spencer (eds), *Environmental Change and Tropical Geomorphology*. London: Allen & Unwin, 39–74.

Downing, A. 1968: Subsurface erosion as a geomorphic agent in Natal. *Transactions of the Geological Society of South Africa*, 71, 131–4.

Dunne, T. and Black, R. D. 1970: An experimental investigation of runoff production in permeable soils. *Water Resources Research*, 6, 478–90.

Edwards, A. M. C. 1973: The variations of dissolved constituents with discharge in some Norfolk rivers. *Journal of Hydrology*, 18 (3–4), 219–42.

Edwards, A. M. C. and Liss, P. S. 1973: Evidence for buffering of dissolved silica in fresh waters. *Nature*, 243, 341–2.

Edwards, K. A. and Blackie, J. R. 1981: Results of East African catchment experiments 1958–1974. In R. Lal and E. W. Russell (eds), *Tropical Agricultural Hydrology*, London: Wiley, 163–89.

Ekern, P. C. 1964: Direct interception of cloud water at Lanaihale, Hawaii. *Proceedings of the Soil Science Society of America*, 28, 419–21.

Ekern, P. C. 1982: Measured evaporation in high rainfall areas, leeward Koolau Range, Oahu, Hawaii. In A. I. Johnson and R. A. Clark (eds), *Proceedings of the International Symposium on Hydrometeorology*, Technical Publication Service TPS-82-1, Denver, CO: American Water Resources Association, 85–9.

El-Swaify, S. A. 1980: Physical and mechanical properties of oxisols. In B. K. G. Theng (ed.), *Soils with Variable Charge*, Lower Hutt: New Zealand

Society of Soil Science, Soil Bureau, Department of Scientific and Industrial Research, 303–24.

Feininger, T. 1969: Pseudokarst on quartz diorite, Colombia, *Zeitschrift für Geomorphologie, NF*, 13, 287–96.

Fontaine, H. 1965: Mode particulier d'érosion dans les formations meubles. *Archive de Géologie du Vietnam*, 7, 34–47.

Foote, D. E., Hill, E. L., Nakamura, S. and Stevens, F. 1972: *Soil Survey of the Islands of Kauai, Oahu, Maui, Molokai and Lanai, State of Hawaii.* US Department of Agriculture, Soil Conservation Service, 130 pp.

Fournier, F. 1960: *Climate et érosion*, Paris.

Fournier, F. 1962: Carte du danger d'érosion en Afrique au Sud du Sahara (fondé sur l'agressivité climatique et la topographie). CEE/CCTA, Bur. Interafr. Sols.

Freeze, R. A. 1972: Role of subsurface flow in generating surface runoff. 2: Upstream source areas. *Water Resources Research*, 8, 1272–83.

Freeze, R. A. and Cherry, J. A. 1979: *Groundwater*. Englewood Cliffs, NJ: Prentice-Hall.

Geiger, R. 1965: *Climate Near the Ground*. Cambridge, MA: Harvard University Press.

Gibbs, R. J. 1967: Amazon river: environmental factors that control its dissolved and suspended load. *Science*, 156, 1734–6.

Gilmour, D. A., Bonell, M. and Cassells, D. S. 1987: The effects of forestation on soil hydraulic properties in the Middle Hills of Nepal: a preliminary assessment. *Mountain Research and Development*, 7 (3), 239–49.

Gladwell, J. S. and Bonell, M. 1988: Hydrology and water management strategies in the humid tropics. *Water International*, 13, 123–9.

Godefroy, J., Muller, M. and Roose, E. J. 1970: Estimation des pertes lixivation des éléments fertilisants dans un sol de bananeraie de basse Côte d'Ivoire. *Fruits*, 25, 403–20.

Goudie, A. S. 1985: *The Encyclopedic Dictionary of Physical Geography*. Oxford: Blackwell.

Grenney, W. J. and Heyse, E. 1985: Suspended sediment-river flow analysis. *Journal of Environmental Engineering*, 3, 790–803.

Gunston, H. and Batchelor, C. H. 1983: A comparison of the Priestly–Taylor and Penman methods for estimating crop evapotranspiration in tropical countries. *Agricultural Water Management*, 6, 65–77.

Gupta, A. 1984: Urban hydrology and sedimentation in the humid tropics. In J. E. Costa and P. J. Fleisher (eds), *Developments and Applications of Geomorphology*, Berlin: Springer.

Hem, J. D. 1970: *Study and Interpretation of the Chemical Characteristics of Natural Water*, 2nd edn. Washington, DC: US Government Printing Office, 363 pp.

Herwitz, S. R. 1988: Buttresses of tropical trees influence hillslope processes. *Earth Surface Processes and Landforms*, 13, 563–7.

Hewlett, J. D. 1961: *Watershed Management*. Asheville, NC: US Department of Agriculture Forest Service, Southeastern Forest Experiment Station, Report for 1961, 61–6.

Hewlett, J. D. and Hibbert, A. R. 1967: Factors affecting the response of small watersheds to precipitation in humid areas. In W. E. Sopper and H. W. Lull (eds), *International Symposium on Forest Hydrology*, New York: Pergamon, 275–90.

Hopkins, B. 1960: Rainfall interception by a tropical forest in Uganda. *East African Agricultural Journal*, 25, 255–8.

Horton, R. E. 1945: Erosional development of streams and their drainage basins: hydro-physical approach to quantitative morphology. *Bulletin of the Geological Society of America*, 56, 275–370.

Huang, W. H. and Keller, W. D. 1972: Organic acids as agents of chemical weathering of silicate minerals. *Nature*, 239, 149–252.

Hudson, N. W. 1971: *Soil Conservation*. London: Batsford.

Hudson, N. W. and Jackson, D. C. 1959: Results achieved in the measurement of erosion and runoff in Southern Rhodesia. *Proceedings of the 3rd Inter-African Soil Conference Dalaba*, CCTA, 573–83.

Hutchinson, Sir J., Manning, M. L. and Farbrother, H. G. 1958: On the characterization of tropical rainstorms in relation to runoff and percolation. *Quarterly Journal of the Royal Meteorological Society*, 84, 250–8.

Iwatsubo, G. and Tsutsumi, T. 1968: On the amount of plant nutrients supplied to the ground by rainwater in adjacent open land and forests (III). On the amount of plant nutrients contained in runoff water. *Bulletin Kyoto University, Forests*, 40, 140–56.

Jackson, I. J. 1975: Relationships between rainfall parameters and interception by tropical forests. *Journal of Hydrology*, 24, 215–38.

Jackson, I. J. 1989: *Climate, Water and Agriculture in the Tropics*. Harlow: Longman.

Jansson, M. B. 1988: A global survey of sediment yield. *Geografiska Annaler*, 70A, 81–98.

Jones, J. A. A. 1981: *The Nature of Soil Piping: a Review of Research*, British Geomorphology Research Group Monograph 3. Norwich: GeoBooks.

Jordan, C. F. 1970: Flow of soil water in the lower montane tropical rainforest. In H. T. Odum and R. F. Pigeon (eds), *A Tropical Rainforest*, Book 3, Springfield, VA: US Atomic Energy Commission, 4.199–4.200.

Jordan, C. F. and Kline, J. R. 1977: Transpiration of trees in a tropical rainforest. *Journal of Applied Ecology*, 14, 853–60.

Kayane, I. 1971: Hydrological regions in Monsoon Asia. In M. Yoshimo (ed.), *Balance of Monsoon Asia – a Climatological Approach*, Honolulu: University of Hawaii Press.

Kesel, R. H. 1977: Slope runoff and denudation in the Rupununi Savanna, Guyana. *Journal of Tropical Geography*, 44, 33–42.

Kirkby, M. J. 1988: Hillslope runoff processes and models. *Journal of Hydrology*, 100, 315–39.

Kirkby, M. J. and Chorley, R. J. 1967: Throughflow, overland flow and erosion. *Bulletin of the International Association of the Science of Hydrology*, 12, 5–21.

Lal, R. 1976: *Soil Erosion Problems on an Alfisol in Western Nigeria and their Control*, IITA Monograph 1. Ibadan: IITA.

Lal, R. 1980: Physical and mechanical composition of alfisols and ultisols with particular reference to soils in the tropics. In B. K. G. Theng (ed.), *Soils with Variable Charge*, Lower Hutt: New Zealand Society of Soil Science, Soil Bureau, Department of Scientific and Industrial Research, 253–79.

Lal, R. 1981: Deforestation of tropical rainforest and hydrological problems. In R. Lal and E. W. Russell (eds), *Tropical Agricultural Hydrology*, London: Wiley, 131–40.

Lal, R. 1983: Soil erosion in the humid tropics with particular reference to agricultural land development and soil management. In R. Keller (ed.), *Hydrology of Humid Tropical Regions with Particular Reference to the*

Hydrological Effects of Agriculture and Forestry Practice, IAHS Publ. 140, London: Wiley, 221–40.
Lamotte, M. and Rougerie, G. 1962: Les apports allochtones dans la genèse des cuirasses ferrugineuses. *Revue de Géomorphologie Dynamique*, 13, 145–60.
Ledger, D. L. 1964: Some hydrological characteristics of West African rivers. In M. F. Thomas and G. Whittington (eds), *Environment and Land Use in Africa*, London: Methuen.
Leigh, C. H. 1978: Slope hydrology and denudation in the Pasoh forest reserve. *Malayan Nature Journal*, 30, 179–210.
Leigh, C. H. 1982: Sediment transport by surface wash and throughflow at the Pasoh forest reserve, Negri Sembilan, Peninsular Malaysia. *Geografiska Annaler*, 64A (3–4), 171–80.
Lewis, L. A. 1976: Soil movement in the tropics – a general model. *Zeitschrift für Geomorphologie, Supplementband, NF*, 25, 132–44.
Livingstone, D. A. 1963: Chemical composition of rivers and lakes. US *Geological Survey Professional Paper 440-G*.
Löffler, E. 1977: *Geomorphology of Papua New Guinea*. Canberra: ANU Press.
Louis, H. 1964: Uber Rumpflächen- und Talbildung in den wechselfeuchten Tropen besonders nach Studien in Tanganyika, *Zeitschrift für Geomorphologie, NS*, 8, 43–70.
Lundgren, L. 1980: Comparison of surface runoff and soil loss from runoff plots in forest and small scale agriculture in the Usumbara Mountains, Tanzania. *Geografiska Annaler* 61A (3–4), 113–48.
Meade, R. H., Dunne, T., Richey, J. E., Santos, U. de M. and Salati, E. 1985: Storage and remobilization of suspended sediment in the lower Amazon river of Brazil. *Science*, 228, 488–90.
Meybeck, M. 1976: Total dissolved transport by world major rivers. *Hydrological Sciences Bulletin*, 21, 265–89.
Meybeck, M. 1979: Concentration des eaux fluviales en éléments majeures et apports en solution aux océans. *Revue de Géographie Physique et de Géologie Dynamique*, 21, 215–46.
Michel, P. 1973: Les bassins des fleuves Sénégal et Gambie. *Etude Géomorphologique*. Mem. ORSTOM 63.
Milliman, J. D. and Meade, R. H. 1983: World-wide delivery of river sediment to the oceans. *Journal of Geology*, 91, 1–21.
Mitchell, H. W. 1965: Soil erosion losses in coffee. *Tanganyika Coffee News*, April–June, 135–55.
Mohr, E. C. J. and Van Baren, F. A. 1954: *Tropical Soils*. The Hague.
Monteith, J. L. 1965: Evaporation and environment. *Symposium of the Society for Experimental Biology*, 19, 205–34.
Monteith, J. L. 1981: Evaporation and surface temperature. *Quarterly Journal of the Royal Meteorological Society*, 107, 1–27.
Morgan, R. P. C. 1972: Observations on factors affecting the behavior of a first order stream. *Transactions of the Institute of British Geographers*, 56, 171–85.
Morgan, R. P. C. 1973: Soil–slope relationships in the lowlands of Selangor and Negri Sembilan, West Malaysia. *Zeitschrift für Geomorphologie*, 17, 139–55.
NEDECCO, 1959: *River Studies – Niger and Benue*. Amsterdam: North-Holland.
Norman, M. J. T., Pearson, C. J. and Searle, P. G. E. 1984: *The Ecology of Tropical Food Crops*. Cambridge: Cambridge University Press, 369 pp.

Nortcliff, S. and Thornes, J. B. 1981: Seasonal variations in the hydrology of a small forested catchment near Manaus, Amazonas, and implications for its management. In R. Lal and E. W. Russell (eds), *Tropical Agricultural Hydrology*, London: Wiley, 37–57.

Nullet, D. and Giambelluca, T. W. 1990: Winter evaporation on a mountain slope, Hawaii. *Journal of Hydrology*, 112, 257–65.

Ogunkoya, O. O., Adejuwon, J. O. and Jeje, L. K. 1984: Runoff response to basin parameters in southwestern Nigeria. *Journal of Hydrology*, 72, 67–84.

Olofin, E. A. 1984: Some effects of the Tiga Dam on valleyside erosion in downstream reaches of the River Kano. *Applied Geography*, 4, 321–32.

Othieno, C. O. 1975: Surface run-off and soil erosion on fields of young tea. *Tropical Agriculture (Trinidad)*, 52 (4), 299–308.

Othieno, C. O. 1980: Effects of mulches on soil water content and water status of tea plants in Kenya. *Experimental Agriculture*, 16, 295–302.

Oyebande, L. 1981: Sediment transport and river basin management in Nigeria. In R. Lal and E. W. Russell (eds), *Tropical Agricultural Hydrology*, London: Wiley, 201–26.

Pathak, P. C., Pandey, A. N. and Singh, J. S. 1985: Apportionment of rainfall in central Himalayan forests (India). *Journal of Hydrology*, 76, 319–32.

Peh, C. H. 1978: Rates of sediment transport by surface water in three forested areas of peninsular Malaysia. Occasional Paper 3, Department of Geography, University of Malaysia.

Penman, H. L. 1948: Natural evaporation from open water, bare soil and grass. *Proceedings of the Royal Society, London*, A193, 129–45.

Pereira, H. C. 1973: *Land Use and Water Resources*. Cambridge: Cambridge University Press.

Priestly, C. H. B. and Taylor, R. J. 1972: On the assessment of surface heat flux and evaporation using large scale parameters. *Monthly Weather Review*, 100, 21–92.

Rapp, A., Axelsson, V., Berry, L. and Murray-Rust, H. 1972: Soil erosion and sediment transport in the Morogoro River catchment. *Tanzania. Geographical Annals*, 54A (3–4), 121–55.

Reading, A. J. 1986: Landslides, heavy rainfalls and hurricanes in Dominica, West Indies. Ph.D. thesis, University of Wales.

Reiger, W. A. and Olive, L. J. 1988: Channel sediment loads: comparisons and estimation. In R. F. Warner (ed.), *Fluvial Geomorphology of Australia*, London: Academic Press.

Rodda, J. C. 1985: *Facets of Hydrology II*. London: Wiley.

Roose, E. J. 1967: *Erosion, Russellement et Lessivage Oblique sous une Plantation d'Hervea en Basse Côte d'Ivoire. III: Resultats des Campagnes 1967, 1968, 1979*. Abidjan: IRCA and ORSTOM.

Roose, E. J. 1970: Importance relative de l'érosion, du drainage oblique et vertical dans la pèdogènes actuelle d'un sol ferralitique de moyenne Côte d'Ivoire. Deux années de mesure sur parcelle expérimentale. *Cahiers ORSTOM, Séries pédologie*, 8 (4), 469–82.

Roose, E. J. 1982: Dynamique actuelle des sols ferrallitiques et ferrugineux tropicaux d'Afrique occidentale; étude expérimentale des transferts hydrologiques et biologiques de matières sous végétations naturelles et cultivées. *Traveaux Document ORSTOM* 130.

Roose, E. J. and Lelong, F. 1976: Les fracteurs de l'érosion hydrique en Afrique tropicale. Etudes sur petites parcelles expérimentales de sol. *Revue de Géographie Physique et de Géologie Dynamique*, 18, 365–74.

Rougerie, G. 1956: Etudes des modes d'érosion et du factionnement des versants en Côte d'Ivoire équatoriale. In *Premier Rapport de la Commission pour l'Etude des versants*, Amsterdam: Union Geographic Internationale, 136–44.

Rougerie, G. 1960: Le factionnement actuel des modèles en Côte d'Ivoire forestière. *Memoire Institut français de l'Afrique Noire*, 58.

Roy, G. B. and Ghosh, R. K. 1984: Infiltration rate at long times. *Soil Science*, 134 (6), 345–7.

Ruxton, B. P. 1967: Slopewash under mature primary rainforest in Papua. In N. Jennings and J. A. Mabbutt (eds), *Landform Studies from Australia and New Guinea*, Cambridge: Cambridge University Press, 85–94.

Sengele, N. 1981: Estimating potential evapotranspiration from a watershed in the Loweo region of Zaire. In R. Lal and E. W. Russell (eds), *Tropical Agricultural Hydrology*, London: Wiley, 83–98.

Simonett, D. S. 1967: Landslide distribution and earthquakes in the Bewani and Torricelli Mountains, New Guinea. In N. Jennings and J. A. Mabbutt (eds), *Landform Studies from Australia and New Guinea*, Cambridge: Cambridge University Press, 64–84.

So, C. L. 1971: Mass movements associated with the rainstorm of June 1966 in Hong Kong, *Transactions of the Institute of British Geographers*, 53, 55–66.

Strakov, N. M. 1967: *Principles of Lithogenesis*, vol. 1. Edinburgh: Oliver & Boyd.

Swan, S. B. St. C. 1970: Piedmont slope studies in a humid tropical region, Johor, Southern Malaya. *Zeitschrift für Geomorphologie, Supplementband*, 10, 30–9.

Takasao, T. and Takara, K. 1988: Evaluation of rainfall-runoff models from the stochastic viewpoint. *Journal of Hydrology*, 102, 381–406.

Talsma, T. and Hallam, P. M. 1980: Hydraulic conductivity measurement of forested catchments, Australia. *Journal of Soil Research*, 18, 139–48.

Temple, P. H. and Rapp, A. 1972: Landslides in the Mgeta area, western Uluguru mountains: geomorphological effects of sudden heavy rainfall. *Geografiska Annaler*, 54A (3–4), 157–93.

Temple, P. H. and Sundborg, A. 1972: The Rufiji River, Tanzania, hydrology and sediment transport. *Geografiska Annaler*, 54A (3–4), 345–68.

Thom, A. S. and Oliver, H. R. 1977: On Penman's equation for estimating regional evaporation. *Quarterly Journal of the Royal Meteorological Society*, 103, 345–57.

Thomas, M. F. 1973: Landforms in equatorial forest areas. In D. Brunsden and J. C. Doornkamp (eds), *The Unquiet Landscape*, Newton Abbott: David & Charles, 141–6.

Thomas, M. F. 1974: *Tropical Geomorphology*. London: Macmillan.

Tricart, J. 1972: *The Landforms of the Humid Tropics, Forests and Savannas*. London: Longman.

Turvey, N. D. 1974: Nutrient cycling under tropical rain forest in central Papua. Occasional Paper 10, Papua New Guinea, Department of Geography.

Turvey, N. D. 1975: Water quality in a tropical rain forested catchment. *Journal of Hydrology*, 27, 111–25.

UNESCO/UNEP/FAO, 1978: *Tropical Forest Ecosystems*. Paris: UNESCO.

Van Bavel, C. H. M. 1966: Potential evaporation: the combination concept and its experimental verification. *Water Resources Research*, 2, 455–67.

Vaughan, R. E. and Wiehe, P. O. 1947: Studies on the vegetation of Mauritius. *Journal of Ecology*, 34, 126–36.

Walling, D. E. 1977: Limitations of the rating curve technique for estimating suspended loads with particular reference to British rivers. *IAHS Publications*, 122, 34–48.
Walling, D. E. 1978: Reliability considerations in the evaluation and analysis of river loads. *Zeitschrift für Geomorphologie*, 29, 29–42.
Walling, D. E. and Webb, B. W. 1983: Patterns of sediment yield. In K. G. Gregory (ed.), *Background to Paleohydrology*, Chichester: Wiley, 69–100.
Walsh, R. P. D. 1980a: Runoff processes in the humid tropics. *Zeitschrift für Geomorphologie*, 36, 176–202.
Walsh, R. P. D. 1980b: Drainage density and hydrological processes in a humid tropical environment: the Windward Islands Ph.D. thesis. University of Cambridge.
Walsh, R. P. D. 1982: Hydrology and water chemistry. In A. C. Jermy and K. P. Kavanagh (eds), *Gunung Mulu National Park, Sarawak: An Account of its Environment and Biota being the Results of The Royal Geographical Society, Sarawak Government Expedition and Survey 1977–1978. Sarawak Museum J.*, 30 (51) (new series), Special Issue 2, Part 1, 121–81.
Walsh, R. P. D. and Howells, K. A. 1988: Soil pipes and their role in runoff generation and chemical denudation in a humid tropical catchment in Dominica. *Earth Surface Processes and Landforms*, 13, 9–17.
Weninger, G. 1968: Etudes hydrobiologiques en Nouvelle Calédonie (Mission 1965). Part 2: Beitrage zum Chemismus der Gerwasser von Neukaledonien (SW-Pazifik). *Cahiers ORSTOM, Series Hydrobiologie*, 2, 35–55.
Weyman, D. R. 1970: Throughflow on hillslopes and its relation to the stream hydrograph. *Bulletin of the International Association of the Science of Hydrology*, 15, 25–33.
Weyman, D. R. 1973: Measurements of the downslope flow of water in a soil. *Journal of Hydrology*, 20, 267–88.
Wiersum, K. F. 1985: Effects of various vegetation layers of an Acaia auriculiformis forest plantation on surface erosion in Java, Indonesia. In S. A. El-Swaify, W. C. Moldenhauer and A. Lo (eds), *Soil Conservation Society of America, Proceedings of the Hawaii Symposium, January 1983*, 79–89.
Wolman, M. G. and Gerson, R. 1978: Relative scales of time and effectiveness of climate in watershed geomorphology. *Earth Surface Processes*, 3, 189–208.
Young, A. 1974: The rate of slope retreat. *Transactions of the Institute of British Geographers*, Special Publication, 7, 65–78.
Zadroga, F. 1981: The hydrological importance of a montane cloud forest area of Puerto Rico. In R. Lal and E. W. Russell (eds), *Tropical Agricultural Hydrology*, London: Wiley, 59–74.

Chapter 8

Adedeji, F. O. 1984: Nutrient cycles and successional changes following shifting cultivation practice in moist semi-deciduous forests in Nigeria. *Forest Ecology and Management*, 9, 87–99.
Aina, P. O. 1984: Contribution of earthworms to porosity and infiltration in a tropical soil under forest and long-term cultivation. *Pedobiologia*, 26, 131–6.

Aitken, S. R. and Moss, M. 1975: Man's impact on the tropical rainforest of Peninsular Malaysia: a review. *Biological Conservation*, 8, 213–29.

Aitken, S. R. and Leigh, C. H. 1992: *Vanishing Rain Forests: The Ecological Transition in Malaysia*. Oxford: Oxford University Press.

Alvim, P. de T. 1978: Agricultural production in the Amazon region. In P. A. Sanchez and L. E. Tergas (eds), *Pasture Production in Acid Soils of the Tropics*, Cali, Colombia: Centro Internacional de Agricultura Tropical (CIAT), 13–23.

Andriesse, J. P. 1987: Monitoring project of nutrient cycling in soils used for shifting cultivation under various climatic conditions in Asia. Research Report, Royal Tropical Institute, Amsterdam.

Andriesse, J. P. and Koopmans, T. T. 1984: A monitoring study of nutrient cycles in soils used in shifting cultivation under various climatic conditions in tropical Asia. III: The influence of simulated burning on form and availability of plant nutrients. *Agriculture, Ecosystems and Environment*, 12, 1–16.

Andriesse, J. P. and Schelhaas, R. M. 1987: A monitoring study of nutrient cycles in soils used for shifting cultivation under various climatic conditions in tropical Asia. III: The effects of land clearing through burning on fertility level. *Agriculture, Ecosystems and Environment*, 18, 311–32.

Arnason, J. T., Lambert, J. D. H., Gale, J., Cal, J. and Vernon, H. 1982: Decline of soil fertility due to intensification of land use by shifting agriculturalists in Belize, Central America. *Agro-Ecosystems*, 8, 27–37.

Aweto, A. O. 1981: Secondary succession and soil fertility restoration in south-western Nigeria. II: Soil fertility restoration. *Journal of Ecology*, 69, 609–14.

Ayanaba, A., Tuckwell, S. B. and Jenkinson, D. S. 1976: The effects of clearing and cropping on the organic reserves and biomass of tropical forest soils. *Soil Biology and Biochemistry*, 8, 519–25.

Baharuddin, K. 1988: Effect of logging on sediment yield in a hill dipterocarp forest in peninsular Malaysia. *Journal of Tropical Forest Science*, 1, 56–66.

Bailly, C., Benoit de Cognac, G., Malvos, C., Ningre, J. M. and Sarrailh, J. M. 1984: Etude de l'influence du couvert naturel et de ses modifications à Madagascar; expérimentations en bassins versants élémentaires. *Cahiers Scientifiques de Centre Technique Forestier Tropicale*, 4, 1–114.

Barois, I., Verdier, B., Kaiser, P., Mariotti, Rangel, P. and Lavelle, P. 1987: Influence of the tropical earthworm *Pontoscolex corethrurus* (Glossoscolecidae) on the fixation and mineralization of nitrogen. In *On Earthworms*, Bologna: Mucchi.

Bartholemew, W. V., Meyer, J. and Laudelout, H. 1953: Mineral nutrient immobilization under forest and grass fallows. INEAC Ser. Sci., 57.

Bersgma, E. 1977: Field boundary gullies in the Serayu river basin, central Java. In *ITC/GUA/VU/NUFFIC Serayu Valley Project, Final Report*, vol. 2, Enschede: ITC, 75–91.

Blong, R. J. 1984: *Volcanic Hazards: a Sourcebook on the Effects of Eruptions*. London: Academic Press.

Brandt, J. 1988: The transformation of rainfall energy by a tropical rain forest canopy in relation to soil erosion. *Journal of Biogeography*, 15, 41–8.

Brass, A. R. 1992: The use of a geographical information system for mapping landslide potential in the West Indies. Ph.D. thesis, University of Reading.

Brinkmann, W. L. and Nascimento, J. C. 1973: The effect of slash and burn on plant nutrients in the Tertiary region of central Amazonia. *Acta Amazonia*, 3, 55–61.

Bristow, C. M. 1977: A review of the evidence for the origin of the kaolin deposits in S. W. England. In E. Galen (ed.), *Proceedings of the 8th International Symposium and Meeting on Alunite*, Rome: Servico de Ministerio de Industrie Energia, Madrid, 1–19.

Bruijnzeel, L. A. 1983: Evaluation of runoff sources in a forested basin in a wet monsoonal environment: a combined hydrological and hydrochemical approach. In R. Keller (ed.), *Hydrology of Humid Tropical Regions with Particular Reference to the Hydrological Effects of Agriculture and Forestry Practice*, IAHS Publ. 140, London: Wiley, 165–74.

Bruijnzeel, L. A. 1990: *Hydrology of Moist Tropical Forests and Effects of Conversion: A State of the Knowledge Review*. Paris: UNESCO, 224 pp.

Bruijnzeel, L. A. and Bremmer, C. N. 1989: Highland–lowland interactions in the Ganges Brahmaputra River Basin: a review of published literature. Occasional paper 11. International Centre for Integrated Mountain Development (ICIMOD), Kathmandu, 136 pp.

Bruijnzeel, L. A. and Wiersum, K. F. 1985: A nutrient balance sheet for *Agathis dammara* Warb. plantation forestry under various management conditions in central Java, Indonesia. *Forest Ecology and Management*, 10, 195–208.

Brunsden, D., Jones, D. K. C., Martin, R. P. and Doornkamp, J. C. 1981: The geomorphological character of part of the Low Himalaya of Eastern Nepal. *Zeitschrift für Geomorphologie, Supplementband, NF*, 37, 25–72.

Burton, I., Kates, R. W. and White, G. F. 1978: *The Environment as a Hazard*. Oxford: Oxford University Press.

Buschbacher, R. J. 1984: Changes in productivity and nutrient cycling following conversion of Amazonian rainforest to pasture. Ph.D. thesis, University of Georgia, 193 pp.

Buschbacher, R. J. 1986: Tropical deforestation and pasture development. *Bioscience*, 36, 22–8.

Buschbacher, R. J. 1987: Deforestation for sovereignty over remote frontiers. Case study no. 4: Government-sponsored pastures in Venezuela near the Brazilian border. In C. F. Jordan (ed.), *Amazonian Rain Forests. Ecosystem Disturbance and Recovery* Berlin: Springer, 9–23.

Carson, B. 1985: Erosion and sedimentation processes in the Nepalese Himalaya. Occasional Paper 1, International Centre for Integrated Mountain Development (ICIMOD), Kathmandu, 39 pp.

Chansang, H. 1988: Coastal tin mining and marine pollution in Thailand. *Ambio*, 17 (3), 223–8.

Colchester, M. 1989: *Pirates, Squatters and Poachers: the Political Ecology of Dispossession of the Native Peoples of Sarawak*. London: Survival International; and Petaling Jaya: INSAN.

Critchley, B. R., Cook, A. G., Critchley, U., Perfect, T. J., Russel-Smith, A. and Yeadon, R. 1979: Effects of bush clearing and soil cultivation on the invertebrate fauna of a forest soil in the humid tropics. *Pedobiologia*, 34, 141–50.

Das, D. C. and Maharjan, P. L. 1988: Terracing in the Hindu Kush-Himalaya: a reassessment. Working paper 10, International Centre for Integrated Mountain Development, Kathmandu (unpublished).

Deitrich, W. E. and Dunne, T. 1978: Sediment budget for a small catchment in mountainous terrain. *Zeitschrift für Geomorphologie, Supplementband, NF*, 29, 191–206.

van Dijk, J. M. and Vogelzang, W. L. M. 1948: The influence of improper soil management on erosion velocity in the Tjiloetoeng basin (Residency of Cheribon, West Java). *Communications, Agricultural Experimental Station, Bogor*, 71, 3–10.

Douglas, I. 1967: Natural and man made erosion in the humid tropics of Australia, Malaysia and Singapore. *International Association of Scientific Hydrologists, Publication 75*, 17–30.

Douglas, I., Greer, R., Wong, W. M., Spencer, T. and Sinun, W. 1990: The impact of commercial logging on a small rainforest catchment in Ulu Segama, Sabah, Malaysia. *International Association of Hydrological Sciences, Publication 192*, 165–73.

Douglas, I., Greer, T., Wong Mai Meng, Kwai Bidin, Waidi Sinun and Spencer, T. 1992: Controls of sediment discharge in undisturbed and logged tropical rain forest streams. In *Proceedings of the 5th International Symposium on River Sedimentation, Karlsruhe.*

Dunne, T. 1979: Sediment yield and land use in tropical catchments. *Journal of Hydrology*, 42, 281–300.

Edwards, R. and Atkinson, K. 1986: *Ore Deposit Geology*. London: Chapman and Hall.

Eswaran, H., Conerera, J. and Sooryanarayanan, V. 1981: Scanning electron microscopic observations on the nature and distribution of iron minerals in plinthite and petroplinthite. In *Lateritisation Processes, Proceedings of the International Seminar on Lateritisation Processes (IGCP-129), Trivandrum, India, 1979*, Oxford and New Delhi: IBH, 335–41.

Euphrat, F. D. 1987: A delicate imbalance: erosion and soil conservation in the Pipal Chuar watershed, Kabhre Palanchok District, Nepal. M.Sc. thesis, University of California, Berkeley, 80 pp.

Ewel, J. J., Berish, C., Brown, B., Price, N. and Raich, J. 1981: Slash and burn impacts on a Costa Rican wet forest site. *Ecology*, 62, 816–29.

Fearnside, P. M. 1987: Causes of deforestation in the Brazilian Amazon. In R. E. Dickinson (ed.), *The Geophysiology of Amazonia*, New York: Wiley; and Tokyo: UNU.

Folster, H., Syed Sofi, S. O. and Tan, P. Y. 1983: A lysimetric simulation of leaching losses from an oil palm field. In Malaysian Society of Soil Science, *Proceedings of the Seminar on Fertilizers in Malaysian Agriculture*, Kuala Lumpur, 45–68.

Fritsch, J. M. and Sarrailh, J. M. 1986: Les transports solides dans l'écosysteme forestier tropical humide en Guyane: les effets du défrichement et de l'installation de pâturages. *Cahiers ORSTOM, Série Pedologie*, 22, 93–106.

Galay, V. 1987: Erosion and sedimentation in the Nepal Himalaya. An assessment of river processes. Report 4/3/010587/1/1 seq. 259, Ministry of Water Resources HMG Nepal, Kathmandu.

Gillman, G. P., Sinclair, D. F., Knowlton, R. and Keys, M. G. 1985: The effect on some soil chemical properties of the selective logging of a north Queensland rainforest. *Forest Ecology and Management*, 12, 195–214.

Gilmour, D. A. 1971: The effects of logging on streamflow and sedimentation in a north Queensland rainforest catchment. *Commonwealth Forestry Review*, 50/143, 39–48.

Goswani, D. C. 1985: Brahmaputra river. Assam, India: physiography, basin denudation and channel aggradation. *Water Resources Research*, 21, 959–78.

Goudie, A. S. 1989: *The Nature of the Environment*. Oxford: Blackwell.

Govind, H. 1989: Recent developments in environmental protection in India: pollution control. *Ambio*, 18 (8), 429–33.

Grove, A. T. and Klein, F. M. G. 1979: *Rural Africa*. Cambridge: Cambridge University Press.

Gupta, R. K. 1983: *The Living Himalayas*, vol. 1, *Aspects of Environment and Resource Ecology of Garhwal*. New Delhi: Today's and Tomorrow's Printers, 378 pp.

Haigh, M. J. 1984a: Landslide prediction and highway maintenance in the Lesser Himalaya, India. *Zeitschrift für Geomorphologie, Supplementband, NF*, 51, 17–37.

Haigh, M. J. 1984b: Ravine erosion and reclamation in India. *Geoforum*, 15, 543–61.

Hase, H. and Folster, H. 1983: Impact of plantation forestry with teak (*Tectona grandis*) on the nutrient status of young alluvial soil in west Venezuela. *Forest Ecology and Management*, 6, 33–57.

Henderson, G. S. and Witthawawatchutikul, P. 1984: The effect of road construction on sedimentation in a forested catchment at Rayong, Thailand. In *Proceedings of the Symposium on Effects of Forest Land Use on Erosion and Slope Stability*, Vienna: IUFRO; and Honolulu: East-West Centre, 247–53.

Herwitz, S. R. 1986: Infiltration-excess overland flow caused by stemflow in a cyclone-prone tropical rainforest. *Earth Surface Processes and Landforms*, 11, 401–12.

Hewitt, K. (ed.) 1983: *Interpretations of Calamity*. London: Allen & Unwin.

d'Hoore, J. 1954: *L'Accumulation des Sesquioxides Libres dans les Sols Tropicaux*. Publications de l'Institut National pour l'Etude Agronomique du Congo, Serie Scientifique, 132 pp.

Horikawa, K. and Shuto, N. 1983: Tsunami disasters and protection measures in Japan. In K. Iida and T. Iwasaki (eds), *Tsunamis: Their Science and Engineering*, Boston, MA: Reidel.

Hurni, H. 1982: Soil erosion in Huai Thung Choa, northern Thailand: concerns and constraints. *Mountain Research and Development*, 2, 141–56.

Jaffre, T. 1985: Composition minérale et stocks de bioéléments dans l'épigée d'écrus forestiers en Côte d'Ivoire. *Acta Oecologia*, 6/20, 233–46.

Jordan, C. F. (ed.) 1987a: *Amazonian Rain Forests. Ecosystem Disturbance and Recovery*. Berlin: Springer.

Jordan, C. F. 1987b: Shifting cultivation. Case study no. 1: Slash and burn agriculture near San Carlos de Rio Negro. In C. F. Jordan (ed.), *Amazonian Rain Forests. Ecosystem Disturbance and Recovery*. Berlin: Springer, 12–23.

Kowal, J. M. and Kassam, A. H. 1978: *Agricultural Ecology of Savanna*. Oxford: Oxford University Press.

Kunstanter, P. and Chapman, E. C. 1978: Problems of shifting cultivation and economic development in northern Thailand. In P. Kunstanter, E. C. Chapman and S. Sabhasri (eds), *Farmers in the Forest: Economic Development and Marginal Agriculture in Northern Thailand*, Honolulu: University of Hawaii Press, 3–23.

Kyuma, K., Tulaphitak, T. and Pairintra, C. 1985: Changes in soil fertility and tilth under shifting cultivation. I: General description of soil and effect of burning on the soil characteristics. *Soil Science and Plant Nutrition*, 31, 227–38.

Lal, R. 1976: *Soil Erosion Problems on an Alfisol in Western Nigeria and Their Control*. Ibadan: International Institute for Tropical Agriculture.

Lal, R. 1988: Effects of macrofauna on soil properties in tropical ecosystems. *Agriculture, Ecology and Environment*, 24, 101–16.

Lam, K. C. 1979: Soil erosion, suspended sediment and solute production in three Hong Kong Catchments. *Journal of Tropical Geography*, 47, 51–62.

Lambert, J. D. H. and Arnason, J. T. 1980: Nutrient levels in corn and competing weed species in a first year milpa, Indian Church, Belize, C. A. *Plant and Soil*, 55, 429–33.

Lanly, J. P. 1982: *Tropical Forest Resources*, FAO Forestry Paper 30. Rome: FAO, 106 pp.

Lavelle, P. 1988: Earthworm activities and the soil system. *Biol. Fert. Soil*, 6, 237–51.

Lavelle, P. and Fragoso, C. 1992: Food-webs in the soils of the humid tropics: importance of mutualistic relationships. In *Food-Webs in Soils*, CRC Press.

Lavelle, P. and Pashanasi, B. 1989: Soil macrofauna and land management in the Peruvian Amazonia (Yurimaguas, Loreto). *Pedobiologia*, 33, 283–91.

Liew, That Chim 1974: A note on soil erosion study at Tawau Hills Forest Reserve. *Malayan Nature Journal*, 27, 20–6.

Low, J. 1836/1972: *A Dissertation on the Soil and Agriculture of the British Settlement of Penang or Prince of Wales Island*. Reprinted as *The British Settlement of Penang*, Singapore: Oxford University Press, 1972.

Macdonald, E. H. 1983: *Alluvial Mining. The Geology, Technology and Economics of Placers*. London: Chapman and Hall, 508 pp.

Mahmood, K. 1987: *Reservoir Sedimentation. Impact, Extent and Mitigation*, World Bank Technical Paper 71. Washington, DC: World Bank.

Malm, O., Pfeiffer, W. C., Souza, C. M. M. and Reuther, R. 1990: Mercury pollution due to gold mining in the Maderia River Basin, Brazil. *Ambio*, 19 (1), 11–15.

Manandhar, I. N. and Khanal, N. R. 1988: Study of landscape processes with special reference to landslides in Lele watershed, central Nepal. Department of Geography, Tribhuvan University, Kathmandu, 53 pp.

Mannion, A. M. and Bowlby, S. B. 1990: *Environmental Issues in the 1990s*. Chichester: Wiley.

Marston, R. A. 1989: *Environment and Society in the Manaslu-Ganesh Region of Central Nepal Himalaya*, Final Report, 1987 Manaslu-Ganesh Expedition. Moscow, ID: University of Idaho, 110 pp.

McFarlane, M. J. 1976: *Laterite and Landscape*. London: Academic Press, 151 pp.

McFarlane, M. J. 1991: Some sedimentary aspects of lateritic weathering profile development in the major bioclimatic zones of tropical Africa. *Journal of African Earth Sciences*, 12 (1–2), 267–82.

McNaughton, S. J. 1976: Serengeti migratory wildebeest: facilitation of energy flow by grazing. *Science*, 191, 92–4.

Megahan, W. F. and Kidd, W. J. 1972: Effects of logging and logging roads on erosion and sediment deposition from steep terrain. *Forestry*, 70, 136–41.

Millington, A. C. 1981: Relationship between three scales of erosion measurement in two small basins in Sierra Leone. Erosion and sediment transport measurement. *International Association of Hydrological Sciences, Publication 133*, 485–92.

Millington, A. C. 1988: Environmental degradation, soil conservation and agricultural policies in Sierra Leone, 1895–1984. In D. Anderson and R. Grove (eds), *Conservation in Africa: People, Policies and Practice*, Cambridge: Cambridge University Press, 229–48.

Mishra, B. K. and Ramakrishnan, R. S. 1983: Slash and burn agriculture at higher elevations in north-eastern India. I: Sediment, water and nutrient losses. *Agriculture, Ecosystems and Environment*, 9, 69–82.

Moberg, J. P. 1972: Some soil fertility problems in the West lake region of Tanzania, including the effects of different forms of cultivar on the fertility of some ferralsols. *East African Agricultural and Forestry Journal*, 37, 35–46.

Moody, K. 1975: Weeds and shifting cultivation. *PANS*, 21, 188–94.

Moody, K. 1982: Changes in weed population following forest clearing in Oyo State, Nigeria. *Tropical Agriculture (Trinidad)*, 59, 298–302.

Morgan, R. P. C. 1986: *Soil Erosion and Conservation*. London: Longman.

Mosley, M. P. 1982: Subsurface flow velocities through selected forest soils, South Island, New Zealand. *Journal of Water Resources*, 55, 65–92.

Mulongoy, K. and Bedoret, A. 1989: Properties of worm casts and surface soils under various plant covers in the humid tropics. *Soil Biology and Biochemistry*, 21, 197–203.

Munchener Ruck 1978: *World Map of Natural Hazards*. Munich: Munchener Ruckversicherungs Gesellschaft.

Nahon, D., Beauvals, A. and Trescases, J. J. 1985: Manganese concentration through chemical weathering of metamorphic rocks under lateritic conditions. In J. I. Drever (ed.), *The Chemistry of Weathering*, Boston, MA: Reidel, 277–91.

Nakano, K. 1978: An ecological study of swidden agriculture at a village in northern Thailand. *Tonan Ajia Kenkyu (South East Asian Studies)*, 16, 411–45.

Newell, C. G. and Self, S. 1982: The volcanic explosivity index (VEI): an estimate of explosive magnitude for historical volcanism. *Journal of Geophysical Research*, 87 (C2), 1231–8.

Nye, P. H. and Greenland, D. J. 1960: Soil under shifting cultivation. Technical Communication 51, Commonwealth Soils Bureau, Harpenden.

Ollier, C. D. 1984: *Weathering*. London: Longman.

O'Loughlin, C. L. 1984: Effectiveness of introduced forest vegetation for protection against landslides and erosion in New Zealand's steeplands. In *Proceedings of the Symposium on Effects of Forest Land Use on Erosion and Slope Stability*, Vienna: IUFRO; and Honolulu: East-West Centre, 275–80.

Pain, C. F. and Bowler, J. M. 1973: Denudation following the November 1970 earthquake at Madang, Papua New Guinea. *Zeitschrift für Geomorphologie, Supplementband, NF*, 18, 92–104.

Pearce, A. J. 1986: Erosion and sedimentation. Working Paper, Environment and Policy Institute, East-West Centre, Honolulu.

Pearce, F. 1988: Cool oceans cause floods in Bangladesh and the Sudan. *New Scientist*, 8 September, 31.

Perry, A. H. 1981: *Environmental Hazards in the British Isles*. London: Allen & Unwin.

Poels, R. 1987: Soils, water and nutrients in a forest ecosystem in Surinam. Ph.D. thesis, Agricultural University of Wageningen, 253 pp.

Prasad, R. C. 1985: The landslide and erosion problems with special reference to the Kosi Catchment. In *Proceedings of the Seminar on Landslides and Erosion Problems with Special Reference to Himalaya Region,*

Gangtok, Sikkim, Calcutta: Indian Society of Engineering Geology, 73–83.

Purata, S. E. 1986: Floristic and structural changes during old-field succession in the Mexican tropics in relation to site history and species availability. *Journal of Tropical Ecology*, 2, 257–76.

Raison, R. J. 1979: Modification of the soil environment by vegetation fires, with particular reference to nitrogen transformations: a review. *Plant and Soil*, 51, 73–108.

Ramakrishnan, P. S. and Toky, O. P. 1981: Soil nutrient status of hill agro-ecosystems and recovery pattern after slash-and-burn agriculture (Jhum) in north-eastern India. *Plant and Soil*, 60, 41–64.

Ramsay, W. J. H. 1986: Erosion problems in the Nepal Himalaya – an overview. In S. C. Joshi (ed.), *Nepal Himalaya. Geoecological Perspectives*, Naini Tal, India: Himalayan Research Group, 359–95.

Ramsay, W. J. H. 1987: Sediment production and transport in the Phewa Valley, Nepal. *International Association of Hydrologyical Sciences, Publication 165*, 239–50.

Rapp, A. and Nyberg, R. 1991: Mudflow disasters in mountainous areas. *Ambio*, 20 (6) 210–18.

Reddy, S. 1989: The 1987 drought in Fiji. *Meteorological Event Report 89/2*, Fiji Met Service.

Rijsdijk, A. and Bruijnzeel, L. A. 1990: Erosion, sediment yield and land use patterns in the upper Konto watershed, East Java, Indonesia. *Konto River Project Communication 18*, 2 vols, Malang, Indonesia: Konto River Project, 58 and 150 pp.

Roose, E. J. 1982: Runoff and erosion before and after clearing depending on the type of crop in West Africa. In *Symposium International sur le Defrichement*, Ibadan: IITA.

Rothwell, R. L. 1978: *Watershed Management Guidelines for Logging and Road Construction in Alberta*, Information Report Nov-X-208. Edmonton: Northern Forest Research Centre, Canadian Forest Service, 43 pp.

Rudra, K. 1979: Causes of floods – some investigations in fluvial geomorphology. *Indian Journal of Power and River Valley Development*, 29, 43–5.

Russell, C. E. 1983: Nutrient cycling and productivity in native and plantation forests at Jari Florestal, Para, Brazil. Ph.D. thesis, University of Georgia, 130 pp.

Ruthenberg, H. 1980: *Farming Systems in the Tropics*. Oxford: Oxford University Press.

Ruxton, B. P. 1967: Slopewash under mature primary rainforest in northern Papua. In J. N. Jennings and J. A. Mabbutt (eds), *Landform Studies from Australia and New Guinea*, Canberra: ANU Press, 85–94.

Saldarriaga, J. G. 1987: Recovery following shifting cultivation. Case study no. 2: A century of succession in the Upper Rio Negro. In C. F. Jordan (ed.), *Amazonian Rain Forests. Ecosystem Disturbance and Recovery*, Berlin: Springer, 24–33.

Sanchez, P. A. 1976: *Properties and Management of Soils in the Tropics*. New York: Wiley.

Sanchez, P. A., Villachica, J. H. and Bandy, E. H. 1983: Soil fertility dynamics after clearing a tropical rainforest in Peru. *Soil Science Society of America Journal*, 47, 1171–81.

Sato, K., Sakurai, Y. and Takase, K. 1984: Soil loss and runoff characteristics of shifting cultivation land in Iriomote island. In C. L. O'Loughlin and A. J. Pearce (eds), *Proceedings of the Symposium on Effects of Forest Land Use on Erosion and Slope Stability*, Vienna: IUFRO; and Honolulu: East-West Centre, 147–53.

Savage, J. M., Goldman, D. P., Janos, D. P., Lugo, A. E., Raven, P. H., Sanchez, P. A. and Wilkes, H. G. 1982: *Ecological Aspects of Development in the Humid Tropics*. Washington, DC: National Academy Press.

Scott, G. A. J. 1987: Shifting cultivation where land is limited. Case study no. 3: Campa Indian agriculture in the Gran Pajonal of Peru. In C. F. Jordan (ed.), *Amazonian Rain Forests. Ecosystem Disturbance and Recovery*, Berlin: Springer, 34–45.

Serrão, E. A. S., Falesi, I. C., Bastos de Veiga, J. and Teixeira Neto, J. F. 1979: Productivity of cultivation on low fertility soils in the Amazon of Brazil. In P. A. Sanchez and L. E. Tergas (eds), *Pasture Production in Acid Soils of the Tropics*, Cali, Colombia: Centro Internacional de Agricultura Tropica, 195–225.

Seubert, C. E., Sanchez, P. A. and Valverde, C. 1977: Effects of land clearing methods on soil properties of an ultisol and crop performance in the Amazon jungle of Peru. *Tropical Agriculture*, 54, 307–21.

Shrestra, T. B. 1988: *Development Ecology of the Arun River Basin in Nepal*. Kathmandhu: International Centre for Integrated Mountain Development, 102 pp.

Smith, D. K. 1989: Natural disaster reduction: how meteorological and hydrological services can help. Publication 722, World Meteorological Organisation, Geneva.

Smith, K. 1992: *Environmental Hazards*. London: Routledge.

Starkel, L. 1972: The role of catastrophic flooding in shaping of relief in the Lower Himalaya (Darjeeling Hills). *Geographia Polonica*, 21, 103–47.

Stromgaard, P. 1984: The immediate effect of burning and ash-fertilization. *Plant and Soil*, 80 (3), 307–20.

Stromgaard, P. 1988: Soil and vegetation changes under shifting cultivation in the miombo of east Africa. *Geografiska Annaler (series B)*, 70 (3), 363–74.

Stromgaard, P. 1991: Soil nutrient accumulation under traditional African agriculture in the miombo woodlands of Zambia. *Tropical Agriculture*, 68, 74–80.

Sutherland, D. G. 1985: Geomorphological controls on the distribution of placer deposits. *Journal of the Geological Society of London*, 142, 727–37.

Tergas, L. E. and Popenoe, H. L. 1971: Young secondary vegetation and soil interactions in Izabal, Guatemala. *Plant and Soil*, 34, 675–90.

Thomas, M. F. 1988: Superficial deposits as resources for development – some implications for applied geomorphology. *Scottish Geography Magazine*, 104 (2), 72–83.

Thompson, R. D. 1981: Incidence and extent of tropical cyclones in the Fiji Islands. *Marine Observer*, L1, 189–95.

Thompson, R. D. 1986: Hurricanes in the Fiji area: causes and consequences. *New Zealand Journal of Geography*, 81, 7–12.

Trenberth, K., Branstator, G. W. and Arkin, P. A. 1988: Origins of the 1988 North American drought. *Science*, 242, 1640–5.

Turvey, N. D. 1974: Nutrient cycling under tropical rain forest in central Papua. Occasional Paper 10, Department of Geography, University of Papua New Guinea.

Uhl, C. 1987: Factors controlling succession following slash-and-burn agriculture in Amazonia. *Journal of Ecology*, 75, 377–407.
Uhl, C. and Jordan, C. F. 1984: Succession and nutrient dynamics following forest cutting and burning in Amazonia. *Ecology*, 65, 1476–90.
Uhl, C., Jordan, C. F., Clark, K., Clark, H. and Herrera, R. 1982: Ecosystem recovery in Amazon caatinga forest after cutting, cutting and burning and bulldozer clearing treatments. *Oikos*, 38, 313–20.
Uhl, C., Buschbacher, R. and Serrao, E. A. S. 1988: Abandoned pastures in eastern Amazonia. *Journal of Ecology*, 76, 663–81.
Valeton, I. 1972: Bauxites. *Development in Soil Science*, vol. 1. Amsterdam: Elsevier, 226 pp.
Velasco, J. R., Vega, M. R. Llena, P. A. and Obien, S. R. 1961: Studies in weed control in upland rice. *Philippines Agriculture*, 44, 373–93.
Virgo, K. J. and Ysselmuiden, I. L. A. 1979: Cultivating soils of tropical steeplands. *World Crops*, 20, 216–21.
Vis, M. 1986: Interception, drop size distributions and rainfall kinetic energy in four Colombian forest ecosystems. *Earth Surface Processes and Landforms*, 11, 59–603.
Vitousek, P. M. 1984: Litterfall, nutrient cycling, and nutrient limitation in tropical forests. *Ecology*, 65, 285–98.
Walling, D. E. 1983: The sediment delivery problem. *Journal of Hydrology*, 65, 209–37.
Walsh, R. P. D. and Reading, A. J. 1991: Historical changes in tropical cyclone frequency within the Caribbean since 1500. *Würzburger Geographische Arbeiten*, 80, 199–240.
Watters, R. F. 1971: *Shifting Cultivation in Latin America*, FAO Forestry Development Paper 17. Rome: FAO.
Whitehouse, I. E. 1987: Geomorphology of a compressional plate boundary, Southern Alps, New Zealand. In V. Gardiner, (ed.), *International Geomorphology 1986*, Part I, Chichester: John Wiley, 897–924.
Whitmore, T. 1975: *Rainforests of the Far East*. Oxford: Clarendon.
Whittow, J. B. 1980: *Disasters: the Anatomy of Environmental Hazards*. London: Allen Lane.
Wiersum, K. F. 1984: Surface erosion under various tropical agroforestry systems. In *Proceedings of the Symposium on the Effects of Forest Land Use on Erosion and Slope Stability*, Vienna: IUFRO; and Honolulu: East-West Centre, 231–9.
Wiersum, K. F. 1985: Effects of various vegatation layers in an Acacia auriculiformis forest plantation on surface erosion in Java, Indonesia. In S. El-Swaify, W. C. Moldenhauer and A. Lo (eds), *Soil Erosion and Conservation*, Ankeny, IA: Soil Conservation Society of America, 79–89.
Wooldridge, R. 1986: Sedimentation in reservoirs: Magat Reservoir, Cagayan Valley, Luzon, Philippines – 1984 reservoir survey and data analysis. Report OD69, Hydraulics Research, Wallingford, 67 pp.
Ziemer, R. R. 1981: Roots and the stability of forested slopes. *International Society of Hydrological Sciences, Publication 132*, 321–61.
Zinke, P., Sabharsi, S. and Kunstadter, P. 1970: Soil fertility aspects of the Lua forest fallow system of shifting agriculture. *Proceedings of the International Seminar on Shifting Cultivation, Chiang Mai, Thailand*, 251–93.
Zulkifli Yusop 1989: Effects of selective logging methods on dissolved nutrient exports in Berembun watershed. Paper presented at the FRIM-IHP-UNESCO Regional Seminar on Tropical Hydrology. Kuala Lumpur, 1989.

Chapter 9

Aitken, S. R. and Leigh, C. H. 1984: A second national park in Peninsular Malaysia? The Endau-Rompin controversy. *Biological Conservation*, 29, 253–76.

Aitken, S. R. and Leigh, C. H. 1992: *Vanishing Rain Forests. The Ecological Transition in Malaysia*. Oxford: Clarendon.

Anderson, J. A. and Spencer, T. 1991: Carbon, nutrient and water balances of tropical rainforest ecosystems subject to disturbance: management implications and research proposals. *MAB Digest*. Paris: UNESCO.

Bach, W. 1972: *Atmospheric Pollution*. New York: McGraw-Hill, 33–9.

Battistini, R. and Verin, P. 1972: Man and the environment in Madagascar. In R. Battistini and G. Richard-Vindard (eds), *Biogeography and Ecology in Madagascar*, The Hague: Junk, 311–37.

Boughton, W. C. 1970: Effects of land management on quantity and quality of available water: a review. Australia Water Resources Council Research Project 68/2, Report 120, University of New South Wales, Australia.

le Bourdière, P. 1972: Accelerated erosion and soil degradation. In R. Battistini and G. Richard-Vindard (eds), *Biogeography and Ecology in Madagascar*, The Hague: Junk, 227–59.

Brookfield, H. C. 1991: Change in land use and the agricultural workforce in peninsular Malaysia, 1966–1982: implications for sustainability. In H. C. Brookfield (ed.), *Transformation and Industrialisation: the Case of Peninsular Malaysia*.

Brookfield, H. C. and Byron, Y. 1990: Deforestation and timber extraction in Borneo and the Malay Peninsula. *Global Environmental Change*, 1, 42–56.

Bruijnzeel, L. A. 1986: Environmental effects of (de)forestation in the humid tropics: a watershed perspective. *Wallaceana*, 46, 3–13.

Bryson, R. A. 1968: All other factors being constant ... a reconciliation of several theories of climatic change. *Weatherwise*, 21, 56–61.

Bryson, R. A. 1971: Climatic modification by air pollution, Preprint, Conference on Environment Future, Helsinki.

Bryson, R. A. and Kutzbach, J. E. 1968: *Air Pollution*. Association of American Geographers Resource Paper 2, 28–31.

Budyko, M. I. 1971: *Climate and Life*. Leningrad: Hydrology Publication House.

Caufield, C. 1987: Conservationists scorn plans to save tropical forests. *New Scientist*, 1566, 33.

Charney, J. 1975: Dynamics of deserts and drought in the Sahel. *Quarterly Journal of the Royal Meteorological Society*, 101, 193–202.

Cole, N. H. A. 1968: *The Vegetation of Sierra Leone*. Njala (Sierra Leone): Njala University College Press, 198 pp.

Cross, A. 1991: *Tropical Deforestation and Remote Sensing: The Use of NOAA/AVHRR over Amazonia*. Final Report to the Commission of the European Communities (D. G. VIII), Article B946/88.

De Vos, A. 1977: Game as food. *Unasylva*, 29 (116), 2–12.

Douglas, J. E. and Swank, W. T. 1975: Effects of management practices on water quality and quantity. Report NE-13:1 for Services, US Department of Agriculture, Washington, DC.

Dunne, N. 1989: Ecuador in $9m debt-for-nature swap. *Financial Times*, 10 April.

Eden, M. J. 1989: *Land Management in Amazonia*. London: Belhaven.

Elliott, S. 1988: Thai forest wins reprieve from dam. *Oryx*, 22, 191–2.

Emanuel, K. A. 1987: The dependency of hurricane intensity on climate. *Nature*, 326, 483–5.

Eyre, L. A. and Gray, C. 1990: Utilization of satellite imagery in the assessment of the effect of global warming on the frequency and distribution of tropical cyclones in the Caribbean, East Pacific and Australian regions. In *Proceedings of the 23rd International Symposium on Remote Sensing of the Environment, Bangkok.*

Fearnside, P. M. 1984: Roads in Rondonia: highway construction and the farce of unprotected reserves in Brazil's Amazonian forest. *Environmental Conservation*, 11, 358–60.

Fearnside, P. M. 1986: Spatial concentration of deforestation in the Brazilian Amazon. *Ambio*, 15, 74–81.

Fearnside, P. M. 1990: Deforestation in the Brazilian Amazonia. In G. M. Woodwell (ed.), *The Earth in Transition: Patterns and Process of Impoverishment*, Cambridge: Cambridge University Press, 211–35.

Fortes, M. D. 1988: Mangrove and seagrass beds of East Asia: habitats under stress. *Ambio*, 17, 207–13.

Frankel, O. H. and Hawkes, G. J. (eds) 1981: *Crop Genetic Resources for Today and Tomorrow*. Cambridge: Cambridge University Press.

Furtado, J. I. 1979: The status and future of tropical moist forest in southeast Asia. In C. McAndrews and C. L. Sien (eds), *Developing Economies and the Environment: the Southeast Asia Experience*, Singapore: McGraw-Hill, 73–120.

Gilmore, D. A. 1971: Effect of rain forest logging and clearing on stream flow and sedimentation in a north Queensland rain forest catchment. *Commonwealth Forest Review*, 50, 38–48.

Gribbin, J. 1986: Temperatures rise in the global greenhouse. *New Scientist*, 110 (1508), 32–3.

Gribbin, J. 1987: An atmosphere in convulsions. *New Scientist*, 116 (1588), 30–1.

Gribbin, J. 1988: *The Ozone Hole*. London: Corgi.

Hamilton, L. S. 1985: Overcoming myths about soil and water impacts of tropical forest land uses. In S. A. El-Swaify, W. C. Moldenhauer and A. Lo (eds), *Soil Erosion and Conservation*, Ankeny, IA: Soil Conservation Society of America, 680–90.

Hansen, J. et al. 1988: Global climatic changes as forecast by Goddard Institute for Space Studies three-dimensional model. *Journal of Geophysical Research*, 93 (D8), 9341–64.

Hecht, S. B., Anderson, A. B. and May, P. 1988: The subsidy from nature: shifting cultivation, successional palm forests, and rural development. *Human Organization*, 47 (1), 25–35.

Ince, M. 1990: *The Rising Seas*. London: Earthscan Publications.

Inglis, A. S. 1988: Rural women and urban men: fuelwood conflicts and forest sustainability in Sussex village, Sierra Leone. Social Forestry Network Paper 6c, Overseas Development Administration, London, 16 pp.

Intergovernmental Panel on Climatic Change, 1990: *Climatic Change: the IPCC Scientific Assessment*. Cambridge: Cambridge University Press.

International Board for Plant Genetic Resources (IBPGR), 1981: *Revised Priorities among Crops and Regions*. Rome: IBPGR.

IUCN, 1980: *World Conservation Strategy*. Morges: IUCN/UNEP/WWF.

JIM (Joint Interagency Planning and Review Mission for the Forestry Sector), 1988: *Cameroon Tropical Forestry Action Plan*. Rome: JIM.

Kartawinata, K. and Vadya, V. P. 1984: Forest conversion in East Kalimantan, Indonesia: the activities and impact of timber companies, shifting cultivators, migrant pepper-farmers, and others. In F. Di Castri, F. W. G. Baker and M. Hadley (eds), *Ecology in Practice*, vol. I, *Ecosystem Management*, Dublin: Tycooly; and Paris: UNESCO.

Kemp, D. D. 1990: *Global Environmental Issues*. London: Routledge, 62–3.

Kumar, R. 1986: *The Forest Resources of Malaysia: Their Economics and Development*. Singapore: Oxford University Press.

Kyaw, U. S. 1987: *National Report: Burma*, Proceedings of an ad hoc FAO/ ECE/FINNIDA Meeting of Experts on Forest Resource Assessment, Kotka, Finland, 26–30 October. Helsinki: Finnish International Development Agency.

Lanly, J. P. 1982: *Tropical Forest Resources*. Rome: FAO.

Lean, G., Hinrichsen, D. and Markham, A. 1990: *Atlas of the Environment*. London: Random Century.

Leysen, M., Conway, J. A. and Sieber, A. J. 1994: Evaluating multi-temporal ENSI-SAR data for tropical forest mapping: regional mapping and change detection algorithms. In *Proceedings of the Second ENS-I Symposium, Hamburg*, esu Sp-361, vol. 1, Noordvijk: European Space Agency, 447–52.

Lugo, A. E. and Brown, S. 1982: Conversion of tropical moist forest: a critique. *Interciencia*, 7, 89–93.

Maher, D. J. 1979: *Frontier Development Policy in Brazil: A Study of Amazonia*. Washington, DC: World Bank.

Malingreau, J.-P. and Tucker, C. J. 1987: Large-scale deforestation in the southeastern Amazon Basin of Brazil, *Ambio*, 17, 49–55.

Malmer, A. 1990: Stream suspended sediment load after clear-felling and different forestry treatments in tropical rainforest, Sabah, Malaysia. *IASH Publication 192*, 62–72.

Mannion, A. M. 1992: Sustainable development and biotechnology. *Environmental Conservation*, 19 (4), 293–306.

Mather, A. S. 1990: *Global Forest Resources*. London: Belhaven.

McDermott, M. J. (ed.) 1988: *The Future of the Tropical Rain Forest*. Oxford: Oxford Forestry Institute.

Melillo, J. M., Palm, C. A., Houghton, R. A., Woodwell, G. M. and Myers, N. 1985: A comparison of two recent estimates of disturbance in tropical forests. *Environmental Conservation*, 12, 37–40.

Millington, A. C. 1988: Environmental degradation, soil conservation and agricultural policies in Sierra Leone, 1895–1984. In D. Anderson and R. Grove (eds), *Conservation in Africa: People, Policies and Practice*, Cambridge: Cambridge University Press, 229–48.

Millington, A. C., Townshend, J. R. G., Saull, R. J., Kennedy, P. A., Prince, S. D. and Madams, R. 1989: *Biomass Assessment*. London: Earthscan Publications.

Millington, A. C., Styles, P. J. and Critchley, R. W. 1992: Mapping forests and savannas in sub-Saharan Africa from advanced very high resolution radiometer (AVHRR) imagery. In P. A. Furley, J. Proctor and J. A. Ratter (eds), *Nature and Dynamics of Forest–Savanna Boundaries*, London: Chapman and Hall, 35–64.

Millington, A. C., Critchley, R. W., Douglas, T. D. and Ryan, P. 1994: *Assessing Woody Biomass in Sub-Saharan Africa*. Washington, DC: World Bank Publications.

Myers, N. 1980: *Conversion of Tropical Moist Forests*. Washington, DC: National Academy of Sciences.

Myers, N. 1988: Tropical deforestation and remote sensing. *Forest Ecology and Management*, 23, 215–25.

Neto, R. B. 1989: Burning continues, slightly abated. *Nature*, 339, 569.

Newell, R. E. 1971: The Amazon forest and atmospheric general circulation. In W. H. Matthews, W. W. Kellog and G. D. Robinson (eds), *Man's Impact on the Climate*, Cambridge, MA: MIT Press, 457–9.

Nortcliff, S., Ross, S. M. and Thornes, J. B. 1990: Soil moisture, runoff and sediment yield from differently cleared tropical rainforest plots. In J. B. Thornes (ed.), *Vegetation and Erosion*, Chichester: Wiley, 419–36.

Norton, D. A. 1985: A multivariate technique for estimating New Zealand temperature normals. *Weather and Climate*, 5, 64–74.

Oldfield, S. 1989: The tropical chainsaw massacre. *New Scientist*, 123, 54–7.

Oyebande, L. 1988: Effects of forest on water yield. In E. R. C. Reynolds and F. B. Thompson (eds), *Forests, Climate and Hydrology: Regional Impacts*, Tokyo: United Nations University, 16–48.

Palaniappan, V. M. 1974: Ecology of tin tailings areas: plant communities and their succession. *Journal of Applied Ecology*, 1, 133–50.

Palca, J. 1987: High-finance approach to protecting tropical forests. *Nature*, 328, 373.

Panturat, S. and Eddy, A. 1989: Some impacts on rice yield from changes in the variance of regional precipitation. Air Group Interim Report to UNEP, University of Birmingham.

Parry, M. 1990: *Climate Change and World Agriculture*. London: Earthscan Publications.

Peneteado, A. R. 1967: *Problemas de Colonizacao e de Uso da Terra na Regiao Bragantina do Estado do Pará*. Belém: Universidade Federal do Pará.

Pereira da Cunha, R. 1989: *Deforestation Estimates through Remote Sensing: The State of the Art in the Legal Amazonia*. Sao Jose dos Campos: INPE.

Perry, A. and Perry, V. 1986: *Climate and Society*. London: Bell & Hyman, 78–9.

Persson, R. 1974: *World Forest Resources*. Stockholm: Royal College of Forestry, 261 pp.

Peters, C. M. and Hammond, E. J. 1990: Fruits from the flooded forests of Peruvian Amazonia: yield estimates for natural populations of three promising species. *Advances in Economic Botany*, 8, 159–76.

Peters, C. M., Gentry, A. and Mendelsohn, R. 1989: Valuation of an Amazonian rainforest. *Nature*, 339, 655–6.

PFMB (Philippines Forest Management Bureau), Department of Environment and Natural Resources, 1988: *1987 Philippines Forestry Statistics*. Manila: Department of Environment and Natural Resources.

Phillips, O. 1993: The potential for harvesting fruits in tropical rainforests: new data from Amazonian Peru. *Biodiversity and Conservation*, 2, 18–38.

Prance, G. T. 1990: Fruits of the rainforest. *New Scientist*, 13 January, 43–5.

Pringle, S. L. 1975: Tropical moist forests in world demand, supply and trade. *Unasylva*, 28, 112–13.

Quy, V. 1988: Vietnam's ecological situation today. *ESCAP Environment News*, 6, 4–5.

Rasoanaivo, P. 1990: Rain forests of Madagascar: sources of industrial and medicinal plants. *Ambio*, 19 (8), 421–3.

Rowntree, P. R. 1990: Estimates of future climatic change over Britain, Part 2: Results. *Weather*, 45 (3), 79–89.

RTFD (Royal Thai Forestry Department), Planning Division, 1986: *Forestry Statistics of Thailand, 1986*. Bangkok: Center for Agricultural Statistics, Office of Agricultural Economics, Ministry of Agriculture and Cooperatives.

Sader, S. A. and Joyce, A. T. 1988: Deforestation rates and trends in Costa Rica, 1940–1983. *Biotropica*, 20, 14.

Secrett, C. 1987: Friends of the Earth UK and the hardwood campaign. In *Proceedings of the Conference on Forest Resources Crisis in the Third World, 6–8 September 1986*, Penang: Sahabat Alam Malaysia, 348–56.

Setzer, A. W., Pereira, M. C., Pereira, A. C., Jr and Almedia, S. A. O. 1988: *Relatório de atividades do projeto IBDF-INE 'SEQE' – Ano 1987*. Sao Jose dos Campos: National Space Research Institute of Brazil.

Shiva, V. 1987: Forestry myths and the World Bank. *The Ecologist*, 17, 142–9.

Sioli, H. 1973: Recent human activities in the Brazilian Amazon region and their ecological effects. In B. J. Meggers, E. S. Ayensu and W. D. Duckworth (eds), *Tropical Forest Ecosystems in Africa and South America: A Comparative Review*, Washington, DC: Smithsonian Institution Press, 321–44.

SMIC, 1971: *Inadvertent Climate Modification*, Report of the study of Man's Impact on Climate (SMIC). Cambridge, MA: MIT Press, 170–9.

Sommer, A. 1976: Attempt at an assessment of the world's moist tropical forests. *Unasylva*, 28, 5–25.

Soussan, J. G. and Millington, A. C. 1992: Forests, woodlands and deforestation. In A. M. Mannion and S. R. Bowlby (eds), *Environmental Issues in the 1990s*, Chichester: Wiley, 79–96.

Spencer, T. 1994: Tropical coral islands – an uncertain future. In N. Roberts (ed.), *The Changing Global Environment*, Oxford: Blackwell, 191–209.

Stoddart, D. R. 1990: Coral reefs and islands and predicted sea-level rise. *Progress in Physical Geography*, 14, 521–36.

Thompson, R. D. 1989: Short-term climatic change: evidence, causes, environmental consequences and strategies for action. *Progress in Physical Geography*, 13, 315–47.

United Nations Environment Programme (UNEP), 1989: Criteria for assessing vulnerability to sea level rise: a global inventory to high risk areas. Draft Report UNEP/Goverment of the Netherlands.

Vohra, B. B. 1987: *Confusion on the Forestry Front*. New Delhi: Advisory Board on Energy.

Walsh, J. 1987: Bolivia swops debts for conservation. *Science*, 237, 597–8.

Warrick, R. A., Gifford, R. and Parry, M. 1986: CO_2, climatic change and agriculture. In B. Bolin et al. (eds), *The Greenhouse Effect, Climatic Change and Ecosystems*, Chichester: Wiley.

van der Weert, R. 1974: The influence of mechanical forest clearing on soil conditions and resulting effects on root growth. *Tropical Agriculture*, 51, 325–31.

Westoby, J. 1989: *Introduction to World Forestry*. Oxford: Blackwell.

Wetterberg, G. B., Ferreria, M., Luiz dos Santos, B. and Campbell de Araujo, V. 1976: *Amazon fauna Preferred as Food*. Brasilia: FAO Technical Report 6.

Whitmore, T. 1975: *Rain Forests of the Far East*. Oxford: Clarendon.

Whitmore, T. C. 1990: *An Introduction to Tropical Rain Forests*. Oxford: Clarendon.

Wiersum, K. F. 1985: Effects of various vegetation layers in an *Acacia auriculiformis* forest plantation on surface erosion in Java, Indonesia. In S. A. El-Swaify, W. C. Moldenhauer and A. Lo (eds), *Soil Erosion and Conservation*, Ankeny, IA: Soil Conservation Society of America, 79–89.

Wilson, E. O. (ed.) 1988: *Biodiversity*. Washington, DC: National Academy Press.

Wood, W. B. 1990: Tropical deforestation. *Global Environmental Change*, 1, 23–41.

Woodroffe, C. 1989: *Maldives and Sea Level Rise*. University of Wollongong.

World Bank, Asia Regional Office, 1988: *Indonesia: Forests, Lands and Water: Issues in Sustainable Development*. Manila: World Bank.

World Resources Institute, 1990: *World Resources 1990–91*. Oxford: Oxford University Press, 383 pp.

Index